Counting the Cost of COVID-19 on the Global Tourism Industry

Godwell Nhamo • Kaitano Dube
David Chikodzi

Counting the Cost of COVID-19 on the Global Tourism Industry

Springer

Godwell Nhamo
Institute for Corporate Citizenship
University of South Africa
Pretoria, South Africa

Kaitano Dube
Department of Ecotourism Management
Vaal University of Technology
Vanderbijlpark, South Africa

David Chikodzi
Institute for Corporate Citizenship
University of South Africa
Pretoria, South Africa

ISBN 978-3-030-56230-4 ISBN 978-3-030-56231-1 (eBook)
https://doi.org/10.1007/978-3-030-56231-1

This Springer imprint is published by the registered company Springer Nature Switzerland AG
The registered company address is: Gewerbestrasse 11, 6330 Cham, Switzerland

Peer Review Process

This book is blind peer-reviewed. Apart from being the international norm, this double-blind peer-review process is mandatory for South Africa–based authors in order to fulfil the requirements of the Department of Higher Education and Training's (DHET) policy for recognised research outputs for subsidy purposes. The authors invested their time to incorporate observations from the blind peer-review process, an aspect that enhanced the quality of the product.

Preface

Tourism is one of the leading economic sectors in both developed and developing countries across the world. The industry accounts for a significant share of the global gross domestic product (GDP) and accounts for close to 10% of the world's employment. As the COVID-19 pandemic emerged, the hopes of continued growth and returns from the tourism and hospitality sector were dashed. With negative travel advice, port closures, isolations, quarantines and social distancing measures and lockdowns, the tourism and hospitality industry got to a virtual standstill. At the peak of the pandemic, air travel and maritime movement was restricted to just essential services, with few cargo planes operating. This book draws the reader's attention to the emerging, untold and yet to be fully determined negative impact of COVID-19 on the tourism and hospitality sector. From the small island developing states of the Pacific, Caribbean and those in Africa to larger economies of China, Japan, Russia, the United States of America, Britain and Germany, the COVID-19 pandemic left and is still leaving a trail of devastation in the tourism and hospitality sector. This book, therefore, creates a new niche in the existing literature, which is mainly in the form of journal articles and not very specific to COVID-19 and tourism. To a larger extent, the work remains novel and will be a huge contribution to the knowledge gap on how the world can address and adjust to emerging global pandemics such as COVID-19.

Pretoria, South Africa — Godwell Nhamo

Vanderbijlpark, South Africa — Kaitano Dube

Pretoria, South Africa — David Chikodzi

Acknowledgements

We, the authors, Prof. Godwell Nhamo, Dr Kaitano Dube and Dr David Chikodzi, wish to thank all the blind peer reviewers for their invaluable inputs during the writing and publishing process of this book. We thank Springer for taking on board this book project and for a quality product. We also wish to thank our families for their ongoing support of our work. The book project was coordinated through the Exxaro Chair in Business and Climate Change at the University of South Africa. The Exxaro Chair is a research Chair funded by the Exxaro Resources (Pty) Ltd Chairman's Fund and hosted by the Institute for Corporate citizenship at the University of South Africa. The Exxaro Chair was established in 2008 and is now in its fourth term running (2018–2022).

The authors also wish to acknowledge the hard work put by the University of South Africa's editing team that include the following: Nkeke Thosago, Dr Malvin Vergie, Jane Franz, Shirley Steenekamp, Debbie Rodrigues, Willem Pretorius, Alexa Barnby, Elmarie Williams and Alfred Nethanani. These individuals worked throughout the hard lockdown to make sure that the quality of the book meets international language standards. In addition, the authors also thank Hlengiwe Precious Kunene and Nthabiseng Mashula for the work they performed as part of our team with internal chapter reviews.

About the Authors

Prof. Godwell Nhamo (PhD) Lead Author and Accounting Officer, is a Full Professor and Exxaro Chair in Business and Climate Change at the University of South Africa (UNISA). He is a National Research Foundation (NRF) C-Rated Researcher undertaking research in the fields of climate change and governance, sustainable tourism, green economy and sustainable development. Prof Nhamo has conceptualised and successfully completed nine book projects. The most recent being: *The Green Building Evolution published by the HSRC Press* (2019); *Scaling up SDGs Implementation: Emerging Cases from State, Development and Private Sectors* published by Springer (2020); *SDGs and Institutions of Higher Education* published by Springer (2020); and *SDG 7 – Ensure Access to Affordable, Reliable, Sustainable and Morden Energy* by Emerald (2020). Prof Nhamo has also published over 80 journal articles. Since 2013, Prof Nhamo has guided 10 PhDs and hosted 10 postdoctoral fellows. Currently, Prof Nhamo is leading two mega research projects on cyclones, Tornados and Floods in the era of SDGs in Southern Africa as well as SDGs for society. Professor Nhamo sits on a number of international and national boards and has received several awards and recognitions for his outstanding work both locally and internationally. Finally, Prof Nhamo is one of the four-member African Union High Level Panel drafting the Green Innovation Framework for the continent. Email: nhamog@unisa.ac.za

Kaitano Dube (PhD) is an Ecotourism Management Lecturer at Vaal University of Technology, South Africa. He is one of Africa's leading tourism geographers researching in the area of tourism, climate change, sustainability and green aviation. He has published in high-impact international peer-reviewed journals with his work receiving global attention. His work has received extensive media coverage including in the National Geographic, Wunderground, Atlasobscura and AFP among others. He has granted a number of international television and radio interviews. Dr Dube holds a PhD and MSc from the University of South Africa. He graduated with a BSc Hons from Midlands State University in Gweru, Zimbabwe. He holds several other qualifications from UNISA, Vaal University of Technology and University of the Witwatersrand Business School in Johannesburg, South Africa. Dr Dube is an

Executive Member of the Tourism Educators Association of South Africa and Tourism Sector Human Resource Development Governance and Institutional Coordination Forum hosted by South Africa's National Department of Tourism. Email: kaitanod@vut.ac.za

David Chikodzi (PhD) is currently a Postdoctoral Fellow in the Exxaro Chair in Business and Climate Change at the University of South Africa. His research interests are in climate change, water resources management, tourism, sustainable development and application of Earth Observation technologies for societal benefit. He has worked for over 10 years in academia at Great Zimbabwe University (Zimbabwe) and has published over 30 journal articles and book chapters. Dr Chikodzi has also taken part in several locally and internationally funded research projects across Southern Africa and has previously worked as a Research Scientist at the Scientific and Industrial Research and Development Centre in Zimbabwe. Dr Chikodzi is a former member of the ISIbalo Africa Young Statisticians and also the Zimbabwe Young Academy of Sciences. Email: chikod@unisa.ac.za

Contents

Part I
Introduction and Background

Chapter 1
The Context: COVID-19, Global Development Agendas and Tourism

Abstract This chapter provides the context in which the book "Counting the Cost of COVID-19 on Global Tourism Industry" is written, including highlights for what follows in the rest of the book. The COVID-19 pandemic came when the world was in the midst of entertaining major global development agendas that had implications on the global tourism value chains. Among such major global development agendas, one could list the 2030 Agenda for Sustainable Development and its aligned 17 Sustainable Development Goals (SDGs), the Paris Agreement, the Sendai Framework and Habitat III's New Urban Agenda. The pandemic also steered geo-political and global economic tensions as key players, including individuals and countries, demanded answers as to what led to such a wild and unstoppable spread of the coronavirus. Accusations of maladministration within the World Health Organization (WHO) and connivance with China emerged. Many countries battled to "flatten" the runaway COVID-19 and economic curves. Many countries and territories went on lockdown, with millions infected and hundreds of thousands dead by the time this book was going into production. Global economic stimulus packages were urgently instituted, and the tourism sector was among the top beneficiaries of such packages. The chapter also presents sections on the importance of the tourism sector to the global economy and how past pandemics affected the sector. It further highlights the overall methodological framework used for data generation and analysis.

Keywords Tourism · COVID-19 · SDGs · Geopolitics · WHO · Methodology · China · Stimulus

1.1 Introduction

On 31 December 2019, the world woke up to the WHO announcement that China had notified it of pneumonia cases of an unknown cause in Wuhan City, Hubei Province of China (WHO 2020a). Within 3 days from the notification to 3 January

G. Nhamo et al., *Counting the Cost of COVID-19 on the Global Tourism Industry*,
https://doi.org/10.1007/978-3-030-56231-1_1

2020, there were 44 case-patients with so-called pneumonia. The source of the infection was identified as the seafood market in Wuhan, with the disease referred to initially as the novel coronavirus (2019-nCoV). As time passed, the 2019-nCoV was renamed COVID-19, with the cause being the SARS-2 coronavirus (SARS-CoV-2) (Backer et al. 2020). The WHO declared the COVID-19 a global pandemic on 11 March 2020, sending huge shock waves across the global community (WHO 2020b). COVID-19 was also known to spread from human to human through droplets and contact, making it a deadly infectious disease (Peng et al. 2020). From the knowledge gathered at the time, COVID-19 was fortunately sensitive to disinfection measures, although it could survive for hours on surfaces in the environment. The main known symptoms were fever, cough and dyspnea. Other symptoms from clinical cases included shortness of breath and chest tightness and/or pain (Deng and Peng 2020). Coughing was also reported for the first case of COVID-19 in Nepal (Bastola et al. 2020) whom on admission to hospital in Kathmandu, the temperature was 37.2 °C, with throat congestion. However, there were no other symptoms. Most of those at high risk of dying included people with hypertension, diabetes, coronary heart disease, cerebral infarction and chronic bronchitis (Deng and Peng 2020).

The Global Preparedness Monitoring Board (GPMB) highlights that while diseases, epidemics and pandemics have been integral parts of human livelihoods, "a combination of global trends, including insecurity and extreme weather" had heightened the risk of outbreaks (GPMB 2019: 6). Disease outbreaks were noted to be on the rise during the past several decades. To this end, one could not discount the possibilities of a spectre of a global health emergency looming large in the future. In almost a prophetic message, the GPMB presented a shock scenario. In its view, "If it is true to say, 'what's past is prologue', then there is a genuine threat of a rapidly moving, highly lethal pandemic of a respiratory pathogen killing 50–80 million people and wiping out nearly 5% of the world's economy" (Ibid.). Hence, experiencing a pandemic of this magnitude would be catastrophic. Not only will it create instability and insecurity, but the world will also not be ready for such a situation. The GPMB (2019) then raises several key points that the world should familiarise itself with in order to deal with future pandemics (Box 1.1).

Global pandemics impact severely on least resourced communities, of which most are in Africa. However, with the spectre of COVID-19, no country was left standing as the pandemic moved across jurisdictions with great and uncontrollable speed. Among the countries that were at the coal face in the early days were China, Hong Kong, Taiwan, Italy and Spain, with the USA, Brazil, Australia and South Africa becoming epicentres in their continents. As of 28 April 2020, there were 3.08 million confirmed cases globally, 213,824 deaths registered (Johns Hopkins University 2020) and the count was rising. The figures were becoming common knowledge! What a traumatised world it became, right through communities, organisations and ultimately families and individuals.

The GPMB (2019) further highlighted the vulnerability of many developing nations that threatened the achievement of the 2030 Agenda for Sustainable Development. Developing nations lacked access to basic food (SDG 2), basic health services (SDG 3), quality education (SDG 4) as well as clean water and sanitation

Box 1.1: Pointers for Future Pandemic Preparedness

- Leaders at all levels hold the key. It is their responsibility to prioritise preparedness with a whole-of-society approach that ensures all are involved, and all are protected.
- For too long, we have allowed a cycle of panic and neglect when it comes to pandemics: we ramp up efforts when there is a serious threat and then quickly forget about them when the threat subsides. It is well past the time to act.
- Heads of government must commit and invest in preparedness.
- All countries must build strong systems.
- Countries, donors and multilateral institutions must be prepared for the worst.
- Financing institutions must link preparedness with financial risk planning.
- Development assistance funders must create incentives and increase funding for preparedness.
- The United Nations must strengthen coordination mechanisms.

Source: authors, based on GPMB (2019: 6–10)

(SDG 6). All these aspects aggravate the spread of infectious pathogens. Other disease amplifiers that added to ever-increasing populations included strains on the environment, climate change and dense urban settlements. Furthermore, air travel had exponentially grown, which increased the risk of both imported and locally transmitted infections from pandemics such as the 2003 SARS pandemic (Breda 2004) and COVID-19, which is the focus of this book. Everyone everywhere would be at risk, including both national and global economies.

1.2 Importance of Tourism to Economies and COVID-19 Impacts

There is no doubt that tourism makes a significant contribution to the global economy. The United Nations World Tourism Organisation (UNWTO) presented positive figures regarding global tourism growth in 2018. There were 1.4 billion international tourist arrivals (up 5% from the same time in 2017), with $1.7 trillion (up 4% from the same period 2017) (UNWTO 2019). This growth was attributed to the emerging middle class, new business models, technology advancement, low-cost travel and the ease of visa facilitation. From the tourism receipts highlighted, it is clear that the tourism sector is a significant player in global economies, creating jobs and wealth as well as changing lives for the better. The tourism sector, therefore, has been assisting in transforming, especially poorer communities. However,

Table 1.1 International tourist arrivals and tourism receipts 2018

	Arrivals		Receipts	
Continent/destinations	Millions	Growth	USD billion	Growth
Africa	67	7%	38	2%
America	216	2%	334	0%
Asia and the Pacific	348	7%	435	7%
Europe	710	5%	570	5%
Middle East	60	5%	73	4%

Source: authors, data from UNWTO (2019: 3)

given the quick growth, the UNWTO noted that there was also an increasing demand for the sector to be more responsible by ensuring effective destination management and reduce negative impacts associated with the sector such as greenhouse (GHG) emissions. The year 2018 was the ninth consecutive year of growth in the tourism sector, and this was a remarkable achievement! With COVID-19 in the mix, predictions from the UNWTO were that international arrivals could drop by between 20% and 30% compared to 2019 (Gössling et al. 2020). The growth in terms of arrivals and receipts for selected continents is reflected in Table 1.1.

Given the widespread negative impacts of COVID-19 on all sectors of the economy, the tourism sector was not spared and became one of the worst affected sectors (Nicola et al. 2020). Global travel restrictions accompanied by stay-at-home orders caused disturbances in the global economy, possibly surpassing damages witnessed during World War II (Gössling et al. 2020). An estimated 90% of the global population was impacted by restrictions on mass gatherings, and this disrupted the tourism and hospitality sector. For example, in the US hotel industry revenue per available room fell by 11.6% for the week ending 7 March 2020 (Nicola et al. 2020). In China, hotel occupancy rates fell by 89% as of 31 January 2020. This resulted in workers having reduced working hours, being put on temporary layoffs and, in extreme cases, being dismissed.

Countries with a heavy dependence on outbound tourists from China including Australia, Fiji, Palau, Samoa, Vanuatu, Thailand, the Philippines, Singapore, Malaysia and Vietnam were on the receiving end of the detrimental effects (ADB 2020; ATIC 2020). The cruise ship business also came under the spotlight with cancellations and port closures (Webeck 2020). Many passengers and crew members got trapped in cruise ships for weeks as they could not disembark. The aviation industry was also badly hit by cancellations and airport closures (Nicola et al. 2020). The cancellations were made abruptly with limited time for a soft landing by the sector. To make matters worse, the travel bans and airport closures were prolonged beyond 2 months in most instances. The sporting sector was not spared either (Stone 2020; Comrades Marathon 2020). There were cancellations and postponements of the Japan Olympics 2020, to golf, tennis, football, athletics, basketball, marathons, rugby, cricket, cycling, boxing, Formula one Grand Prix and ice-skating fixtures and tournaments (PGA Tour 2020; AELTC 2020). Some of the events were moved indefinitely, with other fixed for 2021.

1.3 COVID-19's Landing on Global Development Agendas

Although COVID-19 made a direct landing on SDG 3 (health and wellbeing), the turbulence was intensive and vast, generally resulting in immediate negative impacts on all major global development agendas. Among such major global and regional development agendas were listed the 2030 Agenda for Sustainable Development and the aligned 17 SDGs (United Nations 2015), the Paris Agreement, (United Nations Framework Convention on Climate Change – UNFCCC 2015), the Sendai Framework (United Nations Office for Disaster Risk Reduction 2015) and the Habitat III's New Urban Agenda (United Nations Habitat 2016). Given that the 2030 Agenda for Sustainable Development, COVID-19 and tourism, as well as the Paris Agreement, are discussed later in the book, this section will only highlight some of the take-homes regarding the remaining two global development agendas. The impacts of COVID-19 on the progress to attain the 2030 Agenda for Sustainable Development is overall negative (United Nations 2020), with resources being redirected towards fighting the pandemic across the world (Gössling et al. 2020). However, there were some positive sin terms of reduced carbon emissions that cause global warming, leading to climate change (Quéré et al. 2020).

The Sendai Framework (2015–2030) provides a generic and global platform for dealing with disaster risk reduction (DRR) and management. Central to the Sendai Framework is the DRR cycle that focuses on prevention, response and recovery, as well as the concept of building resilience and building back better (United Nations Office for Disaster Risk Reduction 2015) after disasters such as COVID-19. The Sendai Framework further identifies biological hazards such as pandemics among the key challenges of the twenty-first century. Bringing the reader's attention to the evaluation of the Sendai Framework for the first 3 years of implementation, Bakatsaki and Zampetakis (2020) observed that countries needed to adopt comprehensive and holistic approaches, including sharing responsibilities among key stakeholders and building community resilience (Adekola et al. 2020). The most appropriate leadership style under such disasters should be persuasive as opposed to the directive. These insights remain useful in the case under scrutiny, as the management of the COVID-19 pandemic demanded flexibility and quick decision-making that embraced all stakeholders. Ratner et al. (2020: 1) add elements such as the need to remain "flexible in dynamic situations and embracing discomfort to think bigger about context-specific solutions to collectively build back our systems". The authors also identify the need to accept realities that could not be changed and to look back and learn from disaster situations. Given the scenario being documented in this book, there are incidences where lives were lost, with some businesses folding. These situations could not be changed, but they remained relevant as learning points for the future. Parajuli (2020) presented the concept of disaster science education for effective DRR in developing countries. Such disaster science education would assist in shifting public awareness through local level basic science behind a disaster to minimise exposure, improve preparedness, response and recovery. These efforts were becoming clear across the world as governments channelled COVID-19 disaster science education prioritising washing hands, coughing and sneezing etiquette, self-isolation and social distancing.

The impacts of COVID-19 could also be linked to the New Urban Agenda (United Nations Habitat 2016), that is, in turn, linked to the SDGs and the Paris Agreement. For example, metropolitan transportation systems were brought into the spotlight (Okraszewska et al. 2019) as public transport was rendered useless by lockdowns. However, as the stay-at-home restrictions were relaxed, passenger limitations by public vehicles were also implemented. With high population densities and massive expansion drives in megacities such as Wuhan and Paris (Acuto 2020) that have 11 million-plus dwellers, pandemics such as COVID-19 always present a huge and rapid infection risk. The authors go on to indicate that approximately 43 such high-rise-dense megacities are expected by 2050, with more than 60% of the global population living in them. From the experiences of COVID-19, there was an increased demand for clean water supply and sanitation facilities (SDG 6), access to energy (SDG 7), reliable food supply systems (SDG 2) and other amenities and infrastructure (SDG 9). Apart from being densely populated, megacities host many informal settlements and slums (SDG 11) (Kaika 2017), an element that would present additional complexity to administering the spread of pandemics such as COVID-19. Megacities and other urban spaces are also hosting to most refugees and half of all internally displaced populations (Earle 2016). The municipal officials and other key players were also battling with reductions in carbon emissions, calling for greener building, low carbon transportation systems and settlement densification (Nerini et al. 2019). All this was done in response to the call to reduce carbon emissions as per the Paris Agreement on climate change (UNFCCC 2015) and SDG 13 (United Nations 2015). Hence, it is a cocktail of interventions that COVID-19 brought within the framework on major global development agendas to 2030 and beyond. Drawing from Kaika (2017: 89), however:

> [T]he new call for 'safe, resilient, sustainable and inclusive cities' as enshrined in SDG 11 remains path dependent on old methodological tools (e.g. indicators), techno-managerial solutions (e.g. smart cities), and institutional frameworks of an ecological modernisation paradigm that did not work.

Given the widespread of COVID-19, containment became one of the key intervention measures to "flatten" the curves (McCloskey et al. 2020). The authors noted that historically, major gatherings including pilgrimage and religion, sports, music concert and cinemas remain sources of runaway infections. Most of these events took place in urban areas. Although some major events were held in the past during periods when the WHO had declared public health emergencies such as the Vancouver 2010 Winter Olympics and the 2010 FIFA World Cup in South Africa during the H1N1 influenza pandemic and the 2015 Africa Cup of Nations Football tournament in Equatorial Guinea during the outbreak of Ebola virus (Ibid.), in hindsight, the outbreaks were nowhere near that of COVID-19. Hence, to curb the spread of COVID-19, drastic measures of containment, including mass cancellations and postponements of events involving major gatherings, had to be taken. In addition, mass COVID-19 testing was carried out, and isolations were implemented (Lau et al. 2020), while maintaining essential health services; this was necessary to reduce further loss of lives (Loayza and Pennings 2020). The next section presents an overview of past pandemics and how these impacted on tourism.

1.4 Overview of Past Pandemics and Impacts on Tourism

Following the outbreak of the severe acute respiratory syndrome (SARS) pandemic in 2003, the WHO issued international travel advice to avoid high-risk areas in China (Breda 2004). This was the first pandemic of the twenty-first century (Wilder-Smith 2006). Between March and June 2003, the WHO issued 11 travel advisory notices, with the China National Tourism Administration issuing its own emergency circular making it a top priority to control the spread of SARS by travellers (Breda 2004). The SARS, therefore, remains a good case study on how pandemics can affect tourism. To this end, Beijing International Airport had to install infrared temperature scanners for both inbound and outbound passengers. This measure was not helpful as the WHO issued further travel advisory notices for Beijing and Shanxi Province on 23 April 2003. Although Canada (a country that recorded the highest SARS cases outside Asia) had a travel advisory notice issued by the WHO, this only lasted for a week. The travel advisory of 23 April was followed by two additional recommendations to China, targeting Tianjin, Inner Mongolia and Hebei Province, negatively impacting both out- and inbound tourism.

Tourism became a victim of SARS induced panic (SIP). Several high-profile sporting events were cancelled (Breda 2004), and these included the Women's Ice Hockey World Championship, the Cathay Pacific Squash Open, which were to be held in Hong Kong, and the San Fernando Yacht Race from Hong Kong to the Philippines, and the World Track Cycling Championship was moved by the International Cycling Union from Shenzhen in the Guangdong Province to Germany, and FIFA decided to move the fourth Women's World Cup to the USA. International arrivals globally fell by 1.2% to 694 million in 2003 (Wilder-Smith 2006). The figures were higher in East Asia as arrivals dropped 41% between the first and 21st of April 2003 compared to the same period in 2002. The most affected destinations were China, Hong Kong, Vietnam and Singapore. There was a drop of 12 million arrivals over the months of the SARS outbreak in Asia and the Pacific, comprising a 9% drop compared to 2002. Beijing hotel occupancy fell by 10% (Ibid.). The broad impacts of SIP from SARS on the Hong Kong tourism industry are shown in Box 1.2.

McKercher (2003) ended his paper by asking whether the SIP was warranted, and the author responded to the contrary. This made sense then because there had been breakthroughs in containing SARS, with initial modelled projections to have 80% of Hong Kong infected, which turned out to have been prevented. Only 0.02% of the population had been infected after 2 months, and with the number of new cases between 15 and 30 a day, it would have taken 931 years to infect the entire population. However, with the rapidly spreading and much more contagious COVID-19, taking precautionary measures remained the most viable option. As the

Box 1.2: Impact of SIP on the Hong Kong Tourism Sector

(i) Hotel occupancy across the board was less than 20% with most five-star hotels reporting single-digit occupancies.
(ii) At least six hotels went on sale, with some predicting many more were going into receivership before the end of the (northern) summer.
(iii) More than 40% of all flights in and out of Hong Kong were cancelled with Cathay Pacific alone cancelling 45% of its flights and issuing an unprecedented profit warning.
(iv) Outbound tourism went down by almost 80%.
(v) Inbound tourism went down by almost 80%.
(vi) The loss of the Easter weekend and China 1 May Golden Week period (a total of 2 weeks) cost Hong Kong tourism approximately US$250 million.
(vii) The Hong Kong Restaurant Association warned that up to 5000 restaurants were likely to close.
(viii) The Travel Industry Council warned that 300 travel agents could be forced out of business.
(ix) At least 50,000 jobs were at risk, and many companies had to send their staff home on unpaid leave due to a lack of business.
(x) Economists predicted that Hong Kong would have 0.5% economic growth that year, down from 3% growth forecasted a few months earlier.

Source: authors, based on McKercher (2003: 1)

book was being finalised, Africa remained at the knife-edge as many still believed it was a matter of time. McKercher (2003: 2) ended his story with other observations that were being discouraged and/or promoted to combat the spread of COVID-19. In his view "those of us who live in Hong Kong continue to go to work, continue to shop and continue to lead normal lives. We practice safe SARS by adopting such unusual measures as washing our hands regularly, cleaning our houses, refraining from spitting and seeing doctors if we feel unwell. And yes, we do wear face masks in public". On the question, whether it was safe to travel, the author indicated "Yes", but urged precaution. He also blamed the media for presenting news that sells, some highlighting death to visitors to Asia.

Cooper (2005), focused on the impact of SARS on the Japanese tourism sector, and they discovered that SARS brought instability into the sector. Wilder-Smith (2006) also established an intricate interlinkage between SARS and travel. In their view, while travellers become vectors of SARS, finally travel and tourism becomes the victim. Since SARS anxiety killed travel, governments from both affected and non-affected areas were encouraged to strike a quick and right balance in risk communication. However, this intervention usually came late in many instances when the damage would have already been done.

1.5 The Politics Surrounding COVID-19

Dealing with China in terms of statistics and full disclosure has been problematic in the past. During the novel influenza A (H1N9) outbreak in 2013, the government was accused of hiding information (The Lancet Editorial 2013). While China notified the WHO on 31 March 2013, the first case had been identified on 19 February 2013. This was not the first time China had been accused of concealing information. In 2003 during the SARS outbreak, the country notified the WHO of the pandemic on 11 February 2003, yet the first case had been discovered months earlier on 16 November 2002 (Chan-Yeung and Xu 2003). However, to build back global support, the government fired both the national health minister and the Beijing mayor for poorly handling the situation and China pledging to fully cooperate with the WHO. Since China is a global powerhouse with huge trade linkages with every country on earth, it would have been difficult to immediately close borders to China, and even the USA could not manage it (Stacey and Richards 2020). Yet the COVID-19 pandemic continued spreading like wildfire.

As the COVID-19 pressure grew globally, and particularly in the USA, President Donald Trump on 7 April 2020 announced a move to withhold WHO funding of up to $400 million (Mahase 2020). The President accused the WHO head of incompetence and being biased towards China. The backlash against China seemed to be gaining traction as on 8 April 2020, and Japan announced that $2.2 billion from its stimulus package would be used to incentivise its companies to move out of China back into Japan and also to go elsewhere (Reynolds and Urabe 2020). However, as a typical Chinese show of soft power, interwoven with diplomacy, the comeback fight was on (Knight 2020). As the country was initially accused of covering up on the COVID-19 outbreak, and only notifying the WHO after hundreds had already been infected, the country was rebranding its response as a symbol of good leadership. Several cargo planes were sent all over the world, including the USA, Africa and Europe dropping donated and purchased medical equipment including ventilators, surgical gloves, masks and other protective clothing. As Knight (2020) observed, the Irish Ambassador to China posted a clip thanking Chinese officials who had helped arrange shipment of COVID-19 fighting equipment worth $30 million. China was the first country to reopen closed factories following COVID-19, producing tonnes and tonnes of medical supplies for the world (Bittmann et al. 2020). Assisting the Chinese government's soft power were its mega corporates such as Alibaba that also sent medical supplies and COVID-19 test kits across continents and the USA (Knight 2020).

The US Secretary of State, Mr. Mike Pompeo accused Cuba of "profiting from COVID-19" by exporting doctors to countries such as South Africa and Qatar. The 200 Cuban doctors to South Africa were priced at approximately $24.3 million (BBC 2020a, b). Like China, Cuba uses its medical assistance as a source of soft diplomatic power. However, the Cuban doctors were not welcome in countries such as Brazil, Ecuador and Bolivia (News24 Team 2020). Although the USA claims could have been valid, the USA has for a long time been accusing Cuba of human rights violations.

The COVID-19 political battles spilled over to trade war proxies between China and Australia. Following the passing of a motion to institute an independent international inquiry into the COVID-19 outbreak sponsored by Australia, China immediately instituted an 80% anti-dumping tariff on Australian barley imports (Hurst 2020). This action, if finalised would threaten about $A500 million worth of trade decline annually in the Australian economy for a period of 5 years. China had been investigating a barley anti-dumping complaint since 2018 on the grounds that Australian farmers were being subsidised (Australian Associated Press 2020). To complicate matters, China agreed to a WHO-led investigation into the pandemic outbreak and further pledged $2 billion in aid for 2 years to assist other countries to fight COVID-19 (Shih et al. 2020). Speculation over the pledge grew, with allegations that the pledge was to divert attention from China that needed to give even more towards COVID-19 fight. To this end, China could have succeeded in its pushback as the draft resolution from the 73rd World Health Assembly of the WHO that took place 18–19 May 2020 (supported by over 100 countries) did not mention China or Wuhan (Ibid.).

Devakumar et al. (2020) bring another interesting dimension of racism and discrimination to COVID-19 responses. The authors believe that pandemics create an environment of fear, where racism and xenophobia thrive. To this end, COVID-19 revealed social and political fault lines in communities, with those marginalised usually at the receiving end. There was also increased discrimination against the Chinese people, including violence and blocked entry to certain establishments. COVID-19 policy responses were also found to have been disproportionately affecting people of colour and migrants. Individuals from marginalised communities were seen to have limited healthcare access, at times work in precarious jobs and stay in crowded residential areas and residences. To this end, policy responses such as self-isolation remain inappropriate, presenting the communities with a high risk of infection (Ibid.). Drawing from Latin America, Blofield et al. (2020) brought more evidence of class struggles to COVID-19 responses. The authors highlighted that social distancing measures directly affected the livelihoods of those involved in informal workers who are mainly from lower social classes.

Political leadership in Latin America also came under pressure. Some presidents were "emerging as strong, unifying leaders, while other flounder, in a continent where historically trust in formal institutions, is low" (Blofield et al. 2020: 1). However, Oliver (2020) cautioned, indicating that COVID-19 was not presenting opportunities for premature post mortems on responses and political point-scoring. The author referred to how some national leaders were at odds with the WHO guidance and at times with South East Asian countries that had managed to "flatten" the COVID-19 curve. The reaction came in direct response to the way in which the UK had responded, which was in the author's view, a few weeks late. From South Africa emerged an issue concerning the geography of poverty (Manderson and Levine 2020), where some residential areas in Western Cape remained marginalised and some in the wealthy suburbs clamouring to go ahead with their dog and daily walks.

The African continent has, for a long time, been placed in the spotlight to check compliance with global pandemics such as COVID-19. However, this work revealed several African countries took leadership in certain areas. For example, Nigeria was among the first countries to realise the risk associated with COVID-19 and started planning in early January 2020 (Zumla et al. 2020). In a comprehensive federal effort, the National Coronavirus Preparedness Group was established by the Centre for Disease Control on 7 January 2020. This was just after a week when China first reported the COVID-19 cases to the WHO. Furthermore, within a month, the country had established testing capacity with three functional laboratories. The Centre for Disease Control also put up a national team that met daily to assess the pending coronavirus risk.

The Tanzania and Madagascar presidents also joined the COVID-19 controversies. President John Magufuli fired the national COVID-19 laboratory testing head after undercover samples for a goat, sheep and papaw allocated human names and ages tested positive (Aljazeera 2020). Apart from firing the national COVID-19 testing laboratory head, the President went on to request the military to test the genuineness of the donated COVID-19 tests kits he suspected were faulty. President Magufuli also gave orders for the Madagascar COVID Organics treatment to be bought and used in Tanzania. The COVID Organics is some herbal remedy made out of *Artemisia* plant widely grown in Madagascar and prepared by the Malagasy Institute for Applied Research. Artemisia is also traditionally used to cure ailments that include malaria and the Madagascar President Andry Rajoelina indicated that other countries that took delivery of donated COCVID Organics included Comoros, Guinea-Bissau and Equatorial Guinea (Medical Brief 2020). However, the WHO first dismissed the COVID Organics saying it was not a cure for COVID-19, only to make an about-turn indicating its recognised benefits of traditional and alternative medicines for possible COVID-19 treatment (ANA Reporter 2020). The WHO indicated that the COVID Organics should be tested for efficacy and side effects.

While these events were unfolding, New Zealand became the first country to claim it had managed to stop community transmissions of COVID-19, effectively eliminating the coronavirus (BBC 2020a, b). The announcement came on 27 April 2020. The claim came after the country reported single-digit new cases for several days. However, the Prime Minister, Jacinda Ardern and her officials warned against complacency as the claim did not imply a total freeze in COVID-19 cases. The announcement came soon after the national lockdown was ended with some non-essential business, education and healthcare resuming on Tuesday 28 April 2020. The country had less than 1500 COVID-19 cases and 19 deaths (Ibid.). The success story from New Zealand and other countries is the subject of Chap. 4. The next section focuses on the economic stimulus packages put in place in response to COVID-19.

1.6 Economic Stimulus Packages in Response to COVID-19

This section profiles the emerging global economic stimulus packages in response to COVID-19. The section will not go into detail regarding the tourism specific economic stimulus packages as these are the subject of a whole chapter in the book. Hence, it is only a quick overview and an overview into what transpired until the point the book went into publication.

According to the United Nations Conference on Trade and Development (UNCTAD 2020: 2), in a televised address of 23 February 2020, the President of China, Mr. Xi Jinping highlighted that it was "unavoidable that the novel coronavirus epidemic will have a considerable impact on the economy and society". This view was supported by the Japanese Finance Minister, Taro Aso during the Riyadh G20 meetings in Saudi Arabia on 24 February 2020. Mr. Taro Aso further indicated that COVID-19 was a public health crisis that posed a serious risk to the macroeconomy, and this would be aggravated by drastically reduced production lines, cut-off supply chains, as well as restricting people's movement (Ibid.). PwC (2020) also noted work that was done by the International Monetary Fund (IMF), which concluded that a one percentage point drop in Chinese growth translates into a 0.2 percentage point drop in South Africa's growth. The observations made herein were fulfilled as the global economy needed multiple economic bailouts following the COVID-19 pandemic.

Nicola et al. (2020) identified three economic sectors whose industries were severely affected, namely, those from primary, secondary and tertiary sectors. The primary sectors include agriculture as well as petroleum, oil and gas. The plummeting global demand for foods and beverages, particularly from the hotels and restaurants, resulted in a 20% drop in commodity prices. Brent crude demand went down drastically, hitting a low of around $16 per barrel in April 2020. From the secondary sectors, the manufacturing industry was not spared, with businesses in the UK factoring in decline in revenue until the third quarter if things remained as they were. As for the tertiary sector, education systems were negatively affected, ranging from partial to complete closures of facilities that included schools and universities, with figures of those affected running into billions. The finance sector was another of the tertiary industries affected. Indices of major stock exchanges such as the S&P 500, Dow Jones Industrial Average, Nasdaq, Shanghai Composite, Johannesburg Stock Exchange, UK FUTSE 100, Euro Stoxx 50, Germany's DAX, Japan's Nikkei, Hong Kong's Hang Seng and South Korea's KOSPI all went down drastically following the announcement of COVID-19 as a global pandemic on 11 March 2020 (Ibid.). The central banks had to intervene with drastic repo rate cuts, and the situation as of 30 April 2020 is shown in Fig. 1.1.

A comparison of emerging economic stimulus packages for 2008/2009 and 2020 in logarithmic format for selected countries is shown in Fig. 1.2. What is clear is that the 2008/09 global financial meltdown economic stimulus packages were a fraction of what was required for COVID-19's 2020 economic stimulus packages. For example, South Africa's 2008/2009 economic stimulus packages stood at $3.7 billion compared to $25 billion in 2020. This is an increase of close to sevenfold. For the European Union, the 2008/2009 figure was $255 billion compared to $2.108 trillion for 2020. Similarly, the USA had a stimulus package of $787 billion in 2008/2009 compared to $2 trillion in 2020. The same picture emerged in many countries sampled. Hence, there is no doubt that judging by the amounts of stimulus packages countries were implementing, the COVID-19 impacts were far more severe compared to the impacts of the 2008/2009 global financial crisis. In many instances, countries had to top up their stimulus packages for the third time as the gravity of the COVID-19 pandemic impacted their economies.

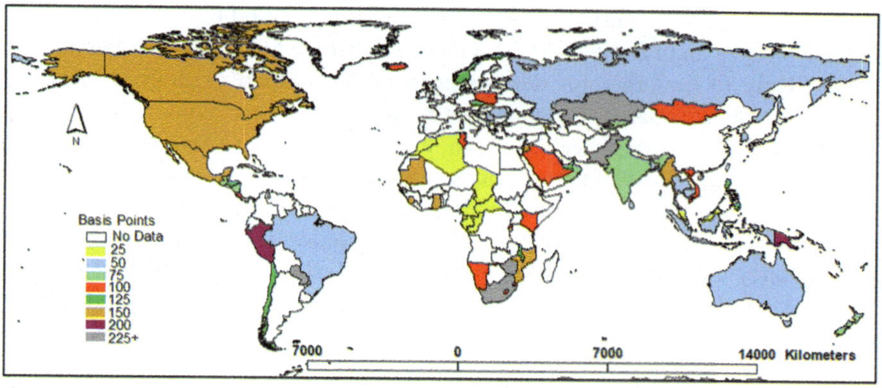

Fig. 1.1 Repo rate cuts by central banks as of 24 May 2020. (Source: authors, data from IMF (2020))

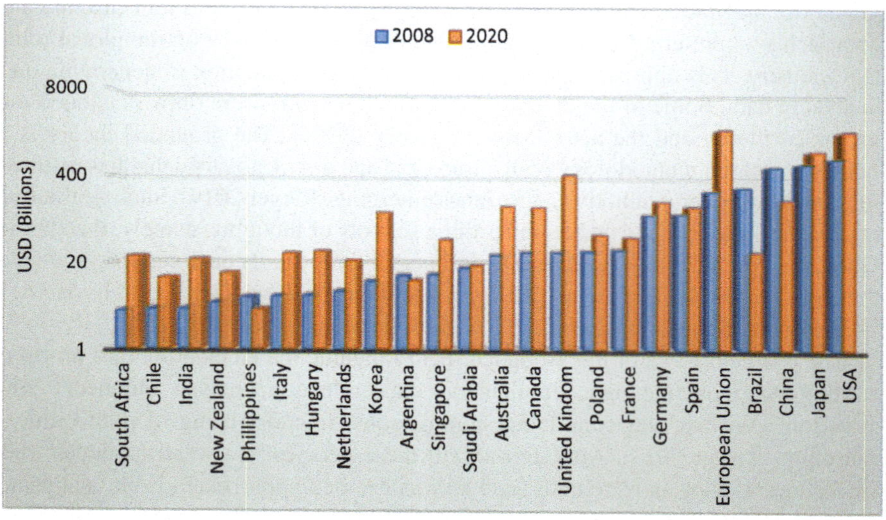

Fig. 1.2 The 2008/2009 and emerging 2020 economic stimulus packages as of 28 April 2020. (Source: authors, data from IMF (2020 online), Alpert (2020 online) and The China Analyst (2009: 4))

Kickbusch et al. (2020) portrayed three linked platforms on which COVID-19 had severe negative impacts, namely, health and wellbeing, social welfare and global economy. Regional and national lockdowns in many other places led to a decrease across all economic sectors and caused large numbers of job losses (Nicola et al. 2020). The World Bank announced immediate support of $12 billion to enable countries to respond to the pandemic, and the IMF announced it was relaxing structural adjustment measures to permit countries and territories to invest in fighting COVID-19 (Kickbusch et al. 2020). The US President Donald Trump announced payroll tax cuts, which usually brought approximately $960 billion to the Federal budget. Loayza and Pennings (2020: 1) warned that the human and economic costs were "likely to be larger for developing countries, which generally have lower health care capacity, larger informal sectors, shallower financial markets, less fiscal space, and poorer governance". As such, the short-term interventions had to focus on avoiding mass layoffs and bankruptcy to stimulate the economy, which was an impossibility under supply-restricting containment measures. Economic stimulus packages were likely to work in the medium- to long-term horizons, with macroeconomic policy focusing on recovery measures involving monetary and fiscal stimulus. Such policies could also not be applied in a blanket manner, as these had to be country-specific.

1.7 Methodology: Theories and Methods Utilised

Given the magnitude and nature of the book, a multi-, inter- and transdisciplinary approach was preferred in documenting the cost of COVID-19 on the global tourism industry. This approach led to a mixed methods application in generating and analysing data. Some of the theories from which insights were drawn included the grounded theory and the actor–network theory (ANT). The grounded theory is a popular research methodology with a traceable and distinct history that has resulted in an array of other qualitative approaches emerging (Rieger 2019). Such qualitative approaches are distinguishable under three schools of thoughts, namely, the classic Glaserian grounded theory, Straussian grounded theory and constructivist grounded theory. The methodology emerged from sociology and was developed by Barney Glaser and Anselm Strauss in the USA in the 1960s (Peiter et al. 2020).

The characteristics of the grounded theory include the elucidation of a process; starting with inductive logic; simultaneous data collection, analysis and theory construction; constant and continuous comparisons; memo writing; and theoretical sampling (Rieger 2019). Although not all these perspectives were applied in this work, several aspects were embraced as was deemed appropriate. From analysing data, a large number of modern-day document and critical discourse analysis components borrow from the grounded theory. Matters such as data coding, category building and formulation of themes resulting in incoherent narratives are borrowed from grounded theory (Tie et al. 2019). Given this set-up, grounded theory has found resonance across many disciplines applying purposive sampling, including tourism studies and general environmental studies. The fundamentals have also been embedded in modern qualitative data analysis software packages such as NVivo and ATLAS.ti.

The ANT has also found application across disciplines and brings additional lenses for tracing and documenting narratives emerging from the interaction of non-human (actants) and humans (actors) within networks (Laasch et al. 2020). The interaction takes place as premised from a quasi-object such as COVID-19 on which certain narratives are built, dismantled and reconstructed. With time, actors recruited to such a narrative may contest perspectives, resulting in some withdrawing from the networks, new actors recruited and so on presenting contested policy domains. As a new phenomenon, narratives and discourses surrounding COVID-19 would naturally bring disagreements and contestations, some as sharp as direct opposites. Technologies including mobile phones, social media and others remain active ingredients in the ANT (Sage et al. 2019).

Organisations such as the United Nations, WHO, World Bank, UNWTO, International Civil Aviation Organization (ICAO), The International Air Transport Association (IATA), EUROCONTROL, the IMF, Airports Council International, the Civil Air Navigation Services Organization, stock exchanges and the International Air Cargo Association become active and crucial in the context of ANT. In addition, everyday influential countries such as China, India, the USA, South Africa, Nigeria, Japan, the European Union, Brazil, Germany, England, Australia and Canada remain in the spotlight and so do their leaders. Top and known philanthropists, such as Bill and Melinda Gates, as well as global multinational corporates such as Alibaba, Facebook, Boeing, Airbus and Nestlé, also grace the discourses, shaping global direction both at national and global levels (Nhamo 2006, 2010). However, it was also important to trace actors and their networks in the context of COVID-19, an aspect that brought together groups of countries according to how they responded to the pandemic. A group of countries that battled to contain the pandemic such as Italy, Spain and the USA emerged. Another group of countries that seemingly did well to manage the pandemic also emerged, and in this group were countries and territories such as Hong Kong, China, Taiwan, Korea, Australia, Canada, South Africa, Rwanda, New Zealand and Germany. There was yet another group of countries that were leading in vaccines and treatment research, which included countries such as Senegal, the USA, China, England, Germany and Australia. In the context of this book, more groups of countries emerged such as the USA, Singapore, Malaysia, Thailand and Australia who had strong tourism economic stimulus packages. These groups are only a few examples of many more that the researchers could identify.

In addition to grounded theory and ANT, the (global) tourism value chain framework, as informed by Michael Porter's value chain model (Porter 1985), and the generic global value chain framework (Antràs 2020) were also utilised. The global tourism value chain framework brought additional methodological underpinnings which guided the researchers in orientating the chapters. Chapters were identified along the global tourism value chain framework to include COVID-19 impacts on the following sectors: air travel, cruise ships, airports networks, restaurants, hotels, sports as well as pilgrimage and religion. Support and cross-cutting chapters were also identified, including the tourism stocks and those dealing with the tourism economic stimulus packages.

With these theoretical framings, several methods that were applicable were identified to generate and analyse the data sets. Figure 1.3 shows the overall methodological orientation as it was applied in the book. Further details on how the methods were applied, including secondary data sources from the World Development Indicators (WDI) database, Johns Hopkins University, the IMF and the UNWTO

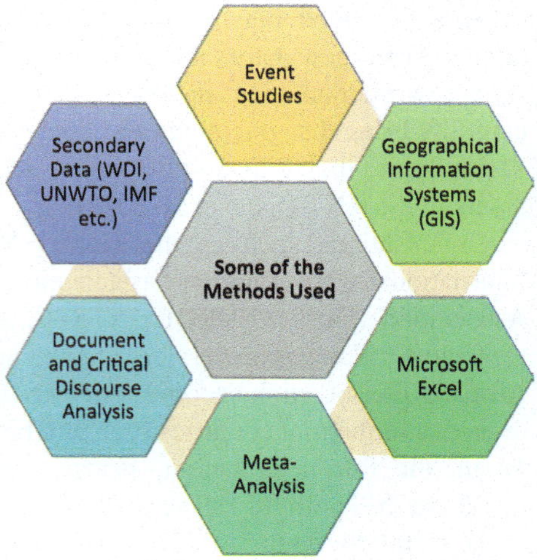

Fig. 1.3 Overall methodological setting. (Source: authors)

portals, are embedded in each of the book's chapters. The authors also made use of announcements, a method that is embedded in events studies (Kumar and Srivastava 2017). The authors recommended that there was a need to carefully select the nature of events that could be subjected to events studies. To this end, the chapters dealing with hotel stocks and shares benefited most, although elements of the methodology were also applied in other chapters as was deemed appropriate. The events studies method has its roots in finance, particularly the analysis of stocks and foreign exchange (Hayward 2018). However, over time the methodology has been applied in other disciplines (Giorgino et al. 2017).

Given that some chapters are location-specific, the geographical information system (GIS) applications were also used. In addition, other generic Microsoft data analysis packages such as Excel were utilised. The GIS remains a suitable tool for tourism studies (Martínez-Hernández and Yubero 2019). Several spatial maps were developed to locate countries and related tourism events that were impacted by the COVID-19 pandemic. Of late, the GIS application has found application in the mapping and spatial analysis of SDGs (Cheikh and Rebai 2019). Reference to the SDGs is made in some chapters as the authors deemed appropriate. Document and critical discourse analysis (D&CDA) (Mendes et al. 2020), as well as meta-analysis, were other methods used in analysing data (Asatani et al. 2020). The D&CDA has been successfully applied to research climate change (Shi et al. 2020; Islam and Kieu 2020) and REDD+ regimes (Mbatu 2020), both of which have direct linkages to the tourism value chains. Both D&CDA and meta-analysis methods were also applied in studying sports tourism (Jiménez-García et al. 2020) and the understanding of environmental perception in tourism (Shi and Sun 2020).

The use of several methods to generate and analyse data further assisted with both methodological and data triangulation (Taswell et al. 2019; Zhang et al. 2019) as some sources of data needed double authentication. There was just too much information generated within very short periods. In addition, there were also updates from the literature, another source of data heavily utilised in the book. In a way, it was like the book was being written live. The next and last section in this chapter presents the book and chapter outlines.

1.8 Book and Chapter Outlines

This book was written in five parts. Part I presents the introduction and background and is made up of a single chapter dedicated to the context of COVID-19, global development agendas and the contribution of tourism to global economies (this chapter). Part II presents the conceptual framework for COVID-19 and tourism. This part comprises two chapters, namely, Chap. 2, that deliberates on the global tourism value chains, the SDGs and COVID-19, and Chap. 3, which documents the preparedness and responses to COVID-19 from selected countries.

Part III is composed of nine chapters and looks at the impact of COVID-19 on (global) tourism value chains. Chapter 4 discusses COVID-19 and its implications for the aviation sector, while the impact of COVID-19 on the global network of airports is the subject matter in Chap. 5. Chapter 6 elucidates the cost of COVID-19 on the global cruise ship industry, with Chap. 7 deliberating on the car rentals, and Chap. 8 focuses on hotel industry and COVID-19 impacts. Chapter 9 profiles restaurants and COVID-19 impacts, with a focus on sustainability and recovery pathways. Chapter 10 presents the impacts of COVID-19 on the global sporting industry and related tourism, while Chap. 11 does the same with a focus on pilgrimage and religious tourism. Chapter 12 looks at COVID-19 and the entertainment industry, while COVID-19 and the stock market with a focus on tourism-related share prices is the focus of Chap. 13.

Part IV is made up of two chapters and is dedicated to philanthropy and tourism economic stimulus packages. Chapter 14 tracks corporate, philanthropic and public donations to dislodge COVID-19, while the tourism economic stimulus packages as a response to COVID-19 is the focus on Chap. 15. Part V, which is made up of another single chapter, draws the book's conclusions and policy recommendations (Chap. 16).

References

Acuto, M. (2020). COVID-19: Lessons for an urban(izing) world. *One Earth*. https://doi.org/10.1016/j.oneear.2020.04.004.

ADB (Asian Development Bank). (2020). *The economic impact of the COVID-19: Outbreak on developing Asia*. Kuala Lumpur: ADB.

Adekola, J., Fischabcher-Smith, D., & Fischabcher-Smith, M. (2020). Inherent complexities of a multi-stakeholder approach to building community resilience. *International Journal of Disaster Risk Science, 11*, 32–45. https://doi.org/10.1007/s13753-020-00246-1.

AELTC (Main Board of the All England Club). (2020). *Cancellation of the championships 2020*. Retrieved from https://www.wimbledon.com/en_GB/news/articles/2020-04-01/cancellation_of_the_championships_2020.html. Accessed 4 Apr 2020.

Aljazeera. (2020). *Tanzania president questions coronavirus kits after animal test.* Retrieved from https://www.aljazeera.com/news/2020/05/tanzania-president-questions-coronavirus-kits-animal-test-200503174100809.html. Accessed 17 May 2020.

Alpert, G. (2020). *COVID-19 government stimulus and financial relief guide.* Retrieved from https://www.investopedia.com/government-stimulus-efforts-to-fight-the-covid-19-crisis-4799723. Accessed 28 Apr 2020.

ANA Reporter. (2020). *WHO says it recognises benefits of traditional, alternative medicine as possible treatments for Covid-19.* Retrieved from https://www.iol.co.za/news/south-africa/who-says-it-recognises-benefits-of-traditional-alternative-medicine-as-possible-treatments-for-covid-19-47544249. Accessed 17 May 2020.

Antràs, P. (2020). *Conceptual aspects of global value chains.* Washington, DC: World Bank Development Economics.

Asatani, K., Takeda, H., Yamano, H., & Sakata, I. (2020). Scientific attention to sustainability and SDGs: Meta-analysis of academic papers. *Energies, 13*, 975. https://doi.org/10.3390/en13040975.

ATIC (Australian Tourism Industry Council). (2020). *Media statement: Tourism industry body thumbs up to federal stimulus combatting COVID-19.* Sydney: ATIC.

Australian Associated Press. (2020). *Australia 'deeply disappointed' after China imposes 80% tariff on barley imports.* Retrieved from https://www.theguardian.com/australia-news/2020/may/19/australia-deeply-disappointed-after-china-imposes-80-tariff-on-barley-imports. Accessed 24 May 2020.

Backer, J. A., Klinkenberg, D., & Wallinga, J. (2020). Incubation period of 2019 novel coronavirus (2019-nCoV) infections among travellers from Wuhan, China, 20–28 January 2020. *European Surveillance, 25*(5). https://doi.org/10.2807/1560-7917.

Bakatsaki, M., & Zampetakis, L. (2020). International trends in managing natural hazards and the role of leadership. In M. Goci'c et al. (Eds.), *Natural risk management and engineering* (pp. 63–87). https://doi.org/10.1007/978-3-030-39391-5_4.

Bastola, A., Sah, R., Rodriguez-Morales, A. J., Lal, B. K., Jha, R., Ojha, H. C., Shrestha, B., Chu, D. K. W., Poon, L. L. M., Costello, A., Morita, K., & Pandey, B. D. (2020). The first 2019 novel coronavirus case in Nepal. *The Lancet, 20*, 278–279.

BBC (British Broadcasting Corporation). (2020a). *Coronavirus: Cuban doctors go to South Africa.* Retrieved from https://www.bbc.com/news/world-africa-52431627. Accessed 30 Apr 2020.

BBC (British Broadcasting Corporation). (2020b). *Coronavirus: New Zealand claims no community cases as lockdown eases.* Retrieved from https://www.bbc.com/news/world-asia-52436658. Accessed 27 Apr 2020.

Bittmann, S., Weissenstein, A., Villalon, G., Moschüring-Alieva, E., Bittmann, L., & Luchter, E. (2020). COVID-19: The importance of adequate personal protective equipment in health-care medical staff. *Journal of Regenerative Biology and Medicine, 2*(2), 1–2. https://doi.org/10.37191/Mapsci-2582-385X-2(2)-021.

Blofield, M., Hoffmann, B., & Llnos, M. (2020). *Assessing the political and social impact of the COVID-19 crisis in Latin America. (GIGA Focus Lateinamerika, 3).* Hamburg: GIGA German Institute of Global and Area Studies. https://giga-hamburg.de/en/system/files/publications/gf_lateinamerika_2003_en.pdf.

Breda, Z. (2004). The impact of Severe Acute Respiratory Syndrome (SARS) on China's tourism sector. *Tourism Research Journal, 1*(2), 5–14.

Chan-Yeung, M., & Xu, R. H. (2003). SARS: Epidemiology. *Respirology, 8*, S9–S14.

Cheikh, Z. B., & Rebai, N. (2019). Mapping and spatial analysis of socio-economic and environmental indicators for sustainable development. In N. Rebai & M. Mastere (Eds.), *Advances in science, technology & innovation* (pp. 3–12). https://doi.org/10.1007/978-3-030-21166-0_1.

Comrades Marathon. (2020). *Comrades Marathon history.* Retrieved from https://www.comrades.com/index.php/home/home-nav-a/history. Accessed 28 Apr 2020.

Deng, S. Q., & Peng, H. J. (2020). Characteristics of and public health responses to the coronavirus disease 2019 outbreak in China. *Journal of Clinical Medicine, 9*, 575. https://doi.org/10.3390/jcm9020575.

Devakumar, D., Shannon, G., Bhopal, S. S., & Abubakar, I. (2020). Racism and discrimination in COVID-19 responses. *The Lancet*. https://doi.org/10.1016/S0140-6736(20)30792-3.

Earle, L. (2016). Urban crises and the new urban agenda. *Environment & Urbanisation, 28*(1), 77–86. https://doi.org/10.1177/0956247815620335.

Giorgino, M. C., Supino, E., & Barnabè, F. (2017). Corporate disclosure, materiality, and integrated report: An event study analysis. *Sustainability, 9*, 2182. https://doi.org/10.3390/su9122182.

Gössling, S., Scott, D., & Hall, C. M. (2020). Pandemics, tourism and global change: A rapid assessment of COVID-19. *Journal of Sustainable Tourism*. https://doi.org/10.1080/09669582.2020.1758708.

GPMB (Global Preparedness Monitoring Board). (2019). *A world at risk: Annual report on global preparedness for health emergencies*. Geneva: World Health Organization. Licence: CC BY-NC-SA 3.0 IGO.

Hayward, R. (2018). Foreign exchange speculation: An event study. *International Journal of Financial Studies, 6*, 22. https://doi.org/10.3390/ijfs6010022.

Hurst, D. (2020). *Why has China slapped tariffs on Australian barley and what can Australia do about it?* Retrieved from https://www.theguardian.com/business/2020/may/20/why-has-china-slapped-tariffs-on-australian-barley-and-what-can-australia-do-about-it. Accessed 24 May 2020.

IMF (International Monetary Fund). (2020). *Policy responses to COVID-19*. Retrieved from https://www.imf.org/en/Topics/imf-and-covid19/Policy-Responses-to-COVID-19. Accessed 28 Apr 2020.

Islam, M. S., & Kieu, E. (2020). Tackling regional climate change impacts and food security issues: A critical analysis across ASEAN, PIF, and SAARC. *Sustainability, 12*, 883. https://doi.org/10.3390/su12030883.

Jiménez-García, M., Ruiz-Chico, J., Peña-Sánchez, A. R., & López-Sánchez, J. A. (2020). A bibliometric analysis of sports tourism and sustainability (2002–2019). *Sustainability, 12*, 2840. https://doi.org/10.3390/su12072840.

Johns Hopkins University. (2020). *COVID-19 dashboard by the Center for Systems Science and Engineering (CSSE) at Johns Hopkins University (JHU)*. Retrieved from https://coronavirus.jhu.edu/map.html. Accessed 28 Apr 2020.

Kaika, M. (2017). 'Don't call me resilient again!': The new urban agenda as immunology … or … what happens when communities refuse to be vaccinated with 'smart cities' and indicators. *Environment & Urbanisation, 29*(1), 89–102. https://doi.org/10.1177/0956247816684763.

Kickbusch, I., Leung, G. M., Bhutta, Z. A., Matsoso, M. P., Ihekweazu, C., & Abbasi, K. (2020). Covid-19: How a virus is turning the world upside down. *The BMJ, 369*, m1336. https://doi.org/10.1136/bmj.m1336.

Knight, W. (2020). *China flexes its soft power with 'Covid Diplomacy'*. Retrieved from https://www.wired.com/story/china-flexes-soft-power-covid-diplomacy/. Accessed 27 Apr 2020.

Kumar, V., & Srivastava, A. (2017). Eventful non-events: Distinguishing an event from a non-event in event studies. *Theoretical Economics Letters, 7*, 1067–1080.

Laasch, O., Moosmayer, D. C., & Arp, F. (2020). Responsible practices in the wild: An actor-network perspective on mobile apps in learning as translation(s). *Journal of Business Ethics, 161*, 253–277. https://doi.org/10.1007/s10551-019-04214-8.

Lau, L. S., Samari, G., Moresky, R. T., Casey, S. E., Kachur, S. P., Roberts, L. F., & Zard, M. (2020). COVID-19 in humanitarian settings and lessons learned from past epidemics. *Nature Medicine*. https://doi.org/10.1038/s41591-020-0851-2.

Loayza, N. V., & Pennings, S. (2020). *Macroeconomic policy in the time of COVID-19: A primer for developing countries*. Kuala Lumper: World Bank Malaysia Hub.

Mahase, E. (2020). Covid-19: Trump threatens to stop funding WHO amid "China-centric" claims. *The BMJ, 369*, m1438. https://doi.org/10.1136/bmj.m1438.

Manderson, L., & Levine, S. (2020). COVID-19, risk, fear, and fallout. *Medical Anthropology*. https://doi.org/10.1080/01459740.2020.1746301.

Martínez-Hernández, C., & Yubero, C. (2019). Explaining urban sustainability to teachers in training through a geographical analysis of tourism gentrification in Europe. *Sustainability, 12*, 67. https://doi.org/10.3390/su12010067.
Mbatu, R. S. (2020). Discourses of FLEGT and REDD + regimes in Cameroon: A nongovernmental organization and international development agency perspectives. *Forests, 11*, 166. https://doi.org/10.3390/f11020166.
McCloskey, B., Zumla, A., Ippolito, G., Blumberg, L., Arbon, P., Cicero, A., Endericks, T., Lim, P. L., & Borodina, M. (2020). Mass gathering events and reducing further global spread of COVID-19: A political and public health dilemma. *The Lancet*. https://doi.org/10.1016/S0140-6736(20)30681-4.
McKercher, B. (2003). SIP (Sars Induced Panic) a greater threat to tourism than SARS (Severe Acute Respiratory Syndrome). *e-Review of Tourism Research, 1*(1). http://ertr.tamu.edu.
Medical Brief. (2020). *Tanzania joins the list of African countries importing Madagascar's COVID-19 'cure'*. Retrieved from https://www.medicalbrief.co.za/archives/tanzania-joins-the-list-of-african-countries-importing-madagascars-covid-19-cure/. Accessed 18 May 2020.
Mendes, R., Fidélis, T., Roebeling, P., & Teles, F. (2020). The institutionalisation of nature-based solutions – A discourse analysis of emergent literature. *Resources, 9*, 6. https://doi.org/10.3390/resources9010006.
Nerini, F. F., Slob, A., Engström, R. E., & Trutneyte, E. (2019). A research and innovation agenda for zero-emission European cities. *Sustainability, 11*, 1692. https://doi.org/10.3390/su11061692.
News24 Team. (2020). *'Profiting from the pandemic' – US politician slams SA for taking Cuban doctors*. Retrieved from https://www.news24.com/SouthAfrica/News/profiting-from-the-pandemic-us-politician-slams-sa-for-taking-cuban-doctors-20200429. Accessed 30 Apr 2020.
Nhamo, G. (2006). Actor/actant-network theory as emerging methodology for environmental policy research in southern Africa. *Southern Africa Journal of Environmental Education, 23*, 34–47.
Nhamo, G. (2010). Dawn of a new climate order: Reading between the USA-India-China partnerships. *Politikon: South African Journal of Political Studies, 37*(2–3), 351–374.
Nicola, M., Alsafi, Z., Sohrabi, C., Kerwan, A., Al-Jabir, A., Iosifidis, C., Agha, M., & Agha, R. (2020). The socio-economic implications of the coronavirus and COVID-19 pandemic: A review. *International Journal of Surgery*. https://doi.org/10.1016/j.ijsu.2020.04.018.
Okraszewska, R., Jamroz, K., Michalski, L., Zukowska, J., Grzelec, K., & Birr, K. (2019). Analysing ways to achieve a new urban agenda-based sustainable metropolitan transport. *Sustainability, 11*, 813. https://doi.org/10.3390/su11030813.
Oliver, D. (2020). David Oliver: Covid-19 – Recriminations and political point scoring must wait. *The BMJ, 368*, m1153. https://doi.org/10.1136/bmj.m1153.
Parajuli, R. R. (2020). Citizen disaster science education for effective disaster risk reduction in developing countries. *Geoenvironmental Disasters, 7*, 12. https://doi.org/10.1186/s40677-020-00150-2.
Peiter, C. C., Santos, J. L. G., Kahl, C., Coelli, F. H. S., Cunha, K. S., & Lacerda, M. R. (2020). Grounded theory: Use in scientific articles published in Brazilian nursing journals with qualis A classification. *Texto Contexto Enferm, 29*, e20180177. https://doi.org/10.1590/1980-265X-TCE-2018-0177.
Peng, X., Xu, X., Li, Y., Cheng, L., Zhou, X., & Ren, B. (2020). Transmission routes of 2019-nCoV and controls in dental practice. *International Journal of Oral Science, 12*, 9. https://doi.org/10.1038/s41368-020-0075-9.
PGA Tour. (2020). *PGA championship announces postponement*. Retrieved from https://www.pgatour.com/news/2020/03/17/pga-championship-harding-park-postponed-pga-tour.html. Accessed 4 Apr 2020.
Porter, M. E. (1985). *Competitive advantage*. New York: Free Press.
PwC (PricewaterhouseCoopers). (2020). *Impact of trade disrupting COVID-19 on South African business*. Johannesburg: PwC.

Quéré, C. L., Jackson, R. B., Jones, M. W., Smith, A. J. P., et al. (2020). Temporary reduction in daily global CO2 emissions during the COVID-19 forced confinement. *Nature Climate Change*. https://doi.org/10.1038/s41558-020-0797-x.

Ratner, L., Martin-Blais, R., Warrell, C., & Narla, N. (2020). Reflections on resilience during the novel coronavirus disease (COVID-19) pandemic: Six lessons from working in resource-denied settings. *American Journal of Tropical Medicine and Hygiene, 102*(6), 1–3. https://doi.org/10.4269/ajtmh.20-0274.

Reynolds, I., & Urabe, E. (2020). *Japan to fund firms to shift production out of China*. Retrieved from https://www.bloomberg.com/news/articles/2020-04-08/japan-to-fund-firms-to-shift-production-out-of-china. Accessed 27 Apr 2020.

Rieger, K. L. (2019). Discriminating among grounded theory approaches. *Nursing Inquiry, 26*, e12261. https://doi.org/10.1111/nin.12261.

Sage, D., Vitry, C., & Dainty, A. (2019). Exploring the organizational proliferation of new technologies: An affective actor-network theory. *Organisation Studies, 41*(3), 345–363. https://doi.org/10.1177/0170840618815524.

Shi, J., & Sun, J. (2020). Meta-understanding of environmental perception in tourism: Implications for China's tourist attractions. *Sustainability, 12*, 1658. https://doi.org/10.3390/su12041658.

Shi, W., Fu, H., Wang, P., Chen, C., & Xiong, J. (2020). Climatechange vs. Globalwarming: Characterising two competing climate discourses on twitter with semantic network and temporal analyses. *International Journal of Environmental Research and Public Health, 17*, 1062. https://doi.org/10.3390/ijerph17031062.

Shih, G., Rauhala, E., Dawsey, J. (2020). *China's Xi backs WGO-led review of COVID-19 outbreak*. Retrieved from https://www.washingtonpost.com/world/asia_pacific/chinas-xi-backs-who-led-review-of-covid-19-outbreak-proposes-aid-for-developing-world/2020/05/18/911a1544-98df-11ea-ad79-eef7cd734641_story.html. Accessed 24 May 2020.

Stacey, S., & Richards, G. A. (2020). Coronavirus epidemic: A South African perspective. *Wits Journal of Clinical Medicine, 2*(1), 1–4. https://doi.org/10.18772/26180197.2020.v2n1a0.

Stone, S. (2020). *Coronavirus: What next for football as cracks begin to appear*. Retrieved from https://www.bbc.com/sport/football/52063085. Accessed 28 Apr 2020.

Taswell, C., Donohue, C., Mastwyk, M. T., Louey, A. G., Giummarra, J., Robertson, J., Darby, D. G., Masters, C. L., & Rowe, C. (2019). Avoiding methodological bias in studies of amyloid imaging results disclosure. *Alzheimer's Research & Therapy, 11*, 52. https://doi.org/10.1186/s13195-019-0495-y.

The China Analyst. (2009). Worldwide stimulus packaged (USD billion). *The China Analyst*, 4.

Tie, Y. C., Birks, M., & Francis, K. (2019). Grounded theory research: A design framework for novice researchers. *SAGE Open Medicine, 7*, 1–8. https://doi.org/10.1177/2050312118822927.

UNCTAD (United Nations Conference on Trade and Development). (2020). *Global trade impact of the coronavirus (COVID-19) epidemic*. Geneva: UNCTAD.

UNDRR (United Nations Office for Disaster Risk Reduction). (2015). *Sendai framework on disaster risk reduction (2015–2030)*. New York: UNDRR Secretariat.

UNFCCC (United Nations Framework Convention on Climate Change). (2015). *Paris agreement*. Bonn: UNFCCC Secretariat.

United Nations. (2015). *Transforming our world: The 2030 agenda for sustainable development*. New York: United Nations Secretariat.

United Nations. (2020). *Shared responsibility, global solidarity: Responding to the socio-economic impacts of COVID-19*. New York: United Nations Secretariat.

United Nations Habitat. (2016). *Habitat III: New urban agenda*. New York: United Nations Habitat Secretariat.

UNWTO (United Nations World Tourism Organisation). (2019). *International tourism highlights: 2019 edition*. Geneva: UNWTO.

Webeck, E. (2020). *Grand Princess Passengers sue cruise line for negligence over COVID-19 outbreak*. Retrieved from http//:www.mecurynews.com. Accessed 27 Apr 2020.

WHO. (2020a). *Coronavirus disease 2019 (COVID-19) situation report – 51*. Retrieved from https://www.who.int/docs/default-source/coronaviruse/situation-reports/20200225-sitrep-36-covid-19.pdf?sfvrsn=2791b4e0_2. Accessed 28 Apr 2020.

WHO (World Health Organisation). (2020b). *COVID-19 solidarity response fund for WHO*. Retrieved from https://covid19responsefund.org/. Accessed 5 Apr 2020.

Wilder-Smith, A. (2006). The severe acute respiratory syndrome: Impact of travel and tourism. *Travel Medicine and Infectious Disease, 4*, 53–60. https://doi.org/10.1016/j.tmaid.2005.04.004.

Zhang, H., Su, Q., Yin, Y., Wu, X., & Xie, W. (2019). Local–migrant interaction in everyday life in an ancient tourism town. *International Journal of Environmental Research and Public Health, 17*, 266. https://doi.org/10.3390/ijerph17010266.

Zumla, A., Kapata, N., Ihekweazu, C., Ippolito, G., & Ntoumi, F. (2020). Is Africa prepared for tackling the COVID-19 (SARS-CoV-2) epidemic? Lessons from past outbreaks, ongoing pan-African public health efforts, and implications for the future. *International Journal of Infectious Diseases, 93*, 233–236. https://doi.org/10.1016/j.ijid.2020.02.049.

Part II
Conceptual Framework for COVID-19 and Tourism

Chapter 2
Global Tourism Value Chains, Sustainable Development Goals and COVID-19

Abstract This chapter presents a conceptual framework and setting for the book. This is informed by the desire to link three critical thematic areas, namely, (i) the global tourism value chains, (ii) COVID-19 and (iii) the 2030 Agenda for Sustainable Development (AfSD) and the 17 embedded sustainable development goals (SDGs). Bringing in the SDGs adds value given that there are three SDGs (SDGs 8, 12 and 14) that make specific reference to tourism. Furthermore, COVID-19 negatively impacted many SDGs leading to governments, civic and private organisations revising budgets to channel resources towards "flattening" both the COVID-19 and economic curves. Understanding the global tourism value chains assists in opening up the complex tourism space and to systematically document COVID-19 impacts along with the industries within the value chain nodes. To this end, this chapter comes across mainly as an essay with heavy dependence on value add from the document and critical discourse analysis, as well as a meta-analysis of secondary data sources. The chapter is useful from both a theoretical and practical application points of view. A section bringing the nexus of the thematic focus areas is slotted in towards the end, with a critique of how the tourism sector should address shortfalls in relation to the SDGs within the COVID-19 pandemic.

Keywords Value chains · Tourism · COVID-19 · SDGs · Conceptual framework

2.1 Introduction

The purpose of this chapter is to demonstrate the nexus of the global tourism value chains, SDGs and the COVID-19 pandemic. This is done in the interests of developing a (conceptual) framework in which the entire book is located. The concept of the value chains is not new, and its current popularity is traceable to Michael Porter's 1985 book titled *Competitive Advantage*, which applied it to the firm level and related follow-up work (Porter 1985, 2001). This firm-level approach to value chains later evolved to wider applications, including the analysis of global value chains (Gereffi et al. 2001) and the creation of shared value by firms (Porter and Kramer 2018). The global value chain analysis comprises four key dimensions that include (1) the input–output structure, (2) geographic scope, (3) governance, and (4)

G. Nhamo et al., *Counting the Cost of COVID-19 on the Global Tourism Industry*,
https://doi.org/10.1007/978-3-030-56231-1_2

institutional context. The governance types include market, modular, relational, captive and hierarchy, while the institutional context looks at how local, national and international policies and conditions shape globalisation across the value chain (Gereffi and Fernandez-Stark 2011). Yilmaz and Bititci (2006) present the existing opportunity to study the tourism industry as a value chain, arguing that this leads to the development of a systemic-oriented performance management and measurement framework. They posit that using this approach and the framework would enable the tourism sector stakeholders to communicate, coordinate and share processes and activities effectively. In this way, the forward and backward linkages in the value chain could be fine-tuned, presenting a (tourism) firm with a competitive edge (Ensign 2001).

In a background paper for the 2020 World Development Report, Antràs (2020) opens up conceptual aspects of the global value chains. The author draws attention to the dying "Made in such and such a county" labels in manufactured goods, as most goods are now effectively "Made in the World". This transformation is informed by the first globalisation production approaches that began and grew rapidly in the period 1870–1914, resulting in a notable increase in international trade that was propelled by the invention of the steamship and the internal combustion engine (Nhamo 2014). Included in traded goods and services is the tourism sector with its service products traversing international boundaries. The movement of people within a globalised world is highlighted among other essentials that include financial capital, data and information (Tang et al. 2020). Although some authors distinguish between tourism and hospitality (Jones et al. 2017), unless stipulated, in this chapter and book, the term "tourism" is used broadly to also include hospitality. Why then the value chain approach? The value chain approach provides a holistic view enabling the identification of challenges, as well as the quick and big wins for a rapid recovery in the entire tourism sector post-COVID-19.

Gereffi and Fernandez-Stark (2011) are of the view that global value chains in sectors such as tourism have significant implications on global trade, production and employment. They also determine how firms from developing countries, producers and workers get absorbed into the global economy. Integration developing countries into global value chains is important for their development. However, integration in itself is not enough. Instead countries need to seek integration through activities that generate high returns. Hence, it is not just a matter of participating in the global economy but how it is done. For example, and from a tourism perspective, Jamaica has advantages in that it has good natural attractions including great beaches, crystal-clear seas and climate, is close to the USA market, has a cultural affinity with both the UK and the USA and has a reputation as a great vacation destination (Fernandez-Stark and Bamber 2018). However, there are set-backs that the country needs to rectify to remain competitive in the global tourism value chains. These set-backs include poor soft and hard transportation infrastructure, high crime rates and general tourist harassment (Ibid.). In addition, the local economy needs to participate and own part of the means that carry and host facilities used by the tourists.

While upbeat about the value of the tourism sector to national and global economies, most notably through contributions such as job creation across gender and age groups, the industry also has its downsides that are exacerbated when and where massive growth is unchecked (Group of Twenty – G20 2019). The negative impacts of the sector's rapid growth include challenges in the conservation of natural resources, general environmental decay, increased greenhouse gas (GHG) emissions leading to global warming and climate change, biodiversity loss, cultural erosion and frequent disputes with local and impoverished communities of many developing countries (Stone 2020; Ren et al. 2019). However, the G20 is also aware of the contribution tourism makes towards the attainment of the 2030 AfSD and the aligned 17 SDGs (United Nations 2015). With the outbreak of COVID-19 and its subsequent proclamation as a pandemic on 11 March 2020 (WHO 2020), the tourism industry was facing both short- and long-term threats (United Nations World Tourism Organisation – UNWTO 2020). The G20 (2019) identified significant roles played by the tourism sector in the attainment of four SDGs, namely, SDG 8 (Decent Work and Economic Growth), SDG 12 (Responsible Consumption and Production), SDG 14 (Life Below Water – Blue/Ocean Economy) and SDG 17 (Partnerships).

Drawing from the foregoing, this chapter develops and presents a conceptual framework for the entire book. It helps to map areas to focus on across the tourism value chain and brings up linkages between COVID-19 and the SDGs in general. Further analysis is done on tourism-specific SDG targets – an aspect that other chapters pick up and expand on later in the book. The methods utilised included document and critical discourse analysis (Perna et al. 2018) and selected elements from meta-analysis (Baldwin and Del Re 2016).

2.2 Understanding Global Tourism Value Chains

This section has three subsections. The first subsection focuses on Michael Porter's value chain propositions, while the second subsection deliberates on the global tourism value chains. The third subsection discusses the concept of the carbon footprint in the global tourism value chain. Each of these subsections will now be considered in turn.

2.2.1 On Michael Porter's Value Chain Propositions

Discussions on global tourism value chains will not make much sense if one does not start with Porter's (1985) generic value chain model. In the model, Porter identifies three main segments, including primary activities, support activities and firm margin (Fig. 2.1). The primary activities involve the physical creation of products and their sale and transfer to the buyer, including sale assistance (Porter 2001). For a firm to maintain and/or grow its profit margin, there is a need to coordinate both internal and external activities. From the model, the external environment deals mainly with inbound logistics and sourcing, whereas the internal linkages imply

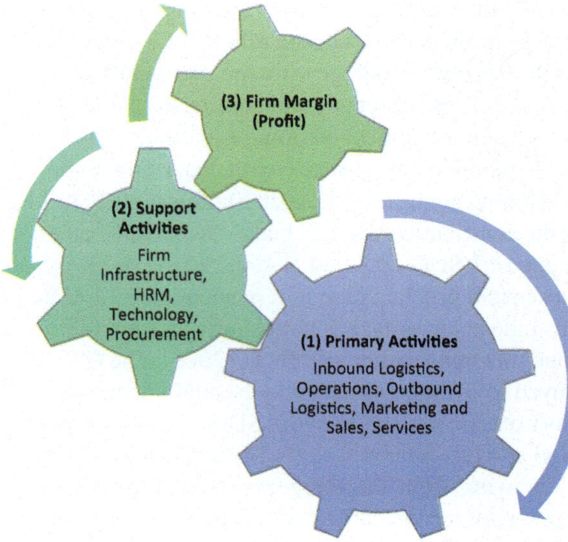

Fig. 2.1 Porter's value chain. (Source: authors, based on Porter (2001: 52))

smooth coordination between and among the support activities and the primary activities. Lui et al. (2020) refer to primary activities as basic value-added activities, and the support activities as auxiliary value-added activities. Drawing from Porter's work, Kaplinsky and Morris (2001) bring up a simple and generic value chain made up of four links, namely, (1) design, (2) production, (3) marketing and (4) consumption and recycling. In the production phase, there is Porter's inward logistics but also additional aspects such as transformation, inputs, packaging and other activities. The value chain is then defined as:

> The full range of activities, which are required to bring a product or service from conception, through the different phases of production (involving a combination of physical transformation and the input of various producer services), delivery to final consumers, and final disposal after use. (Kaplinsky and Morris 2001: 4)

A firm must maintain a competitive advantage that comes from activities it performs "in designing, producing, delivering and supporting its products" (Porter 2001: 50). The firm also looks for the cost advantage that arises from a low-cost physical distribution system, highly efficient assembly processes and/or superior sales utilisation. The value chain, therefore, disaggregates the firm into relevant activities that enable it to understand the behaviour of costs, as well as existing and potential sources of differentiation. Hence, the firm has a competitive advantage if it performs strategic activities cheaper and/or better than its competitors. The value chain for a firm is further embedded into a value system that can be illustrated by a single-industry firm or a diversified firm. Suppliers bring upstream value resulting in a firm's product becoming an integral element of its buyer's value chain. Porter sums up value as "measured by total revenue, a reflection of the price a firm's product commands and the units it can sell" (Ibid.: 52). Hence, although usually portrayed as a vertical chain, there are intrachain linkages that come up in a two-way nature.

For example, a specialised design firm may not only influence the nature of production processes and the manner in which the finished product gets marketed but also constraints in downstream links in the value chain (Kaplinsky and Morris 2001).

Given the foregoing, it is important to note that the tourism global value chain is made up of competing and, at times, collaborating firms, which have been and will still continue to be affected differently by COVID-19. The firms are also bound to respond in ways that at times call for partnerships but mainly in ways that make them retain their competitive advantages through their value chains. Seeing competing companies fold due to COVID-19 is a situation that some firms, which may have good strategies to cope with the COVID-19 storm, may look forward to. A new value for shareholders needs to be created, including through mergers and hostile takeovers. To this end, Porter and Kramer (2018) revisit the value chain within the context of shared value. The authors highlight that opportunities in shared value creation pivot on three things: (1) reconceiving products and markets, (2) redefining productivity in the value chain and (3) enabling local cluster development. Once each firm looks at decisions and opportunities through shared value lenses, this will result in new opportunities that generate greater innovation and growth for companies. The shared value creation approach brings together the long-missing connectivity between the firm's competitive advantage and social concerns (Porter and Kramer 2018). To this end, the configuration of company productivity changes. Company productivity becomes a function of several parameters that include looking at the environmental impact, energy use, water use, employee health, worker safety, employee skills as well as supplier access and viability. The issues raised herein remain significant to the tourism industry.

What Michael Porter did not accurately foresee 30 years later, which was evident during the preparation of this chapter, is the strong advancements in technology, as epitomised by the emergence of the Industry 4.0 (4th Industrial Revolution or 4IR) phenomenon and the centrality of the Smart Factory and Artificial Intelligence (Nhamo et al. 2020; Lee et al. 2018). Industry 4.0 became very visible in Germany in 2014, where the country was battling with ageing workers. Since then, the world has warmed up to the concepts of the Internet of Things, Internet of Services, Internet of Places and Internet of Data (PWC 2016; Kamble et al. 2018), which are all essential in the tourism value chains. As of 2020, the world reached a point where one could accept that it had truly become a global village through industry 4.0. With the Internet of Things, managing the supply chains, in particular, Porter's inbound logistics, has become smarter (Ben-Daya et al. 2017). Furthermore, with Artificial Intelligence, it has become feasible that robots can check guests into hotels and assist in directing other entities. Frederick et al. (2018) come to the party and highlight how the digital economy is influencing industrial equipment. In the authors' view, there are three disruptions in the GVC from the digital economy. These include (1) new industries and stages created in the GVC that change the distribution of value within existing GVCs; (2) significantly different skills requirements, demanding changes to traditional job roles and a greater emphasis on digital skills; and (3) the expansion of company value creation through collaborative ecosystems.

Michael Porter could also not foresee the growing pressure from the environmentalists who demand global stewardship to save the dying planet, particularly addressing climate change (Bramwell et al. 2017; Hall 2019). There is also overwhelming evidence of the increasing frequency of natural disasters and their increasing intensity and propensity to be more destructive. In the context of this book, growing environmentalism meant that the tourism value chain had to do something to protect the environment, particularly through the reduction of carbon footprints (Fang et al. 2018; Dube and Nhamo 2020a). Major global aircraft manufacturers raced to produce low carbon-emitting aircraft and harness biofuels to blend with petrochemicals (Dube and Nhamo 2020a).

Technological advancements also leap-frogged in information communication technologies (ICT). Whereas in the 1990s the most common, quickest and yet expensive mode of communication was the telephone, this changed drastically after a few years. The emergence of the Internet, email, Wi-Fi, mobile phones, social media (Facebook, Twitter, WhatsApp, etc.) and fibre optics all enhanced the global value chains, including in the tourism sector. Tourists can self-book their travel online and can also check in online, joining the travel agents' value chain. Tourism destinations are profiled online and virtually, including on social media, thereby impacting tourists' perceptions (Dube and Nhamo 2020b). Tourists use social media to rate their experiences at certain destinations and accommodation establishments. This impact on the rated establishments (Alonso-Almeida et al. 2019) at times leads to over-tourism where some destinations become overcrowded due to good reviews. Within minutes of incidences occurring, the world can be in the know due to the connectivity that now exists. Across the SDGs, there are several indicators that make reference to ICT, namely, indictors 4.4.1, 4.a.1, 5.b.1, 9.c.1, 17.6.2 and 17.8.1 (IAEG-SDGs 2019). Further details regarding these indicators of importance in the tourism value chains are shown in Box 2.1.

Box 2.1: ICT-Related Indicators of Relevance to the Global Value Chains

- 4.4.1: Proportion of youth and adults with information and communications technology skills, by type of skill
- 4.a.1: Proportion of schools with access to the Internet for pedagogical purposes and/or computers for pedagogical purposes
- 5.b.1: Proportion of individuals who own a mobile telephone, by sex
- 9.c.1: The proportion of the population covered by a mobile network by technology
- 17.6.2: Fixed Internet broadband subscriptions per 100 inhabitants, by speed
- 17.8.1: Proportion of individuals using the Internet (from any location), all individuals (%)

Source: authors, based on IAEG-SDG (2019: 12–38)

Having focused on the generic global value chains, the next section reviews the global tourism value chain. An attempt will be made to develop a simplified global value chain as a framework informing further developments of the book chapters.

2.2.2 Global Tourism Value Chains

Antràs (2020: 5) defines a GVC as "a series of stages involved in producing a product or service that is sold to consumers, with each stage adding value, and with at least two stages being produced in different countries". Gereffi and Fernandez-Stark (2011: 2) see the GVC as focusing on:

> The sequences of value-added within an industry, from conception to production and end-use. It examines the job descriptions, technologies, standards, regulations, products, processes, and markets in specific industries and places, thus providing a holistic view of global industries both from the top-down and the bottom up.

Tourism has become the largest branch of service trade, contributing to the global economy through its value chains (Wang 2018; Daly and Gereffi 2017). A firm integrates in a GVC if it produces at least one stage in a GVC. The GVC provides a framework to understand how global firms are arranged. This is done by evaluating the structure and dynamics of the key actors involved (Gereffi and Fernandez-Stark 2011). In the modern world, there are complex firm interactions to which the GVC methodology assists in tracing the shifting patterns of global production. Christian et al. (2011) observe that tourism is labour-intensive, demanding skills from the lower to the upper range of expertise. Workers are needed in multiple sectors that work together to build the tourism industry and participate in the (global) tourism value chain. In many developing countries, the tourism jobs are of the lower grades such as groundskeepers, housekeeping and cleaning services. Many managerial and skill positions end up in the control of expatriates. This needs to change if tourist attractions in developing countries are to contribute more to the economies of those countries.

For one to understand the global tourism value chain, the entry point is the tourist "footprint" (Christian et al. 2011). The tourist footprint is characterised by the interactions made with firms in the value chain. The firms cover services and products that include transportation distribution, accommodation, excursions and the tourist attractions such as viewing nature, engagement in religious tours, medical tourism, sports tourism and cruise ships (Szpilko 2017). Daly and Gereffi (2017) identify two broad categories of travel tourism: leisure and business (Table 2.1).

As highlighted earlier, several jobs (positions) exist within the global tourism value chain. These include those in distribution, intermediaries and the service providers (Daly and Gereffi 2017). Under the distribution intermediaries are global tour operators, inbound tour operators and travel agents. From the service providers are accommodation establishment managers, airlines agents, accommodation front offices, restaurant/bar staff, retail, local guides, drivers and housekeepers. Jobs in the distribu-

Table 2.1 Definitions and descriptions of leisure and travel tourism

Travel tourism type	Definition and description
Leisure tourism	Means any trip where the tourist travels for pleasure
	Attractions include the sun, sand and surf; environmental or ecotourism; adventure; cultural; etc.
	However, leisure travel does not include visiting friends or relatives
Business tourism	Involves travel for professional reasons, including visits to see clients, scouting trips for potential investment opportunities and travel for conferences
	Conference segments cover meetings, incentives, conferences and exhibitions (MICE)

Source: authors, based on Daly and Gereffi (2017: 40–41)

tion intermediaries require medium to high skill levels. Apart from lodging management staff that may require medium to high skill levels in the service providers segment, most jobs in this category require low to medium skills. Such skills include on-the-job training, technical education and certificate programmes. Figure 2.2 shows a simplified global tourism value chain for both out- and inbound countries.

Drawing from cruise tourism in Shenzhen, China, Lui et al. (2020) identify a typical cruise company value chain. The authors' highlight company infrastructure, human resource management, technological development, procurement and source of tourists as part of the value chain. Covered in firm infrastructure are daily operations management, investment promotion and financial accounting activities. From

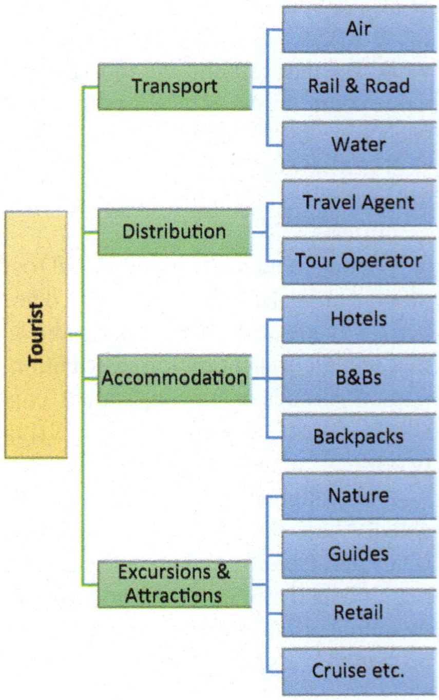

Fig. 2.2 Simplified global tourism value chain. (Source: authors, based on Porter (1985); Christian et al. (2011))

the human resource management angle, activities include recruitment, capacity development and training, as well as labour dispatch. The design and development of cruise lines, research and innovative tourism attractions are included within the technological development segment of the value chain. The procurement segment looks after liaison, negotiation, ordering and maintenance. The tourists' source segment in the cruise company value chain includes customer contact, communication, marketing, planning, operations and after-sales services (Spencer et al. 2014). Daly and Fernandez-Stark (2017) focus on Barbados as it takes part in the cruise tourism global value chain. The travel agents perform outbound activities that include travel package bookings. The cruise ships host hospitality activities, lodging facilities, entertainment and transport onboard. Activities in an inbound country cover port agents, hospitality, shopping and lodging, excursions and domestic transport. A similar cruise tourism global value chain was observed in St. Lucia by the same authors (Daly and Fernandez-Stark 2018). All points of value addition in the global tourism value chain aim to enhance competitiveness at both firm and country level.

From the transport sector, most international tourists use air and cruise services (Christian et al. 2011). Rail and road are also common in other set-ups such as in Europe and Asia. These forms are not common in linking up small island states. To this end, rail and road largely remain applicable to the domestic and regional tourist. However, technological advances such as the underwater train-link between Britain and mainland Europe are changing the reach of rail and road transport. Nevertheless, the international tourist still largely arrives through the airport and thereafter is taken to the local attractions. The attractions include natural site viewing such as the Victoria Falls (shared between Zambia and Zimbabwe), Mount Kilimanjaro of Tanzania and Kruger National Park and Table Mountain of South Africa (Dube and Nhamo 2019a, b). Beach tourism, sailing and wine tourism are other attractions, while religious tourism, including the Hajj and Umrah of Saudi Arabia, are also in the mix. Additional tourism attractions include sports and sites of historical and dark tourism significance such as the 9/11 Twin Towers Memorial Site in the USA and Rwanda's 1994 Genocide monuments (Parry 2018; Potts 2012).

The tourism global value chain entails a country becoming a tourist destination (Antràs 2020). As has been happening the world over, and especially in Mecca, the host city for the Muslin Hajj and Umrah pilgrimages (Ladki and Mazeh 2017), hotel beds, transport and other infrastructure including ICT is developed and expanded by companies, venture capitalists and through Sovereign Wealth Funds to handle growing tourist arrivals (Abuhjeeleh 2019). Relevant to this chapter and book is that all these services and firms act to provide a good experience to the tourist. Any event that impacts any part of this system diminishes this experience. What is of interest to this book from the global tourism value chain perspective is that COVID-19 affected both the mega- and microenterprises. The extent to which these businesses got impacted may have differed, but the bottom line remains, the negative disruptions were massive. In the tourism industry, there is a complex interlink between big corporates on one hand and small, medium and microenterprise (SMMEs) (Christian et al. 2011). Therefore, the ongoing economic stimulus packages have had to routinely and carefully consider the needs of both the SMMEs and the bigger entities.

2.2.3 *Carbon Footprint in the Global Tourism Value Chain*

The UNWTO realised the damage the transport sector was causing to the environment, as goods and services provided within the global tourism value chain result in carbon emissions. The carbon emissions are known to cause global warming that leads to climate change and ultimately causes extreme weather events such as droughts, wildfires, extreme frost and floods (UNWTO 2019). Climate change concerns peaked in 2015 during the United Nations Framework Convention on Climate Change (UNFCCC) 21st Conference of the Parties (COP21) meeting, which concluded the Paris Agreement (UNFCCC 2015). The Paris Agreement placed on the table a number of intervention measures to both mitigate (reduce the emissions of GHG) and build adaptation capacity for nations (learn to live with the changing climate). From a GHG (carbon) emissions perspective, every sector of the economy had to bring up measures to mitigate the emissions. Hence the UNWTO's initiative to audit how firms within the tourism value chain contributes to carbon emissions.

As of 2019, the UNWTO (2019) realised that tourism had grown to represent 10% of total global employment and 10% of global gross domestic product (GDP), thereby assisting in the attainment of the 2030 AfSD. From a baseline on carbon emissions done by the UNWTO in partnership with the United Nations Environment in 2008, it emerged that the tourism sector contributed an estimated 5% of total global emissions, with the transport sector taking up 75% of the sector contributions (UNWTO 2019). A summary of carbon emissions from the tourism sector is shown in Fig. 2.3.

While travel contributes the majority of the industry's carbon emission, other sectors also contribute notably. For instance, the accommodation has a significant share accounting for 21% of the emissions meaning that travel and accommodation make up 96% of the industry GHG emissions.

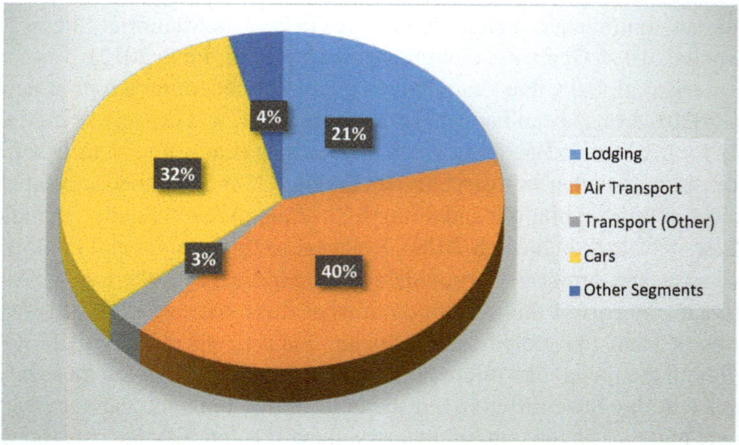

Fig. 2.3 Carbon emissions from the global tourism sector (as of 2005). (Source: authors, data from UNWTO (2019: 12))

In 2016, the UNWTO (2019) narrowed down the profile of carbon emissions in the transport sector. The boundaries applied in calculating the carbon footprint were set to include both passenger transport (car, rail, air) and freight transport (maritime, air, surface). The total emissions stood at 7 230 million tonnes (23% of all global human-induced emissions in the transport sector). Of the tourism sector total, 64% was from passenger transport. What the UNWTO also noted was that although there were improvements in fuel efficiency and advancements in cleaner and greener modes of transport (including electric vehicles), the growth in passenger and freight transport demand would still lead to an increase in carbon emissions by 2030. To this end, carbon emissions from the transport sector of the tourism value chain are expected to grow by 21% compared to 2016. This represents 23% of all global human-induced carbon emissions in the entire transport sector (Ibid.).

The growth in the transport sector carbon emissions mirrors similar findings from the broader tourism sector in one of China's provinces. Ren et al. (2019) conducted a carbon footprint survey for the tourism sector in the Sichuan Province of China. The results show that from 2004 to 2018, the carbon footprint increased 20-fold from about 427.59 million tonnes to about 9.51 billion tonnes. Effectively, this pushed the province to a 2013 carbon deficiency compared to the years 2004 to 2012 when there was a carbon surplus. Following on with air travel, Ştefănică (2017) highlighted that tourism represented 60% of all air travel. To mitigate carbon emissions from air travel, some airlines entered into partnerships to procure sustainable biofuels (Hong et al. 2019). Added to the equation has been the growth in carbon neutral airports, including their accreditation. Airports have been increasing their environmental credentials through commitments to go greener by reducing their carbon emissions. One example is the Brussels Airport in Belgium (Boussauw and Vanoutrive 2019).

Carbon emissions from hotels also prompted the environmental sustainability debate in the tourism industry. Abeydeera and Karunasena (2019) focused on carbon emissions from five hotels in Sri Lanka, arguing that although the subject matter was receiving more attention in developed countries, this had not been the case for developing countries. The results were that each of the case study hotels released more than 7 000 tonnes of GHG emissions annually. Furthermore, purchased electricity, usually from the national grid, contributed the bulk of the carbon emissions. It is a common matter, and many hotels that wish to reduce their carbon footprint end up installing solar energy. Reducing carbon emissions in hotels is but one of the interventions in environmental sustainability. Other interventions include sustainable waste management and water and energy efficiency (Wang et al. 2019). These and other interventions assist in adding value to the hotels and related entities in the tourism value chain and ultimately impact on the firm margin. In Japan, Kitamura et al. (2020) looked into the entire Japanese tourism carbon footprint. The authors found that the sector contributed 136 million tonnes annually, with 56.3% of total emissions from the sector coming from transport. The next section addresses the impact of COVID-19 on the broader 2030 AfSD.

2.3 Impact of COVID-19 on the SDGs

Following the outbreak of COVID-19, the United Nations (2020) presented a quick and consolidated view on how, overall, the pandemic had and was going to negatively impact the attainment of the 2030 AfSD and the SDGs. The anticipated consequences were to be profound, with women, children, the elderly and informal workers being among the most vulnerable stakeholders. However, despite the negative impact, there were also positive outcomes. In the short term, gains were anticipated from rescued GHG emissions as economic activities came to a halt in many countries and territories (Stone 2020). The airline, cruise ship and land-based transportation industries also came to a virtual standstill during the early COVID-19 pandemic era. In China, carbon emissions went down an estimated 18% of normal between early February and mid-March 2020. This was attributed to reductions in the burning of coal to generate electricity against a background of dwindling industrial activity. This represented forgone carbon emissions estimated at 250 million metric tonnes. Reduction of carbon emissions was also witnessed in Europe, which recorded an estimated 400 million metric tonnes, accounting for about 9% of the region's cumulative 2020 emission targets (Ibid.). In one of the publications in the journal, *Nature Climate Change*, Quéré et al. (2020) observe that drastic reductions in energy demand due to the COVID-19 pandemic witnessed daily carbon dioxide emissions reduced by −17% by early April 2020 compared to the 2019 average.

As for the overall impact of COVID-19 on the SDGs, this was mainly negative (United Nations 2020). For example, the loss of employment and income was likely to push more individuals and families below the poverty line (SDG 1). Furthermore, during the initial stages of COVID-19, out-of-pocket expenditure presented an additional burden to the poor, including that of limited health insurance. Food production and distribution were also disrupted (SDG, 2). This was especially so during the first phases of lockdown when factories closed as governments and companies instructed workers to stay home (Nicola et al. 2020). This was eventually relaxed as some countries identified the agriculture sector and related industries as essential industries that could remain open. However, operations remained under strict health and hygiene conditions, including massive use of masks and hand sanitisers. Farmers of specialised seasonal crops, like grapes and those from the horticulture subsector, lost income as there were minimal logistics available, including limited labour. Bottlenecks in the food supply system were likely to aggravate matters of nutrition, thereby directly impacting on the heath (Veldhuizen et al. 2020).

Concerning health (SDG 3), a new challenge was presented to the world that was already under distress from the chronic challenge of HIV/AIDS, TB, malaria, cholera and other diseases (United Nations 2020). Apart from those who were infected and those who passed away, the world was left with a huge health insurance bill. Systems could not cope as the COVID-19 burden threatened to overrun existing capacities. The need for makeshift hospitals was addressed by converting sports and showgrounds into hospitals. This presented a challenge to already strained financial, human and times resources that had to be redirected from other sources specifically to fight the pandemic (Wang and Tang 2020). This effectively reduced the capacity of governments, organised labour, the corporate sector and other social partners to deploy resources towards the attainment of some SDGs.

COVID-19 did not leave the education sector untouched (SDG 4) either. Schools closed, resulting in makeshift online and homeschooling for the privileged (United Nations 2020). However, many from the less developed countries – such as is typical in Africa – had to completely shut doors for months. The impact was felt from the preparatory level, through to primary, secondary and tertiary levels. From a gender perspective (SDG 5), there were increased reports of violence against women during lockdowns (de Lima et al. 2020). Women in rural areas were also identified to be at increased risk, with the main aggressor being the spouse. Campbell (2020) highlights the fact that many countries were reporting increased levels of domestic violence, yet there were no known protocols on how to deal with such incidences during natural disasters like COVID-19. The United Nations (2020) also observed that women made up the bulk of social care and health workers and, as such, were at the forefront of the COVID-19 battle.

The work environment and economic growth (SDG 8) were placed under severe strain too. Economic activities were suspended, and people lost jobs (United Nations 2020). After lockdowns, there was a staggered opening of sectors of the economy, with the tourism industry at the back of the queue (Nicola et al. 2020). Linked to SDG 8 negative impacts are matters pertaining to inequality (SDG 10). As more and more individuals were relegated to unemployment statistics, so did the levels of inequalities rise. Inequalities further rose due to school closures (Armitage and Nellums 2020). From an SDG 11 angle, people living in densely populated areas, including those in informal settlements and refugee camps, remained at high risk (United Nations 2020). Apart from huge numbers, these areas usually have poor water and sanitation facilities. To this list, Ahmed et al. (2020) added prisons, which are usually overcrowded.

What comes out from the discussions herein confirms that the SDGs are intertwined, thus, all in one and one in all (United Nations 2015). As one SDG is negatively impacted, such as was the case with the health SDG from COVID-19, all other SDGs are also impacted in one way or another. Although it was still early for a solid call, it was emerging as a mixed bag for SDG 17 that addresses partnerships. Good partnerships emerged to finance and provide resources for fighting COVID-19 at national, regional and international levels. Yet, threats of cutting funds to the WHO by USA President Donald Trump also emerged (Mahase 2020). The USA president accused the WHO, especially its head of mission, of being "China-centric", thereby conniving to cover up the extent of the potential global damage COVID-19 had caused. Although all member states contribute towards the WHO budget, the USA always contributed a greater portion averaging over 20%. The next section is dedicated to addressing the linkages between tourism and SDGs.

2.4 A Focus on Tourism and the SDGs

In 2019, the Group of Twenty (G20) nations' tourism ministers and some invited countries and organisations – including the UNWTO, Organisation of Economic Cooperation and Development (OECD), International Labour Organisation (ILO) and the World Travel and Tourism Council (TTC) – took notice of the desire to advance tourism's contribution to the SDGs (G20 2019). The G20 also noted that tourism creates jobs for people of all ages and skill levels.

The experiences emerging from across the global tourism industry in response to COVID-19 are not new. The COVID-19 pandemic presents a challenge to the world's quest to attain the 2030 AfSD, with its interlocked 17 SDGs (United Nations 2015). Included among these SDGs are three that make explicit reference to the tourism industry (Table 2.2). Although SDG 14 indicators are not explicit in term of tourism, Target 14.7 mentions tourism. In addition, SDG 17 can be added as this remains generic across all the SDGs. Hence the reason why the G20 (2019) includes it under SDGs to which tourism could make significant contributions is highlighted in the introduction to this chapter.

In its publication on tourism and the SDGs, with a focus on good practices from the Americas, the UNWTO (2018) notes the explicit reference to tourism on the SDGs highlighted in Table 2.2. However, the UNWTO indicates that sustainable tourism should play a pivotal role in delivering solutions across all the 17 SDGs. In the Americas, where 207 million international tourist arrivals were reported in 2017,

Table 2.2 SDGs making explicit reference to tourism

Goal	Goal targets	Indicators
Goal 8. Promote sustained, inclusive and sustainable economic growth, full and productive employment and decent work for all	8.9 By 2030, devise and implement policies to promote sustainable tourism that creates jobs and promotes local culture and products	8.9.1 Tourism direct GDP as a proportion of total GDP and in growth rate
		8.9.2 Proportion of jobs in sustainable tourism industries out of total tourism jobs
Goal 12. Ensure sustainable consumption and production patterns	12.b Develop and implement tools to monitor sustainable development impacts for sustainable tourism that creates jobs and promotes local culture and products	12.b.1 Number of sustainable tourism strategies or policies and implemented action plans with agreed monitoring and evaluation tools
Goal 14. Conserve and sustainably use the oceans, seas and marine resources for sustainable development	14.7 By 2030, increase the economic benefits to small island developing states and least developed countries from the sustainable use of marine resources, including through sustainable management of fisheries, aquaculture and tourism	14.7.1 Sustainable fisheries as a proportion of GDP in small island developing states, least developed countries and all countries

Source: authors, based on IAEG-SDGs (2019: 19–32)

the tourism sector was identified "as a priority to promote economic development and diversification, fight poverty and create sustainable livelihoods" (Ibid.: 8). The focus on poverty points to SDG 1. The UNWTO goes further to identify how tourism could lead to the protection of biodiversity (SDG 15) and act as a catalyst to provide a platform for building peace (SDG 16). Given the foregone, it becomes necessary to elaborate on how tourism contributes (or lack thereof) to the attainment of all the 17 SDGs (Table 2.3).

Table 2.3 Tourism contribution to the SDGs

Goal[a]	Selected key tourism sector contributions
SDG 1: No poverty	Provides income through job creation at local and community levels, with low skills requirement and local recruitment able to empower less favoured groups, particularly youth and women
SDG 2: Zero hunger	Spurs sustainable agricultural growth by promoting the production and supplying to hotels and sales of local products to tourists. Agro-tourism can generate additional income while enhancing the value of the tourism experience
SDG 3: Good health and wellbeing	Tax income generated from tourism can be reinvested in healthcare and services, including improving maternal health, reducing child mortality and preventing diseases
SDG 4: Quality education	Potential to promote inclusiveness and provides opportunities for direct and indirect jobs for youth, women and those with special needs, who should benefit through educational programmes
SDG 5: Gender equality	Empowers women through direct jobs and income generation from SMMEs in tourism and hospitality-related enterprises
SDG 6: Clean water and sanitation	Investment requirement for providing tourism-related utilities play a critical role in achieving water access and security, as well as good hygiene and sanitation for all
SDG 7: Affordable and clean energy	Can accelerate renewable energy shares in the global energy mix, thereby contributing to carbon reductions (SDG 13) and access of energy for all
SDG 8: Decent work and economic growth	One of the top four export earners globally, which provided one in ten jobs worldwide as of 2018 with a strong bias towards youth and women
SDG 9: Industry, innovation and infrastructure	Sector can influence and excite public policy for infrastructure upgrade and retrofit, making them more sustainable, innovative and resource-efficient and moving towards low carbon growth. The Gautrain high-speed train investment in South Africa for the 2010 FIFA Football World Cup remains testimony to that
SDG 10: Reduce inequality	Remains a potentially powerful tool for reducing inequalities if it engages local populations and all key stakeholders in its development. Sector is also an effective pathway for economic integration and diversification
SDG 11: Sustainable cities and communities	Promotes regeneration and preserves cultural and natural heritage assets on which tourism depends. Investment in green infrastructure (more efficient transport, reduced air pollution) should result in smarter and greener cities for all users
SDG 12: Responsible consumption and production	Sector needs to adopt sustainable consumption and production modes, accelerating the shift towards sustainability

(continued)

Table 2.3 (continued)

Goal[a]	Selected key tourism sector contributions
SDG 13: Climate action	Contributes to and is affected by climate change. As such, stakeholders should play a leading role in the global response to climate change by reducing the carbon footprint, particularly in the transport and accommodation sectors
SDG 14: Life below water	Tourism development must be a part of integrated coastal zone management in order to help conserve and preserve fragile marine ecosystems and serve as a vehicle to promote a blue economy, contributing to the sustainable use of marine resources. The sector will also present opportunities for fast-tracking investments in the underdeveloped blue/ocean economy
SDG 15: Life on land	Can play a major role in promoting sustainable management of fragile zones, not only in conserving and preserving biodiversity but also in generating revenue as an alternative livelihood to local communities
SDG 16: Peace, justice and strong institutions	The sector can foster multicultural and interfaith tolerance and understanding, laying the foundation for peaceful societies. It can also consolidate peace in post-conflict societies, particularly at a community level
SDG 17: Partnerships for goals	Has the ability to strengthen private/public partnerships and engage multiple stakeholders from international, national, regional and local scales to mobilise the resources (policy and innovative financing) needed to achieve the 2030 Agenda

Source: authors, based on UNWTO (2018: 20–21)
[a]Full expression of the SDGs can be found from the United Nations (2015: 14)

To demonstrate the contributions the tourism sector may make across the SDGs, the UNWTO (2018) documents some case studies from the Americas. For example, in Chile, there was an initiative to strengthen municipal tourism management, thereby addressing SDG 11 directly. In Colombia, the El Carlos Ecotourism and Archaeological Centre were profiled. The centre was established by ex-guerrilla and ex-paramilitary forces and people with disabilities to address SDG 16 through tourism ventures. Ecotourism projects were also observed and profiled for Panama. In Ecuador, the Karanki Magdalena Community Project is an indigenous establishment that creates deployment (SDG 8) and developed infrastructure (SDG 9) while preserving community culture and traditional livelihoods (SDG 11). In Guyana, the Rewa Eco-lodge provides employment for the Rewa people and has resulted in the village protecting its natural resources (SDG 15). Lodges and natural attractions were discussed earlier in this chapter under the tourism value chain. In Jamaica, the Community Tourism Toolkit has provided community-based tourism businesses with a set of tools on how to establish their enterprises. This answered calls to have locals get involved in SMMEs in the tourism value chain. To harness the blue/ocean economy (SDG 14), the Mayakoba Tourism Development in Mexico focuses on high-level coastal tourism development. Lastly, the Sustainable Destinations Alliance for the Americas brings together 11 destinations in the Caribbean and Central America. These partnered to address the challenges of climate vulnerability and environmental degradation in a bid to attract more arrivals (Ibid.). The next section brings the three thematic areas together.

2.5 The Nexus Between COVID-19, Tourism and the SDGs

This section raises potential red flags and critiques (to revisit and redress as appropriate moving forward). If the tourism sector is to recover quickly and be better in the context of the 2030 AfSD, then there is a need to proactively consider the complex linkages joining the people, planet, prosperity and partnerships (the 4Ps) (United Nations 2015).

Many potential tourists died during the pandemic, and many more were traumatised by caring for the dying and losing their loved ones. There were also issues of economic recession, resulting in job losses, general anxiety and anger and fear with regard to the future (Potash et al. 2020). With economic activities grounded for some major sectors, including tourism and key support industries, projections were clear that both economic and psychosocial recovery would take longer than past pandemics such as SARS and MERS, resulting in even bigger post-traumatic stress. The oil and gas industry was among those hard hit, and it haemorrhaged jobs (Nicola et al. 2020). On Monday 20 April 2020, the oil price crashed to a historical low with West Texas crude crashing to below $0, with contracts getting as low as −$37.63 per barrel, and Brent crude went down further, trading around $25.42 (Omarjee 2020).

Hundreds of thousands were still being infected, with tens of thousands dying, at the time this book was being written. This brings to the fore the need for psychosocial support services for the entire world (Potash et al. 2020) as the COVID-19 effects persisted among affected individuals (Onyeaka et al. 2020). To this end, art therapy remained one of the pathways that could be enrolled in the recovery and post-recovery phases of COVID-19 to benefit the tourism sector (Potash et al. 2020). The tool has the potential to reach out to traumatised communities, secure family connections as well as develop coping and resilience mechanisms, thereby enhancing hope for the future. Art therapy proved to be a very useful tool during past pandemics like SARS and Ebola, especially for those on the frontlines such as medical personnel and security clusters and those in other essential services.

Emerging evidence is pointing to the fact that COVID-19 deaths were mainly associated with the elderly, particularly in Europe. Whereas tourists come from all over the world, Europe remains one of the main suppliers of outbound tourists globally (Macola and Unger 2018). Of the European travellers, senior citizens made up 50% of the total, with 52% out of the total coming from Germany, Switzerland and Austria (Macola and Unger 2018). In Myanmar, for example, Macola and Unger (2018) established that in 2016, about 18.7% of international tourist arrivals were from western Europe, particularly the UK, France, Germany, Italy, Netherlands, Switzerland and Spain. Most of these were badly impacted by COVID-19, with Italy being the European epicentre, both in the numbers of those infected and deaths. Spain was (then) also the second-worst impacted on the continent. Furthermore, an estimated 67.94% of international arrivals came from Asia, especially China and Thailand, with North America coming in as the third highest contributor at 7.2%. China was the original epicentre of COVID-19, and the USA became the global epicentre, taking over from both China and Italy – all important sources of tourists.

On tourism and its role towards the attainment of the SDGs, there are descending voices emerging. Kamp and Mangalasseri (2017) raise a red flag concerning the fact that the concept of tourism for poverty alleviation (SDG 1) and its "trickle-down effect" of having benefits going down to the poor of communities have proved challenging to achieve. Instead, tourism could have increased poverty in many communities due to its inherent inequalities (SDG 10). Deregulation within the tourism value chain and the erosion of workers' rights both lead to loss of employment (Ibid.). There are instances where restaurant workers may largely work for tips, with minimum basic wages, especially if they are migrant labourers. To make matters worse, within this group of the exploited are women and the youth.

With reference to tourism and SDG 2, Zerrudo (2017) is of the view that it remains important to think local, including not only serving the interests of the tourists but also those of the local communities. In the authors' view, local farmers should be empowered to embark on an integrated, holistic food-based approach to tourism. This enables the local farmers to grow nutritional food from sustainable practices, thereby permitting them to earn an income through selling good produce to tourists and thus better sustain their families too. Although there still remain some tourists that demand locally grown and traditional foods, the massive emergence of fast-food chains, including food courts across the world, leads to commercialised food production dominating the supply chain. This has also led to the sourcing of food where it is cheapest across the world. While this practice is sensible in conventional commercial analysis, considerations of sustainability, i.e. the carbon emissions from transporting food, alters many considerations leading to a preference of locally produced and supplied foods albeit at a competitive price.

Regarding the contribution of tourism to health (SDG 3), Jaeger (2017) calls for a comprehensive, cross-sectoral and people-centred approach. Both the local people and tourists' health and safety issues must be considered. The health and safety of tourists and crew members were severely compromised in the cruise ship industry during the COVID-19 pandemic, resulting in infections, deaths and ongoing litigation (Mole 2020). Labour laws are regularly violated in the tourism industry, including lack of or under insurance for medical bills. This suggests a need to ensure that some of the generated revenue from the industry should be invested in healthcare systems for all to be afforded good healthcare (Jaeger 2017).

There are also outstanding matters of education and gender equality in the tourism industry. Novelli and Jones (2017) maintain that tourism development master plans usually overlook education (SDG 4), focusing more on property developments rather than enhancing skills and human capital development needed by the sector. The issue of human capital development was highlighted earlier in the generic value chain as a support activity for the companies. Concerning gender matters, Alarcón (2017) brings to the fore the fact that although many women are involved in the tourism sector, the main issue remains exploitation through low and differentiated salaries. The matter of low and differentiated salaries flies in the face of SDGs 8 and SDG 10 that anticipate decent work and wage equity (Cañada 2017). If tourism's success is to be measured in line with the 2030 AfSD, then it should be evaluated against the manner in which it contributes towards reducing inequalities as enshrined in SDG 10 (Monshausen 2017).

There are also voices opposing the proposition by the UNWTO (2018) that tourism can contribute towards achieving sustainable water and sanitation supplies (SDG 6). Contrary to this view, Jennings (2017) sees tourism leading to water scarcity and further inequalities. In spite of increasing climate-related droughts, destinations overexploit both surface and underground water resources to meet the tourism industry needs as well as local domestic and other industrial uses. In other instances, waste generated in hosting tourists end up in river channels and the sea, mainly plastic waste. This and other issues form a group of matters that require redressing if the tourism sector is going to contribute towards the attainment of SDG 12 on sustainable consumption and production (United Nations 2015; Plüss and Sahdeva 2017) as well as SDG 14.

Many controversial issues emerge when one focuses on tourism, energy consumption (SDG 7) and carbon emission (SDG 13). Many destinations still use dirty, hydrocarbon-based sources of energy including petrol, diesel and jet fuel, leading to huge carbon emissions, especially from the transport sub-chain of the tourism value chain (Eijgelaar and Peeters 2017). In other instances, wood fuel is utilised, resulting in massive deforestation, soil erosion and the silting up of rivers. The negative impacts resulting from carbon emission and general environmental decay have disproportionately affected communities in developing countries. To this end, stakeholders in the tourism value chain are encouraged to localise SDGs as appropriate, leading to the lowering of carbon emissions and footprints (Dube and Nhamo 2020b). Before the onset of the coronavirus infection, this transformation was already in full swing as witnessed by developments in fuel-efficient technologies in the aircraft manufacturing sector and lodging. Many hotels and B&Bs were investing in energy-efficient technologies as well as renewable energy, especially solar (Dube and Nhamo 2020a). Innovation and technological advancements as harnessed in SDG 9 remain at the centre of the tourism sector's contribution to the SDGs too (Kösterke 2017).

With the advent of the revitalised ocean (blue) economy presented in SDG 14, many across the world tour the coastal areas, including floating resorts mainly in the form of the cruise ship subsector. Oceans cover up to 71% of the earth's surface area, supporting millions of people's livelihoods (de Man 2017). Oceans further support biodiversity and life on land as enshrined in SDG 15 (Karschat and Kühhas 2017), including flora and fauna as well as some of the world heritage sites. In the 183 countries with coastlines and 37% of the world's population living in coastal communities, there remains a high risk of tourism development leading to human influxes that could ecologically damage these resorts (de Man 2017). Furthermore, the hotel subsector has encroached on beach fronts and other natural attractions such as nature reserves and parks, breaking building and conservation barriers (Karschat and Kühhas 2017). This once more promotes mass tourism that degrades the biodiversity. These are all red flags to address in 2030. In addition, the shareholders in the tourism value chain need to pay additional attention to small island developing states (SIDS) whose economies are heavily dependent on tourism.

As for good governance (SDG 16) and partnerships (SDG 17), the United Nations (2015) highlighted that there were issues to address leading up to 2030. Although the UNWTO (2018) praised the tourism sector for bringing peace to the Americas through several partnerships, Rutherford (2017) observed that governments and the corporate sector dominate with minimal integration of the local economies in the respective value chains. This tends to result in neglected fundamental human rights. To this end and moving forward, local communities should be empowered to participate in decision-making concerning existing and proposed new tourism developments. There should be effective and transparent monitoring mechanisms at various spatial scales including at local, national and global levels. The corporate sector usually prefers voluntary regulations, and this comes with inherent loopholes (Nhamo and Swart 2012) that require additional verification lenses from non-governmental organisations and governments. The business sector will always try to maximise its profit margins and will always prefer self-regulation. In most instances, self-regulation results in the bare minimum of standards being put in place. Overall, there remains a challenge concerning the choice of SDGs to address by certain entities in the tourism sector as taking on board all 17 SDGs could be problematic (Jones et al. 2017). Entities would need to engage their stakeholders to get assistance and determine which SDGs could be of importance in the tourism value chain. This will eventually enable the stakeholders in the value chain to determine which SDGs are of material value to them. Many big tourism corporates already had sustainability policies and strategies in place. Hence the SDGs and COVID-19 may require the reorientation of such, including building new and relevant human resources capacity.

What is emerging from the foregoing discussion is that, while tourism remains an acknowledged significant economic sector, several red flags demand careful managing to make the sector truly sustainable. For example, debates tend not to factor in the fact that there is strong competition in attracting visitors between and among countries and continents. Furthermore, the sector will remain in the spotlight with government and other bailouts in the mix (Kang and Rhee 2020; Loayza and Pennings 2020).

2.6 Conclusion

The tourism global value chain provided a good framework and setting for the book. The tourism industry and players in the value chain will be seeking ways to bounce back after a disaster and to do so quicker and better after the COVID-19 pandemic subsides. To this end, while the existing working partnerships need to be maintained, new and innovative ways of resuscitating the sector need to emerge. Every segment in the tourism value chain must rise simultaneously because a piecemeal approach may delay recovery. With several key stakeholders involved in industries within the value chain, linkages can be traced between the air travel that is supported by the travel agents and lodging segments, while the lodging segment needs restaurants, retail and the food and beverage sector. It is a complex web. The chapters that follow will draw insights from this chapter and focus specifically on industries within the value chain segments including air travel, cruise ships, airport networks, hotels, restaurants, sports tourism, pilgrimage and religious tourism as well as tourism economic stimulus packages.

References

Abeydeera, L. H. U. W., & Karunasena, G. (2019). Carbon emissions of hotels: The case of the Sri Lankan hotel industry. *Buildings, 9*, 227. https://doi.org/10.3390/buildings9110227.

Abuhjeeleh, M. (2019). Rethinking tourism in Saudi Arabia: Royal vision 2030 perspective. *African Journal of Hospitality, Tourism and Leisure, 8*(5), 1–16.

Ahmed, F., Ahmed, N., Pissarides, C., & Stiglitz, J. (2020). Why inequality could spread COVID-19. *The Lancet*. https://doi.org/10.1016/S2468-2667(20)30085-2.

Alarcón, D. M. (2017). Goal 5: Gender equity. In A. Sud (Ed.), *Transforming tourism: Tourism in the 2030 agenda* (pp. 34–38).

Alonso-Almeida, M. M., Borrajo-Millán, F., & Yo, L. (2019). Are social media data pushing overtourism? The case of Barcelona and Chinese tourists. *Sustainability, 11*, 3356. https://doi.org/10.3390/su11123356.

Antràs, P. (2020). *Conceptual aspects of global value chains*. Washington, DC: World Bank Development Economics.

Armitage, R., & Nellums, L. B. (2020). Considering inequalities in the school closure response to COVID-19. *The Lancet*. https://doi.org/10.1016/S2214-109X(20)30116-9.

Baldwin, S. A., & Del Re, A. C. (2016). Open access meta-analysis for psychotherapy research. *Journal of Counselling Psychology, 63*(3), 249–260. https://doi.org/10.1037/cou0000091.

Ben-Daya, M., Hassini, E., & Bahroun, Z. (2017). Internet of things and supply chain management: A literature review. *International Journal of Production Research*. https://doi.org/10.1080/00207543.2017.1402140.

Boussauw, K., & Vanoutrive, T. (2019). Flying green from a carbon neutral airport: The case of Brussels. *Sustainability, 11*, 2102. https://doi.org/10.3390/su11072102.

Bramwell, B., Higham, J., Lane, B., & Miller, G. (2017). Twenty-five years of sustainable tourism and the Journal of Sustainable Tourism: Looking back and moving forward. *Journal of Sustainable Tourism, 25*(1), 1–9. https://doi.org/10.1080/09669582.2017.1251689.

Campbell, A. M. (2020). An increasing risk of family violence during the Covid-19 pandemic: Strengthening community collaborations to save lives. *Forensic Science International: Reports, 2*, 100089. https://doi.org/10.1016/j.fsir.2020.100089.

Cañada, E. (2017). Goal 8: Decent work and economic growth. In A. Sud (Ed.), *Transforming tourism: Tourism in the 2030 agenda* (pp. 45–51).

Christian, M., Fernandez-Stark, K., Ahmed, G., & Gereffi, G. (2011). *The tourism global value chain: Economic upgrading and workforce development*. London: Centre on Globalization, Governance & Competitiveness, Duke University.

Daly, J., & Fernandez-Stark, K. (2017). *Barbados in the cruise tourism global value chain*. Durham: Centre on Globalization, Governance & Competitiveness (CGGC).

Daly, J., & Fernandez-Stark, K. (2018). *St. Lucia in the cruise tourism GVC*. Durham: Centre on Globalization, Governance & Competitiveness (CGGC).

Daly, J., & Gereffi, G. (2017). Tourism global value chains. In R. S. Newfarmer, J. Page, & F. Tarp (Eds.), *Industries without smokestacks: Industrialisation in Africa reconsidered* (pp. 68–89). Oxford: Oxford University Press.

De Lima, C. A., Alves, P. M. R., De Oliveira, C. J. B., De Oliveira, T. T. N., Barbosa, K. B., Marcene, H. C., & De Oliveira, S. V. (2020). COVID-19: Isolations, quarantines and domestic violence in rural areas. *SciMedicine Journal, 2*(1), 44–45. https://doi.org/10.28991/SciMedJ-2020-0201-7.

De Man, F. (2017). Goal 14: Life below water. In A. Sud (Ed.), *Transforming tourism: Tourism in the 2030 agenda* (pp. 80–85).

Dube, K., & Nhamo, G. (2019a). Climate change and potential impacts on tourism: Evidence from the Zimbabwean side of the Victoria Falls. *Environment, Development and Sustainability, 21*(4), 2025–2041.

Dube, K., & Nhamo, G. (2019b). Evidence and impact of climate change on South African national parks. Potential implications for tourism in the Kruger National Park. *Environmental Development*. https://doi.org/10.1016/j.envdev.2019.100485.

Dube, K., & Nhamo, G. (2020a). Major global aircraft manufacturers and emerging responses to the SDGs agenda. In G. Nhamo, O. Odularu, & V. Mjimba (Eds.), *Scaling up SDGs implementation emerging cases from state, development and private sectors* (pp. 99–113). Cham: Springer International Publishing.

Dube, K., & Nhamo, G. (2020b). Tourist perceptions and attitudes regarding the impacts of climate change on Victoria Falls. *Bulletin of Geography. Socio-economic Series, 47*(47), 27–44.

Eijgelaar, E., & Peeters, P. (2017). Goal 13: Climate action. In A. Sud (Ed.), *Transforming tourism: Tourism in the 2030 agenda* (pp. 75–79).

Ensign, P. C. (2001). Value chain analysis and competitive advantage. *Journal of General Management, 27*(1), 18–42.

Fang, Y., Yin, J., & Wu, B. (2018). Climate change and tourism: A scientometric analysis using CiteSpace. *Journal of Sustainable Tourism, 26*(1), 108–126.

Fernandez-Stark, K., & Bamber, P. (2018). *Jamaica in the tourism global value chain.* Durham: Centre on Globalization, Governance & Competitiveness (CGGC).

Frederick, S., Bamber, P., & Cho, J. (2018). *The digital economy, global value chains and Asia.* Durham: Centre on Globalization, Governance & Competitiveness (CGGC).

G20 (Group of Twenty Countries). (2019). *Advancing Tourism's contribution to the Sustainable Development Goals (SDGs).* Tokyo: G20.

Gereffi, G., & Fernandez-Stark, K. (2011). *Global value chain analysis: A primer.* Durham: Centre on Globalization, Governance & Competitiveness (CGGC).

Gereffi, G., Humphrey, J., Kaplinsky, R., & Sturgeon, T. J. (2001). Introduction: Globalisation, value chains and development. *IDS Bulletin, 32*(3), 1–8.

Hall, C. M. (2019). Constructing sustainable tourism development: The 2030 agenda and the managerial ecology of sustainable tourism. *Journal of Sustainable Tourism, 27*(7), 1–17.

Hong, Y., Cui, H., Dai, J., & Ge, Q. (2019). Estimating the cost of biofuel use to mitigate international air transport emissions: A case study in Palau and Seychelles. *Sustainability, 11*, 3545. https://doi.org/10.3390/su11133545.

IAEG-SDGs (Inter-agency and Expert Group on SDG Indicators). (2019). *Tier classification for global SDG indicators 11 December 2019.* IAEG-SDGs.

Jaeger, L. (2017). Goal 3: Good health and well-being. In A. Sud (Ed.), *Transforming tourism: Tourism in the 2030 agenda* (pp. 21–26).

Jennings, H. (2017). Ensure availability and sustainable management of water and sanitation. In *A. Sud: Transforming tourism: Tourism in the 2030 agenda* (pp. 39–44).

Jones, P., Hiller, D., & Comfort, D. (2017). The sustainable development goals and the tourism and hospitality industry. *Athens Journal of Tourism, 4*(1), 7–18.

Kamble, S. S., Gunasekaran, A., & Sharma, R. (2018). Analysis of the driving and dependence power of barriers to adopt industry 4.0 in Indian manufacturing industry. *Computers in Industry, 101*, 107–119.

Kamp, C., & Mangalasseri, S. (2017). Goal 1: End poverty. In A. Sud (Ed.), *Transforming tourism: Tourism in the 2030 agenda* (pp. 10–15).

Kang, K., & Rhee, C. (2020). *A post-coronavirus recovery in Asia – Extending a "Whatever it takes" lifeline to small businesses.* Retrieved from https://blogs.imf.org/2020/04/23/a-post-coronavirus-recovery-in-asia-extending-a-whatever-it-takes-lifeline-to-small-businesses/. Accessed 24 Apr 2020.

Kaplinsky, R., & Morris, M. (2001). *A handbook for value chain research.* Toronto: International Development Research Centre (IDRC.

Karschat, K., & Kühhas, C. (2017). Goal 15: Life on land. In A. Sud (Ed.), *Transforming tourism: Tourism in the 2030 agenda* (pp. 86–90).

Kitamura, Y., Ihisugi, Y., Karkour, S., & Itsubo, N. (2020). Carbon footprint evaluation based on tourist consumption toward sustainable tourism in Japan. *Sustainability, 12*, 2219. https://doi.org/10.3390/su12062219.

Kösterke, A. (2017). Goal 9: Industry, innovation and infrastructure. In A. Sud (Ed.), *Transforming tourism: Tourism in the 2030 agenda* (pp. 52–57).

Ladki, S. M., & Mazeh, R. A. (2017). Comparative pricing analysis of Mecca's religious tourism. *International Journal of Religious Tourism and Pilgrimage, 5*(1), 20–28. https://doi.org/10.21427/D76Q7Z.

Lee, J., Davari, H., Singh, J., & Pandhare, V. (2018). Industrial artificial intelligence for industry 4.0-based manufacturing systems. *Manufacturing Letters, 18*, 20–23.

Loayza, N., & Pennings, S. (2020). *Macroeconomic policy in the time of COVID-19: A primer for developing countries*. Washington, DC: World Bank.

Lui, Y., Dong, E., Lki, S., & Jie, X. (2020). Cruise tourism for sustainability: An exploration of value chain in Shenzhen Shekou Port. *Sustainability, 12*, 3054. https://doi.org/10.3390/su12073054.

Macola, G., & Unger, L. (2018). *Value chain analysis tourism Myanmar*. Nay Pyi Taw: The Centre for the Promotion of Imports from developing countries (CBI).

Mahase, E. (2020). Covid-19: Trump threatens to stop funding WHO amid "China-centric" claims. *The BJM, 369*, m1438. https://doi.org/10.1136/bmj.m1438.

Mole, B. (2020). *175 now infected with coronavirus on cruise ship, including quarantine officer: Japan is now planning to test everyone aboard*. Retrieved from https://arstechnica.com. Accessed 23 Apr 2020.

Monshausen, A. (2017). Goal 10: Reduced inequalities. In A. Sud (Ed.), *Transforming tourism: Tourism in the 2030 agenda* (pp. 58–62).

Nhamo, G. (2014). Green growth: A game changer ushering in the death of the internal combustion engine? *WIT Transactions on Ecology and the Environment, 186*. https://doi.org/10.2495/ESUS140051.

Nhamo, G., & Swart, R. (2012). A scholarly framework for measuring business responsibility to climate change in South Africa. *Issues in Social and Environmental Accounting, 6*(1&2), 50–71.

Nhamo, G., Nhemachena, C., & Nhamo, S. (2020). Using ICT indicators to measure readiness of countries to implement Industry 4.0 and the SDGs. *Environmental Economics and Policy Studies, 22*, 315–337.

Nicola, M., Alsafi, Z., Sohrabi, C., Kerwan, A., Al-Jabir, A., Losifidis, C., Agha, M., & Agha, R. (2020). The socio-economic implications of the coronavirus and COVID-19 pandemic: A review. *International Journal of Surgery*. https://doi.org/10.1016/j.ijsu.2020.04.018.

Novelli, M., & Jones, A. (2017). Goal 4: Quality education. In A. Sud (Ed.), *Transforming tourism: Tourism in the 2030 agenda* (pp. 27–33).

Omarjee, L. (2020). *Does the oil price crash signal relief for SA consumers?* Retrieved from https://m.fin24.com/Economy/does-the-oil-price-crash-signal-relief-for-sa-consumers-20200421. Accessed 21 Apr 2020.

Onyeaka, H. K., Zahid, S., & Patel, R. S. (2020). The unaddressed behavioral health aspect during the coronavirus pandemic. *Cureus, 12*(3), e7351. https://doi.org/10.7759/cureus.7351.

Parry, T. D. (2018). What is Africa to me now? African-American heritage tourism in Senegambia. *Journal of Contemporary African Studies, 36*(2), 245–263. https://doi.org/10.1080/02589001.2017.1387236.

Perna, L. W., Orosz, K., & Kent, D. C. (2018). The role and contribution of academic researchers in congressional hearings: A critical discourse analysis. *American Educational Research Journal, 56*(1), 111–145. https://doi.org/10.3102/0002831218788824.

Plüss, C., & Sahdeva, N. (2017). Goal 12: Responsible consumption and production. In A. Sud (Ed.), *Transforming tourism: Tourism in the 2030 agenda* (pp. 69–74).

Porter, M. E. (1985). *Competitive advantage*. New York: Free Press.

Porter, M. E. (2001). The value chain and competitive advantage. In D. Barnes (Ed.), *Understanding business: Processes* (pp. 50–68). London: Routledge.

Porter, M. E., & Kramer, M. R. (2018). Creating shared value. In G. G. Lenssen & N. C. Smith (Eds.), *Managing sustainable business: An executive education case and textbook* (pp. 323–346). Amsterdam: Springer.

Potash, J. S., Kalmanowitz, D., Fung, I., Anand, S. A., & Miller, G. M. (2020). Art therapy in pandemics: Lessons for COVID-19. *Art Therapy*. https://doi.org/10.1080/07421656.2020.1754047.

Potts, T. J. (2012). 'Dark tourism' and the 'kitschification' of 9/11. *Tourist Studies, 12*(3), 232–249. https://doi.org/10.1177/1468797612461083.

PWC (PricewaterhouseCoopers). (2016). *Industry 4.0: Building the digital enterprise*. London: PWC.

Quéré, C. L., Jackson, R. B., Jones, M. W., Smith, A. J. P., Abernethy, S., Andrew, R. M., De-Gol, A. J., Willis, D. R., Shan, Y., Canadell, J. G., Friedlingstein, P., Creutzig, F., & Peters, G. P. (2020). Temporary reduction in daily global CO2 emissions during the COVID-19 forced confinement. *Nature Climate Change*. https://doi.org/10.1038/s41558-020-0797-x.

Ren, Y., Zhao, C., & Fu, J. (2019). Dynamic study on tourism carbon footprint and carbon carrying capacity in Sichuan Province. *Journal of Geoscience and Environment Protection, 7*, 14–24.

Rutherford, A. (2017). Goal 16: Good governance. In A. Sud (Ed.), *Transforming tourism: Tourism in the 2030 agenda* (pp. 97–102).

Spencer, J. P., Safari, E., & Dakora, E. A. N. (2014). An evaluation of the tourism value-chain as an alternative to socio-economic development in Rwanda, Africa. *African Journal for Physical, Health Education, Recreation and Dance (AJPHERD), 20*(2:1), 569–583.

Ştefănică, M. (2017). Environmental impact of transportation in the tourism industry – Dimensions and actions. *Studies and Scientific Researches – Economics Edition, 25*, 84–89.

Stone, M. (2020). Carbon emissions are falling sharply due to coronavirus: But not for long. https://www.nationalgeographic.co.uk/environment-andconservation/2020/04carbon-emissions-are-fallin-sharply-due-coronavirus-not-long.

Szpilko, D. (2017). Tourism supply chain – Overview of selected literature. *Procedia Engineering, 182*, 687–693.

Tang, S., Wang, Z., Yang, G., & Tang, W. (2020). What are the implications of globalisation on sustainability? – A comprehensive study. *Sustainability, 12*, 3411. https://doi.org/10.3390/su12083411.

UNFCCC (United Nations Framework Convention on Climate Change). (2015). *Paris agreement*. Bonn: UNFCCC.

United Nations. (2015). *Transforming our world: The 2030 agenda for sustainable development*. New York: United Nations Secretariat.

United Nations. (2020). *Shared responsibility, global solidarity: Responding to the socio-economic impacts of COVID-19*. New York: United Nations Secretariat.

UNWTO (United Nations World Tourism Organisation). (2018). *Tourism and the sustainable development goals: Good practices in the Americas*. Geneva: UNWTO.

UNWTO (United Nations World Tourism Organisation). (2019). *Transport-related CO2 emissions of the tourism sector: Modelling results*. Geneva: UNWTO.

UNWTO (United Nations World Tourism Organisation). (2020). *Supporting jobs and economies through travel & tourism: A call for action to mitigate the socio-economic impact of COVID-19 and accelerate recovery*. Geneva: UNWTO.

Veldhuizen, L. J. L., Giller, K. E., Oosterrveer, P., Brouwer, I. D., Janssen, S., Van Zanten, H. H. E., & Slingerland, M. A. (2020). The Missing Middle: Connected action on agriculture and nutrition across global, national and local levels to achieve Sustainable Development Goal 2. *Global Food Security, 24*, 100336. https://doi.org/10.1016/j.gfs.2019.100336.

Wang, Q. (2018). A comparative study of China-U.S. tourism characteristic industry GVC1 embedment degree. *American Journal of Industrial and Business Management, 8*, 1549–1555.

Wang, Z., & Tang, K. (2020). Combating COVID-19: Health equity matters. *Nature Medicine*. https://doi.org/10.1038/s41591-020-0823-6.

Wang, T. C., Cheng, J. S., Shih, H. Y., Tsai, C. L., TNG, T. W., Tseng, M. L., & Yao, Y. S. (2019). Environmental sustainability on tourist hotels' image development. *Sustainability, 11*, 2378. https://doi.org/10.3390/su11082378.

WHO (World Health Organisation). (2020). *WHO Director-General's opening remarks at the media briefing on COVID-19 – 11 March 2020*. Retrieved from https://www.who.int/dg/speeches/detail/who-director-general-s-opening-remarks-at-the-media-briefing-on-covid-19%2D%2D-11-march-2020. Accessed 20 Apr 2020.

Yilmaz, Y., & Bititci, U. (2006). Performance measurement in tourism: A value chain model. *International Journal of Contemporary Hospitality Management, 18*(4), 341–349. https://doi.org/10.1108/09596110610665348.

Zerrudo, M. R. A. (2017). Gaol 2: Zero hunger. In A. Sud (Ed.), *Transforming tourism: Tourism in the 2030 agenda* (pp. 16–20).

Chapter 3
Preparedness and Responses to COVID-19: A Comparison from Selected Countries

Abstract After witnessing and experiencing the severe acute respiratory syndrome (SARS), the Middle East respiratory syndrome (MERS) and Ebola pandemics and epidemic, one would assume the world was prepared to deal with the COVID-19 pandemic. Alas! This was not to be! Officially, the first cases of COVID-19 were reported to the World Health Organisation (WHO) on 31 December 2019. The impact has been catastrophic, with the entire world affected. Through event study, document and critical discourse analysis methods, the chapter looks at the preparedness, response and recovery measures put in place to deal with COVID-19. It emerged that countries responded differently, some late and some in panic. The main goal was to "flatten" the curve as infections skyrocketed, leading to deaths. The COVID-19 pandemic also placed severe economic burdens to countries across the world. The global epicentre of the pandemic kept shifting, starting in China, moving to Italy, and then to the USA. As the chapter was being completed, Russia had become the second in terms of infections after the USA. The COVID-19 early intervention measures mainly focused on handwashing, promotion of sneezing and coughing etiquette, mask-wearing and social distancing. However, as it became clearer that the pandemic continued spreading, additional measures, including full national lockdowns, were instituted. Since the world was still experiencing lockdowns, amidst other countries moving into the recovery phases, the chapter recommends the sharing of COVID-19 preparedness, response, and recovery strategies and plans. The world is encouraged to consider the COVID-19 within the broader realm of the United Nations Sendai Framework for Disaster Risk Reduction and also work with the WHO as the custodian of global health matters. Lastly, international protocols on certain disease outbreaks with potential to balloon into global pandemics such as coronaviruses must be refined to force countries and territories to notify the WHO earlier after a certain threshold of such diseases' outbreak has been reached.

Keywords COVID-19 · Preparedness · Response · Flatten · China · Italy · WHO · New Zealand · Sendai

G. Nhamo et al., *Counting the Cost of COVID-19 on the Global Tourism Industry*,
https://doi.org/10.1007/978-3-030-56231-1_3

3.1 Introduction

This chapter is situated within the disaster risk reduction (DRR) framework that traces matters from preparedness (readiness and prevention), responses (relief) as well as recovery and reformation (United Nations Office for Disaster Risk Reduction 2015). The DRR framework also brings to the fore the concept of building back better (BBB) after disasters. Natural disasters, particularly pandemics, remain a huge threat to modern societies and in the context of this work, the tourism industry. The Global Preparedness Monitoring Board (GPMB) is of the view that epidemic-prone diseases that include influenza, the severe acute respiratory syndrome (SARS), the Middle East respiratory syndrome (MERS), Ebola, Zika, plague and yellow fever are worrying. These epidemics are "harbingers of a new era of high-impact, potentially fast-spreading outbreaks that are more frequently detected and increasingly difficult to manage" (GPMB 2019: 12). The GPMB further alludes to the fact that poorer countries have suffered and will continue to suffer the most, although all economies remain vulnerable to differing degrees. It is estimated that 10% of the global population suffers from an influenza virus each year (Nickol and Kindrachuk 2019). During the outbreak of the Ebola from 2014–2016, for example, Guinea, Liberia and Sierra Leone lost an estimated combined $ 2.2 billion in gross domestic product (GDP) in 2015. Although similar to both SARS and MERS (Backer et al. 2020), COVID-19 remains different in terms of its swift spread and the number of deaths experienced. The war on these invisible invaders remains difficult to wage (Alexander 2019).

The epidemiology of pandemics is of interest to a broad range of stakeholders, among them the tourism industry, academics, medical authorities, politicians and development practitioners. The demographics of those that get infected, their time of recovery or those that die due to pandemics also remain crucial to many economic sectors. In their study on cruise liners in China's Shenzhen Shekou Port, Lui et al. (2020) find that the major consumer group are the elderly and children. The elderly had enough resources, especially money, leisure time and a strong desire to embark on the cruise. Yet, emerging statistics are pointing to more of the elderly persons succumbing to COVID-19 (Glynn 2020; Onder et al. 2020), an aspect that has the potential to negatively impact the tourism industry. Table 3.1 provides details on some of the known global pandemics and epidemics that have affected the world in the past.

Concerned about the outbreak of the Ebola epidemic in 2014–2016, the United Nations Secretary–General established the Global Health Crises Task Force and Panel to monitor and advocate for readiness in monitoring global health emergencies (GPMB 2019). The taskforce was replaced by the GPMB established in May 2018 by the World Bank and the WHO. The GPMB works independently "to provide the most frank assessments and recommendations possible" around the issue of contagious disease outbreaks (GPMB 2019: 4). In DRR, especially in the case of pandemics, the state of readiness and response determine the extent to which damage can be minimised and at times even avoided.

Table 3.1 Epidemics and pandemics from the past

Year	Pandemic/epidemic	Known estimated deaths
1347–1351	Black Death	200 million
1885	Third plague (China and India)	2 million
1918	Influenza (Spanish flu – H1N1)	50–100 million
1957	"Asian flu" (H2N2)	1–2 million
1968	"Hong Kong flu" (H3N2)	500,000–1 million
1981 to date	HIV/AIDS	25 million-plus (still counting)
2003	SARS	724
2009	Influenza A (H1N1)-swine flu)	200–300,000
2014–2016	Ebola	11,000
2016	MERS	850
2019	COVID-19	165,257[a]

Source: authors, based on Elhakim et al. (2019: 1–2); Martini et al. (2019: E64); Nickol and Kindrachuk (2019: 3); Johns Hopkins Coronavirus Resources Centre (2020)
[a]As of 20 April 2020

This chapter documents the state of readiness and responses to COVID-19 from selected countries. The work further seeks to determine if there were lessons learnt from past pandemic and epidemic outbreaks, among them the 1918 influenza, the SARS and the MERS. Examining and understanding these measures by identifying their strength and weaknesses is critical in explaining the outcomes that will emerge in the post-COVID-19 era. More important, it will inform good response practices relevant to various public and private enterprises as well as other social responses, particularly as these relate to the Sendai Framework on DRR (United Nations Office for Disaster Risk Reduction 2015).

3.2 A Literature Survey

This literature section focuses on preparedness and responses to three pandemics, namely, the 1918 influenza, 2003 SARS and the 2016 MERS. The aim is to draw some insights that may be of help in addressing preparedness and responses to COVID-19.

3.2.1 The 1918 Influenza

There is no better place to start writing on the epidemiology of pandemics than the 1918 influenza, commonly referred to as the Spanish Flu (Jester et al. 2019). In Europe, the 1918 influenza epicentre was Spain (Martini et al. 2019). From there, the influenza spread to France, Great Britain and Italy, severely impacting the military operations during the First World War. From historical records, "Patient Zero" for the Spanish Flu was Albert Gitchel, a cook at Camp Funston in Kansas. He started coughing, having a fever and headaches on 4 March 1918. Later in spring, 75% of French troops and more than half of the British troops fell ill. In total, 100,000 soldiers died. The influenza moved to North Africa in May 1918 and then to Bombay in India in June 1918. In China and Australia, the first cases were reported in July 1918. This first phase of the flu was weaker compared to the second phase that erupted in August 1918 in Spain. The strong flu is suspected of having resulted from a mutated strain of the virus. This new virus was carried from the port city of Plymouth in south-western England by ships bound for Freetown in Sierra Leone (Africa) and Boston (USA). The movements of the armies were also linked to spreads from Boston, Freetown and Brest in France. The second wave lasted about 6 weeks spreading from "North America to Central and South America, from Freetown to West Africa and South Africa in September, and reaching the Horn of Africa in November 1918" (Martini et al. 2019: E64). At the end of September 1918, the Spanish flu had covered entire Europe, including Poland and Russia, with Russia spreading it to northern Asia and getting to India in September.

In the USA, the 1918 influenza's "Patient Zero" was registered in Kansas military camp. It then spread to Boston, New York and Philadelphia in rapid succession (Keeling 2020). In Chicago, 381 people died in a single day, with 1200 others infected. To this day, this event of 17 October 1918 is known as the Black Thursday. In the 2 months from mid-September to mid-November 1918, New York recorded 10,886 deaths, and Philadelphia reported 10,046 deaths. Eventually, 675,000 deaths were recorded in the USA (Alexander 2019). However, these figures were lower compared to 20 million deaths in the Indian subcontinent. Globally, a figure of 50 million deaths is recorded, meaning 40% of these were from the Indian subcontinent. However, other authors put deaths figures at between 50 and 100 million (Nickol and Kindrachuk 2019; Andayi et al. 2019).

In Mexico, where a decade-long revolutionary struggle was ongoing, many died from influenza complications that included pneumonia or bronchitis (Alexander 2019). In October 1918, stories of entire families and peasant populations, miners and urban residents dying started emerging. However, the illness was said to have been transmitted through the railway lines and incoming ships docking at the Atlantic seaports of Tampico and Veracruz. As the pandemic grew, deaths in 1000 s daily were reported, ultimately reaching an estimated total number of half a million. To fight influenza, in typical modern-day language, "flatten" the curve (Strochlic and Champine 2020), the government established sanitary brigades (Alexander 2019). These were sent to every state, and they "enacted a series of behavioural protocols through a national education campaign and established prescriptive regulations on public gatherings and personal mobility, especially on trains" (Alexander 2019: 450). Unpaved streets were irrigated with water, and air was infused with sulphuric compounds. Daily cleaning of frontage property was also enforced. Officials further targeted ships and trains, including quarantine. Curfews were instituted with modest penalties for the poor.

As of December 1918, the greater part of the world was declared influenza-free. Australia then lifted its quarantine measures, and in the austral summer of 1918/1919, the country reported 12,000 new infections. This became the third wave, with New York and Paris experiencing the challenges of this third wave, too (Martini et al. 2019). This third wave ended in May 1919 in the developed Northern Hemisphere, although Japan had the third outbreak at the end of 1919, ending in 1920 (Andayi et al. 2019).

Andayi et al. (2019) argue that the impacts of the 1918 influenza have been under-reported and documented in Africa and specifically in Kenya. The authors focus on the coastal region of Kenya and use historical records as part of their methodology. They found that crude mortality rates and healthcare utilisation increased six- and threefold, respectively, in 1918. They also estimated a pandemic mortality rate of 25.3 deaths/1000 people/year. Paskoff and Sattenspiel (2019) also investigate age-based mortality differences from the 1918 influenza. The authors also make reference to literature advocating for nationally aggregated mortality rates. From this angle, it emerges that from a study of 70 countries and regions, different mortality rates emerged. Some examples are shown in Fig. 3.1. In the Newfoundland island, pregnant mothers were badly affected by the 1918 influenza.

In 1997, scientists in the USA exhumed bodies of indigenous Alaskans that died from the 1918 influenza, which were frozen in the deep permafrost layer of the earth. As a result, the flu strain was determined as H1N1. The H1N1 had an incubation period of 1 to 2 days. On infection, the victims suffered extreme prostration and fever, severe muscle pain, hemoptysis and epistaxis. Temperatures of up to 40 °C were recorded in some patients (Martini et al. 2019). This was accompanied by acute respiratory distress syndrome with mortality levels of between 60 and 70% (Keeling 2020). Death when one starts showing acute respiratory distress was so

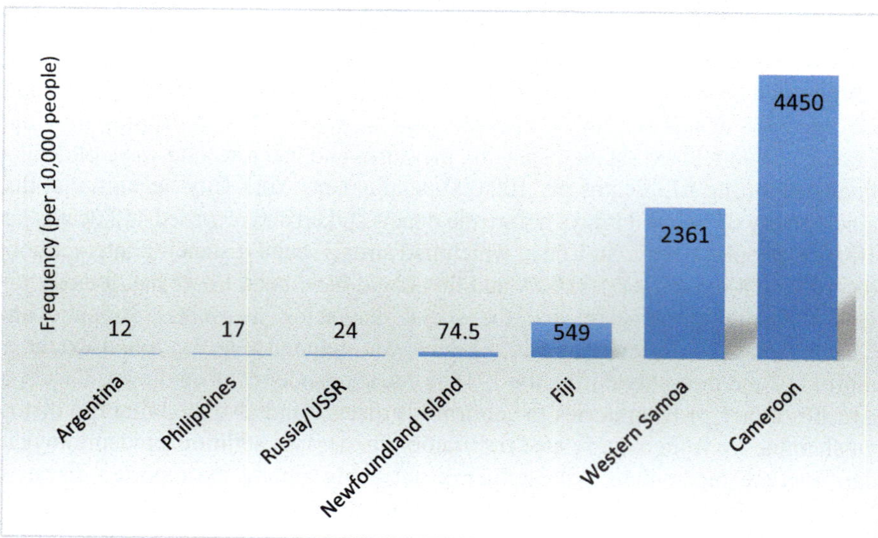

Fig. 3.1 Mortality rates from the 1918 influenza (per 10,000 people). Source: authors, based on Paskoff and Sattenspiel (2019: 1)

rapid, at times taking between 12 and 24 hours, with the majority of deaths being among young people in the age cohort 20–40 (Jester et al. 2019). Young soldiers and pregnant women and premature new-borns were the hardest hit. The mortality curve was rather unusual, as typically, influenza epidemics curves are U-shaped, with most deaths taking place in those that are very young and those that are very old. What is imported to note is that the influenza followed trade routes and shipping lines (Alexander 2019). Van Wijhe et al. (2018) argue that this kind of influenza signature was puzzling. Woo (2019) adds to the equation of young adults' death by indicating that they may have died from an overreaction of their active immune systems, although this explanation has not been satisfactory. This is so given that teenagers also have very active immune systems.

Overall, the 1918 influenza resulted in major public health improvements, including the following (Martini et al. 2019; Andayi et al. 2019): health education, timely isolation, sanitation, surveillance and improved epidemiological knowledge in terms of transmission. Jester et al. (2019), highlight that among some key achievements over the past 100 years since the influenza outbreak are advances in prevention, diagnosis and treatment protocols. Nickol and Kindrachuk (2019) add the rise of vaccines and antivirals from a multitude of scientific and technological advances as well. There has been good access to autopsy samples, with the reconstitution of the influenza pandemic virus having permitted much understanding of how the pandemic virus differs from other seasonal and pandemic influenza virus strains. However, gaps still exist in knowledge regarding the origin of the virus and the molecular mechanisms (be it host and/or viral) underlying differential pathogenesis in contrast to other influenza viruses.

Social distancing proved to be one of the most effective intervention measures to "flatten" the curves (Strochlic and Champine 2020). Scenarios of different approaches to social distancing resulted in different deaths rates from the 1918 influenza in some US cities. Philadelphia (which campaigned against coughing, spitting and sneezing in public) first, and the shortest period of social distancing, recorded 748 deaths per 100,000 people after 24 weeks. This was followed by San Francisco that relaxed social distancing measures and had a second wave of deaths, finally reporting 673 deaths per 100,000 people. New York City began social distancing very early just 11 days before the deaths spiked and recorded 452 deaths per 100,000 people. Lastly, St. Louis, which had strong social distancing interventions recorded 358 deaths per 100,000, and this could have been lower had the city not lapsed and temporarily relaxed the social distancing measures (Strochlic and Champine 2020). Jester et al. (2019), however, contend that the actual extent of morbidity and mortality during the 1918 influenza pandemic is unknown. This is as a result of lack of laboratories to confirm the disease and also the failure to distinguish influenza from other related respiratory diseases. In addition, epidemiological data sets are incomplete, with influenza not being among the reported diseases. Reporting only caught up later when lots of people had already died.

Matters of race often come out when discourses on pandemics feature. Økland and Mamelund (2019) attempt to disentangle this from the 1918 influenza. The authors found that the black population in the USA had lower morbidity. They also had lower mortality during September, October and November 1918. However, there were higher cases of fatality than from the white population. The findings further revealed that the black population had lower influenza morbidity prior to 1918. Although the reasons for lower morbidity among the black population remain unclear, the findings might imply that black people had a lower risk of developing influenza. However, when this happened, there was a higher risk of dying.

3.2.2 *Focus on SARS*

The SARS pandemic broke out in 2002/2003. The SARS was recognised as a new disease in Asia in mid-February 2003, although it had started spreading from November 2002 from Guangdong Province in China. From November 2002 to July 2003, there were over 8090 cases recorded, resulting in 774 deaths in 29 countries and regions. The deaths were also reported on five continents. The most affected country was China, with more than 75% of the cases (Wilder-Smith 2006). Cheng et al. (2007) warn that the SARS coronavirus was an agent of emerging and re-emerging infection.

The first SARS case was detected in Southern China in November 2002 and was exported to Hong Kong on 21 February 2003 (Cherry 2004). Only in March 2003 was a novel coronavirus (SARS-CoV) found to be the causative agent. The viral infection spread to 28 other countries and regions within 11 weeks from 21 February 2003. Cherry summarises the spread of SARS as:

> The mini pandemic peaked during the last week of May 2003, and the last new probable case was on 13 July 2003. There were a total of 8,096 probable cases and 774 deaths. Sixty-six per cent of the cases occurred in China, 22% in Hong Kong, 4% in Taiwan and 3% in both Singapore and Canada. Twenty-one per cent of all cases occurred in healthcare workers (Cherry 2004: 262).

Cherry's narration of the SARS outbreak shows the notable cases of infection of health workers. The high rate of infection of this critical group was a cause for concern, even in the cases of MERS and Ebola cases. The spread was eventually recorded in a total of 28 countries and territories (Fig. 3.2). The highest cases, as indicated earlier, were from China at 5327. This was followed by Hong Kong with 1755. The next three countries and territories were in Taiwan with 346 reported cases, Canada at 251 reported cases and Singapore with 238 recorded cases. These were the countries reporting cases over a hundred. The next country in the rank was Vietnam with 63 cases, with the remainder all reporting 27 or fewer cases. In terms of reported deaths, the top six countries were China (349 deaths), Hong Kong (299), Canada (43), Taiwan (37) and Singapore (33). The other countries reported five or less death cases. South Africa was the only country in Africa affected by the SARS with one case that resulted in death.

Cooper (2005) observes that nearly 3 months after SARS outbreak, the Chinese Ministry of Health notified the WHO of 300 cases in Guangdong Province. The numbers eventually rose to 5327 for the country. Expectedly, the Chinese

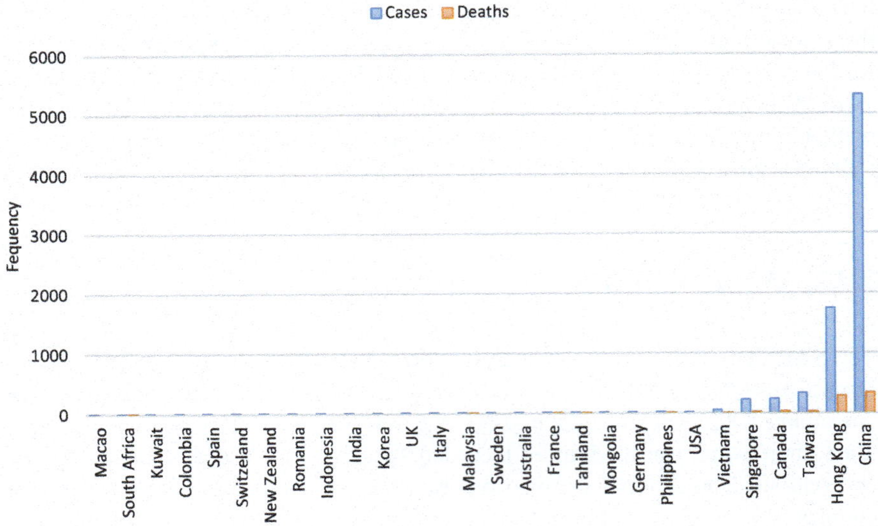

Fig. 3.2 Spread of SARS pandemic into countries and territories. Source: authors, data from Cooper (2005: 120)

government faced harsh international criticism for not initially and comprehensively reporting the full scale of SARS outbreak to the WHO. From the British Broadcasting Corporation (BBC) report of July 2004, the first case of SARS was discovered on 16 November 2002 and an official report from the Chinese Ministry of Health was only made in 11 February 2003 (BBC 2004). In this case, Chan-Yeung and Xu (2003) accuse China of imposing a SARS information blackout. This resulted in the dismissal of both the national health minister of China and the Beijing mayor based on an accusation of poorly handling the situation. Subsequently, China pledged to fully cooperate with the WHO. After 26 March 2003, Beijing immediately put in place stringent control measures such as the isolation and quarantining of infected individuals as well as tracing those who have been in contact with individuals with confirmed infection. This practice resulted in the rapid decline of cases countrywide. Ultimately, the fatality ratio of 7% was realised. Nevertheless, China's record in terms of reporting first case epidemics was tainted, and with each new outbreak from China, the world remains suspicious in terms of accuracy of the reported figures.

In a classical publication focusing on SARS as an agent of emerging and re-emerging infection, Cheng et al. (2007) conclude with a question: "Should we be ready for the re-emergence of SARS?" The authors argue that despite huge successes recorded in containing SARS within a short period of time, there remained gaps and areas requiring further attention, including broader understanding of managing similar epidemics. This assertion is informed by the fact that coronaviruses are known to undergo genetic recombination, with the potential to new genotypes and new pandemic outbreaks (Cheng et al. 2007). This was enhanced by the realities of the presence of SARS-CoV-like viruses in horseshoe bats, civets and growing culture of eating exotic mammals in Southern China. This scenario resembled a ticking time bomb for the next global pandemic outbreak, which in the focus of this work was confirmed by COVID-19 in December 2019.

Hsieh et al. (2005) focus on SARS incidences in Taiwan. The authors highlight that about 151,270 people were put on quarantine, with 668 cases finally testing positive for SARS-CoV and 181 (27%) of them succumbing to the disease (Chen et al. 2005). The initial SARS screening in the country commenced on 14 March 2003 with the pandemic experienced right through to 30 July 2003 (Chen et al. 2005). The screening involved potentially infective persons for quick diagnosis and hospitalisation. This assisted in cutting off many potential transition pathways. The full-scale quarantine measures were instituted on 28 April 2003 following "Patient Zero" confirmation on 8 March 2003 (20 days later) (Hsieh et al. 2005). The declaration of full-scale quarantines and the period it took remains important for COVID-19 as a good learning point. Other matters related to hygiene, especially handwashing, and these issues are of interest to COVID-19 relate to hygiene.

Fung and Cairncross (2006) review selected studies that sought to determine the effectiveness of handwashing in breaking the transmission pathways of SARS. Their analysis shows that nine out of the ten studies confirmed handwashing with soap as an effective SARS infection management measure. Apparently, this approach is applicable to COVID-19 intervention campaigns as well. In addition to hygiene, the matter of communication is also critical. Bowen and Heath (2007: 73) posit that the SARS epidemic taught the world the need to respond with "rapid, factual, and honest narratives and an ethical dedication to communicate on behalf of the public interest to prevent the needless spread of disease and loss of life". In the era of massive social media communication networks, facts on COVID-19 are pivotal.

Drawing from SARS experience, Cherry and Krogstad (2004) observe that the successes in defeating the pandemic also came from "unusual" international cooperation enabled by accurate electronic media coverage. The media augmented scientific data. To this end, Wilder-Smith (2006) is of the view that the international spread of diseases demands strong global public health systems. Such systems should have excellent reporting mechanisms that are coupled with robust health service infrastructure and expertise. There is a need to mobilise resources quickly across borders to match the spread of these global pandemics. The SARS gave rise to the revision of the International Health Regulations. This was a great positive indeed.

Finally, the World Health Organisation (WHO) organised an inaugural meeting on SARS on 16–17 May 2003 in Geneva. The objective was to have a consensus document on the understanding of the epidemiology of SARS and continue to inform public health practice (WHO 2003). The deliberations zoomed into seven key topics, namely, incubation period; infectious period; case-fatality ratios; routes of transmission, exposure dose and risk factors for transmission; the presence and significance of subclinical infection; reproduction number in different transmission settings and under different control strategies; and the animal and environmental reservoirs.

3.2.3 Lessons from the MERS

South Korea was one of the countries severely impacted by MERS. As such, some lessons can be drawn in the manner the country handled the epidemic. The MERS experience assisted in refining tracking protocols and the rolling out of early testing in South Korea (Kim 2020). "Patient Zero" during the MERS outbreak was a 68-year-old businessman who returned to Seoul after a 10-day business trip to Bahrain and other countries in the Middle East on 4 May 2015. He was only diagnosed with MERS on 20 May 2015 (15 days later). By this time, 28 patients at the hospital had already been infected from the movement of the businessman (Cho et al. 2016). Another patient – Patient 14, a 35-year-old later diagnosed with MERS – had already infected 82 people resulting in the country suffering the largest outbreak outside the Middle East. The total eventually was 186 confirmed cases with 38 deaths, while 16,000 were quarantined (Kim 2020).

The key lesson from the South Korean MERS case is that timely interventions are critical to restrict the rapid spreading of infectious diseases (Cho et al. 2016). Key in the interventions is the rapid identification of the source of their infection and infected individuals and tracking those they could have infected as well. In this mix, hospitals have been identified as key transmission points, as health workers are not aware of the infection or lack the means to effective protective measures. Furthermore, it is important to have the means to diagnose an infection. In the case of MERS in South Korea, there was a lack of early diagnosis of MERS. The aforementioned businessman went from hospital to hospital with symptoms, but the MERS infection was not detected. Then, the testing procedure was long and cumbersome. Lessons from this case resulted in the government enacting a new law after MERS in 2016, permitting laboratories to use unapproved in vitro diagnostic kits in the case of a public health emergency. As will be encountered in this chapter, this law assisted the case of COVID-19 greatly (Kim 2020). However, from earlier studies and following the outbreak of MERS, Ng and Tan (2017) maintain their worry on the lack of progress in developing vaccines to counter similar future outbreaks of coronavirus. Despite this, the authors remain positive regarding the progress made in terms of the surveillance and the preparation of possible future outbreaks.

3.3 Methodological Orientation

The objective of this chapter is to determine how selected governments and territories responded to COVID-19 prior and after the so-called Patient Zero. This brings to fore the elements of DRR starting from preparedness (readiness), disaster striking, response (relief) as well as recovery and reformation. The chapter also seeks to document the lessons for future similar pandemics in order to improve on efficiency in dealing with such. Given the nature of the pandemic, event study methodology, employing the case study approach in some instances, was employed. This is a method used by many authors, including Bimha and Nhamo (2017) and MacKinlay (1997). To this end, events surrounding COVID-19 were traced since the first announcement by the WHO on 31 December 2019 to the time when the chapter went into publication in June 2020. An effort was made to plot geographical information (GIS) maps on lockdowns and hotspot countries in terms of infections and deaths. Other official publications were also retrieved and included emergency declarations, national disaster declarations, decrees, directives, regulations and daily updates on statistics. The event study methodology was complemented by the document and critical discourse analysis (D&CDA) method (Antwi-Agyei et al. 2018; Cooper 2005). Few countries were purposefully selected, including global powerhouses China and the USA. The findings from the study are now considered in the next section.

3.4 Presentation of Data and Discussion of Findings

This section comes in three subsections that include preparedness, response and recovery and reformation. Each of these subsections will now be considered in turn in the next paragraphs.

3.4.1 Preparedness for COVID-19

The Sendai Framework for Disaster Risk Reduction (2015–2030) is clear that there is a need for "enhanced work to reduce exposure and vulnerability, thus preventing the creation of new disaster risks, and accountability for disaster risk creation are needed at all levels" (United Nations Office for Disaster Risk Reduction 2015: 10). The United Nations goes further to identify pandemics and epidemics among the disasters that require attention. The Sendai Framework remains the overarching platform for globally coordinated disaster risk reduction. The policies and practices for disaster risk reduction and management should be embedded in understanding disaster risk from all dimensions. These dimensions include vulnerability, capacity, exposure of persons and assets, hazard characteristics and the environment (Ibid.). Hence, the Sendai Framework presents the foundation for countries and organisations to prepare for disasters such as COVID-19 and any eventualities. What remains clear, however, is that when the disaster strikes, the impacts will be felt differently, at different times across the world. This gives some countries more time to prepare and other lesser time to prepare. The extent of impact from the disaster will also be a function of the already prevailing vulnerability. For example, least developed countries and small island states will, by nature, remain more vulnerable to any major global disasters such as COVID-19. Hence, more vulnerable and poor countries may not be able to deal with pandemics such as COVID-19 without external financial, technical, human resources and other assistance. The New Zealand COVID-19 Alert Systems remains of interest to the world (Table 3.2).

Table 3.2 New Zealand COVID-19 Alert System – Levels 1 and 2 (Prepare and Reduce)

Alert level	Outcome/state	Risk assessment	Selected measures at local or national level
1 (prepare)	The disease is contained in New Zealand	COVID-19 is uncontrolled overseas. Isolated household transmission could be occurring in New Zealand	Border entry measures to minimise risk of importing COVID-19 cases Intensive testing for COVID-19 Rapid contact tracing of any positive case Self-isolation and quarantine required Schools and workplaces open and must operate safely Physical distancing encouraged No restrictions on gatherings Stay home if you're sick, report flu-like symptoms Wash and dry your hands, cough into your elbow, don't touch your face No restrictions on domestic transport – avoid public transport or travel if you're sick
2 (reduce)	The disease is contained, but the risk of community transmission remains	Household transmission could be occurring. Single or isolated cluster outbreaks	Physical distancing of 1 metre outside the home including on public transport Gatherings of up to 100 people indoors and 500 outdoors allowed while maintaining physical distancing and contact tracing requirements Sport and recreation activities are allowed if conditions on gatherings are met, physical distancing is followed and travel is local Public venues can open but must comply with conditions on gatherings and undertake public health measures Health services operate as normally as possible Most businesses open, and business premises can be open for staff and customers with appropriate measures in place. Alternative ways of working are encouraged, such as remote working, shift-based working, physical distancing, staggering meal breaks, flexible leave Schools and early childhood education centres open, with distance learning available for those unable to attend school, such as people self-isolating People advised to avoid non-essential interregional travel People at high risk of severe illness such as older people and those with existing medical conditions are encouraged to stay at home where possible and take additional precautions when leaving home. They may choose to work

Source: authors, based on Government of New Zealand (2020)

Nkengasong and Mankoula (2020) called upon the African continent to be vigilant to the looming threat of COVID-19. The authors made this call when Egypt reported its first confirmed case on 14 February 2020. Noting the high volumes of air traffic and other modes of transportation of goods and services to and from China into the country, and of course the world over, the authors made a call for an Africa-wide focus to manage the then impending pandemic. Nkengasong and Mankoula (2020: 841) highlight that Africa remained vulnerable due to its weak health systems, poverty and inequality as well as "inadequate surveillance and laboratory capacity, scarcity of public health human resources, and limited financial means." Models on the importation of COVID-19 to the African continent direct from China identified Algeria, Egypt and South Africa as having the highest importation risk given their trade relations. Based on this criterion, countries with moderate risk include Angola, Ethiopia, Ghana, Kenya, Nigeria, Sudan and Tanzania. The models indicated that the risk would come from Guangdong, Fujian and Beijing. As a measure of preparedness, the African Union Africa Centres for Disease Control and Prevention (AUCDC) and WHO, in partnership with member states, established the Africa Taskforce for Coronavirus Preparedness and Response (AFTCOR). The partnership established six linked work streams (Fig. 3.3). Despite this effort, what was worrying though is that at the beginning of February 2020, only two countries in Africa had the diagnostic capacity to test for COVID-19. Fortunately, this number increased to more than 40 countries by 25 February 2020 (Ibid.).

The case of Africa and COVID-19 presented thus far raises many challenges. Most notable is that the absence of early confirmed COVID-19 cases in some African may not necessarily have been to no infections. Rather, this could have been due to the limited capacity to roll out massive tests on COVID-19 in various jurisdictions. This concern resulted in the WHO, the African Union and partners convening urgent health ministers' meeting with 55 member states represented. The meeting confirmed a commitment to swiftly act against COVID-19 spread (Nkengasong and Mankoula 2020). This is when the aforementioned AFTCOR Task force was formed, with an Africa-wide COVID-19 strategy endorsed. The strategy sought to ensure that African countries acted accordingly well before COVID-19 crossed into their jurisdictions.

The key in this strategy was access to resources. To this end, the strategy called for the immediate release of resources for investment in COVID-19 preparedness activities. The donors and development partners had to commit and release funds before the crisis landed in Africa to "address supply chain management, mapping, and stockpiling of COVID-19 response needs. Key focus issues in this space include large quantities of personal protective equipment i.e. gloves, surgical masks, coveralls, and hoods, and medical countermeasures like antiviral agents" (Ibid.: 842). Nkengasong and Mankoula (2020) ended by raising matters pertaining to the whole government approach, with governments urged to urgently put in place proper quarantine and infection control protocols. Other measures included developing some of the protocols such as social distancing through banning mass gathering and the closure of public facilities and building capacity by training medical staff at major hospitals to be able to handle and deal with individuals or at-risk of COVID-19 infection.

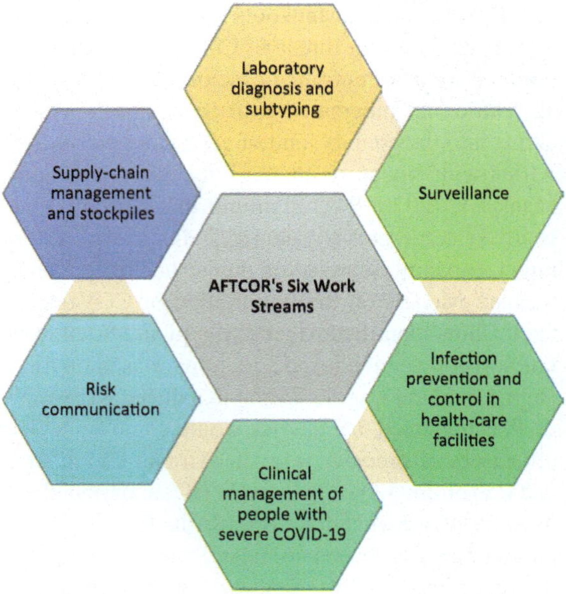

Fig. 3.3 Africa Taskforce for Coronavirus Preparedness and Response. Source: authors, based on Nkengasong and Mankoula (2020: 842)

3.4.2 COVID-19 Response

During the response phase, the Sendai Framework for Disaster Risk Reduction emphasises the need to increase public education and awareness of disaster risk, as well as bringing up rapid response measures (United Nations Office for Disaster Risk Reduction 2015). The foregoing brings one to that old folktale many grandmothers used to tell around the evening fires. The folktale presented the case of a cat, mice, warning bell and a tying string. A key question among the mice was who among them was going to tie the bell around the cat's neck? This had to be so for the entire mice kingdom to be alerted of the movements of the cat and hide on time and be safe and free. No mice did! As a result, the mice kingdom remained in danger to the cat to this generation. Likewise, it took time for the world to impose stringent travel bans and doing business with China ("the cat") as this could have resulted in big economic ramifications after the COVID-19 epidemic subsided. Hence, the world remained lukewarm to the pending COVID-19 pandemic emanating from Wuhan, in Hubei, a Province of China. As indicated in Chap. 1, Australia tried it through just sponsoring a resolution for an international independent inquiry into COVID-19 outbreak and paid dearly through a 5-year anti-dumping tariff on its barley by China.

The range of the values for the incubation period remains important in epidemiology case definitions. These help in the determination of appropriate periods of quarantine and the effectiveness of entry screening and contact tracing (Backer et al. 2020). However, in the absence of incubation periods, the SARS and MERS coronavirus ranges were used during COVID-19 initial cases. Drawing from a sample of 88 confirmed COVID-19 cases detected outside Wuhan, the authors estimated the average incubation period to be 6.4 days (95% credible interval). This, however, ranged from 2.1 to 11.1 days at 2.5th to 97.5th percentile. The progression of symptoms reported in "Patient Zero" in Taiwan was similar to those from China and the US cases. Chen et al. (2020) reported fever on day 5 of illness and dyspnea with mild hypoxemia and radiographical pneumonia on day 9 of illness. As the first case of COVID-19 in Taiwan remained under observation, there was no known effective medical treatment. After quarantine, "Patient Zero" remained positive on day 15 and only turned negative on day 17. Jalava (2020) then urged the world to find the exact mechanism of initial transmission events. The SARS-CoV-2 transmission can follow pathways that include either human-to-human or animal-to-human (Amodio et al. 2020). The WHO also presented COVID-19 scenarios, as reflected in Fig. 3.4.

What also came up in the response phase is the concept of moving COVID-19 epicentres (Fig. 3.5). Between the COVID-19 official outbreak in China on 31 December 2019, three global epicentres emerged starting with China and then Italy as of 20 March 2020 and the USA from 27 March 2020. However, there were other regional and continental epicentres including Australia, Brazil, Iran and South Africa. Another fascination was how Spain had the potential to be the new epicentre for Europe. Drawing from the Johns Hopkins University Coronavirus Resource Centre data, a number of deductions were made (Johns Hopkins University 2020).

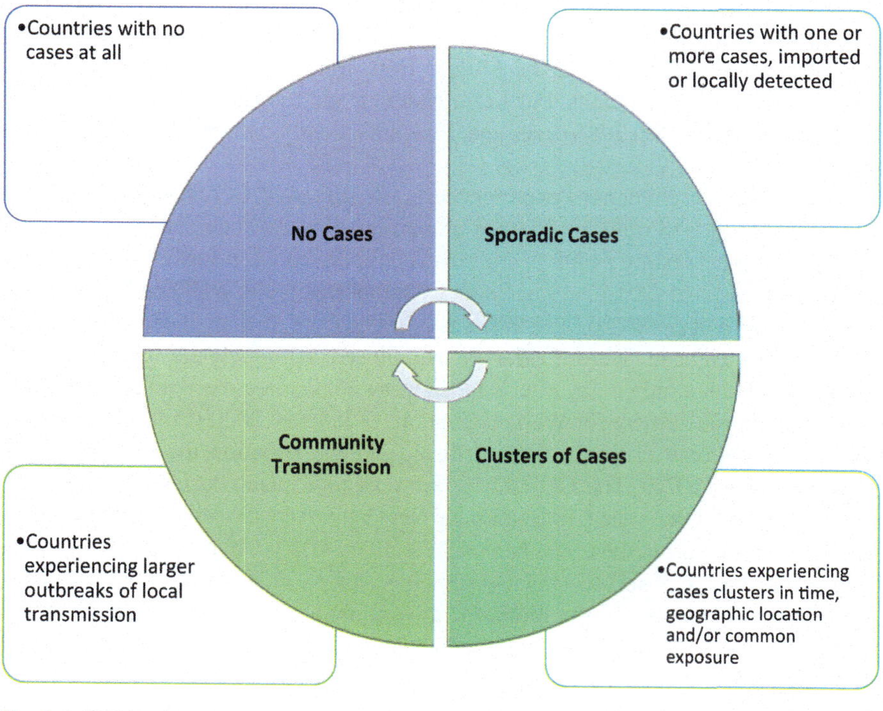

Fig. 3.4 WHO's four COVID-19 scenarios. Source: authors, based on WHO (2020)

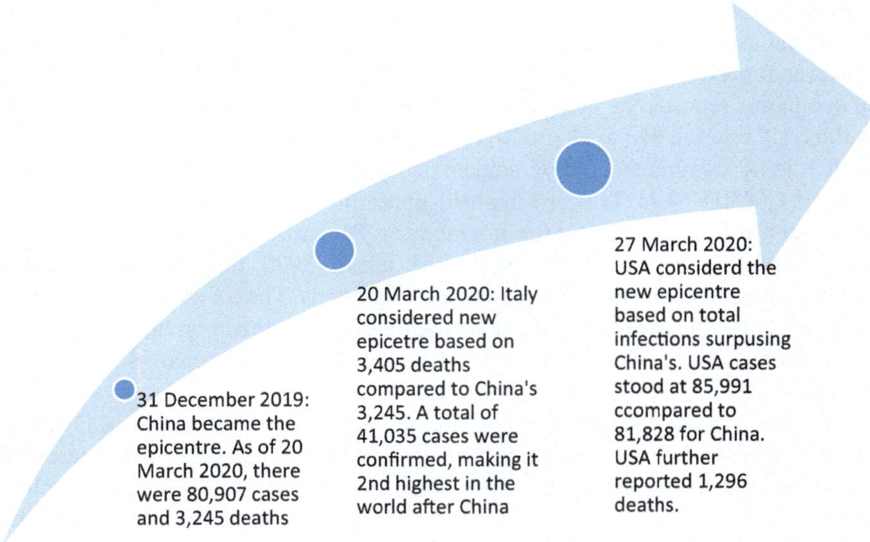

Fig. 3.5 The moving COVID-19 global epicentres. Source: authors, data from Johns Hopkins University Coronavirus Resource Centre (2020)

As of 27 March 2020, the country had 57,789 cases and 4365 deaths. Spain was closely followed by Germany with 47,278 and 281 deaths and France with 29,567 cases and 1698 deaths. Australia was the epicentre in East Asia and the Pacific with 3143 cases and 13 deaths. In the Middle East, Iran remained the epicentre with 29,406 case and 2378 deaths. For Latin America and the Caribbean (LAC), Brazil attracted attention with 2985 cases and 77 deaths. Lastly, South Africa was in focus with 927 cases and 2 deaths.

One of the key challenges experienced in the spread of COVID-19 Coronavirus has been church and other pilgrimage gatherings. Several such gatherings resulted in large numbers testing positive, and this is deliberated upon further in Chap. 10.

The first cases of COVID-19 in China were reported to the WHO on 31 December 2019 (Amodio et al. 2020), and this was coming from Wuhan. The following day the Wuhan's Huanan Seafood Market was closed. Although the initial key pronouncement came only a day after on 1 January 2020 when the Wuhan market was shut, the main lockdown for Wuhan came on 23 January 2020 (Ibid.). China came under heavy global criticism for not disclosing the situation to the public early (Nishiura et al. 2020). The COVID-19 outbreak took place during the most celebrated time in China – the Chinese Lunar New Year. An estimated 5 million people had left Wuhan before the travel ban on 23 January 2020 (Chen et al. 2020). Indeed Chinese residents observed social distancing, stayed at home and put on masks while in public places. Furthermore, the government extended the Lunar New Year holidays, with social marketing and media being central to the success rate (Li et al. 2020). The New Zealand's COVID-19 Alert System for recovery became relevant for other countries (Table 3.3).

Table 3.3 New Zealand COVID-19 Alert System – Levels 3 and 4 (Prepare)

Alert level	Outcome/state	Risk assessment	Selected measures at local or national level
3 (restrict)	High risk as the disease is not contained	Community transmission might be happening. New clusters may emerge but can be controlled through testing and contact tracing	People instructed to stay home in their bubble other than for essential personal movement – including to go to work, school if they have to or for local recreation The physical distancing of 2 metres outside home including on public transport or 1 metre in controlled environments like schools and workplaces Bubbles must stay within their immediate household bubble but can expand this to reconnect with close family/whānau, or bring in caregivers, or support isolated people. This extended bubble should remain exclusive Schools between years 1 and 10 and early childhood education centres can safely open but will have limited capacity. Children should learn at home if possible People must work from home unless that is not possible Businesses can open premises, but cannot physically interact with customers Low-risk local recreation activities are allowed Public venues are closed. This includes libraries, museums, cinemas, food courts, gyms, pools, playgrounds, markets Gatherings of up to ten people are allowed but only for wedding services, funerals and tangihanga. Physical distancing and public health measures must be maintained Healthcare services use virtual, non-contact consultations, where possible Interregional travel is highly limited to, for example, essential workers, with limited exemptions for others People at high risk of severe illness such as the elderly and those with existing medical conditions are encouraged to stay at home where possible and take additional precautions when leaving home. They may choose to work

Table 3.3 (continued)

Alert level	Outcome/state	Risk assessment	Selected measures at local or national level
4 (lockdown)	Likely that disease is not contained	Community transmission is occurring. Widespread outbreaks and new clusters	People instructed to stay at home in their bubble other than for essential personal movement Safe recreational activity is allowed in the local area Travel is severely limited All gatherings cancelled and all public venues closed Businesses closed except for essential services, such as supermarkets, pharmacies, clinics, petrol stations, and lifeline utilities Educational facilities closed Rationing of supplies and requisitioning of facilities possible Reprioritisation of healthcare services

Source: authors, based on Government of New Zealand (2020)

In the USA, Patient Zero was reported on 21 January 2020 (USA Centre for Disease Control and Prevention 2020). The patient from Washington DC had returned from Wuhan on 15 January 2020. Prior to this, several measures had been implemented including issuing of the first alert to clinics to be on the lookout for potential COVID-19 cases on 8 January 2020. Guidelines for clinics were developed, and the implementation of public health entry screening at the San Francisco (SFO), New York (JFK) and Los Angeles (LAX) airports had been instituted on 17 January 2020. The Atlanta (ATL) and Chicago (ORD) airports were added the same week.

In South Africa, Patient Zero was a 38-year-old male who had travelled to Italy with his wife. They were part of a group of ten people who arrived back home on 1 March 2020 (South Africa Department of Health 2020a). This was after the inaugural tests were conducted on 95 persons on 19 February 2020, all of which tested negative for the virus (South Africa Department of Health 2020b). The first case in Rwanda was a male Indian patient from Mumbai, India, who entered Rwanda on 8 March 219 (Rwanda Ministry of Health 2020). In Italy, the first two cases were confirmed on 31 January, being two Chinese tourists in Rome (Tega and Massa 2020). Table 3.4 presents details regarding the chronology of reporting from selected jurisdictions, some which will be discussed in depth in the chapter.

On tracking COVID-19, Singapore, Taiwan and Hong Kong are heralded as success stories in their strategies (Beech 2020). In Singapore, for example, medical teams were given 2 hours to uncover how the patient contracted the coronavirus. They are also supposed to identify the nature of people the patient might have infected. The four quick screening questions the medical teams ask include the following: (1) Did they travel abroad? (2) Do they have a link to one of the five clusters of contagion identified across the city-state? (3) Did they cough on someone in the street? and (4) Who are their friends and family, their social and prayer partners? The contact tracing in Singapore unearthed a group of avid singers who spread the

Table 3.4 Reported first COVID-19 cases[a] from selected national jurisdictions

Country	Estimated population 2019[b]	Day patient entered country	1st case(s) reported	Days it took to confirm Patient Zero	Days to partial/full lockdown since Patient Zero (date)
China	1.4 billion	N/A	31 December 2019	Unknown[c]	33 (3 February 2020)
Italy	60.3 million	31 January 2020	31 January 2020	1	7 (31 January 2020)
South Korea	51.8 million	19 January 2020	20 January 2020	1	30 (21 February 2020)
Rwanda	12.6 million	8 March 2020	14 March 2020	1	7 (21 March 2020)
South Africa	58.6 million	1st March 2020	5 March 2020	4	18 (23 March 2020)
Thailand	69.6 million	13 January 2020	22 January 2020	9	63 (26 March 2020)
The USA	329.5 million	15 January 2020	21 January 2020	6	59 (20 March 2020, California)

Source: Authors
[a]All imported apart from China
[b]As per the World Bank 19 September 2019
[c]Li et al. (2020) claim the first case could have been on 8 December 2019, but WHO was only notified 31 December 2020

virus to their families, transferred the coronavirus to a gym and then a church. This became the largest concentration of cases in the country. Some of the measures put in place in Singapore included building specialised epidemics clinics and official messaging that encouraged the public to wash their hands and/or sneeze in tissues during flu seasons. There are also freely available and accessible testing facilities and such facilities enabled the country to race against the COVID-19 spread. With a population of about 5.7 million people, Singapore had the capacity to test 2000 people daily. This was in sharp contrast to many countries and bigger states like Washington DC whose public laboratories could process up to 400 samples daily.

As for Hong Kong, government workers were ordered to work from home, with many private companies following suite. The schools were closed in January and scheduled to reopen on 31 April 2020 (Beech 2020). Borders were also tightened. The case of Taiwan is different and interesting in that the territory receives direct flights from Wuhan (Cheng et al. 2020). The territory started screening passengers from Wuhan late December 2019 well before China admitted COVID-19 coronavirus was spreading between people (Beech 2020). Despite this screening, the first case in the territory was reported on 21 January 2020. Similar to Hong Kong, the government response was based on lessons and practices developed during SARS infection period. The National Health Command Centre was key in the response operations. As of 31 January, Taiwan had suspended flights from China against advice from WHO. After detecting COVID-19 coronavirus during the Diamond Princess Cruise ship stopover, text messages were sent to every mobile phone on the island. The text messages listed each restaurant, tourist site and destination that the ship's passengers had visited during their time in Taiwan. Another notable measure was a call for self-quarantine of arrivals from 20 countries and 3 American States (Beech 2020). The efficient and effective monitoring systems in Hong Kong and Taiwan were built in response to SARS. This was in sharp contrast to the USA that had disbanded its pandemic response unit in 2018 (Beech 2020).

"Patient Zero" in Italy was reported on 21 February 2020, with the northern part of the country responding by imposing a lockdown in 11 towns. The affected population attempted to disregard government directives, and the government threatened to imprison residents who broke the rules (Spina et al. 2020). The case of Italy showed that the epicentre of COVID-19 had shifted from Wuhan to Europe. This realisation served as a global wake-up call that sent European health ministries on a quest to urgently search for COVID-19 prevention measures. While this is commendable, our lived experiences as authors suggest that this could have been late. The COVID-19 horse had bolted out of the stable and the world was on fire.

Germany's COVID-19 response attracted admiration globally. The country recorded its first case on 27 January 2020 in Munich (Carrel and Poltz 2020). However, its methodical medical manhunt to trace, widely test and isolate patients seemed to have borough early breakthroughs in COVID-19 containments. The move further gave Germany to buy time in building more effective defence systems. Bostock (2020) attributed the success to three key factors: (1) the testing programme witnessed Germany testing between 120,000 and 500,000 weekly in March 2020; (2) the country was prepared as it spends more on healthcare per capita compared to many countries in Europe and had the second most critical-care beds per capita in Europe; and (3) Germany enforced a lockdown on much earlier on 22 March 2020. Germany may also have been fortunate in that as of 27 March 2020, the average age of those infected was 46 years compared to Italy's 63 (Ibid.).

Immediately after the confirmation of "Patient Zero" in Rwanda on 15 March 2020, the following day the Ministry of Health issued a statement pronouncing six specific actions. The six were (1) closure of places of worship as of 15 March 2020; (2) closure of schools and institutions of higher educations as of 16 March 2020; (3) employees to work from home as appropriate in consultation with employers; (4) postpone large gathering including weddings and sporting events and minimise the number of people attending burials (5) at least 1 metre social distancing to be observed in business and restaurants whole continuing to operate; and (6) no overcrowding in public transport and need to avoid unnecessary movements (Rwanda Ministry of Health 2020). The proclamation indicated handwashing with soap or an alcohol-based sanitiser and social distancing as the most effective methods of constricting and/or blocking the pathways of COVID-19 transmission. The Ministry of Health further indicated that the use of masks was not necessary for the general public but recommended it for patients and those in direct contact with such patients. Lastly, the ministry highlighted the symptoms of COVID-19 that include a dry cough and high fever. In addition, the ministry advised those who had such symptoms and suspected that could be COVID-19 infected stay at home and telephone a specific response centre. Calls to this centre and consultations with community health workers were without a charge (Ibid.).

After “Patient Zero” was confirmed in South Africa on 5 March 2020, government response came 4 days later on 15 March 2020 in the form of the Presidential Statement outlining measures to combat COVID-19 (Republic of South Africa 2020a). Drawing from e-lessons and the history of COVID-19 spread elsewhere, the President clearly stated that South Africa, like all the other countries in the world, was not immune to the rapidly spreading pandemic. The President stated that this was probably the greatest calamity facing South Africa since democracy in 1994. At the time of this statement, South Africa had 61 confirmed cases of COVID-19. Against this background, the President stated expectations of higher cases of infection in the future. While all confirmed cases then were imported from Italy, growth was largely expected from cases of local transmissions. As such, the statement articulated that situation called “for an extraordinary response; there can be no half measures” (Republic of South Africa 2020a: 1). The envisaged response was to be adequate and robust to manage COVID-19, protect the residents and minimise the negative impacts of the pandemic on society and the economy.

Notably, the President declared a National State of Disaster in terms of the National Disaster Management Act of 2002 (Act No. 57 of 2002) (Republic of South Africa 2002) as the Head of the National Disaster Management Centre had classified the COVID-19 pandemic a national disaster on 15 March 2020 (DCOGTA 2020a, 2020b). In the interim, a number of measures were announced (Republic of South Africa 2020a) that included limiting contact with people who may be infected and travel ban on foreign nationals from high-risk countries – Italy, Iran, South Korea, Spain, Germany, the USA, the UK and China as from 18 March 2020. Control measures also included the revoking of valid visas held by visitors from these countries. In addition, the issuing of visas to travellers from these countries was suspended. Furthermore, South African citizens were advised to immediately refrain from all forms of travel to or through the European Union countries, the USA, the UK and other identified high-risk countries. Other measures proclaimed that citizens returning from high-risk countries would be tested and advised to self-isolate upon return to the country. In addition, travellers from low-risk countries like Portugal, Hong Kong and Singapore were to be screened, and all travellers that entered South Africa from high-risk countries since 15 February were to present themselves for COVID-19 testing at designated points.

The President also announced intervention measures linked to the ports of entry. Notable here was the strengthening of COVID-19 surveillance, screening and testing measures in the major ports of entry specifically at OR Tambo, Cape Town and King Shaka International Airport (Republic of South Africa 2020a). A total of 72 ports of entry into the country were identified, and 35 of the 53 land ports of entry were closed from 16 March 2020. In addition, two of the seaports were closed for passengers and crew changes. All non-essential government official travel was suspended with immediate effect, and non-essential local travels, especially by air, rail, taxis and buses were discouraged. To enforce social distancing, additional measures were announced including banning gathering of more than 100 people in mass celebrations of national holidays such as the Human Rights Day, funerals, weddings and other events. Schools closed from Wednesday 18 March and were then directed to remain closed until the Easter weekend. In addition, visits to correctional centres (prisons) were suspended for 30 days with immediate effect, and the Minister of Higher Education, Science and Innovation consulted with the Vice-Chancellors of universities to find an appropriate response.

Matters of hygiene came into the fore as well too. Businesses were called upon to ensure necessary measures to enhance hygiene within their premises (Republic of South Africa 2020b). This directive covered shopping malls entertainment and amusement centres and other places accommodating large numbers of people. As these measures were being affected, the government began the process of identifying isolation and quarantine sites in all districts and municipalities covering even the lowest levels of government dealing with disaster risk reduction. Developments on this front included the designation of COVID-19 hospitals in various places in the nine provinces of the country. There was also a noteworthy exercise seeking to enhance the process of both potential infected and infectious individuals. This was being done in partnership with the private sector, with a mass communication campaign on good hygiene underway. Summing up, these measures were a call for residents to change common behavioural practices from all residents. For example, there was a call for elbow greeting to replace the traditional handshaking practice.

Since the outbreak of COVID-19 in South Africa, the institutional set-up of managing had involved an Inter-Ministerial Committee chaired by the Minister of Health. However, as of 15 March 2020, the government announced the establishment National Command Council (NCC) chaired by the President (Republic of South Africa 2020a). The NCC included, among others, members of the Inter-Ministerial Committee. Meeting three times a week, the role of the NCC was to coordinate all aspects of COVID-19 extraordinary emergency responses. A number of developments emerged from this group. Lastly, the President discouraged the spread of fake news and hinted that the COVID-19 national emergency "demands cooperation, collaboration and common action" and also "requires solidarity, understanding and compassion" (Republic of South Africa 2020a: 6).

The Italian COVID-19 outbreak provides some learning points to the world. The infections and deaths spiralled shockingly in that country. Silver et al. (2020) identified two key lessons from Italy: (1) the importance of rapidly imposing strict epidemic management rules with accompanying measures to sanction those who derogate these rules and (2) the importance of clear and consistent communication. In Italy, Prime Minister resisted demands and calls from his government and opposition leaders for strict containment intervention measures at the onset of the pandemic in the country. As the first lockdown loomed in the north of the country, media leaks caused confusion and panic, leading to thousands of residents leaving the region. The national lockdown was eventually proclaimed on 9 March 2020.

Initial sanctions for resident disregarding the lockdown rules was a fine of 200 Euros (€200) imposed on those roaming streets without legitimate reasons. The figures were revised upwards at the peak of the pandemic to between €400 and €3000 as of 25 March 2020. The State of Emergency declared on 31 January 2020 mainly sought to activate the national civil protection system. With this law, the Head of the Department of Civil Protection accelerated procurement of resources to manage the epidemic. A key development during this time was the suspension of the payment of bank loans. (Order No. 642) gave clarity on the scope and content of decrees adopted by the President of the Council of Ministers (Order No. 646) and extended to the whole nation some measures initially limited (Order No. 648) (Tega and Massa 2020: 1–4).

The lockdown of 23 February 2020 marked Phase 2 of interventions in Italy (Tega and Massa 2020). It provided the basis for a series of decrees from the President of the Council of Ministers (DPCMs) that included the following: DPCM 8 March 2020 (partial lockdown), DPCM 9 March 2020 (national lockdown) and DPCM 11 March 2020 (closing of all shops, apart from those dealing in food and essentials). The third phase focuses on curing Italy, and a Decree Law No. 18 was passed on 17 March 2020. The law provides for boosting the health system and hiring of health workers as well as police and military forces. A raft of state-funded economic relief measures was also proclaimed. These included the suspension of fiscal dates and social security contributions, tax credits and incentives for enterprises to face the emergency. Work dismissals for COVID-19 reasons were suspended for 60 days, and businesses could resort to the redundancy fund to pay unused workforce. An amount of 25 billion euro was set aside for economic relief to be borrowed from the market, and overall, 350 billion euro was targeted (Tega and Massa 2020).

The restrictions on people's movement became more common as COVID-19 outbreaks spiralled the world over. Countries started with general restrictions on movements, moving to a partial city and regional lockdowns and eventually into full-blown national lockdowns. Figure 3.6 shows the state of people's movement restrictions as of 5 May 2020.

The COVID-19 data as of 5 May 2020 revealed that the Northern hemisphere countries seemed to be in greater trouble, both in terms of infections and deaths from the top 22 selected (Fig. 3.7). The USA topped the list of both infections and deaths with 1,181,885 infections and 69,079 deaths as of 5 May 2020. This was

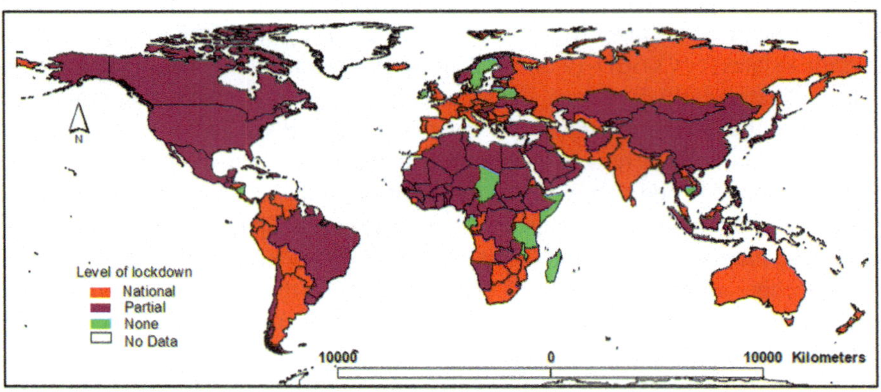

Fig. 3.6 COVID-19 lockdown record up to 17 May 2020. (The records do not show relaxations in lockdowns. Rather, the map shows the extent to which countries went in as far as lockdowns were concerned.). Source: authors

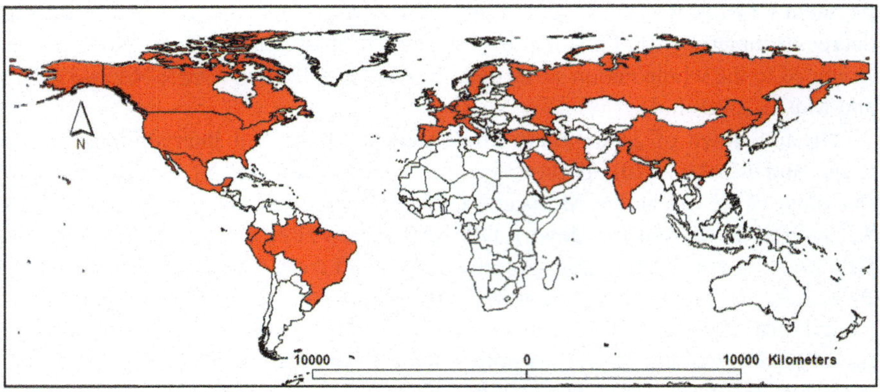

Fig. 3.7 Top 22 countries in terms of COVID-19 infections and deaths (5 May 2020). Source: authors, data from John Hopkins University (2020)

followed by Spain (218,011), Italy (213,013), the UK (196,238) and France (169, 583) in terms of infections and a similar pattern in terms of death with the UK (29,502), Italy (29,315), Spain (25,428) and France (25,204) ranking 2nd to 5th in that order. Those at the bottom of the top 20 packs in terms of infections were Saudi Arabia (30,251), Switzerland (30,009), Portugal (25,702), Mexico (24,905) and Sweden (23,216). Twelve of the top 22 countries were from Europe, 2 apiece from Asia and the Middle East, 4 from Latin America and 2 from North America.

One possible explanation for the picture regarding infections and deaths for the top 22 countries could be the mobility within the Northern Hemisphere that remains very high as facilitated through thousands of (cheap) flights daily, compared to the southern hemisphere. This is also linked to high volumes of trade with China. Secondly, there is the possibility of countries from the southern hemisphere, especially the African continent having a longer time to prepare for the pandemic with the first case reported a month and a half later in Egypt and being an imported case. As revealed earlier, the African Union Centre for Disease Control prepared the member states for COVID-19 outbreak in partnership with the WHO. Thirdly, other aspects, particularly for Africa, could have been to do with the re-emerging Ebola outbreaks that regularly prepare the countries, especially those in West Africa for epidemics. Fourth, there is evidence of slow reactions from countries such as the USA and Italy. In addition, it seems the Spanish Flu lessons had been forgotten as there were huge impacts in some of the countries caught up in COVID-19 pandemic including Italy, Spain, the UK and the USA. Spain was the epicentre then and almost became the epicentre during COVID-19. Lastly, the cities in the Northern Hemisphere remain with very high population densities that are favourable for COVID-19 human-to-human transmissions.

The COVID-19 response from South Korea is a good case for reflection. As it reported 7513 COVID-19 cases on 10 March 2020, the country had 54 deaths (0.6%), which was the lowest in the world (Kasulis 2020). Indeed this was low in comparison to 5% and 4.1% in China and Italy, respectively. So, what did South

Korea do right? From Kasulis' (2020) account, there were a number of key interventions put in place. The country acted very fast to combat the spread of infection. After the confirmed first case of a Chinese woman on 20 January 2020 and a South Korean businessman who had gone to Wuhan 3 days later, there was a plethora of aggressive measures sought to minimise the spread of infections. More important is that there was thorough and screening of all travellers from Wuhan (Kim 2020). However, with 11 cases confirmed by 31 January, anti-Chinese sentiment grew, with over half a million people signing a petition on the Presidential Blue House website calling for a ban on all Chinese travellers (Kasulis 2020). Despite these measures, the number of confirmed cases rose to more than 6000 in less than a week, and the first death was reported on 20 January 2020.

Third, from 26 February 2020, in addition to hundreds of testing clinics, 50 free "drive-through" testing centres began operating, with results texted in less than 3 days. By then, an estimated 46,127 people had been tested, compared to Japan's 1846 and the USA's 426 cases (Kim 2020). As of 10 March 2020, over 210,000 people had taken tests (Kasulis 2020). Fourth, authorities worked together to precisely document the movement of infected people down to the minute from testimonies, closed-circuit television (CCTVs) and smartphone GPS data (Kim 2020). The success in testing took advantage of the post-MERS reform and authorised an unlicensed COVID-19 test. Fifth, as from 5 March 2020, the government boosted the distribution of masks to medical staff and the public and banned face masks exports. Sixth, to prevent price-gouging and shortages, all face mask producers were required to report sales volumes. The government supplied 4.5 million masks with each person limited to a maximum of five. The masks were also sold at reduced prices of less than US$1 (Kasulis 2020).

However, the success story did not go without challenges pertaining to mass gatherings. Between 20 January 2020 and 17 February 2020, there were only 30 known COVID-19 cases. Then came a first case (Patient 31) reported in Daegu, the third largest city of about 2.5 million. Patient 31 attended a service at the Shincheonji Church of Jesus and spread COVID-19 to 14 people in that church gathering, with 3150 cases eventually reported and 17 deaths as of 29 February (Kim 2020). Investigations confirmed that some cases were as a result of the church's major event held in Gwacheon City, Gyeonggi Province on 16 February where about 10,000 members attended. By 1 March 2020, there were 3526 confirmed cases of which an estimated 60% of the confirmed cases were linked to the church. Another church-related event resulted in 46 people infected on 8 March 2020 well after the government pronounced social distancing measures. The service had been conducted in Seongnam, a city close to Seoul by the River of Grace Community Church (Kim 2020).

The measures in South Korea are in sharp contrast to those in the USA. President Donald Trump sought to play down the challenge indicating that the 15 cases that had been reported as of 27 February 2020 could fall to zero in a "few days" (Farrer 2020). The President was basing his thinking on past successes and even had the audacity to point out to Italy and other countries he thought had real challenges. Later events in the USA proved Trump's thinking was very wrong. The cases in the

USA increased by huge numbers. As of 26 March 2020, the confirmed COVID-19 cases in the USA were at 69,197 with death cases at 1046 people (News24 Team 2020).

Drawing from mobile phone surveillances experiences elsewhere, for example, in South Korea, Taiwan, Germany and Israel, the South African government and mobile network providers agreed to use mobile phone data to fight COVID-19 (De Wet 2020). The data was used to estimate the number of contacts an infected individual may have had. In a typical replica of COVID-19 church-related infections in South Korea, South Africa reported 50 new cases from a church as the national figures spiked to 709 on 25 March 2020 from 554 cases reported on 24 March 2020 (a 28% increase) and went over the 1000 marks to 1170 on 27 March 2020 (New24 Wire 2020). The Jerusalem prayer breakfast, hosted by the Divine Restoration Ministries in Bloemfontein, Free State Province, had hosted COVID-19 positive tourists from the USA, Israel and France. The Department of Health announced it was now tracing 300 people that attended the service. Among those that attended the church service were two politicians and a known pastor.

Eventually, massive COVID-19 testing programmes were instituted in many countries across the world. Furthermore, handwashing, face mask usage and other necessary hygiene measures became part and parcel of a package of COVID-19 response measures. Countries increased their bed capacities as well, although both the COVID-19 vaccines and treatment solutions were not yet found 5 months after the first case was reported. The medical field and other related professionals were trying various potential vaccines and treatments that fall outside the scope of this chapter.

3.4.3 Recovery from COVID-19: The Battle to "Flatten" Both Curves

The concept of building back better (BBB) is emphasised in the Sendai Framework for Disaster Risk Reduction (United Nations Office for Disaster Risk Reduction 2015). BBB comes in the post-disaster recovery, rehabilitation and reconstruction phase after a disaster such as COVID-19 and is necessary to prevent the creation of and to reduce disaster risk. Although it was too early to make bold proclamations on COVID-19 recoveries, there were signs of countries coming out of the COVID-19 lockdown hibernations. These countries included China, New Zealand, Australia, Germany, South Korea, South Africa, Italy and Spain, among others. The recovery pathways were being unveiled from carefully planned phased approaches from lockdowns, although not all countries went on lockdowns. Some countries that had not to go on lockdowns at the time of writing included Madagascar, Sweden and Tanzania. Recovery further included economic stimulus packages, which are a subject of discussion in other chapter seen in this book. The battle from lockdowns (whether full or partial) had been the continuously falling economic curve. Hence, the arguments shifted from "flattening" the COVID-19 curve to "flattening" both the COVID-19 and economic curves.

China was the first country to go back to business after the COVID-19 outbreak. On 27 March 2020, Knight (2020) reported that the country's big corporates were recalling workers, with malls and restaurants opening up to crows. This was, however, done with strict safety measures. Some precautionary measures included having workers in shifts that were staggered in order to minimise the risk of COVID-19 infections. Other measures included the installation of infrared cameras and security staff that checked the temperatures. Workplaces were also disinfected thrice daily, and workers had to wear their face masks all the time. Only a maximum of six people were allowed in elevators, with face-to-face meetings discouraged and so were group gatherings for tea or lunch meals. Travel bans in Hubei were also lifted. Senior students in Wuhan opened schools during the week the first week of May, although the preventive measures for factories also applied, including additional measures for students to sit with desks 1 metre apart (AFP 2020). In South Korea, life was also getting back to normal, with workers getting back to work and libraries and museums opening (Ibid.).

South Africa started its COVID-19 recovery journey 35 days after a full national lockdown (26 March to 30 April 2020). The country devised five levels of COVID-19 response and recovery. These are presented in Fig. 3.8. The scope for COVID-19 response and recovery levels was also being used in some countries including Namibia and Zimbabwe, although this did not follow the same five levels. The levels were accompanied by respective regulations specifying the exact activities that would be allowed at any level. Level 4 also meant that all people had to wear compulsory face masks. However, these did not come without challenges. In South Africa, for example, the government was challenged for banning tobacco and alcohol sales under Level 4 (Steyn and Klopper 2020).

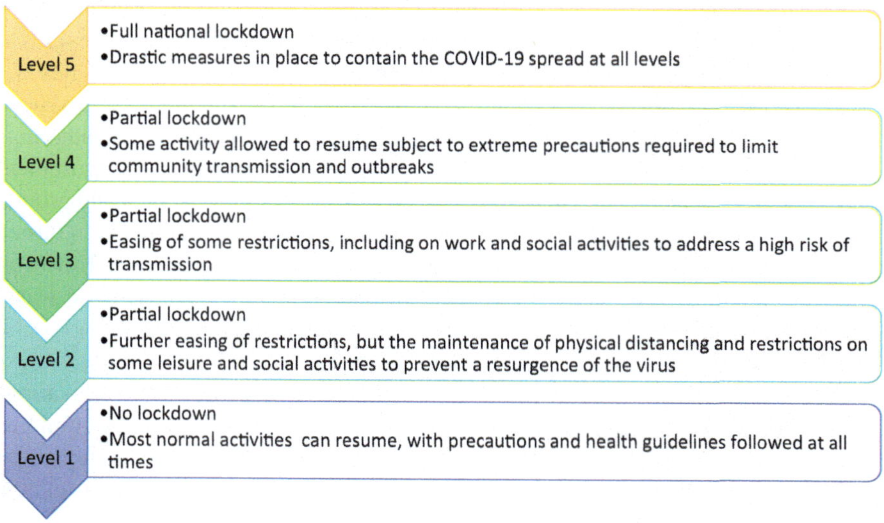

Fig. 3.8 South Africa's COVID-19 response and recovery levels. Source: authors, based on the Republic of South Africa (2020c)

European countries warned up to COVID-19 recovery, mainly after witnessing the "flattening" of curves in Italy and Spain. Austria and Italy opened shops on 14 April 2020 (BBC 2020). Other recovery activities included schools opening in Denmark, businesses allowed to return in Spain and Poland gradually lifting restrictions. Italy opened bookshops and clothes stores for young children, although bookshops were not permitted to open in Lombardy. There was also further use of thermo-scanning and screening robots in certain retail and other establishments. Austria was the second country in Europe to follow Italy's strict lockdowns. On return to normalcy, the country allowed shops of $400m^2$ to open alongside hardware and garden centres. People had to wear compulsory facemasks. Germany was due to allow schools and all shops to open in May as well as restart their top-flight football league the Bundesliga (AFP 2020). The AFP also revealed that car sales went down 61% in April 2020.

New Zealand became the first country to proclaim some form of COVID-19 eradication after close to 5 weeks of lockdown since 26 March 2020 (Hollingsworth 2020). The government reported that COVID-19 new infections had dropped to single digits over several days on 28 April 2020. The country had recorded 19 deaths out of 1472 infections. With this success story, about half a million workers returned to work after the alert level was lowered to 3 (Withers 2020) on 27 April 2020. This level permitted workers to return to construction sites and factories, with take-away food outlets also opened. Hospitality and retail outlets were to maintain strict criteria on physical distancing for customers. As highlighted earlier, New Zealand developed a four-tier COVID-19 Alert System as follows: Alert Level 1, Prepare; Alert Level 2, Reduce; Alert Level 3, Restrict; and Alert Level 4, Lockdown (Government of New Zealand 2020).

In Australia, a three-step pathway for easing restrictions was developed and code-named the roadmap to a COVIDSafe Australia (Australian Government 2020). Under Step 1, five visitors were now allowed at home, with ten in businesses and public places. Working from home was still encouraged, with restaurants, cafes and shops allowed to operate. Libraries, community centres and boot camps were allowed to resume as well as local and regional travel. Under Step 2, gatherings numbers increased to 20, with gyms, hair salons, cinemas, galleries and amusement parks opening. Selected interstate travel was allowed as well as caravan and ground camping. Under Step 3, gatherings numbers were increased to 100; people returned to work, nightclubs, food courts; and all interstate travel also opened. However, Australians had to keep maintaining the 1.5 metre social distancing, hand wash, practice safe sneezing, download the COVIDSafe app and put in place COVIDSafe Plans at workplaces.

As the chapter was being finalised and getting into production, more relaxations of lockdowns were being unveiled. Furthermore, additional economic stimulus packages were also being put in place. The world in COVID-19 hibernations was coming back to life. However, no one or country for that matter was clear as to how the relaxation of COVID-19 restrictions were going to unveil. It was a huge gamble and a massive global experiment, with the whole world being a COVID-19 laboratory.

3.5 Conclusions

The COVID-19 coronavirus pandemic dictated extraordinary measures to be put in place before and especially after "Patient Zero". From the South Africa Government, new institutions and special vehicles to manage the pandemic were instituted including the Presidential Command Council, Inter-Ministerial Committee on COVID-19 and the Solidarity Fund. Several Decrees were raised in Italy too. Lockdowns became common sites, and these were instituted in several counties under review, with the army deployed to support the police in maintaining law and order. In terms of quick response, Rwanda remains the pathfinder, having instituted its first set of measures after "Patient Zero" within a day and only 7 days to lockdown. Everything was done in 8 days, with the lockdown becoming effective the day it was announced. South Korea provided a good example in terms of free testing, contact tracing, distribution of face masks and controlling their prices. Citizen activism was also great, with reforms in law following the MERS assisting inefficiencies in the government systems. In addition, intervention measures should be in place long enough to avoid a recurring of infections, which was a huge worry for Wuhan. A raft of regulations were promulgated to institute travel bans from the air, land and sea as well as to protect consumers from price spikes and manipulations. While some countries declared the COVID-19 as state of emergencies, many others opted to declare the pandemic a state of national disaster. All these were done using applicable legislation, especially disaster management acts. There was also a huge challenge in the manner in which China handled the first cases as it delayed notifying the World Health Organisation (WHO). This was not the first time China did it, having done so during the SARS outbreak in 2002/2003. What became clear is that should similar pandemics happen, the source country must notify the WHO early so that the disease outbreak can be managed quickly and possibly contain early. COVID-19 also brought insights into the manner in which the epicentres migrated. It was first China (Asia), then Italy (Europe), the USA (North America) and South Africa (Africa).

There was a basket of measures encouraged to break COVID-19 infections. Many of these measures were from experiences with past pandemic outbreaks and including social distancing, testing with self-isolation and regulatory quarantines, wearing of masks and lockdowns. Social distancing is an intervention measure from the 1918–1919 influenza pandemic when the New York City Department of Health enforced such, with other complementary measures. This resulted in the mega-city registering the lowest death rate. The road to COVID-19 recovery enhanced through the gradual and staged phasing out of lockdown and other restrictions. These measures were put in place from across the world from Austria, Australia, Italy, New Zealand and South Africa. Australia, for example, put in place the roadmap to a COVIDSafe Australia with three stages, New Zealand had a COVID-19 Alert System covering all the stages from preparedness, through response and then recovery. South Africa had a five-stage COVID-19 Response and Recovery Plan. Although there was still talk on the extent of COVID-19 in Africa, particularly, the low infections, early indications were that the pandemic had not manifested itself in a manner it did in other areas, especially in the Northern Hemisphere. This probably could have been due to the fact that the continent had adequate time to prepare for COVID-19.

Looking ahead, there is definitely no guarantee that the world will not experience yet another pandemic of similar manifestation and magnitude, with a high potential that China could be the source given the hosts of the coronavirus in that country. Ultimately, if the world is to make lasting progress in trying to prevent the coronaviruses that are mainly zoonotic in nature, there is a great need for eating behavioural changes. The success stories embedded in SARS and MERS history for South Korea, Singapore, Taiwan and Hong Kong means every nation should take the COVID-19 as a learning platform to better prepare for the future. In addition, there was inconclusive evidence in terms of the effectiveness of lockdowns. The cases of countries like Sweden, Tanzania and Madagascar, among others, need to be revisited and documented further. However, from the historical experience of the 1918 influenza, US states that had shorter lockdowns experienced more cases and deaths compared to those that had longer lockdowns showing the effectiveness of lockdowns in controlling the pandemic. The world was shaken and will remain so for a while!

References

AFP. (2020). *EU faces historic recession as China, US spar over virus origins*. Retrieved from https://www.afp.com/en/news/3954/eu-faces-historic-recession-china-us-spar-over-virus-origins-doc-1r08k34. Accessed 30 May 2020.

Alexander, R. M. (2019). The Spanish flu and the sanitary dictatorship: Mexico's response to the 1918 influenza pandemic. *The American*, 443–465. https://doi.org/10.1017/tam.2019.34.

Amodio, E., Vitle, F., Cimino, L., Casuccio, A., & Tramuto, F. (2020). Outbreak of novel coronavirus (SARS-Cov-2): First evidences from international scientific literature and pending questions. *Healthcare, 8*, 51. https://doi.org/10.3390/healthcare8010051.

Andayi, F., Chaves, S. S., & Widdowson, M. A. (2019). Impact of the 1918 influenza pandemic in coastal Kenya. *Tropical Medicine and Infectious Diseases, 4*, 91. https://doi.org/10.3390/tropicalmed4020091.

Antwi-Agyei, P., Dougill, A. J., Agyekum, T. P., & Stringer, L. C. (2018). Alignment between nationally determined contributions and the sustainable development goals for West Africa. *Climate Policy, 18*(10), 1296–1312. https://doi.org/10.1080/14693062.2018.1431199.

Australian Government. (2020). *Roadmap to a COVIDSafe Australia: A three-step pathway for easing restrictions*. Sydney: Australian Government.

Backer, J. A., Klinkenberg, D., & Wallinga, J. (2020). Incubation period of 2019 novel coronavirus (2019-nCoV) infections among travellers from Wuhan, China, 20–28 January 2020. *Euro Surveillance, 25*(5). https://doi.org/10.2807/1560-7917.

BBC (British Broadcasting Corporation). (2004). *Timeline: Sars virus*. http://news.bbc.co.uk/2/hi/asia-pacific/2973415.stm. Accessed 22 Mar 2020.

BBC (British Broadcasting Corporation). (2020). *Coronavirus: Austria and Italy reopen some shops as lockdown eased*. Retrieved from https://www.bbc.com/news/world-europe-52275959. Accessed 6 May 2020.

Beech, H. (2020). *Tracking the coronavirus: How crowded Asian cities tackled an epidemic*. Retrieved from https://www.nytimes.com/2020/03/17/world/asia/coronavirus-singapore-hong-kong-taiwan.html?action=click&module=Spotlight&pgtype=Homepage. Accessed 24 Mar 2020.

Bimha, A., & Nhamo, G. (2017). Sustainable development, share Price and carbon disclosure interactions: Evidence from South Africa's JSE 100 companies. *Sustainable Development, 25*, 400–413. https://doi.org/10.1002/sd.1670.

Bostock, B. (2020). *Germany is reopening with one of the world's lowest death rates. Here's how it barely missed a trick*. Retrieved from https://www.businessinsider.co.za/coronavirus-germany-reopening-low-deaths-testing-lockdown-2020-5?r=US&IR=T. Accessed 17 May 2020.

Bowen, S. A., & Heath, R. L. (2007). Narratives of the SARS epidemic and ethical implications for public health crises. *International Journal of Strategic Communication, 1*(2), 73–91. https://doi.org/10.1080/15531180701298791.

Carrel, P. & Poltz, J. (2020). *It was the saltshaker: How Germany meticulously traced its coronavirus outbreak*. Retrieved from https://www.weforum.org/agenda/2020/04/it-was-the-saltshaker-how-germany-meticulously-traced-its-coronavirus-outbreak. Accessed 17 May 2020.

Chan-Yeung, M., & Xu, R. H. (2003). SARS: Epidemiology. *Respirology, 8*, S9–S14.

Chen, K. T., Twu, S. J., Chang, H. L., Wu, Y. C., Chen, C. T., Lin, T. H., Olsen, S. J., Dowell, S. F., & Su, I. J. (2005). SARS in Taiwan: An overview and lessons learned. *International Journal of Infectious Diseases, 9*, 77–85.

Chen, S., Yang, J., Yang, W., Wang, C., & Bärnighausen, T. (2020). COVID-19 control in China during mass population movements at new year. *The Lancet, 395*. https://doi.org/10.1016/S0140-6736(20)30421-9.

Cheng, V. C. C., Lau, S. K. P., Woo, P. C. Y., & Yuen, K. Y. (2007). Severe acute respiratory syndrome coronavirus as an agent of emerging and reemerging infection. *Clinical Microbiology Reviews, 20*(4), 660–694. https://doi.org/10.1128/CMR.00023-07.

Cherry, J. D. (2004). The chronology of the 2002–2003 SARS mini pandemic. *Paediatric Respiratory Reviews, 5*, 262–269. https://doi.org/10.1016/j.prrv.2004.07.009.

Cherry, J. D., & Krogstad, P. (2004). SARS: The first pandemic of the 21st century. *Pediatric Research, 56*(1), 1–5.

Cho, S. Y., Kang, J. M., Ha, Y. E., Park, G. E., Lee, J. Y., Ko, J. H., et al. (2016). MERS-CoV outbreak following a single patient exposure in an emergency room in South Korea: An epidemiological outbreak study. *The Lancet, 388*, 994–1001. https://doi.org/10.1016/S0140-6736(16)30623-7.

Cooper, C. (2005). Japanese tourism and the SARS epidemic of 2003. *Journal of Travel & Tourism Marketing, 19*(2–3), 117–131. https://doi.org/10.1300/J073v19n02_10.

DCOGTA (South Africa Department of Co-operative Governance and Traditional Affairs). (2020a). *Disaster management act, 2002 (act no. 57 of 2002): Classification of a National Disaster*. Pretoria: Department of Co-operative Governance and Traditional Affairs.

DCOGTA (South Africa Department of Co-operative Governance and Traditional Affairs). (2020b). *Regulations issued in terms of section 27(2) of the disaster management act, 2002*. Pretoria: Department of Co-operative Governance and Traditional Affairs.

De Wet, P. (2020). *South Africa will be tracking cellphones to fight the COVID-19 virus*. Retrieved from https://www.businessinsider.co.za/south-africa-will-be-tracking-cellphones-to-fight-COVID-19-2020-3. Accessed 26 Mar 2020.

Elhakim, M. M., Kandil, S. K., Elaziz, K. M. A., & Anwar, W. A. (2019). Epidemiology of Severe Acute Respiratory Infection (SARI) cases at a sentinel site in Egypt, 2013–15. *Journal of Public Health*, 1–9. https://doi.org/10.1093/pubmed/fdz053.

Farrer, M. (2020). *Coronavirus: Trump says US risk 'very low' as Australian PM warns pandemic is 'upon us'*. Retrieved from https://www.theguardian.com/world/2020/feb/27/coronavirus-trump-says-risk-to-us-is-very-low-as-australia-enacts-emergency-plan. Accessed 25 Mar 2020.

Fung, I. C. H., & Cairncross, S. (2006). Effectiveness of handwashing in preventing SARS: A review. *Tropical Medicine and International Health, 11*(11), 1749–1758. https://doi.org/10.1111/j.1365-3156.2006.01734.x.

Glynn, J. R. (2020). Protecting workers aged 60–69 years from COVID-19. *The Lancet*. https://doi.org/10.1016/S1473-3099(20)30311-X.

Government of New Zealand. (2020). *COVID-19 Alert System: New Zealand's 4-level Alert System specifies measures to be taken against COVID-19 at each level*. Retrieved from https://covid19.govt.nz/alert-system/covid-19-alert-system/. Accessed 6 May 2020.

GPMB (Global Preparedness Monitoring Board). (2019). *A world at risk: Annual report on global preparedness for health emergencies*. Geneva: World Health Organization. Licence: CC BY-NC-SA 3.0 IGO.

Hollingsworth, J. (2020). *How New Zealand 'eliminated' Covid-19 after weeks of lockdown*. Retrieved from https://edition.cnn.com/2020/04/28/asia/new-zealand-coronavirus-outbreak-elimination-intl-hnk/index.html. Accessed 6 May 020.

Hsieh, Y. H., King, C. C., Chen, C. W. S., Ho, M. S., Lee, J. Y., Liu, F. C., Wu, Y. C., & JulianWu, J. S. (2005). Quarantine for SARS, Taiwan. *Emerging Infectious Diseases, 11*(2), 278–282.

Jalava, K. (2020). First respiratory transmitted food borne outbreak? *International Journal of Hygiene and Environmental Health, 226*, 113490. https://doi.org/10.1016/j.ijheh.2020.113490.

Jester, B., Uyeki, T. M., Jernigan, D. B., & Tumpey, T. M. (2019). Historical and clinical aspects of the 1918 H1N1 pandemic in the United States. *Virology, 527*, 32–37. https://doi.org/10.1016/j.virol.2018.10.019.

Johns Hopkins Coronavirus Resources Centre. (2020). *Coronavirus COVID-19 Global Cases by the Centre for Systems Science and Engineering (CSSE) at Johns Hopkins University (JHU)*. Retrieved from https://coronavirus.jhu.edu/map.html. Accessed 27 Mar 2020.

Kasulis, K. (2020). *Timeline: What the world can learn from South Korea's COVID-19 response*. Retrieved from https://thegroundtruthproject.org/timeline-how-south-korea-reacted-to-the-COVID-19-outbreak/. Accessed 5 Mar 2020.

Keeling, A. W. (2020). The 1918 influenza pandemic: Lessons from the past for a global community. *J-STAGE*, 1–2. https://doi.org/10.24298/hedn.2019-SP00.

Kim, H. (2020). *South Korea learned its successful COVID-19 strategy from a previous coronavirus outbreak: MERS*. Retrieved from https://thebulletin.org/2020/03/south-korea-learned-its-successful-COVID-19-strategy-from-a-previous-coronavirus-outbreak-mers/. Accessed 26 Mar 2020.

Knight, W. (2020). *China goes back to work as the coronavirus rages on elsewhere*. Retrieved from https://www.wired.com/story/china-goes-work-coronavirus-rages-elsewhere/. Accessed 6 May 2020.

Li, H., Chen, X., & Huang, H. (2020). The novel coronavirus outbreak: What can be learned from China in public reporting? *Global Health Research and Policy, 5*, 9. https://doi.org/10.1186/s41256-020-00140-9.

Lui, Y., Dong, E., Lki, S., & Jie, X. (2020). Cruise tourism for sustainability: An exploration of value chain in Shenzhen Shekou Port. *Sustainability, 12*, 3054. https://doi.org/10.3390/su12073054.

MacKinlay, A. C. (1997). Event studies in economics and finance. *Journal of Economic Literature, 35*(1), 13–39.

Martini, M., Gazzaniga, V., Bragazzi, N. L., & Barberis, I. (2019). The Spanish influenza pandemic: A lesson from history 100 years after 1918. *Journal of Preventive Medicine and Hygiene, 60*, E64–E67. https://doi.org/10.15167/2421-4248/jpmh2019.60.1.1205.

News24 Team. (2020). *More than 2.6 billion worldwide told to observe lockdowns*. Retrived from https://m.news24.com/SouthAfrica/News/coronavirus-all-the-latest-news-about-COVID-19-in-south-africa-and-the-world-20200312. Accessed 30 May 2020.

Ng, O. W., & Tan, Y. J. (2017). Understanding bat SARS-like coronaviruses for the preparation of future coronavirus outbreaks — Implications for coronavirus vaccine development. *Human Vaccines and Immunotherapeutics, 13*(1), 186–189. https://doi.org/10.1080/21645515.2016.1228500.

Nickol, M. E., & Kindrachuk, J. (2019). A year of terror and a century of reflection: Perspectives on the great influenza pandemic of 1918–1919. *BMC Infectious Diseases, 19*, 117. https://doi.org/10.1186/s12879-019-3750-8.

Nishiura, H., Jung, S., Linton, N. M., Kinoshita, R., Yang, Y., Hayashi, K., Kobayashi, T., Yuan, B., & Akhmetzhanov, A. R. (2020). The extent of transmission of novel coronavirus in Wuhan, China, 2020. *Journal of Clinical Medicine, 9*, 330. https://doi.org/10.3390/jcm9020330.

Nkengasong, J. N., & Mankoula, W. (2020). Looming threat of COVID-19 infection in Africa: Act collectively, and fast. *The Lancet, 395*, 841–841.

Økland, H., & Mamelund, E. E. (2019). Race and 1918 influenza pandemic in the United States: A review of the literature. *International Journal of Environmental Research and Public Health, 16*, 2487. https://doi.org/10.3390/ijerph16142487.

Onder, G., Rezz, G., & Brusaferro, S. (2020). Case-fatality rate and characteristics of patients dying in relation to COVID-19 in Italy. *The Journal of the American Medical Association*. https://doi.org/10.1001/jama.2020.4683.

Paskoff, T., & Sattenspiel, L. (2019). Sex- and age-based differences in mortality during the 1918 influenza pandemic on the island of Newfoundland. *American Journal of Human Biology, 31*, e23198. https://doi.org/10.1002/ajhb.23198.

Republic of South Africa. (2002). *Disaster management act, 2002 (act no. 57 of 2002)*. Cape Town: South African Government.

Republic of South Africa. (2020a). *President Cyril Ramaphosa: Measures to combat Coronavirus COVID-19 epidemic*. https://www.gov.za/speeches/statement-president-cyril-ramaphosa-measures-combat-COVID-19-epidemic-15-mar-2020-0000. Accessed 21 Mar 2020.

Republic of South Africa. (2020b). *Statement by President Cyril Ramaphosa on escalation of measures to combat the COVID-19 epidemic*, Union Buildings, Tshwane. Retrieved from http://www.thepresidency.gov.za/speeches/statement-president-cyril-ramaphosa-escalation-measures-combat-COVID-19-epidemic%2C-union. Accessed 23 Mar 2020.

Republic of South Africa. (2020c). *Statement by president Cyril Ramaphosa on South Africa's response to the coronavirus pandemic*. Pretoria: The Presidency.

Rwanda Ministry of Health. (2020). *Statement on new measures to prevent COVID-19 coronavirus transmission*. Kigali: Ministry of Health.

Silver, V., Migliaccio, A., & Follain, J. (2020). *A preview from Italy: What life is like in lockdown*. https://www.fin24.com/Opinion/a-preview-from-italy-what-life-is-like-in-lockdown-20200324. Accessed 25 Mar 2020.

South Africa Department of Health. (2020a). *First case of COVID-19 coronavirus reported in South Africa*. Retrieved from http://www.nicd.ac.za/first-case-of-COVID-19-coronavirus-reported-in-sa/. Accessed 21 Mar 2020.

South Africa Department of Health. (2020b). *COVID-19 update*. Pretoria: Department of Health.

Spina, S., Marrazzo, F., Migliari, M., Stucchi, R., Sforza, A., & Fumagalli, R. (2020). The response of Milan's emergency medical system to the COVID-19 outbreak in Italy. *The Lancet, 365*, e49–e50.

Steyn, F. & Klopper, H. (2020). *Cigarette ban backfiring with 7 million smokers against government decision*. Retrieved from https://www.iol.co.za/capeargus/opinion/cigarette-ban-backfiring-with-7-million-smokers-against-government-decision-47545171. Accessed 5 May 2020.

Strochlic, N. & Champine, R.D. (2020). *How some cities 'flattened the curve' during the 1918 flu pandemic*. Retrieved from https://www.nationalgeographic.com/history/2020/03/how-cities-flattened-curve-1918-spanish-flu-pandemic-coronavirus/. Accessed 28 Mar 2020.

Tega, D. & Massa, M. (2020). *Fighting COVID-19: Legal powers and risks: Italy*. Retrieved from https://verfassungsblog.de/fighting-COVID-19-legalpowers-and-risks-italy/. 25 March 2020.

UNDRR (United Nations Office for Disaster Risk Reduction). (2015). *Sendai framework on disaster risk reduction (2015–2030)*. New York: UNDRR Secretariat.

USA Centre for Disease Control and Prevention. (2020). *First travel-related case of 2019 novel coronavirus detected in United States*. Washington: Centre for Disease Control and Prevention.

Van Wijhe, M., Ingholt, M. M., Andreasen, V., & Simonsen, L. (2018). Loose ends in the epidemiology of the 1918 pandemic: Explaining the extreme mortality risk in young adults. *American Journal of Epidemiology, 187*(12). https://doi.org/10.1093/aje/kwy148.

WHO (World Health Organisation). (2003). *Consensus document on the epidemiology of severe acute respiratory syndrome (SARS)*. Geneva: WHO.

WHO (World Health Organisation). (2020). *Health topics: Epidemiology*. Retrieved from https://www.who.int/topics/epidemiology/en/. Accessed 21 Mar 2020.

Wilder-Smith, A. (2006). The severe acute respiratory syndrome: Impact of travel and tourism. *Travel Medicine and Infectious Disease, 4*, 53–60.

Withers, T. (2020). *New Zealand economy gets back to work as lockdown is eased*. Retrieved from https://www.bloomberg.com/news/articles/2020-04-27/new-zealand-economy-gets-back-to-work-as-ardern-eases-lockdown. Accessed on 6 May 2020.

Woo, G. (2019). Age-dependence of the 1918 pandemic. *British Actuarial Journal, 24*(e3), 1–16. https://doi.org/10.1017/S1357321719000023.

Part III
Impact of COVID-19 on Industries Within (Global) Tourism Value Chains

Chapter 4
COVID-19 and Implications for the Aviation Sector: A Global Perspective

Abstract In recent years we have witnessed several disasters and extreme events that have carried a variety of negative socio-economic and environmental impacts. These events and disasters threaten the attainment of the ambitious targets outlined in the 2030 Agenda for Sustainable Development. While intense tropical cyclones, extreme floods, fires, droughts and record polar ice melting characterised 2019, the year 2020 ushered in COVID-19, a global pandemic that has left a trail of untold suffering and economic disruption. Although the pandemic is known to have shaken citizens around the globe, to date, there is very little knowledge of how this has affected various economic sectors. Nevertheless, the tourism industry appears to have suffered a serious blow. This study investigated how one subsector of tourism, namely, aviation, was affected by and responded to the COVID-19 pandemic. Using critical document analysis, the study found that the aviation sector was brought to a halt, costing billions of dollars in losses around the world. COVID-19 also resulted in job losses along the entire aviation industry's value chain and in related sectors; these were due to losses in passenger revenue, cargo revenue, air ticket refunds, fixed costs and increased parking fees for grounded aircraft, among other things. The study identified various strategies that were used to provide relief, including private and public funding through bailouts, tax deferment, waiver of individual taxes, provision of grants and low-interest loans. They also recommend additional documentation of impacts as they continue unfolding, thus providing valuable lessons for handling any future disasters.

Keywords COVID-19 · Aviation · Disaster · Tourism · IATA · Airline bailouts

4.1 Background and Introduction

Globalisation and continued aviation efficiency mean that many people can reach distant places within a relatively short period. It is established that global mobility efficiency has been a critical factor in supporting human civilisation and commerce for a better world (Dube and Nhamo 2020). As such, the aviation industry is credited with being central to human development and its capacity to bring people together. The sector employs 165 million people across the world, and its net worth is approximately US$ 2.7 trillion (International Air Transport Association (IATA) 2020a).

G. Nhamo et al., *Counting the Cost of COVID-19 on the Global Tourism Industry*,
https://doi.org/10.1007/978-3-030-56231-1_4

Despite substantial crucial economic contribution made by the aviation sector, there are some challenges posed by the aviation industry, including the industry's contribution to greenhouse gas (GHG) emissions, which contribute to climate change (Dube and Nhamo 2019). The aviation sector is also blamed for contributing to the spreading of pandemics and diseases (Grais et al. 2003), such as COVID-19. Hall (2006) argues that, for some time, travel and trade have been a significant transmission vehicle for infectious diseases, as they facilitate the movement of infected people. With 8.3 billion passenger trips in 2018 (International Airport Review 2019) alone, the risk of people's spreading diseases as they move across the world remains very high.

On 1 April 2020, the United Nations warned that the impact of COVID-19 was likely to surpass all previous pandemics and trigger a worse economic recession than that of 2008–2009; indeed, it was likely to have a greater impact than that of World War II and plunge the world into an economic recession (United Nations 2020a). They also expressed a real concern that such a development would have a drastic impact on the attainment of the Sustainable Development Goals (SDGs). As a result, the United Nations (2020b) proposed a set of measures to assist in defeating the world pandemic and building a better world. As part of that proposal, the United Nations called for an examination of the impacts of the coronavirus at the regional level to start the process of resource mobilisation for reconstruction.

This study is, therefore, a response to the United Nations' call and examines the impact of COVID-19 on the global aviation industry.

4.2 Literature Review

Disease outbreaks are often problematic and disruptive to businesses, leading to disruption of mobility between destinations (Joo et al. 2019). Such a phenomenon is most evident in the aviation industry and the tourism industry. This is particularly true when the pandemic restricts people's movement, thereby crippling the aviation industry in several ways. The industry depends on people, cargo movement and other travel-related products to generate the bulk of its revenue (Wolf et al. 2016; Konttinen 2017). Ordinarily, the aviation industry has huge fixed and semi-fixed expenses that they incur, even when the industry is grounded. Some of these expenses include depreciation, insurances, leases, station expenses, ticketing, administrative costs, crew costs and commercial aircraft maintenance, repair and overhaul (MRO). According to IATA (2020b), the fixed and semi-fixed costs constitute about 49% of an airline budget, while variable costs for fuel, user charges and passengers constitute about 51%. Hence, any disruption to traffic movement has dire consequences for the budget and viability of commercial airlines.

It could be argued, therefore, that the aviation industry is a fickle industry that is sensitive to internal and external shocks (Senbeto and Hon 2020), whether these are natural or human-induced. Hyndman and Hyndman (2016) distinguish and categorise various threats that face the aviation industry by pointing out that a crisis is an internal threat to the business, while a disaster is an external environmental threat.

Inasmuch as crises, pandemics and disasters affect people's movement, evidence suggests that various tourist typologies are affected differently. From Fuchs and Reichel's (2011) perspective, first-time visitors are usually sensitive to health-related issues. However, repeat visitors are sensitive to financial meltdowns. Women, on the other hand, are said to be more sensitive to health and food issues than men (Bodosca et al. 2014). Consequently, each disaster would shape the travel patterns of tourists differently. Given that such disasters are different in terms of extent and nature, there is often a need to evaluate and understand each on its own merits and demerits with regard to impacts. Senbeto and Hon (2020) argue that regardless of the frequency and magnitude of disaster events, such typically bring about fear and cause tourists to avoid affected areas and destinations.

The outbreak of pandemics often affects the aviation business negatively, including the share price and financial performance. This has severe implications for shareholders. Loh (2006) notes that the airline stocks were more sensitive to the severe acute respiratory syndrome (SARS) outbreak than were other stocks. The reason for this could be that human beings tend to adopt defensive mechanisms, such as avoidance of certain activities, as they face personal threats. Such defence mechanisms include, but are not limited to, journey postponements and/or trip cancellations. All these also affect revenue streams for tour operators. A study by Fenichel et al. (2013) found that tourists had foregone air tickets to the value of approximately $50 million through cancellations during the outbreak of SARS, as they feared to contract the deadly disease. SARS thus had a devastating impact on tourist arrivals. Such shocks to the system can result in the industry's taking months to recover.

The year 2020 started with the pronouncement of a deadly virus that originated in Wuhan, China, in December 2019. Given the virulence and rate of spread of the disease, which came to be known as 2019-nCoV (COVID-19) acute respiratory disease, the World Health Organization (WHO) declared the outbreak a public health emergency. This was after 19 countries had confirmed infections among 9826 citizens, with the bulk of them being in China; of those infected with COVID-19 at that time, a staggering 9720 were in China (WHO 2020a). The WHO declared COVID-19 a pandemic on 11 March 2020 (2020b). At that stage, there were 118,319 people infected and 4292 deaths across the world.

Following this announcement, there was a frenzy of activity, marked by confusion, misinformation and – at times – half-hearted attempts to contain the disaster. However, by 31 March 2020, much of the world was in lockdown, as both death and infection figures skyrocketed: globally, the infections had soared to 823,626 and 40,598 deaths were reported (WHO 2020c). Following the pandemic declaration, many countries closed their borders and ports of entry, thus imposing lockdowns that threatened the tourism and travel industry. Many states and corporate organisations were warning that inaction would collapse national and global economies, as the lockdowns posed the threat of an economic recession. There were also additional calls in many countries for social distancing as a measure of curbing the spread of the disease (IATA 2020c; the United Nations World Tourism Organization (UNWTO) 2020a).

Tourism was subsequently singled out as one of the sectors that were being – and would be – affected severely by the pandemic; this would result in negative economic growth for the sector itself, as well as at national, regional and global level. In what could be considered a conservative estimate, the United Nations World Tourism Organization (UNWTO 2020b) stated that there would be between a 20% and 30% annual decline in tourist arrivals due to the impact of COVID-19. This decline was estimated to lead to a decline in international tourism receipts (exports) of between US $300 and $450 billion. The United Nations International Labour Organization (ILO) estimated that the pandemic would result in between 5 to 25 million jobs being lost and a further loss of labour income of between $860 billion and $3.4 trillion by the end of 2020 (ILO 2020). The next section looks at the research methodology applied in the chapter.

4.3 Research Methodology

Given the context and nature of this research, critical document analysis was selected as the most appropriate methodology because it is fast becoming a preferred method in the era of big data and the Internet of Things. Most public and private organisations generate data in the form of reports, articles, media statements and pictures, which can be used in answering specific research questions. Critical document analysis has been in use for a long time as a qualitative way of conducting research and offering numerous advantages to the researcher, as it is a rich source of data (Bowen 2009). In this research study, data generated by various authoritative aviation companies and organisations were therefore retrieved for use.

The researchers used reports from 290 airlines in 120 countries, airport companies, regional and international aviation organisations (such as IATA), the International Civil Aviation Organization (ICAO) – a United Nations agency for aviation – and the European Organisation for the Safety of Air Navigation (EUROCONTROL), comprising 41 member states. Additional data were retrieved from the International Airport Review, Airport Council International (ACI), Uniting Aviation, the Collaborative Arrangement for the Prevention and Management of Public Health Events in Civil Aviation (CAPSCA) and flight tracking systems, such as Flightradar24. Critical document analysis provided additional advantages, given that the aviation industry is heavily regulated, and obtaining primary data would have taken a long time. However, the aviation industry remains one of the most organised industries, with much of the necessary data being archived and available publicly on Internet databases. This method was also preferred because it was cheaper and convenient for the researchers.

The data that was sought for the study include departures, costs and volumes between 2019 and 2020. In addition, multiplatforms were utilised to triangulate and corroborate the data. This was necessary to ensure reliability, authenticity and credibility in the generated data and findings. The qualitative data were analysed through content and thematic analysis, while the numerical data were analysed using the Microsoft Excel Analysis ToolPak.

4.4 Presentation of Data and Results

Given the sensitivity of the aviation industry, it emerged that the industry was closely monitoring the disease from very soon after it received media attention. IATA presented weekly and monthly updates, providing insight into the likely impact of COVID-19 on the aviation industry. These updates were given by using previous disease outbreaks, such as SARS, as the basis for their projections. The impact was subsequently revised upwards to give credence to observations as the situation unfolded. Many aviation regulatory organisations created dedicated website panels for COVID-19, where updates were given as and when these were deemed necessary.

China subsequently implemented travel restrictions on 23 January 2020. According to IATA (2020d), this led to the slowest air traffic in nearly a decade. Consequently, the passenger demand for January 2020 was 2.5%, compared with 3.7% recorded for the same month in 2019. Much of the impact was caused by less demand for air traffic in the Asia Pacific, which recorded a growth rate of only 2.5%, compared with a rate of 3.9% in 2019. In January and February 2020, China – the COVID-19 epicentre – reported a decline of 6.8% in domestic air travel due to restrictions that were imposed. Between late January 2020 and early February 2020, the Chinese Ministry of Transport reported that they had witnessed an 80% decline in air traffic volumes.

4.5 Impact of COVID-19 on Global Air Traffic

The study found that the disease outbreak had a significant impact on flight operations (Figs. 4.1 and 4.2). On 11 March 2020, when the WHO declared COVID-19 a pandemic, Italy and other European countries seemed to take the lead in the number of infected people and recorded deaths. As a result, on 13 March 2020, the US government suspended all aircraft from Europe, with the exception of traffic from Britain and Ireland. This was done to protect its territory, as most infected people that were being reported at the time had a history of travel either to China or Europe. These restrictions were later expanded to include the two countries that had initially been excluded, namely, Britain and Ireland. This meant that the whole of Europe was prevented from flying to the USA to curb any further spread of the disease. By that time, only 50 people had died from the disease in the USA, and there were 2200 confirmed cases of people with coronavirus. This marked the beginning of extensive restrictions across the world as various countries affected inward travel bans to protect their territories.

The flight restrictions into the USA had a devastating global impact, but most prominently to the European market given the significance of the size of the USA market. For instance, the IATA that the travel restrictions between the USA and Schengen region plus the UK and Ireland had impacted approximately 5.5 million passengers. In this mix, the USA–Schengen travel route alone accounts for about

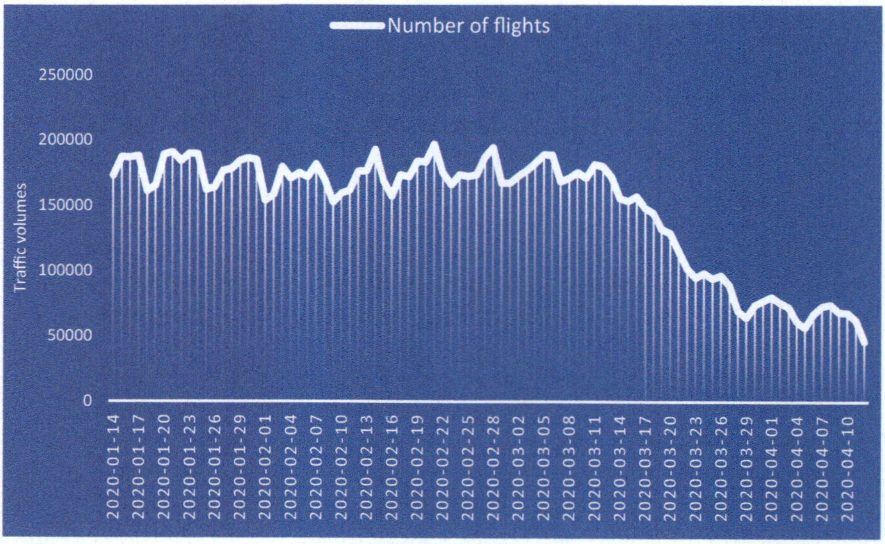

Fig. 4.1 Global total number of observed daily flights. Source: authors; data from Flightradar24

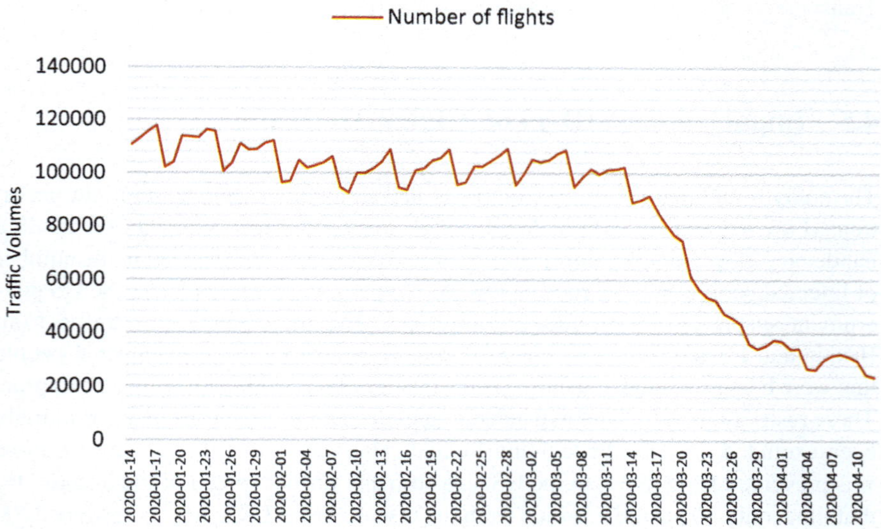

Fig. 4.2 Global total number of observed commercial flights a day. Source: authors; data from Flightradar24

550 flights per day roughly, which is equivalent to 125,000 travellers a day. The UK alone has a market size of 47,000 passengers per day. Broadly this region has an annual revenue market to US$10,4 billion (IATA 2020e). As such, a disturbance in traffic has direct implications on revenue and a knockdown effect on supporting industries such as airport companies, air traffic navigation companies and shops that

operate at airports. The adverse impact on the aviation industry worsened as more countries imposed inward bound passenger air traffic. The ban led to a series of diversion (Fig. 4.3) and rescheduling, all of which were costly to the industry (IATA 2020f). This led to a decline in flights, and as restrictions increased, more and more flights were cancelled in Europe and across the world – as can be seen by the sharp decline in air traffic in Figs. 4.1 and 4.2 but especially in Fig. 4.3. According to data from EUROCONTROL, the industry continued a downward spiral, from a decline rate of about 20% to about 86% at the end of February.

As more countries imposed travel restrictions, passenger aircraft were operating, within the African airspace being almost emptied by the beginning of April 2020. Only a few cargo flights, distributing mainly medical supplies, could be observed on the radar by the beginning of April 2020. The restrictions and grounding of many airlines resulted in distress signals from several airlines, which feared collapse due to the unexpected grounding that cut off ticket sales as tourists became increasingly anxious about the situation.

Table 4.1 highlights how some of the major airline companies in Europe responded to COVID-19 restrictions. According to reports from IATA, the increased number of cases of COVID-19 at the end of February 2020 led to an 8% loss of passenger capacity, particularly among those countries that had a stronger connection with China. By March 2020, the pandemic had affected almost the entire world. This led to a reduction of passenger traffic, not only among countries that had reported early infections but the entire aviation region, with the industry witnessing a 33% decline in passenger capacity.

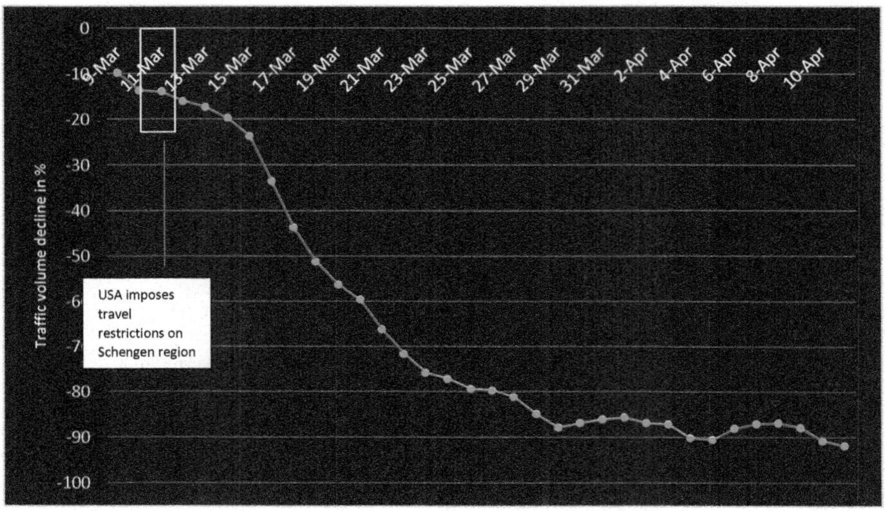

Fig. 4.3 Air traffic deviation from 2019 base traffic. Source: authors; data from EUROCONTROL

Table 4.1 State and airline response to COVID-19 as of 3 April 2020

Airline	Response to COVID-19 restrictions	Airline market share in 2019
Lufthansa Group	Adapted flight schedules for returnees between 23 March and 19 April: ~ 40 daily connections Grounded 700 out of its 763 aircraft fleet Lufthansa Cargo was still flying its regular programme, except to mainland China Retired 6 Airbus A380 double-deckers, 7 Airbus A340–600, 5 Boeing 747–400 and 3 Airbus A340–300, which were operated by Lufthansa CityLine	11.1%
Ryanair Group	Cut 80% of flights between 18 and 24 March Suspended all flights from 24 March until April/May RYR has a limited schedule: 13 routes from IE and 12 routes from GB (27 March–9 April) Instituted 50% pay cut for its staff	8.5%
IAG	Cut capacity by a minimum of 75% in April and May, with effect from 16 March	7.3%
easyJet	Suspended all flights on 30 March	6.2%
Lufthansa	Reduced hours for 60% of staff from 2 April	5.1%
Turkish Airlines	All international flights suspended between 27 March and 17 April. Domestic flights operated between 13 airports only, with effect from 29 March to 17 April As of 28 March, flights down by 93%	4.3%
Air France	Flights' capacity cut by 90% from 23 March for 2 months Nationalisation proposed on 18 March All A380 grounded on 16 March	
British Airways	Suspended all flights to and from London Gatwick Airport as of 30 March On 3 April, flights were down by 89% 80% of staff were suspended on 2 April Furloughed 30,000 of the airline's workers for the months of April and May; a modified version of the UK's COVID-19 Job Retention Scheme grants furloughed employees 80% of their pay and specific allowances. The deal allows workers to divert pension contributions to their pay	3%
SAS	Flights down by 87%	3%
KLM	Cut 90% of flights between 29 March 29 and 3 May Grounded its 5 B747s on 16 March	2.4%
Air New Zealand	Made 387 of its new and young pilots redundant (which is 30%) Slushed its capacity by 95% Negotiated a debt funding agreement of US$538 with its government, which owns 52% of the share value	

Source: authors; data from EUROCONTROL

4.5.1 *COVID-19 Impacts on Global Passenger and Cargo Markets*

Given the anxiety of passengers and the ensuing restrictions by various governments, the pandemic has had severe implications for global passenger movement. This has negatively affected the revenue and operations of airlines. After lower-than-expected passenger numbers for January 2020, particularly in the Asia Pacific region, the industry witnessed a record low performance in February 2020, which was only seen previously following the September 2011 attack in the USA. The industry-wide revenue passenger kilometres (RPKs) contracted by 14.1% on a year-on-year basis in February 2020, with industry-wide capacity (ASKs) declining by 8.7% year on year. The industry performance for February 2020 was far worse than the January 2020 performance. This was primarily as a result of a double-digit decline in RPKs for the Asia Pacific region, which declined −30.4%, and in Latin America, which reported infections in December 2019 and January 2020. At that time, the epicentre of the pandemic was China. Figure 4.4 shows the impact of the pandemic in January and February 2020. There was a severe impact of COVID-19 on RPK growth in both months as the epicentre of the virus which started in January and extended all the way to February 2020. Evidence shows that the main driver for the losses was the Asia Pacific region and China. The domestic arrivals were down nearly 84% at the backdrop of lockdown and restricted movements in that country. As a result of the lockdown and restricted movements in China, domestic arrivals decreased by nearly 84%. China recorded declines in both January and February 2020.

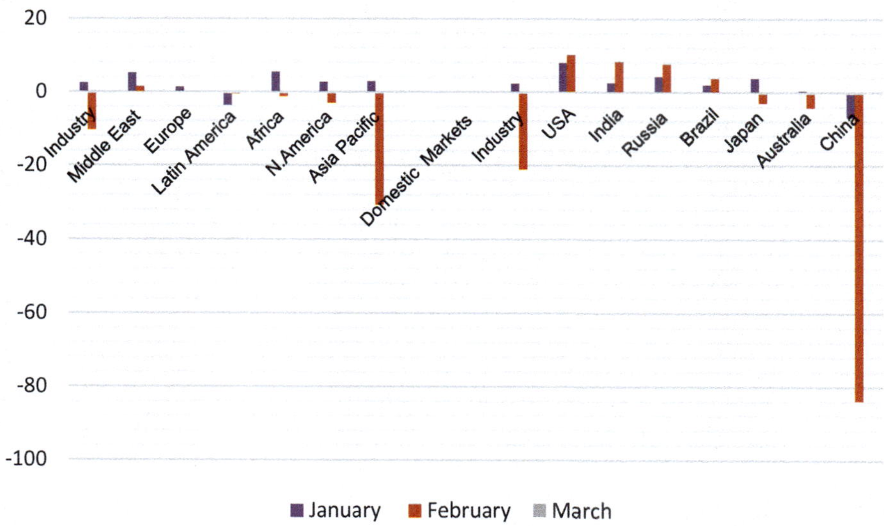

Fig. 4.4 Performance of regional and major domestic markets in response to COVID-19. Source: authors; data from IATA

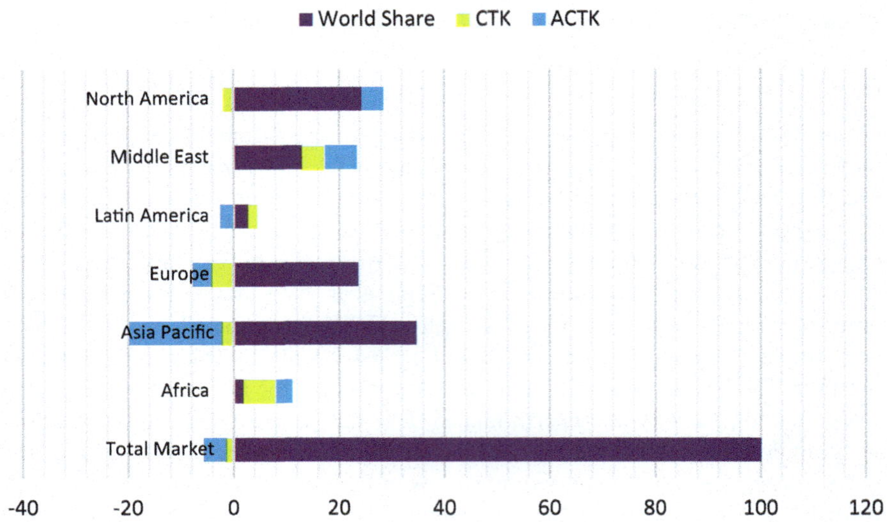

Fig. 4.5 The impact of COVID-19 on air cargo for 2019. Source: authors

As mentioned earlier, the aviation industry is critical for trade and industry. This industry transports about 52 million metric tonnes of goods annually, which accounts for about 35% of global trade by value. Daily, the aviation industry transports $18.6 billion worth of goods. According to reports from IATA, the pandemic has also harmed air cargo demand. The decline in cargo demand has been caused by three factors, namely, (i) a decline in manufacturing production in mainland China due to company closures and travel restrictions; (ii) falling global orders due to a contraction in the Purchasing Managers Index (PMI); and (iii) the reduction in airline passenger operations due to travel restrictions as a measure to contain COVID-19.

The developments have resulted in a decline in cargo tonne kilometres (CTKs) of 1.4, compared with the same period in 2019, as illustrated in Fig. 4.5. The cargo capacity (measured in available cargo tonne kilometres (ACTKs)) dropped by 4.4% on a year-on-year basis in February 2020. Fig. 4.5 shows that the highest declines were recorded in Asia Pacific, Europe, Latin America and North America, which account for the most significant portion of global decline.

4.5.2 *Impact of COVID-19 on Weather and Climate Observations*

The aviation industry plays a pivotal role in providing weather data that are crucial in weather forecasting and making high-resolution observations through the Aircraft Meteorological Data Relay Programme (AMDAR). The World Meteorological

Organization (WMO) (2020) noted that the AMDAR observation system typically produces over 700,000 high-quality observations per day, consisting of air temperature and wind speed and direction, together with the required positional and temporal information. The decline in air traffic has negatively affected their data collection capability, which has implications for weather forecasting. This could put certain countries in hydrometeorological danger, requiring areas where climate monitoring could not take place due to shortage of aircraft monitoring to increase their national hydrometeorological stations' monitoring, as continued disruption of aviation traffic will have an impact on the quality of weather forecasting.

4.5.3 *Impact of COVID-19 on Aviation Employment*

Apart from the pandemic's posing a threat to the liquidity of several airlines, another challenge for airline executives was how to sustain a large number of the aviation industry and related sector employees, totalling 65.5 million (IATA 2020i). Owing to constrained liquidity globally, most aviation employees found themselves facing an uncertain future. Several airlines laid off their staff, reduced working hours and suspended staff – with and or without pay in some instances – and also introduced (forced and voluntary) salary cuts (Table 4.1). Other airlines outside of the Eurozone instituted similar measures to contain their company expenses. In Africa, Air Ethiopia gave its employees 90 days' paid leave and suspended services of all the outsourced contract workers for 3 months. Emirates instituted a temporary pay cut of between 25% and 50% for its employees (Emirates 2020). However, junior employees' salaries were left unchanged, while the president of Emirates took a 100% pay cut for 3 months. In all these scenarios, the employees' allowances remained unchanged. The staff affected the most were the contract and casual employees.

4.6 Discussion

The study showed that both from a passenger and a cargo perspective, the pandemic has had negative impacts. The magnitude of the impact will only be discernible when the pandemic is under control. However, one of the most immediate challenges was that of fiscal liquidity in the industry.

Given that about 49% of the airlines' expenses are fixed, this presented a challenge for a number of airlines. Grounding meant airlines were burning cash without generating any income. At best, most airlines were operating below capacity or were grounded. Under normal circumstances, airlines ordinarily require at least 3 months of cash flow for them to be considered to have a reasonable safety buffer. A report by IATA (2020g) shows that most airlines across the world had a median cash flow ranging between 1 and 2 months, with airlines in Europe and the Asia Pacific taking severe strain. This meant that the airlines in those regions and globally had to rely on an external cash injection if they were to survive the pandemic. The pandemic also hurt investor returns and those along aviation's supply chain. These findings are in line with observations made by Loh (2006), who found that disasters had a negative impact on airline stocks.

Of real concern, however, was the impact of COVID-19 on the return on invested capital (ROIC) for those airlines that had not been performing well in general. Such airlines were particularly evident in Africa, Asia Pacific, Latin America and the Middle East, as they have a tendency to damage investor value on aggregate. The ROIC for airlines reached a peak in 2015 when fuel prices declined. While the bulk of the airline industry had modelled their business on a $100-per-barrel price mark, the decline in fuel price during COVID-19 was expected to provide much relief. The price of jet fuel dropped from the beginning of 2020 and was at about $31.01/bbl at the beginning of April 2020. IATA argued that, at that price, it would have translated to an expenditure reduction of $36.9 billion for global aviation (IATA 2020h). Nonetheless, on 9 April 2020 – around the time this article was being finalised – the Organisation for Petroleum Exporting Countries (OPEC) and other countries were in a meeting to review an oil pumping war between Saudi Arabia and Russia, which resulted in a significant cutback on production. Oil prices were expected to rise again, although not to the pre-COVID-19 levels in short to medium term.

The year 2020 had seen record aircraft deliveries for the European market. By March 2020, Europe had taken delivery of 105 narrow-body aircraft and 318 wide-body jets; other aviation regions had also received a significant number of aircraft. On the other hand, by April 2020, two of the largest air manufacturers of Boeing had made 30 deliveries (Boeing 2020), while Airbus had made 55 deliveries (Airbus 2020), pointing to the huge capitalisation by airlines. This meant that some airlines had incurred enormous expenses in acquiring new and efficient airlines, which had a bearing on their future liquidity. Most importantly, the pandemic affected orders for 2020 – and will most likely also affect those for 2021. This could be dire for Boeing, which had been dealing with the fallout of the 737 MAX aircraft type. Consequently, there is a possibility that the company would request a bailout or file for bankruptcy.

Also worrying is that some airline was expecting to receive new aircraft from manufacturers. With current closures in many production firms, it was highly unlikely that, with quarantines and social distancing, the air manufacturing companies would be able to reduce their backlog. Puget Sound facility, which accounts for 80% of Boeing's overall commercial and defence aircraft production activity, had to close for 14 days in March 2020 owing to COVID-19. As a result, they anticipated that the delivery backlog would increase by the end of 2020, given the disruptions caused by the pandemic.

At the beginning of 2020, Airbus had a backlog of 7482 aircraft, or about 7 to 8 years of production (Bellamy 2020). In the same period, Boeing had a backlog of 5351 jets (Hemmerdinger 2020). As of February 2020, some airlines had started cancelling their orders, which might have indicated the financial squeeze they were already facing or anticipated would follow. One of the cancellations for the orders for both Airbus and Boeing was from Avon, a leading leasing company that had a fleet of 549 aircraft at that time. Avon cancelled its order of 75 Boeing Max and four Airbus Neo aircraft, citing the challenges associated with COVID-19 as the reason.

Furthermore, easyJet – a company that operates a fleet of 340 aircraft – argued that the coronavirus was a *force majeure* and they were pushing for a cancellation of their $5.5 billion order with Airbus to ensure that they had adequate liquidity to withstand the impact of the pandemic.

On a different note, the grounding of a large fleet from airlines presented an unusual problem, namely, a shortage of parking space. The usual requirement is that aircraft be checked every 1 to 2 months if grounded. Regular checks of avionics, hydraulics and some other operating systems had to be conducted during the process. This meant additional storage and parking costs for the already struggling airlines. Most affected airports were in North America, Europe and other regions that have a large number of flights. Under normal circumstances, parking fees account for about 2% of an airport's revenue. However, the parking problem led some companies to contemplate putting their old fleet on early retirement. IATA reports singled out American, Delta and KLM as some of the airlines that were contemplating retiring their old Boeing 747 fleet.

In addition to parking, there was also a challenge of maintenance as the hangers were not sufficient to handle all the aircraft in these facilities. The usual requirement is that the aeroplane is checked on a constant basis if grounded. Regular checks of avionics, hydraulics and some other operating systems had to be conducted during the process.

4.6.1 Aviation and Relief Packages

As the aviation industry suffered huge losses due to flight cancellations by no-show passengers, border closures and other restrictions, airlines left many customers stranded when they shut down operations or cancelled routes. This also meant that the airlines held about $35 billion in unutilised customer tickets that needed to be refunded (IATA 2020g). Most airlines were simply reluctant to refund passengers, although in most cases they did not have the means or capacity to provide the service that passengers had paid for. Other airlines offered free booking date changes of between 6 months to about a year, with a few extending the period to 2 years. In March, IATA made two global financial appeals that were revised in less than a month, as the situation changed rapidly, requesting governments and corporates to extend a helping hand to the struggling airline industry.

In order to ensure sustainability, there was a need to find the means to provide liquidity for the ailing airline industry, since the industry was critical in the provision and distribution of medicine and protective equipment across the world at the height of the pandemic. Some passenger aircraft were being turned into a cargo fleet, although the cargo fleets were not as severely affected as the passenger airlines in terms of revenue losses. Most countries allowed cargo airlines access into their territories. The sustenance of the aviation industry and its continuity was crucial for the global developmental agenda in facilitating industry and commerce. While grounding aircraft was one of the strategies that were used by the aviation industry to avoid so-called ghost flights, that measure alone was not adequate to safeguard the industry's future. According to IATA, the airline industry needed about $252 billion in bailouts (IATA 2020b).

After that call by the global aviation body (IATA), several governments and private entities made an effort to respond to the appeal to assist the aviation industry to survive the fallout from the pandemic. On 7 April 2020, when about 90% of the fleet of most airlines were grounded, EUROCONTROL announced €1.1 billion to defer air traffic control fees in support of the airline industry (EUROCONTROL 2020). The Australian government also indicated that it was putting together a $430 million relief package for the aviation industry and introducing a range of waivers aimed at providing relief for the ailing aviation sector (Australian Government 2020). Other significant relief packages included a package from the US government's Coronavirus Aid, Relief, and Economic Security Act (H.R. 748). This provided $61 billion to assist specific airlines to navigate their businesses in the post-COVID-19 era. It remains to be seen how far this amount will go in providing the much-needed relief. Reports indicated that American Airlines had withdrawn more than $2.7 billion from a credit account they had created in 2014 (Arnold 2020), but the airline also had plans to apply for $12 billion government aid, which was approved by Congress. Fortunately, the American bailout had conditions attached, namely, that the airlines maintain their labour force. This provided some much-needed financial relief for airline employees.

Some of the weaknesses in the airline industry included leased aircraft, as airlines had obligations to pay according to their lease agreements. Therefore, there was a need for grounded airlines to seek and negotiate terms of payment that did not cripple the airline. By April 2020, Avon Leasing had received requests from more than 80% of its clients for relief from payment obligations. Such requests included short-term rent deferrals, which seemed likely to be granted.

Various airlines responded to the distress call by IATA and provided different interventions, such as the nationalisation of the airline, as illustrated in Table 4.1. Thus it would seem that state-owned airlines were protected somewhat from the financial consequences of the pandemic since there was a strong likelihood that they would receive a state bailout. The three largest airlines in the Middle East, namely, Qatar Airways, Emirates and Etihad, were also promised state bailouts. Furthermore, Emirates indicated its intention to seek private funding in the form of a loan. Other airlines that received funding from their government included Norwegian Air Shuttle, which was set to receive $270 million, Finnair (€600 million), SAS AB ($300 million) and Alitalia ($647 million) (Kammel 2020). Although state intervention was welcome, there were fears among industry players that the crisis would see the industry's reversing the privatisation efforts that had transformed the industry over a number of years.

In addition to the global cash burn of $61 billion in the second quarter of 2020 by the US government, there was a need for all role players to come on board to assist the airlines to survive the pandemic phase. The government could assist by wavering/forgoing certain taxes or deferring the payment of taxes to ensure that airlines have enough liquidity to operate. The private sector could also help by extending loans at reduced interest rates to ensure viability.

Table 4.2 The potential number of job losses resulting from COVID-19

Region	Job losses in millions
Asia Pacific	11.2
Europe	5.6
Latin America	2.9
North America	2
Africa	2

Source: authors; data from IATA

IATA projected huge direct job losses in the aviation and tourism industries, amounting to 25 million jobs and a cost of $252 billion to the industry (IATA 2020i). (See the details in Table 4.2.) A loss of jobs in the tourism and aviation industries would have a ripple effect on the market. Therefore, there was a need for the aviation industry to consider the plight of employees carefully.

The subject of offering relief to airlines – or any other business – is contested in business literature. While some authors advocate state bailouts as a measure to ensure business viability during turbulent times, Nulsch (2014) argued that state bailouts of private companies were controversial. At best, these serve to postpone the failure of affected companies; therefore, Nulsch advocated a long-term restructuring of firms instead. De Rugy and Leff (2020) strongly opposed the proposed state bailout of airlines in the USA, claiming that the airlines had sufficient leeway to raise money independently, without relying on state payouts, as this was not sustainable in either the short or long term. These authors argued that the airline industry had the capacity to raise cash from financial institutions, such banks, with whom they have credit card deals. Besides, they alleged that any airline seeking a bailout should have gone through the bankruptcy process, as required by law in the USA; the authors claimed that companies that had gone through this process in the past had emerged stronger, citing American Airlines, Delta Airlines and United Airlines as examples. They rebutted the argument that failed airlines would have a knock-on effect on the economy by pointing out that no failure of an airline had an adverse knock-on effect on the economy. Other critics of state bailouts argued that it was not prudent to use the job losses narrative, as some companies had already shut down, job losses were not unique to airlines and any measure to assist employees should not be done through such companies since aviation companies had enough assets to sell to weather the COVID-19 pandemic.

De Rugy and Leff (2020) further argued that there are challenges associated with state bailouts; for instance, they distract airline executives from focusing on their role of making prudent management decisions in favour of pleasing politicians as a way of creating relations for future bailouts. In addition, they claimed that some of the conditions attached to bailouts were detrimental to the future efficiency of airlines and could result in airlines' investing in technologies that would lead to job losses in any case. Conditions such as setting up the minimum wage and curbing executive salaries were not regarded as the best way to ensure operational efficiency. Any bailout offered by the government must, therefore, be the last resort that is given in the form of a loan or loan guarantee.

4.7 Conclusion and Recommendations

This study aimed to examine the impact and extent of COVID-19 on global aviation, as one of the tourism sectors significantly affected. The study found that the pandemic was one of the most devastating disasters to have affected the aviation industry in living memory. The impact of COVID-19 was global and affected every aviation region, thus affecting both international and domestic routes. The closure of borders and restrictive measures that were put in place, including airport closures and lockdowns, had a devastating impact on the aviation industry. This resulted in record cash burn, threatening more than 25 million of the 65.5 million jobs in the aviation industry and supporting sectors. The pandemic threatened the survival of most airlines across the world and resulted in demands for bailouts to cover the fixed costs. The unprecedented nature of the disaster resulted in order cancellations, thereby negatively affecting those in the supply chain too.

Given the impacts of COVID-19 on aviation, the researchers believe that there is a need to rethink how the aviation industry operates. This could include the revision of policy on the median cash availability, which – if allowed to be as low as the levels were in 2020 – would prove to be unsustainable in the future. While a 2-month cash threshold could be adequate for regular times, the COVID-19 pandemic shows how this cover is inadequate when lengthy and severe challenges emerge. Thus suggests a need to extend the financial safety threshold. Noting, cash flow challenges in the industry in the industry demands innovative financial solutions. This is important so that government concentrates on other social issues instead of bailing out private enterprises. It is important that these safety measures also cover the welfare of employees. This removes the argument that state interventions seek to manage unemployment. Lessons and practices from such actions could inform policy and practice in the entire tourism industry.

The scope and form of intervention will definitely differ in countries and regions as we anticipate the impacts of COVID-19 and as indeed adverse impacts of past pandemics have and will vary across regions and nations, to this end country- and region-specific empirical impact studies.

References

Airbus. (2020). *The month in review: February 2020.* Airbus. Retrieved April 6, 2020, from https://www.airbus.com/aircraft/market/orders-deliveries.html

Arnold, K. (2020). *American Airlines pulls $2.7 billion from credit lines as it prepares for lean months ahead.* The Dallas Morning News. Retrieved April 9, 2020, from https://www.dallasnews.com/business/airlines/2020/04/01/american-airlines-pulls-27-billion-from-credit-lines-as-it-prepares-for-lean-months-ahead/

Australian Government. (2020). *Aviation.* Department of Infrastructure, Transport, Regional Development and Infrastructure. Retrieved April 9, 2020, from https://www.infrastructure.gov.au/aviation/

Bellamy, W. (2020). *Airbus, Boeing make production decisions amid COVID-19 Pandemic*. Avionics International. Retrieved April 6, 2020, from https://www.aviationtoday.com/2020/03/24/airbus-boeing-make-production-decisions-amid-covid-19-pandemic/

Bodosca, S., Gheorghe, G., & Nistoreanu, P. (2014). Tourist consumption behaviour before and after the crisis from 2008. In *Proceedings of the 21st international economic conference* (pp. 77–87). Sibiu: Procedia Economics and Finance.

Boeing. (2020). *Current year deliveries through February 2020*. Retrieved April 6, 2020, from http://www.boeing.com/commercial/#/orders-deliveries

Bowen, G. A. (2009). Document analysis as a qualitative research method. *Qualitative Research Journal, 9*(2), 27.

De Rugy, V., & Leff, G. (2020). *The case against bailing out the airline industry*. Mercatus Center George Mason University. Retrieved April 8, 2020, from https://www.mercatus.org/system/files/de-rugy-leff-airline-bailouts-mercatus-v1.pdf

Dube, K., & Nhamo, G. (2019). Climate change and the aviation sector: A focus on the Victoria Falls tourism route. *Environmental Development, 29*, 5–15.

Dube, K., & Nhamo, G. (2020). Major global aircraft manufacturers and emerging responses to the SDGs agenda. In G. Nhamo, O. Odularu, & V. Mjimba (Eds.), *Scaling up SDGs implementation emerging cases from state, development and private sectors* (pp. 99–113). Cham: Springer International Publishing.

Emirates. (2020). *The Emirates Group's business response to COVID-19*. Emirates. Retrieved April 6, 2020, from https://www.emirates.com/media-centre/the-emirates-groups-business-response-to-covid-19-updated/

EUROCONTROL. (2020). *EUROCONTROL States agree a €1.1 billion deferral package to assist airlines*. EUROCONTROL. Retrieved April 9, 2020, from https://www.eurocontrol.int/press-release/eurocontrol-states-assist-airlines-11bln-deferral

Fenichel, E. P., Kuminoff, N. V., & Chowell, G. (2013). Skip the trip: Air travelers' behavioral responses to pandemic influenza. *PLoS One, 8*(3), 1–10.

Fuchs, G., & Reichel, A. (2011). An exploratory inquiry into destination risk perceptions and risk reduction strategies of first time vs. repeat visitors to a highly volatile destination. *Tourism Management, 32*(2), 266–276.

Grais, R. F., Ellis, J. H., & Glass, G. E. (2003). Assessing the impact of airline travel on the geographic spread of pandemic influenza. *European Journal of Epidemiology, 18*(11), 1065–1072.

Hall, C. M. (2006). Tourism, disease and global environmental change: The fourth transition? In S. Gössling & C. Hall (Eds.), *Tourism and global environmental change* (pp. 173–193). Routledge.

Hemmerdinger, J. (2020). *Boeing's order book declines as customers cancel and modify new-jet orders*. FlightGlobal. Retrieved April 6, 2020, from https://www.flightglobal.com/airframers/boeings-order-book-declines-as-customers-cancel-and-modify-new-jet-orders/137193.article

Hyndman, D., & Hyndman, D. (2016). *Natural hazards and disasters*. Boston: Cengage Learning.

IATA. (2020a). *What can we learn from past pandemic episodes?* IATA. Retrieved April 3, 2020, from https://www.iata.org/en/iata-repository/publications/economic-reports/what-can-we-learn-from-past-pandemic-episodes/

IATA. (2020b). *COVID-19 Cashburn Analysis*. IATA. Retrieved April 2, 2020, from https://www.iata.org/en/iata-repository/publications/economic-reports/covid-19-cash-burn-analysis/

IATA. (2020c). *IATA Economics' Chart of the Week Liquidity is crucial for airlines to overcome COVID-19 pandemic*. IATA. Retrieved April 2, 2020, from https://www.iata.org/en/iata-repository/publications/economic-reports/liquidity-is-crucial-for-airlines-to-overcome-covid-19-pandemic/

IATA. (2020d). *IATA economics' chart of the week COVID-19 pandemic puts employment at risk*. International Air Transport Association. Retrieved April 02, 2020, from https://www.iata.org/en/iata-repository/publications/economic-reports/covid-19-pandemic-puts-employment-at-risk/

IATA. (2020e). *Evolution of COVID-19 travel restrictions 12–23 March 2020.* IATA. Retrieved April 5, 2020, from https://www.facebook.com/iata.org/photos/a.647512395314902/3027979637268154/?type=3&theater

IATA. (2020f). *Potential Impact of US-Schengen Travel Ban.* IATA. Retrieved April 3, 2020, from https://www.iata.org/en/iata-repository/publications/economic-reports/infographic-impact-of-the-us-travel-ban/

IATA. (2020g). *Liquidity is crucial for airlines to overcome COVID-19 pandemic.* IATA. Retrieved from https://www.iata.org/en/iata-repository/publications/economic-reports/liquidity-is-crucial-for-airlines-to-overcome-covid-19-pandemic/

IATA. (2020h). *Jet Fuel Price Monitor.* IATA. Retrieved April 6, 2020, from https://www.iata.org/en/publications/economics/fuel-monitor/

IATA. (2020i). *25 Million Jobs at Risk with Airline Shutdown.* IATA. Retrieved April 7, 2020, from https://www.iata.org/en/pressroom/pr/2020-04-07-02/

ILO. (2020). *COVID-19: Protecting workers in the workplace.* International Labor Organization. Retrieved April 2, 2020, from https://www.ilo.org/global/about-the-ilo/newsroom/news/WCMS_738742/lang%2D%2Den/index.htm

International Airport Review. (2019). *In the balance: Global air transport demand in 2018 and 2019.* International Airport Review. Retrieved March 31, 2020, from https://www.internationalairportreview.com/article/101952/world-airport-traffic-report-preview-aciworld/

Joo, H., Maskery, B. A., Berro, A. D., Rotz, L. D., Lee, Y. K., & Brown, C. M. (2019). Economic impact of the 2015 MERS outbreak on the Republic of Korea's tourism-related industries. *Health Security, 17*(2), 100–108.

Kammel, B. (2020). *Desperate airlines turn to European Governments for support.* Bloomberg Businessweek. Retrieved April 9, 2020, from https://www.bloomberg.com/news/articles/2020-03-25/desperate-airlines-turn-to-european-governments-for-support

Konttinen, J. (2017). How can airports and airlines better integrate their retail business in the future? *Journal of Airport Management, 11*(3), 238–242.

Loh, E. (2006). The impact of SARS on the performance and risk profile of airline stocks. *International Journal of Transport Economics/Rivista internazionale di economia dei trasporti, 33*(3), 401–422.

Nulsch, N. (2014). *Is Subsidizing Companies in Difficulties an Optimal Policy? An Empirical Study on the Effectiveness of State Aid in the European Union (No. 9/2014).* IWH discussion papers. Retrieved April 8, 2020, from https://www.econstor.eu/handle/10419/98743

Senbeto, D. L., & Hon, A. H. (2020). The impacts of social and economic crises on tourist behaviour and expenditure: An evolutionary approach. *Current Issues in Tourism, 23*(6), 740–755.

UN. (2020a). *Shared Responsibility, Global Solidarity: Responding to the socio-economic impacts of COVID-19.* United Nations. Retrieved April 2, 2020, from https://www.un.org/sites/un2.un.org/files/sg_report_socio-economic_impact_of_covid19.pdf

UN. (2020b). *UN launches COVID-19 plan that could 'defeat the virus and build a better world'.* UN. Retrieved April 2, 2020, from https://news.un.org/en/story/2020/03/1060702

UNWTO. (2020a). *COVID – 19: Putting people first.* UNWTO. Retrieved April 2, 2020, from https://www.unwto.org/tourism-covid-19-coronavirus

UNWTO. (2020b). *International tourist arrivals could fall by 20–30% in 2020.* UNWTO. Retrieved April 2, 2020, from https://www.unwto.org/news/international-tourism-arrivals-could-fall-in-2020

WHO. (2020a). *Novel Coronavirus(2019-nCoV) Situation Report – 10.* World Health Organization. Retrieved March 29, 2020, from https://www.who.int/emergencies/diseases/novel-coronavirus-2019/situation-reports

WHO. (2020b). *Coronavirus disease 2019 (COVID-19) Situation Report – 51.* World Health Organization . Retrieved March 29, 2020, from https://www.who.int/emergencies/diseases/novel-coronavirus-2019/situation-reports

WHO. (2020c). *Coronavirus disease 2019 (COVID-19) Situation Report – 72.* World Health Organization. Retrieved April 2, 2020, from https://www.who.int/docs/default-source/coronaviruse/situation-reports/20200401-sitrep-72-covid-19.pdf?sfvrsn=3dd8971b_2

Wolf, H., Forsyth, P., Gillen, D., Hüschelrath, K., & Niemeier, H. M. (2016). À la carte pricing to generate ancillary revenue: The case of Ryanair. In H. Wolf, P. Forsyth, D. Gillen, K. Hüschelrath, & H.-M. Niemeier (Eds.), *Liberalization in aviation* (pp. 207–216). Routledge. https://doi.org/10.4324/9781315592305-18.

World Meteorological Organization (WMO). (2020). *WMO is concerned about impact of COVID-19 on observing system.* World Meteorological Organization. Retrieved April 7, 2020, from https://public.wmo.int/en/media/press-release/wmo-concerned-about-impact-of-covid-19-observing-system?fbclid=IwAR2y3GHOk-H4Xd_9DwX7_QgZC37iUb-BqG6sg9XigA_seDX_lsk2gYUyBlQ

Chapter 5
Impact of COVID-19 on the Global Network of Airports

Abstract Airports have grown in number, size and function in the recent past, leading to vast economic spinoffs to many communities. They are critical global nodes that link the world and facilitate economic development. In 2020, the global aviation industry was brought to a near standstill by the COVID-19 pandemic, with devastating global economic and associated impacts. This study evaluated the impact of COVID-19 on the airports' sector. Using archival and secondary data from authoritative sources such as the International Civil Aviation Organization (ICAO), EUROCONTROL and Flightradar24, it emerged that COVID-19 had a negative impact on airport operations mainly affecting their revenue streams. Travel restrictions and border closures had a debilitating impact on the airports and the aviation component of the tourism industry. With airlines grounded, the airports converted into parking lots and ghost towns. With this, there were drastic declines in receipt in landing, departure fees and non-aeronautical revenue. The remaining limited operations that continued in many airports related to service and facilitate the delivery of medical and other essential supplies, including food and critical raw materials. These could not raise adequate revenues to sustain many airports. Globally, the closure of airports also affected more than 6 million people directly employed in these airport-related services, including shops, car rental companies, airport restaurants and that of air navigation systems. The study recommends that the same support offered to the aviation industry in the form of grants, tax exemptions, preferential loans and so on be extended to airports. Airports play a central role in the global economy, tourism economy and future recovery efforts.

Keywords Airport · Shopping · COVID-19 · Economic impact · Aviation relief · Disease control

5.1 Introduction

Airports have witnessed increased use and function over the past decades. This is as a result of increased global air traffic mainly driven by increased levels of tourism development. Besides acting as important transport nodes in the past, airports have

G. Nhamo et al., *Counting the Cost of COVID-19 on the Global Tourism Industry*,
https://doi.org/10.1007/978-3-030-56231-1_5

become economic hubs offering additional services, including shopping malls, car rental facilities, conference facilities, lodging and logistics for exports and imports. Evidence shows that airports are large cities with multiple functions.

A record of 8.8 billion travellers passed through the world's airports in 2018, marking a 6.4% increase from the previous year (International Airport Review 2019). This showed that there was an increased demand for air travel. Indications were that trend was likely to continue in the foreseeable future as the demand for business and leisure travel increases. Figure 5.1 shows how traffic at airports grew and evolved over the past decade, with some significant events impacting aviation such as the economic recession and outbreaks of swine flu, Middle East respiratory syndrome (MERS) and Ebola. Data from the International Air Transport Association (IATA) show that the distance between travels is becoming shorter and shorter (IATA 2019). Hence, people are travelling more frequently today than ever before. As a consequence, there is a higher demand for airport usage, and the demand was expected to grow in the short and long term.

In 2019, airports contributed significantly to the global economy through aeronautical, non-aeronautical and nonoperating revenue. According to the Airports Council International (ACI), globally airports made US$178.2 billion, of which 59.9% came from aeronautical services, 39.2% from non-aeronautical services and the remainder from nonoperating revenue (ACI 2020a). The non-aeronautical revenue came from retail concessions (28.9%), car parking (20.4%), property and real estate (14.9%) and other services such as car rentals, advertising and food and beverages accounted for 35.8%. The data show the importance of different components of the airport ecosystem.

Given the number of people who use airports daily, the establishments are also part of the global disease transmission and control channels, with some of the

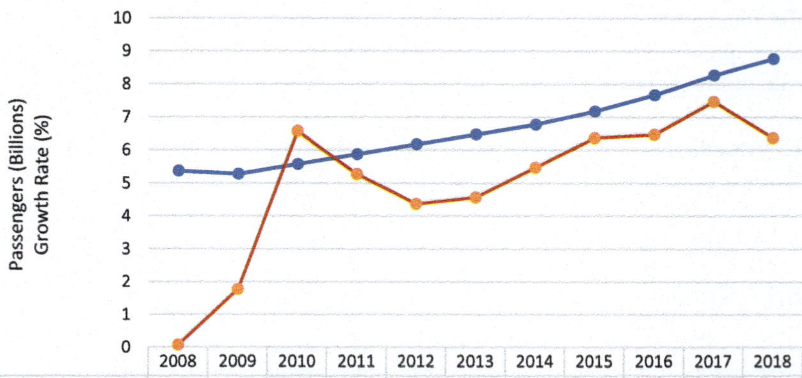

	2008	2009	2010	2011	2012	2013	2014	2015	2016	2017	2018
Passengers in billions	5.4	5.3	5.6	5.9	6.2	6.5	6.8	7.2	7.7	8.3	8.8
Growth Rate %	0.1	1.8	6.6	5.3	4.4	4.6	5.5	6.4	6.5	7.5	6.4

Fig. 5.1 Number of travellers making use of airports and annual passenger growth rate (2008–2018)
Source: Authors; data from ACI (2020b)

biggest airports facilitating millions of tourists per annum. Chung (2015) argues that the rapid commercialisation of the aircraft industry paved the way for disease proliferation at a faster rate. Given the history of pandemic and infectious diseases, the world has learned over time that the only plausible measures of controlling such diseases are screening, quarantines and isolations (Gensini et al. 2004; Safi and Gumel 2011; Hou et al. 2020). One of the greatest cities, the New York islands, is famously known for having been a smallpox quarantine site (Chung 2015). Isolation, social distancing and quarantine were also used in the COVID-19 outbreak, which led to restrictions in people moving across the world (Wilder-Smith and Freedman 2020; Adams et al. 2020). The travel limitations included ground, water, rail and air travel restrictions. A disruption in air traffic has implications for traffic flow, which affects revenue collection for airports. Airports make the bulk of their money from landing and departure fees, and airport taxes charged to customers.

Experience has shown that the aviation and tourism industries are vulnerable to political, economic and natural disasters. A review of the history of aviation revealed that several historical events had an impact on the aviation industry, including the oil crisis in 1973, the Iran–Iraq war in the 1980s, the Gulf War in the early 1990s, the Asian crisis of the late 1990s, the September 2011 (9/11) terrorist attack in the USA, the Severe Acute Respiratory Syndrome (SARS) outbreak in 2003, the global financial crisis in 2008 and 2009, the avian flu outbreak in 2013 and the MERS outbreak in 2016. The impact of some of the outbreaks is presented in Fig. 5.1. Looking at the trend, it appears that the number of shocks affecting the travel industry is increasing. Of interest is the increase in the frequency of outbreaks, epidemics and pandemics. The threat of epidemics and pandemics on the aviation industry led to the formation of a board tasked with coordinating international aviation's response to public health risks – the Collaborative Arrangement for the Prevention and Management of Public Health Events in Civil Aviation (CAPSCA). CAPSCA has since been formalised as a branch of the International Civil Aviation Organization (ICAO) (CAPSCA 2020). As a consequence, airport authorities have learned some invaluable lessons on how to manage disease pandemics from countries such as Japan (Fujita et al. 2011), Australia (Eastwood et al. 2010), Canada (Johanis 2007) and Singapore (Wilder-Smith et al. 2003).

The COVID-19 pandemic did not only threaten public health and wellness but also had an impact on the global economy. The International Monetary Fund (IMF) officially announced in March 2020 that the world had entered a recession (IMF 2020). An economic recession has an adverse impact on economic performance in every sector, including the aviation industry, and indeed traffic at various airports. Harvey and Turnbull (2009) documented the adverse impact of the 2008 global economic recession on aviation labour, which led to significant retrenchments and loss of income. However, Chung (2015) notes that in as much as disease epidemics harm airport economics, there is a lack of literature on the impacts of previous diseases. This is despite the fact that airports are often the last line of defence against exported/imported pandemics and other biosecurity risks (Warren et al. 2010; Melly and Hanrahan 2020; Yadav et al. 2020).

This study examined the impact of the COVID-19 pandemic on the global socio-economic network of airports. It also sought to document responses of various airports across the world to the challenges presented by the pandemic.

5.2 Literature Review

Besides their central economic role in various metropolitan areas across the world, airports are critical strategic national key points for trade and development. They play an essential role as biosecurity areas that can assist in screening tourists entering a country (Warren et al. 2010). Regardless of the responsibility placed on airports to control and eliminate the entry of disease and make (tourist) travel safe, measures put in place have often failed to curb the entry and spread of diseases (Malone et al. 2009). This resulted in diseases spreading faster due to the rapid development of the aviation industry. In a few minutes, hours or overnight, a disease could be spread from one national location to another, one country to another and one continent to another. The need to understand how disease spreads through airports and other transport networks has been of interest as academics try to model and understand the spread of epidemics and pandemics (Nicolaides et al. 2012).

Understanding how diseases spread through airports and airline networks is a critical aspect of addressing the spread (Briand et al. 2011). This is crucial in efforts to curb the spread of pandemic diseases, which appear to be on the increase and have an increased cost on the tourism and travel industries. When airports fail to detect diseases from arriving hosts through the use of mainly thermal scans, governments and authorities often resort to quarantines and border closures to limit the human and economic costs of a pandemic (Ulrich and Mariner 2018; Nussbaumer-Streit et al. 2020). Due to such restrictions, people cannot move freely, and this negatively impacts the airport business.

Since screenings are conducted at airports at destinations (Nishiura and Kamiya 2011), this usually discourages people from moving from their places of origin, leading to declining traffic from risky areas. Tourists and even residents are often afraid of being quarantined, particularly in a distant airport out of touch with family and relatives if they are to test positive to a pandemic disease (Blendon et al. 2006). The effectiveness and efficiency of thermal scanners have recently come into question amid the high rate of exportation of COVID-19 from the disease's hotspots areas (Gostic et al. 2020).

As indicated earlier, airports are economic hubs with a significant impact on the neighbouring community and sphere of influence. However, they are equally vulnerable to economic downturns. The airports in the USA were affected by the economic recession of 2008/2009, resulting in some getting bailouts under the American Recovery and Reconstruction Act of 2009 (Bilotkach 2018). While the IMF declared that the COVID-19 pandemic had triggered the worst economic recession in history, projections were that the recovery would assume a "U recovery shape" (Nicola et al. 2020) – with a decline followed by some flattening at the bottom before upward growth – as opposed to a V-shaped recovery (a sharp drop and a sharp rise). The study predicted that most of the impact would be felt in the tourism and aviation industries.

Preliminary studies had already shown the ravaging impact of the pandemic on the tourism industry. Rapid initial impact assessment carried out by Gössling et al. (2020) and another study by Brouder (2020) had found an imagined impact of the pandemic on travel, hospitality and attraction sectors of the tourism economy. A study by Dube et al. (2020) found that the restaurant and hospitality industry was almost decimated by the pandemic across the world. Concern was also being raised over the ramifications of labelling of Chinese nationalities as coronavirus carriers on the tourism economy (Zheng et al. 2020). The outbreak of the pandemic presented new scenarios for the tourism industry (Chen et al. 2020) and a possible opportunity to reset (Nepal 2020). The tourism reset, as proposed by Romagosa (2020), was likely to have implications for the aviation industry.

5.3 Research Design

The study took a global view that looked at multiple cases to draw a real narrative of the events. The basic foundation of the study was archival data from various aviation official sources. Jones (2010) and Webb et al. (2017) argue for the use of archival data in social research, outlining that it offers numerous advantages. According to Welch (2000:197), "using archival material provides longitudinal data, generates new theories about business networks and improves the validity and reliability of findings". Since the data are available and have been collected periodically, such as in the aviation industry, the data are very rich, of high quality and authentic. Given the scope of the research, archival data were found to be inexpensive on the researcher's part. Since there was a need for historical data, the researchers could not go back in time and start researching. Most organisations opened up access to some restricted data in the public interest and to foster an understanding of how the COVID-19 pandemic was affecting various sectors. The researchers took advantage of this publicly available data for this study.

The first port of call was data from the United Nations regulatory body the ICAO. The ICAO consolidates data on COVID-19 from sector organisations such as the ACI, Civil Air Navigation Services Organization (CANSO), IATA, International Air Cargo Association and World Health Organization. Data from the COVID-19 Airport Status app was used to track daily departures of aircraft across the world. The data was gathered from the COVID-19 NOTAMS (notice to airmen) and could be accessed via application programming interfaces (APIs) through ICAO's API Data Service. The COVID-19 cases per day were sourced from the European Centre for Disease Prevention and Control, and this was updated daily.

There was also a specific focus on grounded aircraft in the Eurozone. To this end, data from the EUROCONTROL area were sought and analysed. The data were obtained from a collation made by EUROCONTROL and archived on their database. The data excluded military aircraft, circular flights, unknown operators and special operators such as Airbus. The data on grounded aircraft were based on the identification of aircraft that last landed somewhere in the EUROCONTROL area between 15 February and 7 April 2020 and have been inactive since then. The researchers further looked at aircraft departures from China, as the initial epicentre of the pandemic, using data from Flightradar24 web archives.

Additional (secondary source) data were gathered from authoritative sources such as aviation media outlets and airport reports. The qualitative data were analysed through content and thematic analysis after Vaismoradi et al. (2013). Microsoft Excel Analysis ToolPak was used to analyse the quantitative (Dube and Nhamo 2019).

5.4 Results and Discussion

This section is divided into four subsections. The first subsection is dedicated to presenting the impact of the COVID-19 pandemic on the global airport networks in the various ICAO aviation regions, while the second subsection is reserved for examining the various interventions undertaken at various airports to deal with the fallout of COVID-19. The third subsection is dedicated to the impact of COVID-19 on the global airport retailing market, with the fourth subsection focusing on airport employment and economies.

5.4.1 Impact of COVID-19 on Global Airport Departures

The declaration of COVID-19 as a pandemic seems to have marked a turning point in departures. At that point, record lows in departures were reported. Observations revealed that the other wave of declines in departures, particularly in March 2020, was triggered by at least three incidences, including the USA imposing travel restrictions on Europe on 13 March 2020. By presidential proclamation, the USA initially imposed travel restrictions on 26 European countries that form part of the Schengen region. Britain and Ireland, which were at first excluded, were later included, and this led to a drastic decline in departures to and from the USA given the region's share of traffic globally. Several countries later followed suit, which led to a global downscale of departures.

The study revealed that the pandemic drastically affected departures of aircraft across the world. The disruptions covered air traffic movement in ICAO's seven aviation regions, namely, (i) Eastern and Southern Africa; (ii) Western and Central Africa; (iii) the Middle East; (iv) Central America and Caribbean; (v) Asia and Pacific; (vi) Europe and North Atlantic; and (vii) South America. An analysis of 6 months' traffic data between October 2019 and April 2020 shows that at a global level, the maximum number of departures were over 33,500 sometime in October 2019 and the departures dropped to less than 2000 departures at the end of April 2020 (Fig. 5.2).

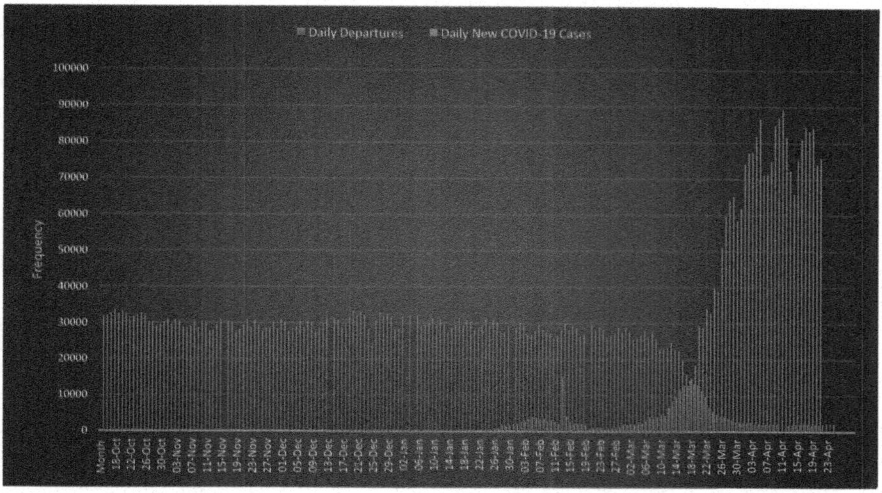

Fig. 5.2 Daily traffic departures at all the airports and statistics on daily new COVID-19 infections Source: Authors; data from ICAO (2020)

The outbreak of COVID-12 or its announcement resulted in a decline in global departures, with impacts starting to show at the end of January 2020. The impacts were more pronounced in early February 2020. The hardest knock was observed on 12 March 2020, as reflected in a marked global downward spiral of departures. This resulted in a decrease of about one-third of aircraft departures. In about 10 days after the pandemic declaration on 11 March 2020, departures plummeted to less than 10,000 aircraft globally. By the end of April 2020, there were fewer than 2000 departures on average. Globally an estimated 95% of the departures were lost. The bulk of aircraft departures and landing at airports mainly comprised cargo and domestic flights (particularly in the USA and China). Primarily, cargo aircraft was delivering medical supplies across the world, and the few passenger flights were as a result of slight relaxation-imposed travel restrictions, which also resulted in poor demand for air transport services.

It emerges that as new infections were reported globally, traffic was on the decline. CANSO data confirm these findings. Using measuring hours, the data show a drastic decline in flying hours across regions between January and April 2020. According to CANSO (2020), flying hours declined by 56% in North America and by 76% in Asia, with other regions recording much higher declines totalling as high as 90%.

The decline in departures across the global airport network could be attributed to several factors that took place in various aviation regions and affected aviation globally. A slight decline in departures in January and February 2020 could be attributed to the ban on people to leave or enter Wuhan during what is the busiest travel period in China. The travel bans led to a sharp decline of traffic at several airports, with airlines reporting traveller declines of between 85% and 90% for the 2 months year on year comparison (Chua 2020). The Wuhan ban resulted in the cancellation of air flights and trains.

The slump in departures presented challenges for airports that generate up to half of their revenue from landing and departure fees. Ordinarily, airports make close to half of their revenue from the following revenue streams:

(i) Landing fees (determined by weight and origin)
(ii) Aircraft parking fees (determined by duration and aircraft size)
(iii) Passenger service charges (in most cases, these are charged only on departures) (ACI 2020b)

While airports rely on landing and departure fees, it has other revenue sources that are linked to landing and departure. Most notable is revenue from aviation fuel surcharges. Revenue from this source was not affected by reduced sales, relevant losses compounded by notable lower fuel prices because of an oil price slump emanating from squabbles Organization of the Petroleum Exporting Countries (OPEC).

The trend from both African regions (Figs. 5.3 and 5.4) shows similar patterns. However, the African regions were affected rather later, in mid-March 2020, and the departures dropped more abruptly than in any other part of the world. After the disease was declared a pandemic, there was a sharp decline in departures owing to movement restrictions imposed by various states. The few departures that were observed show that few airlines were conducting stranded citizen rescue and relief missions, especially at the end of March and early April 2020. As in other parts of the world, the bulk of emanating flights were largely cargo aircraft medical supplies and other essentials that were needed.

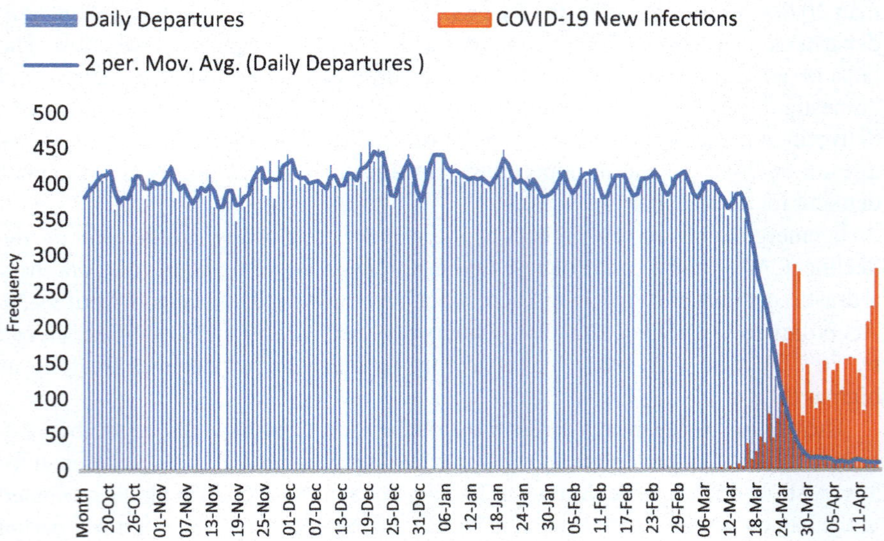

Fig. 5.3 East and South Africa – impact of COVID-19 on daily aircraft departures
Source: Authors; data from ICAO (2020)

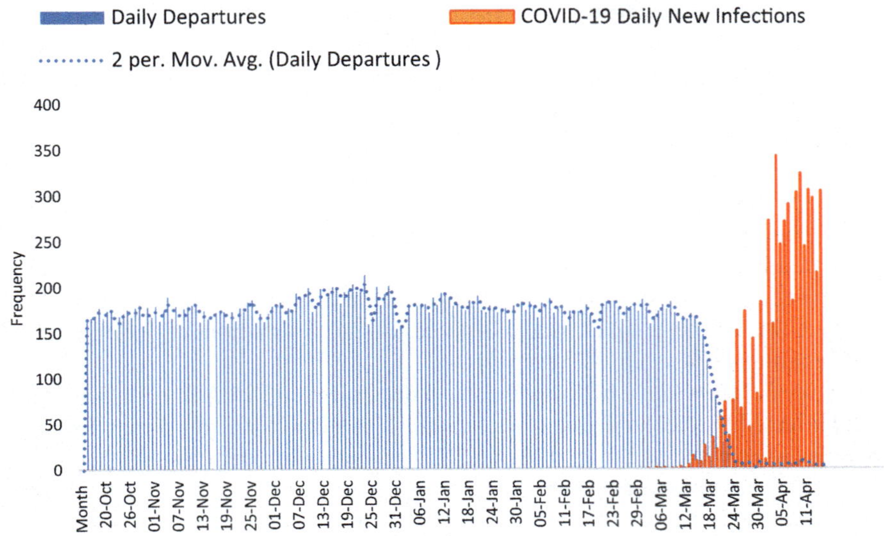

Fig. 5.4 Western and Central Africa – impact of COVID-19 on daily airport departures
Source: Authors; data from ICAO (2020)

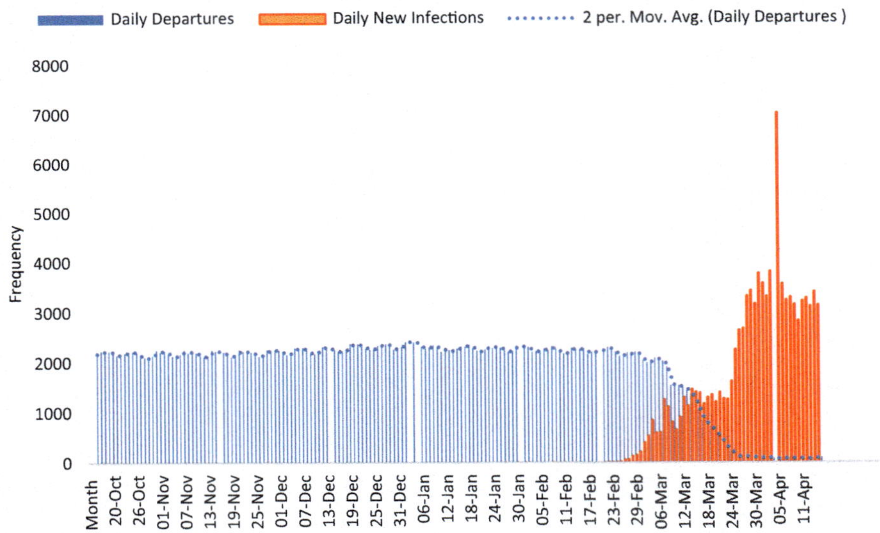

Fig. 5.5 Middle East – impact of COVID-19 on daily aircraft departures
Source: Authors; data from ICAO (2020)

The Middle East is one of the areas with a growing influence on aviation. It hosts some of the most prestigious airlines in the world, including Qatar Airways, Etihad Airways and Emirates. Figure 5.5 shows that the region was affected earlier, around February 2020, when it reported its first cases (Fig. 5.4). This could be attributed to its wide network, which resulted in exposure to the virus. The downward spiral of airport departures started around 6 March 2020, 6 days earlier than the other regions. By the end of March 2020, very few aircraft departed from the region, with about 90 departures from a peak of around 3300 in February 2020. As in other regions, the 90 aircraft departures pertained to cargo flights. Interestingly some airlines turned their passenger flights into cargo airlines.

The USA market showed some resilience, with serious departure declines being witnessed only after the travel restrictions to the European market in early March 2020. The market has a robust domestic airline industry, hence the resilience that witnessed the market taking a bit longer than other markets to get to the bottom. However, it is important to note that the international and domestic policies adopted by the USA had huge significance for the local and international aviation markets and departures (Figs. 5.6, 5.7 and 5.8).

Figure 5.7 shows how both the European departures and the American departures were affected by the pandemic from January 2020. However, significant departure declines were witnessed from 6 March 2020 onwards. Europe suffered a huge blow when the USA banned travel to the region at the height of Italy's massive infection rates and deaths. Cancellations affected flights between the two regions (the USA and the Eurozone).

Since COVID-19 emerged in China, the Asia-Pacific region recorded the first cases of both infection and death. The region witnessed a significant drop in departures, as can be seen in both Figs. 5.9 and 5.10. Departure declines were reported as far back as at the end of January 2020, with the highest infection spike on 17 February 2020. Daily new infections showed highs and lows since January 2020. In line with the global trend, there were very few departures at the airports in the

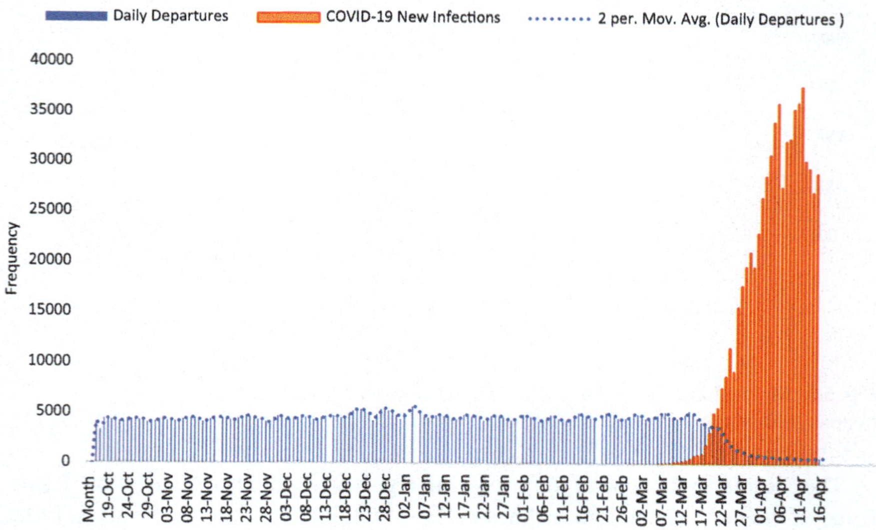

Fig. 5.6 North America and the Caribbean – impact of COVID-19 on daily aircraft departures
Source: Authors; data from ICAO (2020)

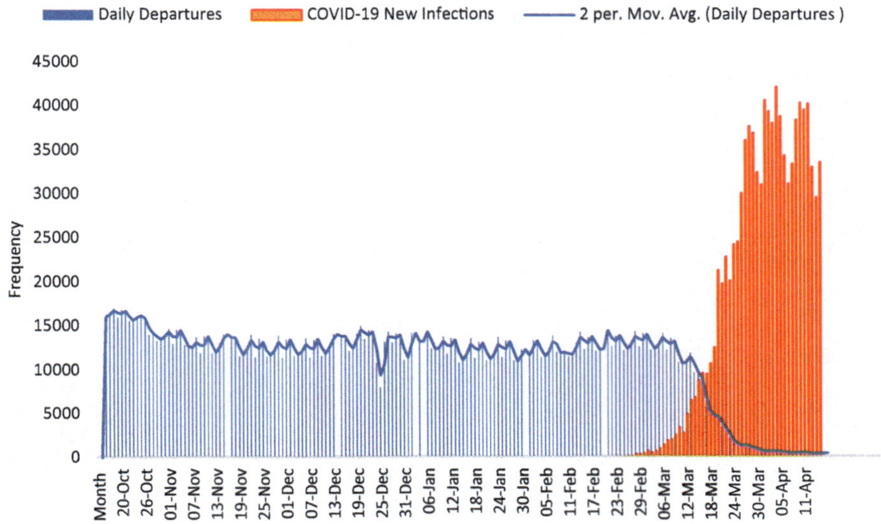

Fig. 5.7 Europe and North America – impact of COVID-19 on daily airport departures
Source: Authors; data from ICAO (2020)

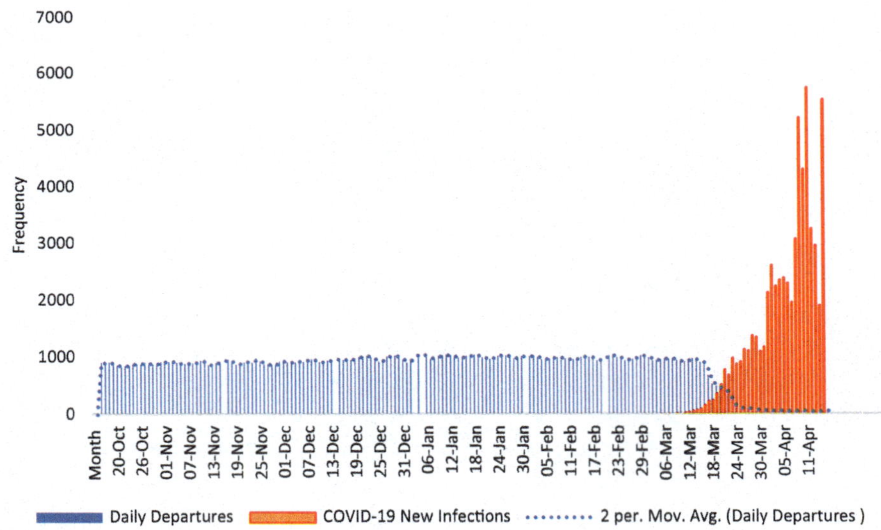

Fig. 5.8 South America – impact of COVID-19 on daily airport departures
Source: Authors; data from ICAO (2020)

Asia-Pacific region. As a consequence, the regional board for airports (the ACI Asia-Pacific) reported that instead of the predicted revenue of US$12.4 billion in the first quarter of the year, they were expecting a revenue loss of US$3 billion (ACI 2020b).

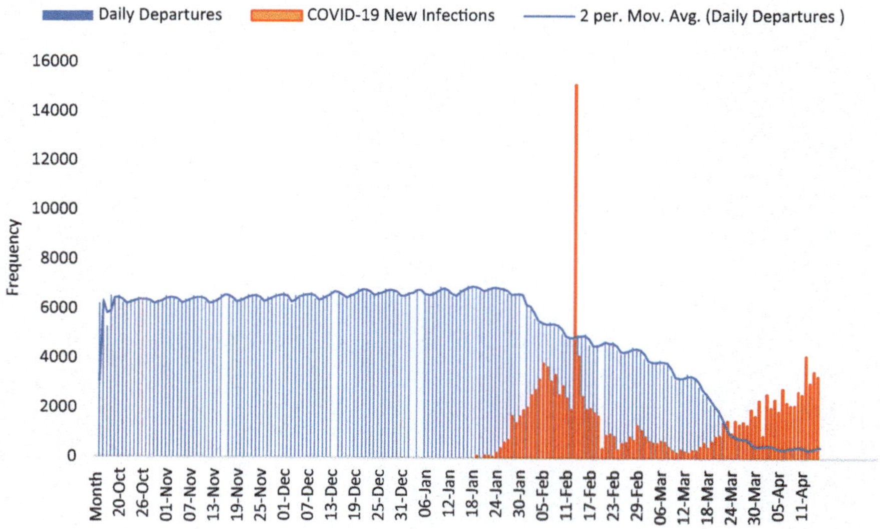

Fig. 5.9 Asia Pacific – impact of COVID-19 on daily airport departures
Source: Authors; data from ICAO (2020)

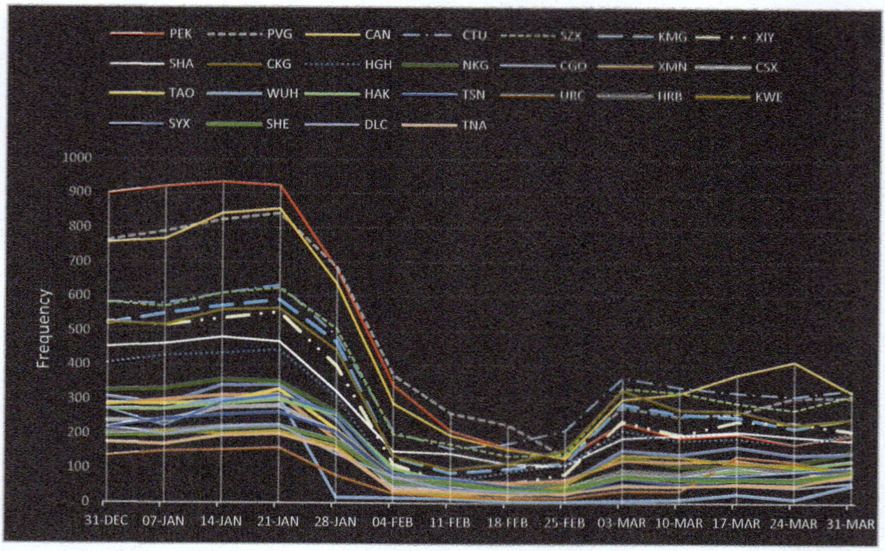

Fig. 5.10 Impact of COVID-19 on departures at China's 25 most busy airports
Source: Authors; data from Flightradar24 (2020)

They blamed the revenue loss on the suspension of the 80/20 slot rule by IATA, which triggered airline cancellations and led to a decline in "traffic and passenger demand, airports' aeronautical revenues and non-aeronautical revenues" (ACI 2020a:1b).

On the busiest day in January 2020, China's 25 busiest airports recorded more than 10,000 departures on average on a single day. This accounted for about one-third of global departures. China has 220 airports across the country. The busiest airport is Beijing Capital International Airport, which at the peak had over 900 departures per day. This figure reached an all-time low of 159 departures on 10 February 2020 due to restrictions on travel rules implemented to control the spread of COVID-19. COVID-19 severely disrupted the operations of the giant airport that services 89 international airlines (Beijing Capital Airport 2020). Figure 5.10 shows the trend at China's 25 most busy airports, where an 80% decline in departures were observed over the period in question due to the impact of COVID-19.

5.4.2 COVID-19 and Airport Parking Challenges and Economies

In as much as the grounding of many airlines presented a huge financial blow from declining departure and related fees overall, there was an unforeseen opportunity for airports to generate revenue from parking fees. With many grounded flights, most airports struggled to find parking space, with some ending up using part of their runways as parking space. This led to the closure of some runways to accommodate the unprecedented number of aircraft piling up in several airports across the world. Figure 5.11 shows a situation playing out across several airports in the world, with aircraft parked on pavements and runways. Across the world, aircraft were

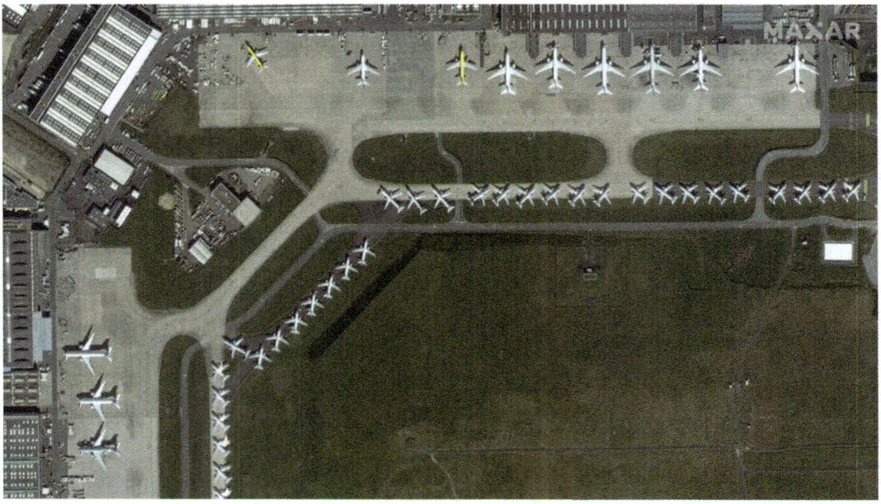

Fig. 5.11 Aerial photograph of aircraft parked at Paris' Charles de Gaulle Airport as a result of groundings due to COVID-19
Source: MAXAR Technologies (2020)

grounded due to COVID-19 travel restrictions. This led to pavement damage at some airports.

Among the most affected airports were those in countries with large fleets of aircraft such as the United Arab Emirates. The next section focuses on the airport situation in Europe (Figs. 5.12, 5.13, 5.14 and 5.15). Aircraft were piling up at various airports in Europe since February 2020, with the most significant number of parked aircraft witnessed mid-March 2020 (Fig. 5.12). Ryanair, Lufthansa and Turkish Airways had the highest number of parked aircraft, with a total of 915 (Fig. 5.13). The airports with the highest number of parked aircraft were in Madrid, Istanbul, London and Vienna (Fig. 5.14). On 17 April 2020, there were 5472 parked aircraft at 303 airports across Europe. As can be seen in Fig. 5.15, narrow-bodied aircraft formed the bulk of the parked aircraft, and a significant number were very large and wide-bodied aircraft.

The USA has a huge aircraft fleet and some airlines. Airport resorted to using closed airports or airports not frequently used as parking sites for grounded aircraft. These airports include the Victorville airport, California, and Marana, Arizona. Delta, FedEx and Southwest Airlines were some of the airlines that used this less frequented airport, probably to evade the high cost at other busy airports. The airport is also advantageous in that it has subhumid conditions. The dry conditions ensure that aircraft are rust free, which helps in lowering maintenance cost. Faced with substantial parking costs, probably for a more extended period, some airlines decided to retire their bigger aircraft. Airlines such as Lufthansa retired a sizeable number of wide-bodied aircraft.

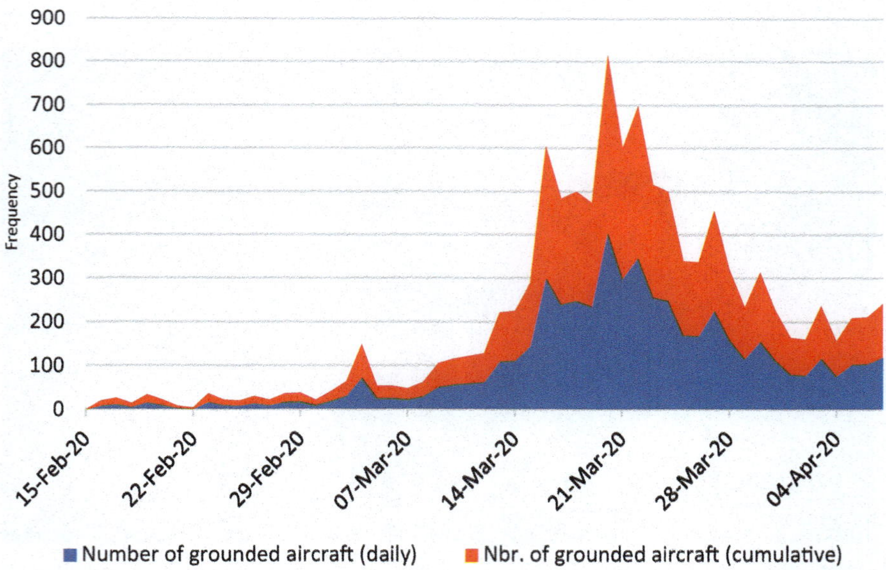

Fig. 5.12 Trend in airport parking due to COVID-19 in the Europe region (March–April 2020) Source: Authors; data from EUROCONTROL (2020)

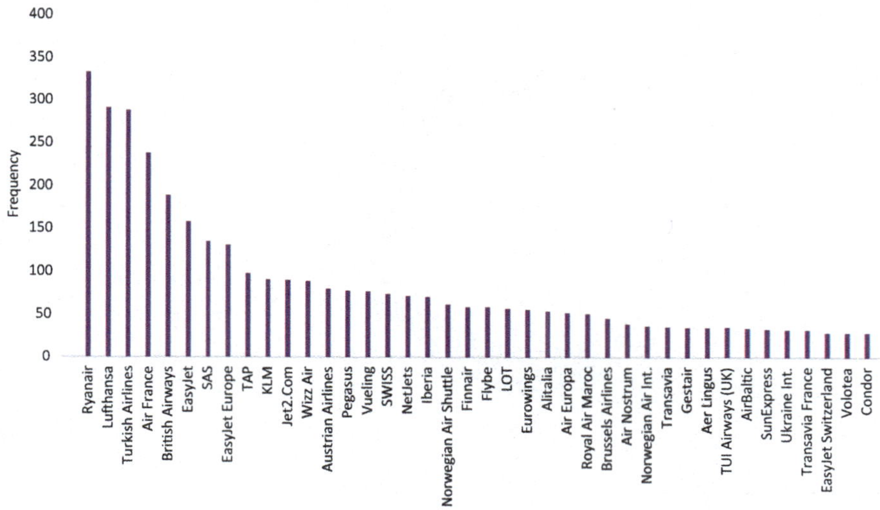

Fig. 5.13 Selected airlines with over 50 parked aircraft due to COVID-19
Source: Authors; data from EUROCONTROL (2020)

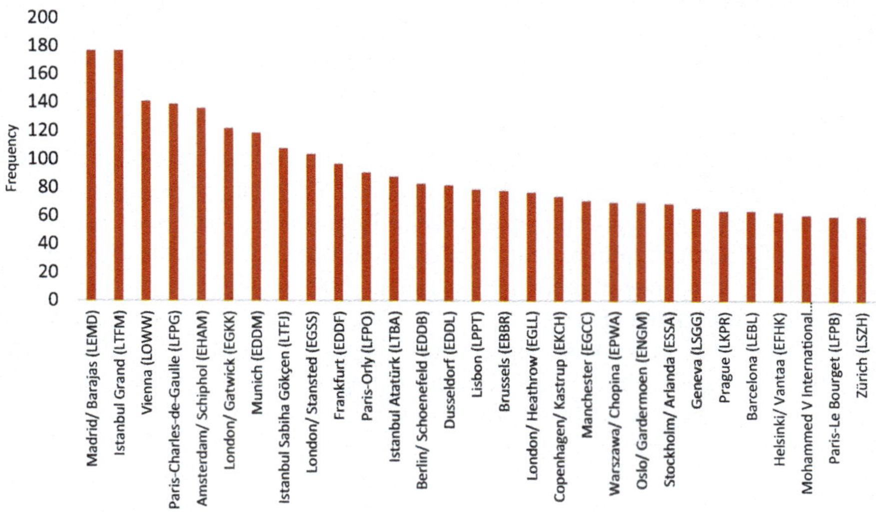

Fig. 5.14 Airport with more than 60 parked aircraft due to COVID-19 travel restriction
Source: Authors; data from EUROCONTROL (2020)

As already indicated, the piling up of aircraft presented several challenges for airports across the world. In Europe, there was a call to cut airport parking fees, which effectively reduced revenues. As such, the airports were also incapable of cashing in on the troubled airlines. Parking an aircraft at an airport is particularly expensive, especially if no business is going on. In India, for example, parking fees

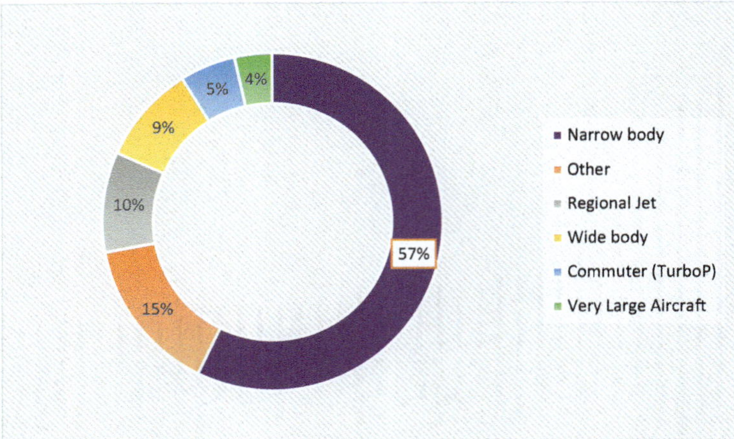

Fig. 5.15 Size of parked aircraft due to COVID-19 in the Eurozone (n = 5472)
Source: Authors; data from EUROCONTROL (2020)

per day can be as high as US$1000 (Heathrow 2019a). Parking fees for a wide-bodied aircraft at Heathrow Airport are £59.35 and that for narrow-bodied aircraft £24.73 every 15 minutes, with the costs waived between 22:00 and 05:59 UTC (Ibid). While the parking costs could have been reduced by government and airport authorities through concessionaries and waivers, there were serious financial implications for airlines with wide-bodied aircraft and those with a large fleet (Figs. 5.13 and 5.15).

Due to the increased demand for parking, some airports had challenges resulting from damaged airport infrastructure. Istanbul had a capacity of parking space for 500 aircraft and could accommodate the large Turkish fleet. The airport was built at the cost of US$16.65 billion, and a significant portion of this amount was sourced through a build-operate-transfer process and a (US$5bn) loan from several banks, including Ziraat Bank, Halkbank and VakıfBank (Airport Technology 2019). The 90-million-passengers capacity airport that employs 31,000 employees (iGA 2020) reduced operations, which had a devastating impact on its employees. Thus, the disruptions caused by COVID-19 could hamper the period of the rate of return and the repayment process. Several airports across the world faced similar challenges.

The Southern Hemisphere saw the construction of many airports, with some not realising their capital investment. Observations revealed that in Southern Africa, there were new airports or major airport upgrades at Lanseria International Airport (South Africa), Victoria Falls International Airport (Zimbabwe), Cape Town International Airport (South Africa) and OR Tambo International Airport (South Africa). China reported the completion of facilities worth U$11.7 billion at Beijing Daxing International Airport, with a capacity to handle more than 100 million passengers per annum (Keju 2019). In the USA, there were 10 airport projects with a budget of over US$1.3 billion, with the highest having a budget of US$2.6 billion (Statista 2020) which were likely to be affected by the COVID-19 pandemic liquidity crunch. The liquidity crunch that affected airports is feared to have reversed the progress of carbon–neutral projects that were on the increase globally.

While airports could have modelled their investments in projected traffic figures based on financial gains and observations from trends in recent years, the pandemic upset that scenario and changed airport companies' financial outlook. This had dire consequences for financially insecure airports, particularly private airports, with limited access to government funding. The pandemic led to a wave of disturbances throughout the tourism value chain and had a devastating impact on the airline industry. The placement under administration and/or ultimate collapse of a number of airlines have implications for the viability of airports going forward. No doubt the disturbances at Virgin Australia with its fleet of 120 aircraft (Flightradar24 database) may have a significant impact on the revenue of airports on its routes if it is allowed to fail. The airliner has been put under administration and is battling a credit bill of US$4.4 billion. The development has seen Perth Airport attaching some of the airline's property. There are fears that many airlines that were financially fragile will go under after the pandemic, with implications for airports across the world.

The airline industry on the African continent has been identified as vulnerable to shocks owing to poor governance, restrictive taxes and poor route network planning (Dube and Nhamo 2019). The pandemic has had an adverse impact on South African Airways, which have been in trouble for years and depends on government bailouts. The pandemic might signal its demise as the government denied the perennial loss – making airline further bailouts due to a restricted fiscal space (Makinana 2020). With South African Airways one of the leading regional and continental airlines in sub-Saharan Africa, its liquidation and that of SA Airlink which made a loss of R365 million in November 2019 (Ramalepe 2020) and SA Express can cripple some of the airports in South Africa and the Southern African Development Community (SADC) region where the two airlines are dominant. These airlines are the only airlines servicing some of the airports in the SADC region, and their demise is likely to have negative implications for jobs and the sustainability of the affected airports. If these airlines eventually close down, major routes to tourist attractions via Kruger Mpumalanga International Airport, Victoria Falls International Airport, Livingstone International Airport (LVI, also known as Harry Mwanga Nkumbula International Airport), Skukuza Airport and Bisho Airport amongst others could be compromised, if not closed down altogether. This situation is not unique to South Africa.

5.4.3 *Impact of COVID-19 on the Global Airport Retailing Market*

Apart from getting revenue from aeronautical services, several airports get substantial revenue from retail, car rentals, property leases and ground transport. For example, Sydney Airport made AUS$1637, nine million in 2019 (SYD 2019). Figure 5.16

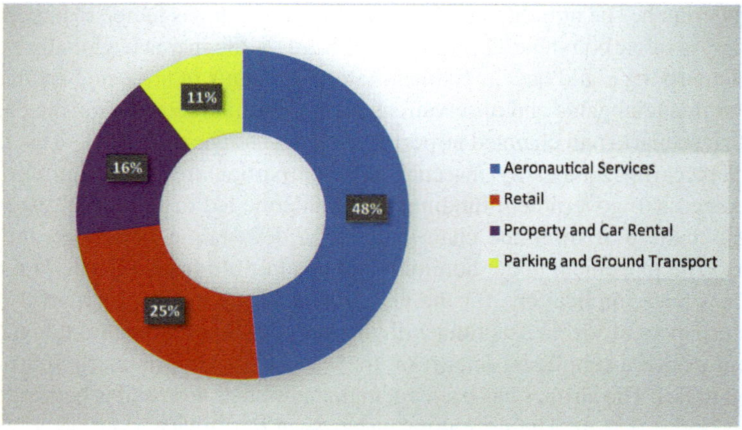

Fig. 5.16 Revenue collection by airport business units at Sydney Airport in 2019 (Total revenue $1637.9 million)
Source: Authors; data from SYD (2019)

shows the contribution of each airport business unit to total revenue collected by the airport.

With millions of passengers using the airports and using various services and facilities, the closure of various retail units as a result of the grounding of airlines led to a loss of substantial revenue from various business units. Closed services in the space include airport restaurants, bars, airport lounges and car rental and ground transport facilities, all of which generate significant amounts of revenue for airport-owning companies.

Besides retail, some airports rent or lease hotel space all adding another revenue stream. With aircraft grounded, most shops (particularly restaurants, duty-free shops, electronic shops and souvenirs shops) and airport lounges shut down. The global airport retail value in 2017 was pegged at US$40 billion, with the figure expected to reach US$58.4 billion by 2022 (News Wire 2020). In 2019, the global retail market grew by 5%, with profit margins at 5.6% (Research and Markets 2019). It was estimated that the loss in potential revenue could have easily exceeded US$5 billion and came at a considerable cost to employees in the retail sector.

5.4.4 *Impact of COVID-19 on Employment and Economies of Airports*

Airports indirectly provide employment to a lot of people who provide cleaning services for airlines and airport facilities, security services and ancillary services such as baggage handling and airport shuttle services. Given that most of these employees are either semi-skilled or employed on a temporary basis, indications

Table 5.1 Impact of COVID-19 on selected airport labour and employment

Name of airport/ground handling company	Number of dismissed/furloughed employees
Leeds Bradford Airport	Up to 250 people directly employed by the airport were laid off
Houston Airport	Swissport temporarily laid off at least 50 full-time employees of its airport staff
	United Airlines, which uses 60% of the airport and employs nearly 14,000 people in the Houston area, indicated the intention to lay off airport staff
Charlotte Douglas Airport	112 airport workers lost their jobs
	More than 600 HMSHost employees lost their jobs after the airport closed restaurants and bars
John Menzies Axes	17,500 employees in 200 countries were dismissed globally, which was 50% of its workforce (BBC 2020)
Airports in India	35,000 ground handling agents were placed on forced leave without pay (Simhan 2020)
Heathrow Airport	Security guards and firefighters accepted a 10% reduction in their pay for 9 months
	Other airport staff were forced to take a 15% salary reduction after business fell by 90% (Partridge 2020)
	In 2019, the airport had, on average, 7827 monthly employees (Heathrow 2019b)
Bristol Airport	Furloughed 80% of staff (Gogarty 2020)

Source: Authors

were that the bulk of them were furloughed or dismissed due to reduced operations or the closure of airports across the world. Table 5.1 shows some of the cases where airport employees were either furloughed or dismissed.

While the aviation industry was advocating for bailouts from the share of emerging economic stimulus packages from governments to save tens of millions of jobs, it was clear that the employees were on the receiving end of the coronavirus attack. Using baseline data from the previous year (ACI 2020b) at the beginning of April 2020, the total loss for airports due to decreased passenger demand was estimated to be in the region of US$76 billion. This amount is significant as it equates to the financial budget of several developing countries. ACI (which represents 668 members operating 1979 airports in 176 countries) indicated that in a bid to protect viability, the airports had adopted a wide variety of 6+ measures (highlighted in Fig. 5.17).

It has to be stated that in most cases, the government were devoid of support to airports. If anything, governments and authorities instituted measures that put airports at a financial disadvantage. Although understandable, the call for fee reductions and suspension of slots increased the financial strain on airports across the world. With very little coming from aeronautical services, the airports' other revenue streams (discussed earlier) were incapacitated because they were shut down as a control measure due to COVID-19.

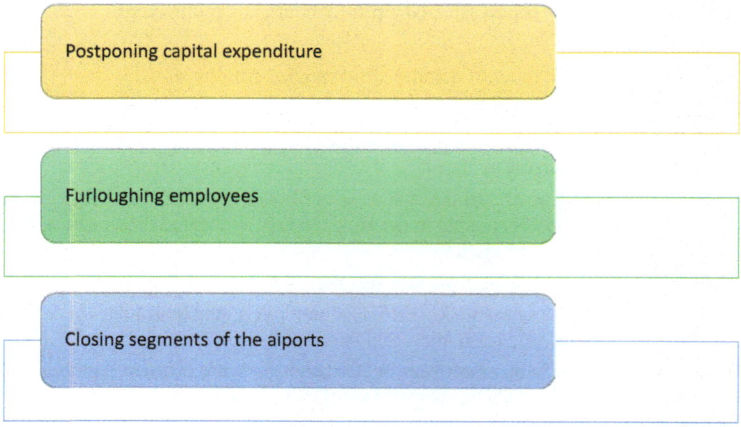

Fig. 5.17 COVID-19 airports mitigation strategies
Source: Authors; data from ACI (2020a)

The COVID-19 pandemic also brought under the spotlight the effectiveness of thermal scanners in the screening of infected passengers. Thermal scanners were often used previously to control the outbreak of diseases such as SARS (John et al. 2005). Quilty et al. (2020) observe that out of 100 COVID-19 cases, an estimated 47% was not detected, 44% was detected at exit screening, 1% was detected as severe on the flight and 8% was detected at entry screening. This points to the need for continuous research and innovation to improve the efficiency and effectiveness of thermal scanners at airports. In most cases, however, screening is not done on departure and arrival, with experience showing that on departure airports hardly conduct any screenings.

5.5 Conclusions and Recommendations

The study investigated how COVID-19 pandemic affected the airport industry across the world. The study found that the industry, like any other industry, was adversely affected by the pandemic. This led to a loss of traffic around the world owing to plummeting departures. Initial assessment shows that the industry was set to lose about US$76 billion in the first few months of the pandemic. The loss in revenue due to reduced traffic had an adverse impact on individual airport revenues, which in turn, resulted in revenue losses for airports and air navigation companies. The airports responded by cutting back on capital expenditure, furloughing employees and closing sections of the airport as cost containment measures. Regardless of the challenges, some airports with bigger financial capacity took the opportunity to upgrade and conduct critical maintenance in preparation for the post-pandemic period.

There were fears that the impact of airlines closing down would seriously affect airports, particularly in Africa. This was because some airports and the tourism industry would be crippled by the closure of individual airlines. With the airport industry expected to take a U-shaped (instead of a V-shaped) recovery path, countries with weak domestic travels were likely to have the worst affected airports. Due to revenue losses, the pandemic drastically affected loan repayments for airports whose projects were premised on forecasted revenue instead of cash balance. With several airports having capital projects lined up, some aimed at improving and ensuring seamless travel and promoting green airports, the postponement of some projects might compromise and reverse the gains made in ensuring green travel.

The study further found that there were still challenges in terms of the effectiveness and efficiency of thermal scanners at airports. This requires continued investment and attention to improve the apparatus used to ensure biosecurity at airports going forward, thereby protecting global economies. The ACI called for support for airports and aeronautical companies across the world in the form of providing financial relief packages to enable service sustainability post COVID-19 or when the economy is restarted. Other support measures could come through the protection of airport charges and revenues, the provision of tax relief, a concession fee waiver, continuity of air cargo operations and a comprehensive relief package that includes offering the industry concessionary grants, loan grants at preferential rates and subsidies to ensure the sustainability of airports.

The study highlights a need to support the sector to ensure the continued existence of affected airports and associated operations. There is also a need to protect airport employees from the impact of the COVID-19 economic shock, the magnitude of which had never been seen in history. The study recommends a post-pandemic impact assessment of airlines, air navigation systems and airports in countries and regions to better understand the impact of COVID-19 on the sector.

References

ACI. (2020a). Economic analysis shows COVID-19 is an existential threat to airport business. *Airport Council International*. Retrieved April 24, 2020, from https://aci.aero/news/2020/04/01/economic-analysis-shows-covid-19-is-an-existential-threat-to-airport-business/

ACI. (2020b). Airport economics at a glance. *Airports Council International*. Retrieved April 24, 2020, from https://store.aci.aero/wp-content/uploads/2020/02/Cover-image_Airport-Economics-Infographic_For-website.jpg

Adams, S., Lindeque, G., & Soma-Pillay, P. (2020). Bioethics and self-isolation: What about low-resource settings? *South African Medical Journal, 110*(5). https://doi.org/10.7196/SAMJ.2020.

Airport Technology. (2019). Istanbul New Airport, Black Sea Coast. *Airport Technology*. Retrieved April 22, 2020, from https://www.airport-technology.com/projects/istanbul-new-airport-black-sea-coast/

BBC. (2020). John Menzies axes 17,500 jobs amid travel slump. *BBC News Scotland*. Retrieved April 23, 2020, from https://www.bbc.com/news/uk-scotland-scotland-business-52070491

Beijing Capital Airport. (2020). *Airlines at Beijing capital airport (PEK)*. Beijing: Beijing Capital Airport. Retrieved April 26, 2020, from https://www.beijing-airport.com/pek-airlines.

Bilotkach, V. (2018). Political economy of infrastructure investment: Evidence from the economic stimulus airport grants. *Economics of Transportation, 13*, 27–35.

Blendon, R. J., DesRoches, C. M., Cetron, M. S., Benson, J. M., Meinhardt, T., & Pollard, W. (2006). Attitudes toward the use of quarantine in a public health emergency in four countries: The experiences of Hong Kong, Singapore, Taiwan, and the United States are instructive in assessing national responses to disease threats. *Health Affairs, 25*(Suppl1), W15–W25.

Briand, S., Mounts, A., & Chamberland, M. (2011). Challenges of global surveillance during an influenza pandemic. *Public Health, 125*(5), 247–256.

Brouder, P. (2020). Reset redux: Possible evolutionary pathways towards the transformation of tourism in a COVID-19 world. *Tourism Geographies*, 1–7. https://doi.org/10.1080/14616688.2020.1760928.

CANSO. (2020). CANSO-aireon data driven insights. *CANSO*. Retrieved April 24, 2020, from https://www.canso.org/canso-aireon-data-driven-insights?platform=hootsuite

CAPSCA. (2020). CAPSCA. *CAPSCA*. Retrieved April 24, 2020, from http://www.capsca.org/IntroandObjectivesWeb.pdf

Chen, H., Huang, X., & Li, Z. (2020). A content analysis of Chinese news coverage on COVID-19 and tourism. *Current Issues in Tourism*, 1–8. https://doi.org/10.1080/13683500.2020.1763269.

Chua, A. (2020). Chinese carriers' February traffic collapse amid pandemic hit. *FlightGlobal*. Retrieved April 19, 2020, from https://www.flightglobal.com/air-transport/chinese-carriers-february-traffic-collapse-amid-pandemic-hit/137390.article

Chung, L. H. (2015). Impact of pandemic control over airport economics: Reconciling public health with airport business through a streamlined approach in pandemic control. *Journal of Air Transport Management, 44*, 42–53.

Dube, K., & Nhamo, G. (2019). Climate change and the aviation sector: A focus on the Victoria Falls tourism route. *Environmental Development, 29*, 5–15.

Dube, K., Nhamo, G., & Chikodzi, D. (2020). COVID-19 cripples global restaurant and hospitality industry. *Current Issues in Tourism*, 1–4. https://doi.org/10.1080/13683500.2020.1773416.

Eastwood, K., Durrheim, D. N., Butler, M., & Jones, A. (2010). Responses to pandemic (H1N1) 2009, Australia. *Emerging Infectious Diseases, 16*(8), 1211.

EUROCONTROL. (2020). Daily Traffic Variation - States. EUROCONTROL. Retrieved May 1, 2020, from https://www.eurocontrol.int/Economics/DailyTrafficVariation-States.html

Flightradar24. (2020). AIR TRAFFIC AT CHINA'S BUSIEST AIRPORTS NOW IN SLOW RECOVERY. Retrieved May 1, 2020, from https://www.flightradar24.com/blog/air-traffic-at-chinas-busiest-airports-down-80-since-the-beginning-of-the-year/

Fujita, M., Sato, H., Kaku, K., Tokuno, S., Kanatani, Y., Suzuki, S., & Shinomiya, N. (2011). Airport quarantine inspection, follow-up observation, and the prevention of pandemic influenza. *Aviation, Space, and Environmental Medicine, 82*(8), 782–789.

Gensini, G. F., Yacoub, M. H., & Conti, A. A. (2004). The concept of quarantine in history: From plague to SARS. *Journal of Infection, 49*(4), 257–261.

Gogarty, C. (2020). *Bristol Airport furloughs 80 per cent of staff and may make difficult decisions*. Bristol: BristolLive. Retrieved April 24, 2020, from https://www.bristolpost.co.uk/news/bristol-news/airport-furlough-flights-coronavirus-expansion-4064450.

Gössling, S., Scott, D., & Hall, C. M. (2020). Pandemics, tourism and global change: A rapid assessment of COVID-19. *Journal of Sustainable Tourism*, 1–20. https://doi.org/10.1080/09669582.2020.1758708.

Gostic, K., Gomez, A. C., Mummah, R. O., Kucharski, A. J., & Lloyd-Smith, J. O. (2020). Estimated effectiveness of traveller screening to prevent international spread of 2019 novel coronavirus (2019-nCoV). *medRxiv*, 1–3. https://doi.org/10.1101/2020.03.12.20034793.

Harvey, G., & Turnbull, P. (2009). The impact of the financial crisis on labour in the civil aviation industry (No. 994476973402676). *International Labour Organization*. Retrieved from https://ideas.repec.org/p/ilo/ilowps/994476973402676.html

Heathrow. (2019a). Heathrow Airport Limited conditions of use including Airport charges from 1 January 2019. *Heathrow Airport Limited*. Retrieved April 21, 2020, from https://www.heathrow.com/content/dam/heathrow/web/common/documents/company/doing-business-with-heathrow/flights-condition-of-use/conditions-of-use-documents/Heathrow-Airport-Limited-Conditions-of-Use-2019.pdf

Heathrow. (2019b). *Annual report and financial statements 2019*. London: Heathrow (SP) Limited. Retrieved April 24, 2020, from https://www.heathrow.com/content/dam/heathrow/web/common/documents/company/investor/reports-and-presentations/annual-accounts/sp/Heathrow_SP_ARA_2019.pdf.

Hou, C., Chen, J., Zhou, Y., Hua, L., Yuan, J., He, S., & Zhang, J. (2020). The effectiveness of the quarantine of Wuhan city against the Corona Virus Disease 2019 (COVID-19): Well-mixed SEIR model analysis. *Journal of Medical Virology*. https://doi.org/10.1002/jmv.25827.

IATA. (2019). WATS world air transport statistics 2019. *World Air Transport Statistics 2019*. Retrieved April 15, 2020, from https://www.iata.org/contentassets/a686ff624550453e8bf0c9b3f7f0ab26/wats-2019-mediakit.pdf

ICAO. (2020). Global COVID-19 Airport Status. ICAO. Retrieved May 1, 2020, from ICAO Safety: https://www.icao.int/safety/Pages/COVID-19-Airport-Status.aspx

iGA. (2020). This is where we are building This is where we are building Istanbul New Airport. *iGA*. Retrieved April 22, 2020, from https://web.archive.org/web/20180328103248/http://igairport.com/en/istanbul-new-airport/construction/about-the-construction

International Airport Review. (2019). In the balance: Global air transport demand in 2018 and 2019. *International Airport Review*. Retrieved April 24, 2020, from https://www.internationalairportreview.com/article/101952/world-airport-traffic-report-preview-aciworld/

International Monetary Fund -IMF. (2020). Opening remarks at a press briefing by Kristalina Georgieva following a Conference Call of the International Monetary and Financial Committee (IMFC). *International Monetary Fund*. Retrieved April 9, 2020, from https://www.imf.org/en/News/Articles/2020/03/27/sp032720-opening-remarks-at-press-briefing-following-imfc-conference-call

Johanis, D. (2007). How Toronto Pearson International Airport applied lessons from SARS to develop a pandemic response plan. *Journal of Business Continuity & Emergency Planning, 1*(4), 356–368.

John, R. K., King, A., De Jong, D., Bodie-Collins, M., Squires, S. G., & Tam, T. W. (2005). Border screening for SARS. *Emerging Infectious Diseases, 11*(1), 611.

Jones, C. (2010). Archival data: Advantages and disadvantages for research in psychology. *Social and Personality Psychology Compass, 4*(11), 1008–1017.

Keju, W. (2019). Construction of new Beijing airport finished. *China Daily*. Retrieved April 22, 2020, from http://global.chinadaily.com.cn/a/201907/29/WS5d3e4d6ca310d83056401745.html

Makinana, A. (2020). Government refuses SAA business rescuers' request for R10bn. *Times Live*. Retrieved April 26, 2020, from https://www.timeslive.co.za/politics/2020-04-14-government-refuses-saa-business-rescuers-request-for-r10bn/

Malone, J. D., Brigantic, R., Muller, G. A., Gadgil, A., Delp, W., McMahon, B. H., & Mihelic, F. M. (2009). US airport entry screening in response to pandemic influenza: Modeling and analysis. *Travel Medicine and Infectious Disease, 7*(4), 181–191.

MAXAR Technologies. (2020). Satellites reveal striking impact of COVID-19 on people and air quality. Space News. Retrieved May 28, 2020, from https://spacenews.com/satellites-reveal-striking-impact-of-covid-19-on-people-and-air-quality/

Melly, D., & Hanrahan, J. (2020). Tourism biosecurity risk management and planning: An international comparative analysis and implications for Ireland. *Tourism Review*. https://doi.org/10.1108/TR-07-2019-0312.

Nepal, S. K. (2020). Travel and tourism after COVID-19–business as usual or opportunity to reset? *Tourism Geographies*, 1–5. https://doi.org/10.1080/14616688.2020.1760926.

News Wire. (2020). Global Airport retailing market report 2018: Sales reached US$40.0 billion in 2017, and is expected to reach $58.4 billion by 2022. *Research and Markets*. Retrieved April 21, 2020, from https://www.globenewswire.com/news-release/2019/03/04/1746147/0/en/Global-Airport-Retailing-Market-Report-2018-Sales-Reached-US-40-0-Billion-in-2017-and-is-Expected-to-Reach-58-4-Billion-by-2022.html

Nicola, M., Alsafi, Z., Sohrabi, C., Kerwan, A., Al-Jabir, A., Iosifidis, C., & Agha, R. (2020). The socio-economic implications of the Coronavirus and COVID-19 pandemic: A review. *International Journal of Surgery*. https://doi.org/10.1016/j.ijsu.2020.04.018.

Nicolaides, C., Cueto-Felgueroso, L., González, M. C., & Juanes, R. (2012). A metric of influential spreading during contagion dynamics through the air transportation network. *PLoS One, 7*, 7.

Nishiura, H., & Kamiya, K. (2011). Fever screening during the influenza (H1N1-2009) pandemic at Narita International Airport, Japan. *BMC Infectious Diseases, 11*(1), 111.

Nussbaumer-Streit, B., Mayr, V., Dobrescu, A. I., Chapman, A., Persad, E., Klerings, I., & Gartlehner, G. (2020). Quarantine alone or in combination with other public health measures to control COVID-19: A rapid review. *Cochrane Database of Systematic Reviews*. https://doi.org/10.1002/14651858.CD013574.

Partridge, J. (2020). Heathrow Airport staff threatened with sack if they reject "voluntary" pay cuts. *The Guardian*. Retrieved April 24, 2020, from https://www.theguardian.com/uk-news/2020/apr/08/heathrow-workers-voluntary-pay-cuts-airport-coronavirus

Quilty, B. J., Clifford, S., Flasche, S., & Eggo, R. M. (2020). Effectiveness of airport screening at detecting travellers infected with novel coronavirus (2019-nCoV). *Eurosurveillance, 25*(5), 1–5.

Ramalepe, P. (2020). Over 1 700 jobs at risk at Airlink due to SAA withholding revenue, court hears. *fin24*. Retrieved April 23, 2020, from https://www.fin24.com/Companies/Industrial/over-1-700-jobs-at-risk-at-airlink-due-to-saa-withholding-revenue-court-hears-20200212

Research and Markets. (2019). Airport retailing trends. *Research and Markets*. Retrieved April 21, 2020, from https://www.researchandmarkets.com/reports/4773556/airport-retailing-trends#rela2-4755693

Romagosa, F. (2020). The COVID-19 crisis: Opportunities for sustainable and proximity tourism. *Tourism Geographies*, 1–5. https://doi.org/10.1080/14616688.2020.1763447.

Safi, M. A., & Gumel, A. B. (2011). Mathematical analysis of a disease transmission model with quarantine, isolation and an imperfect vaccine. *Computers & Mathematics with Applications, 61*(10), 3044–3070.

Simhan, T. (2020). Coronavirus impact: Hundreds of ground handling agents of international airlines told to go on leave. *BuisnessLine*. Retrieved April 23, 2020, from https://www.thehindubusinessline.com/economy/logistics/coronavirus-impact-hundreds-of-ground-handling-agents-of-international-airlines-told-to-go-on-leave/article31089563.ece#

Statista. (2020). Top airport projects in the United States 2019–2023. *Statista*. Retrieved April 23, 2020, from https://www.statista.com/statistics/1012620/airport-projects-united-states/

SYD. (2019). *Annual report 2019*. Sydney: Sydney Airport. Retrieved April 25, 2020, from https://assets.ctfassets.net/v228i5y5k0x4/4VyuoCbo3sqHVBggCxV7h3/5ad8f884f3ac89516391d8ea459d50ff/SYD_Annual_Report_2019_FINAL.pdf.

Ulrich, M. R., & Mariner, W. K. (2018). Quarantine and the federal role in epidemics. *SMUL Review, 71*, 391.

Vaismoradi, M., Turunen, H., & Bondas, T. (2013). Content analysis and thematic analysis: Implications for conducting a qualitative descriptive study. *Nursing & Health Sciences, 15*(3), 398–405.

Warren, A., Bell, M., & Budd, L. (2010). Airports, localities and disease: Representations of global travel during the H1N1 pandemic. *Health & Place, 16*(4), 727–735.

Webb, E. J., Campbell, D. T., Schwarz, R. D., & Sechrest, L. (2017). The use of archival sources in social research. In M. Bulmer (Ed.), *Sociological research methods* (pp. 113–130). New York: Routledge.

Welch, C. (2000). The archaeology of business networks: The use of archival records in case study research. *Journal of Strategic Marketing, 8*, 197–208.

Wilder-Smith, A., & Freedman, D. O. (2020). Isolation, quarantine, social distancing and community containment: Pivotal role for old-style public health measures in the novel coronavirus (2019-nCoV) outbreak. *Journal of Travel Medicine, 27*(2), taaa020.

Wilder-Smith, A., Goh, K. T., & Paton, N. I. (2003). Experience of severe acute respiratory syndrome in Singapore: Importation of cases, and defense strategies at the airport. *Journal of Travel Medicine, 10*(5), 259–262.

Yadav, M. P., Singh, R. K., & Malik, Y. S. (2020). Emerging and transboundary animal viral diseases: Perspectives and preparedness. In Y. Malik, R. Singh, & Y. M. (Eds.), *Emerging and transboundary animal viruses* (pp. 1–25). Singapore: Springer.

Zheng, Y., Goh, E., & Wen, J. (2020). The effects of misleading media reports about COVID-19 on Chinese tourists' mental health: A perspective article. *Anatolia, 31*(2), 337–340.

Chapter 6
Counting the Cost of COVID-19 on Global Cruise Ship Industry

Abstract The cruise ship industry is one of the fastest-growing segments of the tourism sector. In 2019, the industry transported over 30 million passengers across the world, creating over 1.11 million jobs. In as much as the industry plays such a significant role to global and regional economies, it remains vulnerable to a cocktail of challenges, including disease outbreaks. The chapter documents the impacts of the COVID-19 pandemic on the global cruise ship industry. Secondary data and document analyses were the main forms of data collection used, while content and thematic analyses were used to analyse the data. The results showed that the COVID-19 outbreak imposed both economic and human costs to the industry. The outbreak decimated the industry, which saw the value of shipping companies' stock falling by up to 80% in March 2020, with major sources of revenue shut due to the suspension of cruises. With no bailout packages in many instances, the industry faced bankruptcy and imminent collapse. COVID-19 also left the industry facing several criminal investigations and lawsuits due to passenger infections on board. Some ships were denied access to ports and medical facilities on land due to fear of spreading the virus. Those that got access were not allowed disembarkations, with passengers often quarantined for days or weeks. This resulted in additional infections, trauma and psychological stress of the passengers. Probably the biggest loss the industry has incurred is reputational damage, and this has had the potential to impact negatively on its ability to recover after the COVID-19 shock. Through improved public health standards, aggressive marketing and the offering of massive discounts, the industry has the potential to rebound from the COVID-19 disaster. These stand out as the chapter's key recommendations.

Keywords COVID-19 · Cruise ships · Port of call · Cruise ship industry · Tourism

6.1 Introduction

The cruise ship industry is an important segment of the global tourism industry providing leisure, employment and revenue across the whole world. After the emergence of COVID-19, the United Nations (2020) observed that its rapid spread and the measures put in place to try and curb it presented more than just a health crisis

G. Nhamo et al., *Counting the Cost of COVID-19 on the Global Tourism Industry*,
https://doi.org/10.1007/978-3-030-56231-1_6

but also had massive socio-economic implications, including attacking societies at their core. According to González (2018), the United Nations World Tourism Organisation (UNWTO) conceptualises cruise ships as floating resorts as opposed to being merely a means of transport. From these floating resorts, recreational activities alternate both on the ship and onshore. The cruise ship industry has been the fastest-growing segment in the travel tourism industry between 1980 and 2019. The Business Research and Economic Advisors (BREA) report the average annual growth rate in the number of cruise passengers worldwide to have been above 8.4% over the past 20 years (BREA 2018). In 2019 alone, over 30 million passengers were transported on cruise lines globally. Records from the Cruise Lines International Association (CLIA) assessed that the industry created over 1.11 million jobs worth over US$45.6 billion in wages and salaries in 2018 (CLIA 2019). The industry made over 20,000 calls at ports and deployed assets to all continents in 2017 (BREA 2018).

Notably, as the cruise ship industry expanded and gained popularity, so too have outbreaks of communicable diseases on cruise ships been more frequent. Crockett (2020) highlighted the increased prevalence of communicable diseases mostly associated with respiratory, enteric and vaccine preventable pathogens on cruise ships, noting more than 353 gastrointestinal or norovirus outbreaks since 2010. The modes of transmission associated with cruise ship outbreaks comprise foodborne, waterborne, environmental and person-to-person transmission (Stock et al. 2015). Outbreaks of the norovirus, shigellosis, measles, Legionnaires' disease, influenza and varicella have been reported on cruise ships around the world. The European Centre for Disease Prevention and Control (ECDC) observed that as early as 2009, there were conscious efforts towards enhanced surveillance of communicable diseases on cruise ships (ECDC 2010). Apart from the increased frequency of disease cases and outbreaks, the most compelling reasons for such initiatives were the increasing popularity of cruise ships in almost all regions of the world, against a background of inadequate surveillance of diseases on ships, and the variability of ship inspection findings regarding sanitation and hygiene.

In December 2019, an epidemic caused by a new coronavirus (COVID-19) emerged and broke out in Wuhan, China. It quickly spread around the world with the World Health Organization (WHO) declaring it a pandemic on 11 March 2020 (Gan et al. 2020). This was when the disease had spread to 114 countries in the world. It was not long until fear of the disease began to show in world markets and sensitive industries such as tourism. As observed by Djalante et al. (2020), COVID-19 rapidly transformed into unprecedented health, economic and geopolitical crisis for the entire world. With accumulating evidence showing that COVID-19 could spread faster in confined settings such as cruise ships (National Institute of Infectious Disease 2020), it became only a matter of when, and not if, the contagion was going to impact on the industry.

Given the foregoing, this chapter determines and evaluates the impact of the COVID-19 pandemic on the global cruise ship industry. This is important as it helps to bring to the forefront plausible strategic interventions to minimise the spread and impact of similar future infections on the industry given its increasing popularity.

6.2 Literature Survey

The cruise ship industry has witnessed dynamic growth over the past three decades in all areas of the world (BREA 2018). This extraordinary growth can be attributed to the industry's development of new and diverse products, including new ships, itineraries and ports well embraced by their clientele (González 2018). The success story could also be attributed to the conversion of traditional land-based resort guests into cruise passengers; delivery of a high level of passenger satisfaction, leading to repeat customers; the creation of a business model adaptable to changing market conditions; and efficient cruise line control of competition, operational costs and revenue streams (Micallef 2018). In short, the industry managed to sustain its growth due to its well-organised and managed structures. Brida and Aguirre (2008) observed the positive economic impact of the cruise ship industry to a port of call as emanating from the industry's value chain. This includes the impact from the spending by cruise passengers and crew and shore side staffing by the cruise lines for their headquarters, marketing and tour operations. It also involves expenditures by the cruise lines for goods and services necessary for cruise operations, spending by the cruise lines for port services, and expenditures by cruise lines for maintenance. There is also the contribution to the fiscus. The next subsection draws the reader's attention to further details regarding the economic impact of the cruise ship industry.

6.2.1 Economic Importance of the Cruise Ship Industry

The economic impact of the global cruise ship industry can be quantified through the number of passengers that it ferries, the number of deployments around the world, the number of ports of call and the total number of those employed by the industry. As observed by the CLIA, the cruise industry has enjoyed dynamic growth, which saw the increase in passengers from 17.8 million in 2009 to over 30 million in 2019 (CLIA 2019). Figure 6.1 shows the global increase in passengers of cruise ships from 2009 to 2019. The strong annual increase proves the popularity of this segment of tourism in the world and as it grew, so too did the direct and indirect economic impacts of the industry.

The capacity deployed by the cruise ship industry, as measured by bed days, rose from 125.8 million bed days in 2012 to 170.6 million in 2017 (BREA 2018). Figure 6.2 shows the cruise ship deployments around the world. The Caribbean was the most popular cruise destination for passengers sourced from North America, with 59.3 million bed days deployed in 2017 – translating to 34.4% of the overall market. These figures show the importance of this industry in the tourism mix of the receiving Caribbean countries.

From Fig. 6.2, it emerges that the cruise industry is indeed a worldwide industry with passengers coming from all over the world and also cruise itineraries destined for countries and ports around the globe. The dynamism of the industry can be

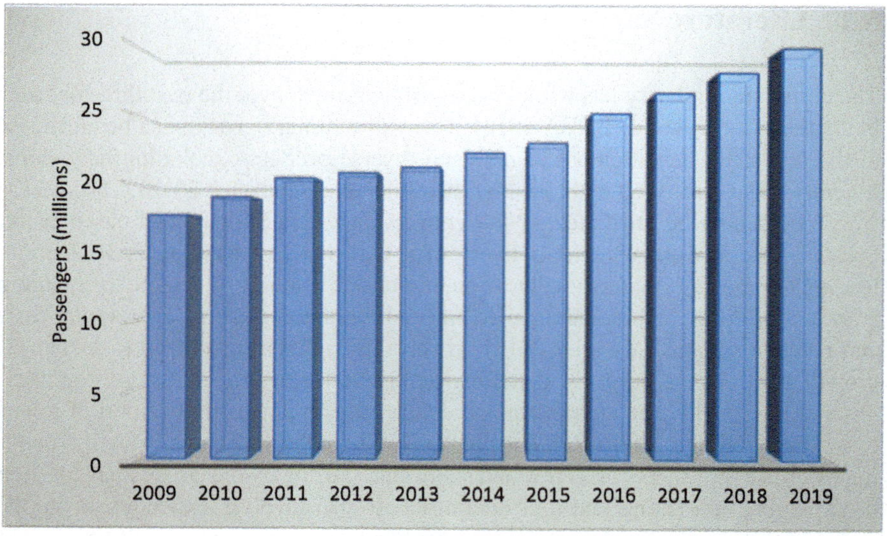

Fig. 6.1 Global number of passengers from cruise ships
Source: Authors, data from CLIA (2019)

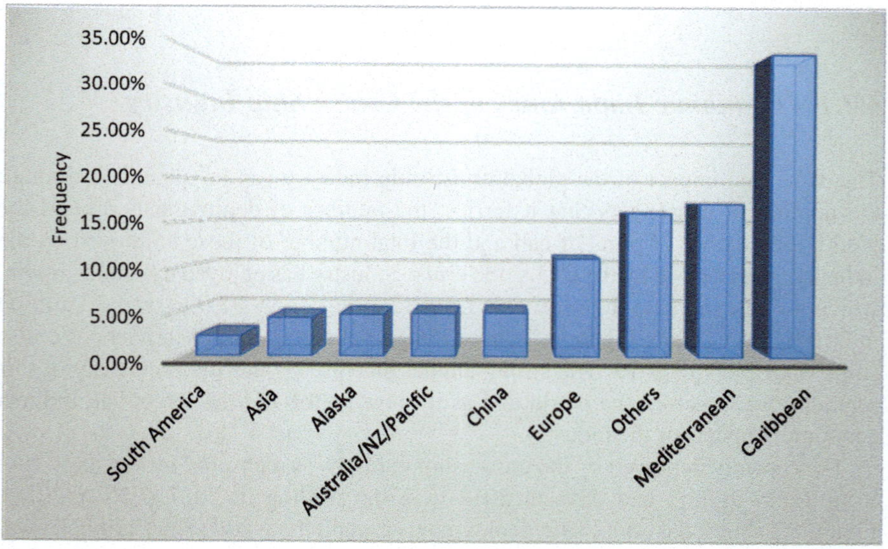

Fig. 6.2 Cruise ship deployment in 2017
Source: Authors, data from BREA (2018)

shown by its 69% global growth rate over the past decade, which by far exceeds that of land-based tourism at 42% over the same period (BREA 2018). The growth of the industry also has a spatial dimension, which saw significant growth in Europe, Asia and Australia. In the end, cruise ship tourism has impacted on the global economy, creating jobs, income and tax revenues in all regions of the world.

Table 6.1 Total global economic contribution of the cruise sector in 2017

Category	2017	2016	% Change
Passenger and crew onshore visits (Million)	136.86	129.38	5.8%
Total direct expenditures (US$ Billion)	$61.02	$57.93	5.3%
Total output contribution (US$ Billion)	$133.96	$125.96	6.3%
Total income contribution (US$ Billion)	$45.57	$41.09	10.9

Source: Authors, CLIA (2019)

Table 6.1 shows that in 2017, about 136.9 million onshore visits by passengers and crew generated $61.02 billion in direct cruise sector spending at source markets and destinations around the globe. This figure also includes the direct expenditures of the cruise lines for goods and services in support of their cruise operations. These payments generated a total global output of $134 billion through direct, indirect and induced transactions related to the cruise ship value chain (CLIA 2019).

Caribbean Tourism (2017) views the Caribbean economies among the world's most tourism dependent. The countries had close to 26.3 million stay-over tourists and 24.5 million cruise ship arrivals in 2014. This was a 5.3% rise from 2013, with trends continuing on an upward trend to 2019. These visitors spent US$29.2 billion, which provided key revenue streams that supported livelihoods and economic development in the area (Caribbean Tourism 2017). The Caribbean is also the principal cruise destination for passengers sourced from North America. Its share of the cruise industry's global deployment has remained relatively constant from 2012 to 2017 at about 35% (BREA 2018). It also remained the largest destination market with 59.3 million bed days deployed in the region during 2017.

The Asian region has continued with the biggest growth margins in the cruise industry, with Japan the most preferred destination (CLIA 2019). The deployed capacity in Asia rose by 18% in 2017 to 17.8 million bed days (BREA 2018). Since 2012, Asia's deployed capacity has risen from 3.2 million to 17.8 million bed days, representing a 453%. Despite the fact that the growth of only 2.4% in 2017, Australia, New Zealand and the Pacific has experienced significant growth since 2012, increasing from 5.5 million to 10.2 million bed days in 2017. This was a rise of 86% (Ibid.). Although passengers of cruise ships were sourced from all over the world, in 2017, North America accounted for 49% of the 26.75 million global cruise passengers that stood at 13.12 million (CLIA 2019). Europe followed with 26% of the total passengers. The rest of the world accounted for the remaining 25% (6.67 million passengers). The rest of the world has good potential for growth in the industry as it accounts for about 82% of the world's population, yet only 25% of world cruisers.

There were thousands of port calls from cruise ships in the USA in 2019. The USA is the major source market. Florida was the leading state with over 3000 port calls (Vera 2020). The distribution of other calls is presented in Fig. 6.3.

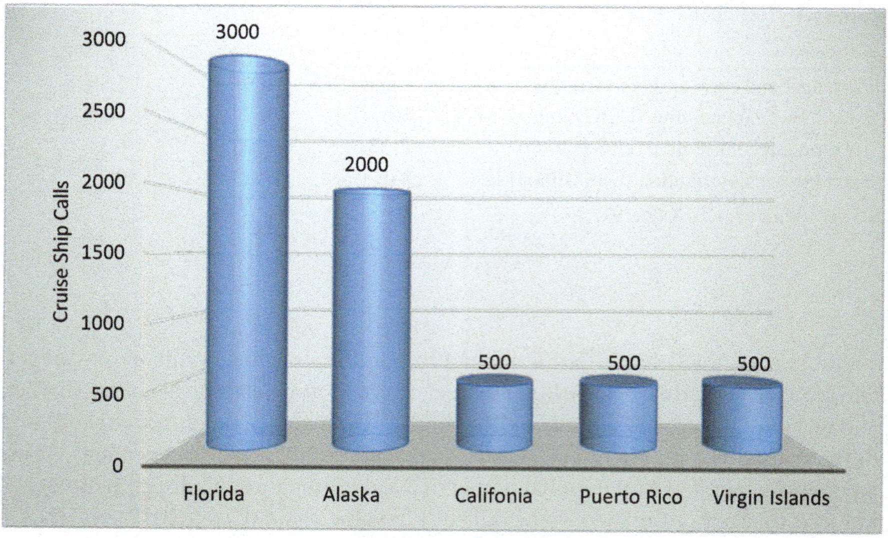

Fig. 6.3 Cruise ship calls for different parts of the USA
Source: Authors, Vera (2020)

Some 7154 port calls were made throughout Asia in 2019 at 306 different destinations (CLIA 2019). This constituted an increase from the 288 different destinations in 2018. Destinations that saw the most calls in 2019 were in Japan, which recorded 2681 calls. The distribution of the other calls is shown in Fig. 6.4. Each of these stops brought with it tourism-related business and employment to the local economy in addition to the docking fees and other related services provided to the industry. Ship chandlers are another significant employer in port areas, as they supply navigation-related parts and equipment as well as food, beverages, linen and other passenger-related supplies (Vera 2020). The more the calls a port receives, the more the people employed directly or indirectly by the industry at the ports of call.

Globally, in 2019, the industry created over 1.11 million jobs, which paid US$45.6 billion in wages and salaries. This makes the industry a very significant employer globally (CLIA 2019). While most of the major cruise lines were incorporated and/or domiciled in foreign countries, most were headquartered in the USA. In addition, most of their clients were from the USA, and so too were the office staff who handle issues of advertising, sales, bookings and vessel planning (Crockett 2020). Miami has the largest contingent of workers employed by the cruise ship industry, with Los Angeles and Washington State as secondary centres of employment. The cruise industry supports more than 421,000 jobs in the USA and annually contributes nearly $53 billion to the economy (Vera 2020).

In South Africa, for example, when a cruise ship carrying an average of 2000 passengers docks, the spending at the destination equates to ZAR2[1] million per day,

[1] The exchange rate averaged US$1 = ZAR19 in April 2020.

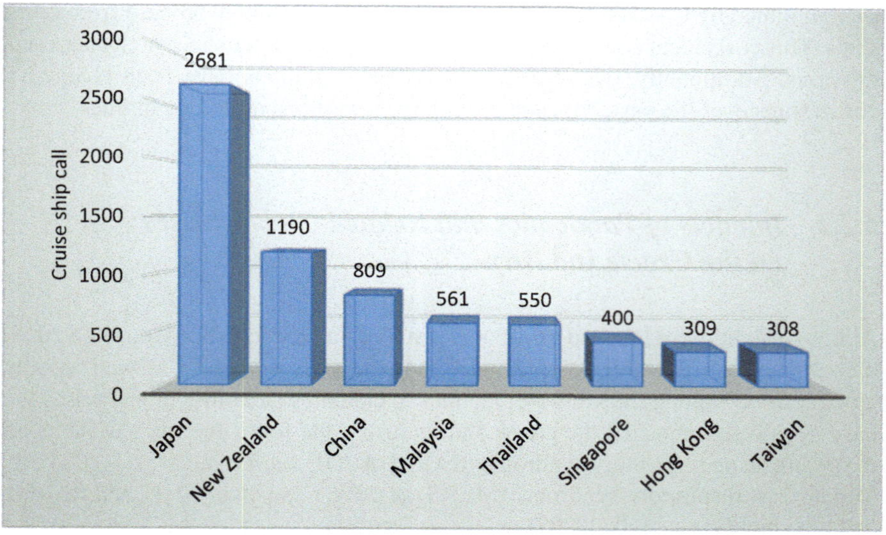

Fig. 6.4 Cruise ship calls in Asia and the Pacific
Source: Authors, data from CLIA (2019)

with tourists spending between ZAR500 and ZAR1,000 each day (Holmes 2020). This excludes the cost of their accommodation. For every 12 tourists visiting the country, one job is created. Hence the drive to attract cruise ship calls at mainly the Durban and Cape Town ports. This saw the industry growing to create demand for expansion of the number of present homeport and port of call facilities in Durban and Cape Town. These offer the most compelling cruise homeport opportunities for cruise lines with others such as Richard's Bay, Port Elizabeth and Mossel Bay being lucrative ports of call candidates for the country. Over the mid to long-term periods, the cruise ship industry capacity growth in South Africa hopes to stimulate expansion into new market regions, such as Walvis Bay in Namibia, Port of Mombasa in Kenya, Beira in Mozambique and Mauritius (South African Tourism 2009).

With continued interest in cruising, all over the world, some local governments continue to invest large amounts of money in high-quality infrastructure to service these massive ships and thousands of passenger arrivals. A question then arose from Brida and Aguirre (2008) querying whether the benefits of attracting cruises to a tourism destination were not higher than the costs of developing and maintaining such infrastructure. Some studies, for example, González (2018) and Brida and Aguirre (2008), noted that the costs of maintaining cruise ship ports were way higher than the revenue realised from arrivals. This was due to competition between destinations and cruise lines. Cruise ships were destinations in themselves, generating almost no impact on the local economy (González 2018). Furthermore, the amount of cruise passenger expenditures was higher than expenditures of normal tourists who have to pay for hotels and meals. Brida and Aguirre (2008) observed that, on average, about US$1690 is spent by a cruiser-based tourist compared to US$1180 spent by a land-based tourist in a week. Other problems the industry faces as highlighted by Crockett (2020) include passengers falling overboard from cruise ships. This costs local and rescue authorities millions of dollars in search and rescue missions. Additionally, over 500 environmental violations have been charged to cruise ships over the past 25 years, costing them millions of dollars in fines.

6.2.2 *Impacts of Pandemics and Related Global Crises on the Cruise Industry*

Although some have labelled cruise tourism as a pandemic itself (González 2018), the industry is vulnerable to epidemics and pandemic outbreaks as well as other global crises. Among the known epidemics, pandemics and other global crises that have negatively impacted the cruise ship industry, the following may be included: the severe acute respiratory syndrome (SARS) of 2003 (Perez and Mckay 2003), the Middle East respiratory syndrome (MERS) of 2016 (Cho et al. 2016) and the 2008 financial meltdown (Belke 2009).

As highlighted earlier, the biggest problem of cruise ships is the high possibility of disease outbreak during voyages, and the global nature of the industry makes it a possible agent of spreading diseases the world over. This is a history that has repeated itself over the years. Crockett (2020) noted that there had been over 353 gastrointestinal or norovirus outbreaks on cruise ships since 2009. The year 2019 alone, the US Centers for Disease Control and Prevention (CDC) reported 10 outbreaks on cruise ships. Stock et al. (2015) observed that for centuries, ships had transmitted infections or vectors that cause diseases with norovirus being the leading cause of illness on cruise ships. It accounts for over 90% of infections on board and one of the leading causes of diarrheal outbreaks on cruise ships (CDC 2014). This makes disease spread from cruise ships a two-way matter, with great worries of onboard passengers taking the diseases to land.

Due to closed living quarters, shared dining areas as well as large passenger turnovers, controlling norovirus outbreaks on cruise ships can be challenging (Stock et al. 2015). Norovirus can be brought on board when ships dock and its ability to persist on surfaces means that it can infect passengers on consecutive cruises (CDC 2014). The most common mode of transmission in cruise ships is person-to-person contact. However, disease spread through contaminated food or water and contact with contaminated surfaces can also occur. Reoccurrence of infections on successive cruises leads to new, susceptible passengers boarding at different ports along the voyage getting infected, thereby prolonging the outbreaks and transferring such to several other cruises (Stock et al. 2015). The disease transmission is exacerbated because cruise ships move from country to country, where different standards of hygiene exit, with the quality and levels of disease surveillance practices varying (Hadjichristodoulou et al. 2013). The average cruise lasts 7 days, a common period within disease incubation and transmission of many infectious diseases. After their cruise vacation, travellers disperse and may further spread acquired infections after leaving the cruise ship (Miller et al. 2000). Approximately a third of cruise ship passengers are senior citizens above 70 years. These elderly cruisers and other persons with underlying comorbidities are at increased risk of complications from infection by norovirus or other common diseases on cruise ships such as influenza (Stock et al. 2015). Since cruise ships are a popular travel destination for the elderly who are at the highest risk of complications, early recognition of outbreaks has been observed to minimise respiratory complications (Hadjichristodoulou et al. 2013). However, outbreaks that occur in closed and crowded settings and have multiple routes of transmission have proven particularly challenging to identify and control on cruise lines (Kuusi et al. 2002).

A lot has been done to try and contain outbreaks on ships and also to prevent them from spreading diseases to land. International biosafety and health regulations require cruise operators to report the presence of any notifiable and contagious diseases on their ship before they enter any port (Stock et al. 2015). Responsibility has been put on ship owners to make these reports and keep their crews updated on these regulations, especially quarantine laws and the international health regulations, which are regularly updated to make them more responsive to current and emerging infectious diseases (Ferson and Ressler 2005). These regulations come with enhanced reporting and response requirements for port health authorities, cruise ship companies and other stakeholders in the event of an outbreak (Stock et al. 2015). There must also be the availability of medical assistance on board, including the capacity for medical consultation with a physician, well-equipped medical facilities and enhanced systems of disease surveillance (Hung et al. 2018).

However, each time disease outbreaks have occurred on cruise ships, there have been concerns regarding the level of preparedness for such occurrences given that repeatedly cruise companies have been caught seemingly unprepared (Stock et al. 2015). In situations where quarantining was deemed necessary, there was always reported poor coordination between responding government agencies and cruise ship companies, as well as inadequate support systems for the infected in the event of a pandemic (Hung et al. 2018). Jamal and Budke (2020) highlighted that given past experiences with previous outbreaks, hospitality-related stakeholders must be better prepared. Such preparedness was expected during the COVID-19 outbreak. This preparedness should have involved proactive, coordinated crisis response and management planning, which unfortunately was still poor in the sector.

In the past, the outbreak of diseases has been met with a corresponding decline in the amount of cruise ship passengers. This is because high and increasing levels of visitor arrivals increase the potential risk for visitors and locals transmitting or acquiring diseases from each other. In the Caribbean during SARS, the H1N1 pandemic of 2009, Chikungunya in 2013 and Zika outbreaks, there was significant cancellation of bookings and declines in arrivals by both air and cruise ships (Caribbean Public Health Agency 2017). The Zika virus outbreak led to more than 80% losses in the Caribbean tourism market in 2015, including the cruise ship segment. This was attributed to heightened concern, alerts and specific measures for travellers from regional and international health agencies like the WHO and the CDC. The consequences of reduced tourist numbers for the millions of people whose livelihoods depended on the sector were serious, contributing to the adverse economic and social burden on the Caribbean governments and potentially damaging the tourism destination reputation of these countries (Caribbean Public Health Agency 2017).

During the SARS outbreak, the heaviest impact on the industry was on Asian cruises. This was because they were located at the epicentre of the disease. There were massive cancellations of bookings by passengers who had earmarked the region, with some companies cancelling or reconfiguring their itineraries to other regions of the world (Perez and Mckay 2003). The largest operators went further by freezing the hiring of Asian workers during the outbreak and also discouraging Asian crew already on board from going back home when on leave. The workers were told that they risked not being allowed back to work if they went home to their affected countries. Some cruise companies refused to board passengers who had travelled to China, Hong Kong and Singapore during the previous 10 days and all those who had come into close contact with SARs patients and SARS health providers. Reduced demand resulted in cruise companies reducing prices and also offering discounts in order to try and entice passengers on board (Perez and Mckay 2003). Effectively, these measures dented the bottom lines of the cruise ship and related industries in the tourism value chain. The next section considers the methodology applied in generating and analysing data for the chapter.

6.3 Materials and Methods

This chapter sought to document the impacts of the COVID-19 pandemic on the global cruise ship industry. The objectives were to identify and evaluate the impacts of COVID-19 on the cruise ship industry and bring to the forefront the need to devise strategic interventions to minimise the impacts in the future should similar occurrences happen. Figure 6.5 shows the major source markets and deployment of

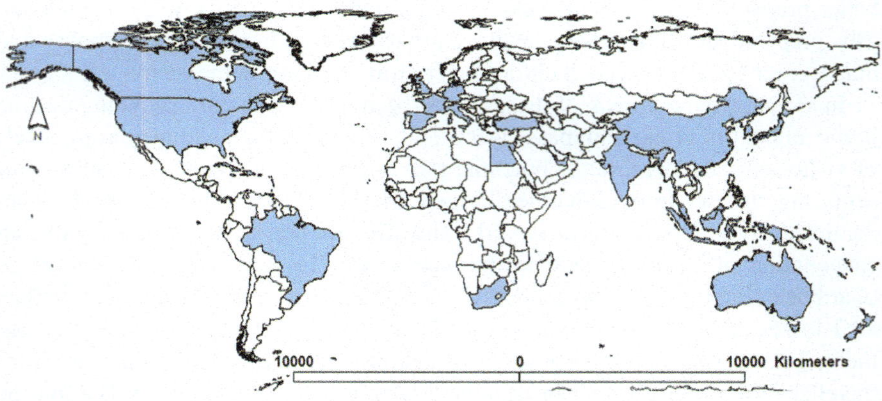

Fig. 6.5 Main cruise ship source markets and deployments – 2019
Source: Authors

cruise ships around the world in 2019. The industry is truly global, impacting on every continent of the world from Australia, through to Asia, Africa, Europe and the two Americas.

Document analysis is depicted by Kumar et al. (2020) as a process of systematically arranging, classifying, analysing content and deriving outcomes from documents. Document analysis was performed from data derived from reports, scientific publications and news publications from different authentic media houses. A document was selected for use in the study only if it offered sufficient insights into the objective of the chapter, which were to examine the impacts of COVID-19 outbreak on the cruise ship industry, specifically documenting the costs. The method was chosen because as it amalgamated information from different sources, which were similar in view of the impacts of the outbreak on the cruise ship industry. Data collection and analysis were done in three stages which were electronic database search, gathering and thematising the obtained content and then the final write up and design of graphics based on the derived themes.

6.4 Presentation of Data and Discussion of Findings

As the COVID-19 outbreak was peaking in March and April 2020, there were massive cancellations of cruise ship bookings by the guests. Eventually, the entire industry voluntarily cancelled planned voyages. This was mainly due to measures put in place worldwide to limit the spread of COVID-19, the outbreak of the disease on some of the cruise ships and also advisories from different national health authorities against travelling on cruise ships. A combination of these events brought the cruise ship industry to a standstill, negatively impacting on their revenue streams. The following subsection highlight some of the impacts COVID-19 imposed on the cruise ship industry.

6.4.1 Economic Cost of COVID-19

The COVID-19 pandemic was a disaster for the whole travel industry but turned out to be a perfect storm for the cruise ship industry. Figure 6.6 shows the drop in share prices of some of the world's biggest cruise liner companies as of the second week of April 2020. The industry was almost decimated by the coronavirus, with share prices for some of the largest players in the industry plunging. Some of these companies include Carnival Cruise, which dropped −63.40%, the Royal Caribbean that fell by −80.80% and the Norwegian Cruise, which dropped −80% (Forbes 2020). These loses signified the worst stock performance on record for the industry. The COVID-19 outbreak, therefore, made the industry lose its appeal to stock buyers and investors, a situation that was bound to continue as long as the coronavirus persisted and vessels could not take to the sea waters.

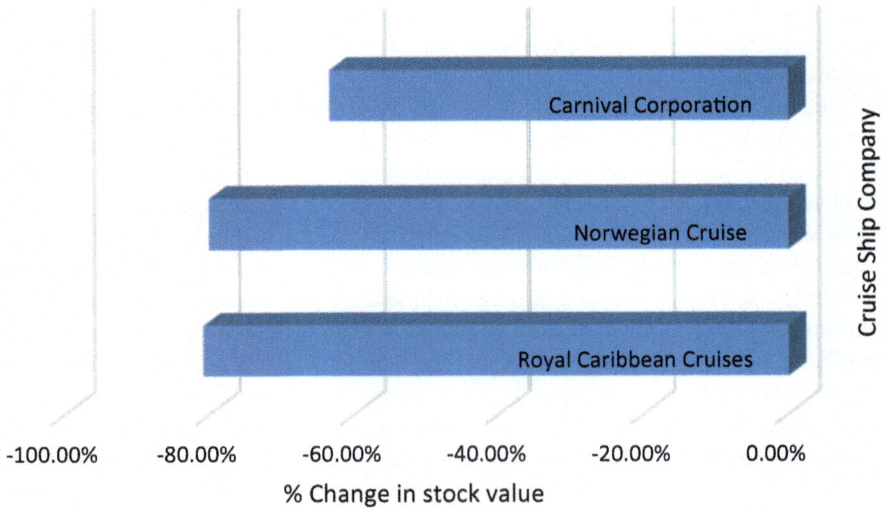

Fig. 6.6 Change in stock value of cruise ship companies
Source: Authors, data from Forbes (2020)

To put it into actual figures, Carnival which owned 122 ships valued at over $55 billion saw its share price plunge from $51.90 on the 17th of January to $8.80 as of the 1st of April 2020 (Swift 2020). This plummeting value of shares had the effect of hampering the raising of money to sustain even the skeletal operations during the period of the pandemic. This free fall in share prices led some like Ebbs (2020) to speculate the beginning of an end for the industry as the bankruptcy was certainly on the table. The cruise ship industry was observed by Bloom (2020) to have gone from being a multibillion dollar industry, with over 30 million passengers per year, to a total standstill industry in just 2 months. As for the Carnival Corporation, the cruise ship shutdown costed the company, which ran 105 ships at least $1billion per month in lost potential revenue (Ebelthite 2020).

Initially, some cruise lines attempted to weather the storm of falling stocks and mass cancellations by trying to sell tickets at all costs, with some even telling customers that COVID-19 did not exist in tropical countries (Crockett 2020). However, the unprecedented fall of the cruise ship companies was made worse by the fact that the United States CDC and other authorities around the world issued no sail orders and advisories to their citizens against going on cruise trips during the COVID-19 pandemic (CDC 2020). This was buttressed by the lockdowns instituted in most countries popular with cruise ships such as China, Caribbean, Spain, Italy, Australia and South Africa. This led to all cruise ship companies cancelling their planned voyages till the pandemic was put under control. The questions that many would ask is if the cruise companies would make it to the other end of the COVID-19 crisis and for how long they could afford to stay in business without any form of revenue as the industry had been brought to a complete stop. Hence with no bailout package for them, most of these companies faced an uncertain future (Ebbs 2020).

To make matters worse, the reputational damage done by COVID-19 was certainly going to make it difficult for the industry to get back on its feet. Cruise companies were being accused of putting the lives of their passengers at risk for monetary gain. Hence, continued operations even when it was clear that COVID-19 was spreading globally (Crockett 2020). Cruise lines also endangered the lives of passengers and crews by refusing to cancel their itineraries and refunding passengers at the earlier stages of the crisis. Instead, the companies continued to cram passengers on to capacity packed ships under conditions, which some Japanese health experts called "coronavirus mills". Others further classified cruise ships as "floating petri dishes" because of the proficiency with which they managed to spread the illnesses (Ebbs 2020).

Decisions to continue operations amid the COVID-19 pandemic left the industry, trapped in unprecedented nightmare scenarios all over the world. At the time of writing most of the cruise companies, for example, Carnival Cruises had up to 6000 passengers stranded at sea due to the coronavirus incidences on board the ships (Woodyard and Hines 2020). This was so because, at one point, no country was authorising cruise ships to dock at their ports fearing the spiking incidences of COVID-19 in the host communities. The Diamond Princess Cruise ship saw 542 of its 3500 passengers falling ill to the coronavirus, leading to the passengers being quarantined in the ship. At the time of writing, another ship, the Coral Princess, was heading towards Florida in the USA with an unknown number of sick passengers. The same applied to the Holland America Zaandam, which had four fatalities on board and had requested entry into Florida (McCormick et al. 2020). Fortunately, the officials in Florida agreed but only to accept Florida residents from the infected ships (Ebbs 2020). The other passengers and crew on the ship were unfortunately left to fend for themselves, a matter that may have led to further preventable deaths.

The logic of the Florida Authorities was that having worked very hard to make sure that they had adequate hospital space in the event of a COVID-19 surge, they were not prepared to release the beds to foreign-owned cruise ships. The US Coast Guard further directed cruise ships with infected passengers to stay offshore indefinitely and even instructed foreign-flagged vessels to attempt the evacuation of passengers through their countries of registration (Ebbs 2020). The affected countries included Liberia, Panama and Bahamas. Memories of such incident are likely to scare future tourists causing them to avoid sea cruises as part of their holiday and leisure activities. These scenarios also led to some questioning the future of cruise ships post COVID-19 (Ebbs 2020).

To add insult to injury, major cruise ship companies were not included in many economic stimuli, and bailout packages rolled out by countries to support their productive sectors. This was as a result of the cruise ships companies avoiding paying taxes in the past, registering in tax haven countries such as Bermuda, Liberia, Panama and the Bahamas (Singh 2020). They also registered offshore and as such would not conform to the USA and European labour laws. To this end, cruise ship companies employed more workers from the developing countries, overworked them and pay them less (Bloom 2020). These foreign-registered companies, for example, were not included in the USA's $2 trillion stimulus package. With the chances of getting bailouts in the USA close to zero. Over the last 3 years, some of the cruise ship companies had been performing very well and profiteering, with the Royal Caribbean having paid out US$1.8 billion in dividends and having spent US$900 million on buybacks (Singh 2020). Some would argue that the companies had enough rainy day funds to survive the COVID-19 downturn. However, their management decided to spend it on wrong priorities, hence being seen to be bailing out the rich would not look very good for the different governments. Given the coronavirus driven shutdown of the industry, which made it impossible for to obtain revenue and service their debt, the poor prospects of a bailout in sight, bankruptcy looked inescapable for most players in the industry (Singh 2020).

Bloom (2020) further observed that no one was eager to bail out the cruise ship industry because of the negative impacts that they impose on their destinations and the environment. The industry is noted to contribute little if anything to the local economy of the destinations they visit. They are known for depositing thousands of crowding tourists who spend very little. These tourists further move around carrying packed lunches from their "floating hotel", after which they return for dinner and sleeping. This arrangement was designed to leave as little as possible at their destinations and bring as much as possible to the cruise liners. Cruise lines generally pay a small head tax that ranges from $4 to $15 per passenger to call on a port. However, as observed by Crockett (2020), most of these countries spend more on maintaining facilities for cruise ships than they make through the fees. Further, they promise to boost the local economies where they call, but the cruise companies are noted to take up to 70% of the onshore revenue with studies indicating that local populations at foreign ports do not get much out of such initiatives. Hence some argue that they were little missed if at all at their usual destinations during and after the COVID-19 pandemic, and they also had few friends in high places hence not included in rescue packages.

6.4.2 Impact on Employment

Employment remains a common thread through the cruise ship tourism segment value chain. Although most European, American and Australian nationals have been employed in the onshore business segment of the cruise ship industry, the offshore segment (on ships), foreign workers, fill most positions. Almost all ocean-going cruise ships calling at US ports are foreign owned (Frittelli 2020). The COVID-19 pandemic saw a loss in the employment of these foreign workers, most of whom were from the Philippines, Thailand, Indonesia and India. While passengers were evacuated during the COVID-19 outbreaks, most of the crew members remained on board. It is reported the more than 90,000 cruise crew members all over the world were left stranded on the cruise ships – at times without pay – because their schedules had been cancelled. This was concerning because just like their passengers, they also suffered due to the spread of the virus, especially those who were not getting paid anymore. However, for some time, cruise line workers were not allowed off the cruise ships due to stringent repatriation conditions put in place by health authorities from different countries (Woodyard and Hines 2020).

The industry therefore still had an ongoing obligation for the welfare, care and safety of its employees during such periods. However, as Woodyard and Hines (2020) observed, the obligation was becoming complicated and harder to fulfil. Some of the workers – like those who work in ship casinos, entertainers, performers and waiters – automatically become classified as nonessential crew members as soon as there are no passengers aboard. As such, they immediately ceased getting paid. To mitigate this scenario, these workers continued to stay on board, off the payroll but received free lodging, meals, medical care and Internet services to maintain contact with loved ones. The repatriation of cruise ship employees was further complicated by the CDC, and many health authorities around the world that recommended that anyone returning home, whether sick or fit, had to get into chartered or private aircraft. This meant that cruise ship companies needed to charter planes for their employees, instead of simply issuing airline tickets for them (Woodyard and Hines 2020). A further complication was that by the time employees got on shore, and most places had been placed under lockdown, and the movement of people was strictly monitored. It is no wonder why most of the cruise ships ordered their employees to remain on board, while they worked out crew repatriation plans. This traumatic experience is likely to linger in the minds of many who were trapped.

For those who remained at work as part of the core and skeletal cruise ship staff for maintenance and securing the cruise ships during their idle period, most had to take substantial salary cuts and observe reduced working hours. The Norwegian, for example, moved its corporate employees to a 4-day working week. It also implemented a pay cut of up to 20% for the shore side employees, with those being paid hourly no longer allowed to work overtime (Peterson et al. 2020). The Royal Caribbean cruise company laid off and/or furlough over 26% of its workforce in the USA. This translated to over 1300 of their 5000 employees there (Reuters 2020). The industry therefore significantly shed both the onshore and offshore employees in order to limit their financial commitments during the COVID-19 downturn.

6.4.3 Emergence of COVID-19 Cruise Ships Lawsuits

In addition to the bad publicity the cruise ship companies got during tours in the midst of the COVID-19 pandemic, the industry has had to contend with potentially costly lawsuits and criminal investigations from all over the world. As each day passed, more lawsuits were filed across the world. In April 2020, Australia launched a criminal investigation into the Ruby Princess cruise ship to establish how the ship managed to disembark 2700 passengers in Sydney (Boseley 2020). About 600 of the passengers later tested positive for COVID-19, ten of whom died. The investigation aimed to establish if the country's biosecurity laws had been broken in the process and if so, to punish those liable. By law, vessels were allowed to dock and disembark passengers only if the captain of the ships assured the local authorities that their ship was free from contagious disease. There was, therefore, a noted discrepancy on what the law said and the information that was provided by the shipping company (Ibid.). In the USA, Florida's Attorney General announced an investigation into whether sales representatives from cruise companies spread misinformation and/or sought to downplay concerns about COVID-19 from their customers. In some instances, there were allegations that emails were sent informing concerned customers that the coronavirus could only survive in cold temperatures, as such, the Caribbean as a cruise destination was not affected by the virus (Crockett 2020).

Passengers who were infected and/or affected by COVID-19 on the cruise ships also launched individual lawsuits against the companies. They argued that their cruises should never have set sail given that by March 2020 the global cruise industry was well aware of the two Princess cruise ships that experienced massive outbreaks of the virus with numerous deaths (Davies and Butler 2020). The key question in Australia, for example, is why the Ruby Princess chose to run a cruise given that the previous cruise had 158 instances of sickness with the ship turning around in a few hours. Hence, the need for answers if it had been sanitised from top to bottom as required. Furthermore, if the company had disclosed anything about the sickness during the previous trip, would the passengers have risked their lives by boarding the ship? The passengers are therefore seeking millions from cruise companies through damages for distress caused by being infected with COVID-19 and living under the threat of being infected (Davies and Butler 2020). In California, a group of former passengers accused Princess Cruises of negligence in its response to the COVID-19 outbreak on the Grand Princess. As such, they have filed several lawsuits against it. They argue that the company chose to place profits before the safety of its passengers, crew and the general public. To this end, the company may have acted negligently by proceeding with the cruise, yet it knew the risk posed to the cruise ship if an infected person was to get on board. Plaintiffs, in this case, are asking for damages of up to US$5million each and also intend to hold the company responsible for implementing new safety measures on its ships (Webeck 2020).

However, as highlighted by Kelley (2020), old laws could potentially shield the cruise industry from huge payouts in the COVID-19 lawsuits. For example, the High Seas Act of the USA, which was passed in 1920 and has been amended only once since then, prevents survivors from suing for emotional distress or suffering even when one loses their loved ones at sea. Other laws also limit the liability of ship companies to the value of their vessel. Cruise ships, just like many industries, are legally not obliged to ensure passenger safety even if they are required to take reasonable steps to keep their passengers from harm. As such, what is reasonable can be debatable in different circumstances. Kelley (2020) also observed the ticket contracts of cruise lines to be "legal works of art". Included in the terms and conditions of the tickets is a waiver to the right of class-action lawsuits against the companies, and emotional distress is not considered an injury. Judging from history, one may reasonably foresee very limited legal recourse against the cruise industry. This is worrying given that Western governments have very limited oversight of this sector. This is mainly due to the fact that most cruise ships are foreign registered and do not have to comply with the stringent safety regulations of the European Union, Japan, Australia and the USA. The challenges emerging herein are those for which the corporate world prepared a legal route out years ago. The risks are well calculated, with the profit margins secured in most instances.

6.4.4 The Human Cost

The COVID-19 pandemic also came with traumatic experiences for both the passengers and the workers of the cruise lines. In February 2020, the cruise ship, Diamond Princess, carrying about 3700 passengers, temporarily became the largest cluster of COVID-19 cases outside China (Mole 2020). The cruise ship docked at the Yokohama Bay, Japan, on 3 February 2020. The passengers were subjected to quarantine for 2 weeks. In the process, at least 705 passengers tested positive for the virus, including about 189 who were asymptomatic and who had initially tested negative (Inui et al. 2020). The total included infections among passengers and crew members, plus one case of a Japanese quarantine officer working on the vessel and four fatalities (Mole 2020). The ship becomes an occupational hazard for the cruise workers as they were required to keep working on board the vessel, despite the risk of infection, to ensure the smooth running of the cruise. As such, many of them – especially those who delivered food to the cabins – became infected (Inui et al. 2020).

During the quarantine period, concerns were raised as to whether quarantining both the sick and in the healthy on one ship was the best thing to do. The other issues were quarantining people on cruise ships, which, by the set-up, could lead to the acceleration of the infections on board. By default, there is limited sanitation on board, as well as restricted water and food supplies (Sawano et al. 2020). Furthermore, given that the majority of passengers were the elderly and vulnerable, issues emerged surrounding how best to care for them. There were more than 200 such elderly over 80 years on Diamond Princess (Mole 2020). There were also those with pre-existing medical conditions who were at bigger risk not just from COVID-19 but also from the physical and psychological stress that could make their existing ailments worse. With increased international criticism on the poor management of those on board, the Japanese government eventually allowed some elderly passengers to leave the ship on 14 February 2020 before the end of the quarantine period on 19 February 2020. After the quarantine period lapsed, 36 severe cases of COVID-19 were reported among the passengers, requiring treatment in land-based intensive care units. However, those fit to travel were repatriated to their countries, where again they were subjected to another 14 days or more of either self-quarantine or other regulated quarantines under lockdowns (Sawano et al. 2020).

Another cruise ship, the Dutch liner, Westerdam, sailed out of Hong Kong on 1 February 2020. The ship had on board 1455 passengers, including 91 from the Netherlands. It was turned away from ports in the Philippines, Taiwan, Korea, Japan, Thailand and the USA territory of Guam. This was in spite of the fact that there were no confirmed cases on board (Pieters 2020). The ship was finally allowed to dock in Sihanoukville, Cambodia, after 13 days at sea. Fearmongering on infection by people on board cruise ships, which had been likened to "floating petri dishes" of the coronavirus, had been heightened the world over by the Diamond Princess saga in Japan. After this traumatic experience, the passengers were able to disembark within a few days after being subjected to medical checks by the Cambodian authorities (Pieters 2020). There were also increasing reports of workers on cruise ships being shunned and harassed by a fearful public because of their occupation and being perceived to have the coronavirus when they returned home (Koh 2020). These remain the real psychosocial matters that will remain in the tourism, and particularly the cruise industry, for a long.

The problem of quarantining people on cruise ships was made worse by the fact that most modern liners have relatively small cabins that are in sync with the industry's economic model. The economic model is hell-bent on getting as many passengers as possible to spend money in the ship's restaurants, shops, bars and spas (Bloom 2020). In the end, thousands of innocent cruise ship guests had to endure being marooned on board in cruise ship cabins that were beset by outbreaks of the coronavirus. Even those who were free of the virus were shunned by normally welcoming ports of call around the world. The CDC (2020) *listed* 21 ships in which travellers tested positive for the coronavirus, either while aboard or after their cruises. Table 6.2 shows some of the ships that had COVID-19 incidences and/or were turned away from their scheduled port of call during their voyages.

Bloom (2020) observed that the cruise ship industry had always been prepared to deal with illnesses on board since they all had resident doctors, nurses and also clinics. However, in normal periods, when a person felt sick, the plan would be to stabilise the patient, get to the nearest port, get everyone off and then sanitise the ship and continue with the voyage. This standard practice was, however, not applicable during the COVID-19 outbreak as ports turned the ships away. The cruise lines were left to fend for their sick guests, with limited resources on board. It can be argued that the industry was caught off guard and unprepared for the impacts of the pandemic, hence imposing a huge toll on its passengers. The health toll of the COVID-19 pandemic on the passengers of cruise ships was higher compared to other forms of transport and most probably will take long periods of time for the psychological trauma to heal. The untold suffering of these innocent cruise line guests was exacerbated mostly by misinformation, panic and fear among the authorities in different countries who denied cruise ships docking even for scheduled stops and when there were no confirmed cases of COVID-19. Where there were emergency cases that necessitated the ill passengers to be evacuated from the ships, they were (in many cases) not allowed to disembark from the ships and left to fend for themselves leading to preventable deaths. The psychological wounds from such experiences meant that cruise ship companies had to do more to ensure that their former passengers had access to psychological counselling where possible. This bad publicity of the industry will be difficult to shake off in the post-COVID-19 period. It will likely suffer prolonged negative impacts more than other segments of the leisure travel and tourism industry.

Table 6.2 Selected cruise ships that had COVID-19 incidences and/or were turned away from ports of call

Cruise ship	Passengers and crew	Receiving/rejecting port
The Grand Princess	2422 passengers on board	Oakland
	At least 21 out of 46 coronavirus tests administered on board were positive	
MSC Meraviglia	4500 passengers and 1600 crew members	Rejected by Jamaican and the Cayman Islands ports
		Finally allowed to dock in Cozumel, Mexico, before being moved to Miami
Costa Fortuna (Italy)	Nobody on board tested positive for COVID-19	Denied permission to dock in Thailand and Malaysia
		Allowed to dock and disembarked in Singapore
A Sara (Nile River)	33 passengers tested positive for COVID-19	Luxor
MS Braemar	At least five of the over 600 passengers aboard the cruise tested positive for COVID-19	Denied docking in multiple Caribbean ports
	25 guests and 27 crew members showed flu-like symptoms	Anchored 8 km off the coast of Cuba; only docked after charter flights arrived
The Golden Princess	2600 passengers and 1100 crew	New Zealand, not allowed to disembark
	Three passengers quarantined	
Pacific Princess	No one tested positive for COVID-19	Having difficulty finding a place to dock; all planned stops refused it entry, including Australia
	Was still on a global voyage for 111 days	
Greg Mortimer (Antarctic cruise)	132 passengers and 85 crew	Anchored 20 km off Uruguay; not allowed to dock
	128 of them tested positive for COVID-19	
	Ship's doctor fell ill with a fever and was left unable to perform his duties	
AIDAmira	1720 passengers and crew	Cape Town
	No one tested positive for COVID-19, but passengers not allowed to disembark	

Source: Authors, based on Chang (2020)

6.5 A Way Forward

History shows that tourists have short memories when it comes to bad experiences. Good examples include the Paris and Brussels terror attacks, Bali bombings and the southeast Asian tsunami. All these had about 3 months of a fall in tourist numbers before going back to normal (Bloom 2020). As soon as travel restrictions are removed, the industry needs to launch massive advertising campaigns and significantly reduce prices to lure customers back. However, the irony is that the ships need to be full in order to make profits, and discounting may impact on their capacity to break even. The good thing is that oil prices, one of the major fixed costs of the industry, have also been going down during the COVID-19-induced downturn, and cruise companies need to take advantage by buying fuel futures.

Munarriz (2020) observed that it would take several months after gaining control of COVID-19 for these cruise ships to be sailing again and start generating revenue. It will take at least a couple of years before customer interest in cruise vacations returns with the same appetite as before. Furthermore, when one factor in the inevitable world economic recession, which will follow the COVID-19 outbreak, the situation looks very gloomy for the industry. In a survey done by the Carnival Cruise Company, 45% of the guests contacting the company on cancelled sailings were taking up sweetened future cruise packages, with 55% of the displaced customers taking refunds (Munarriz 2020). For Norwegian cruise line passengers, about 50% were opting for compensation in the form of future cruise credit rather than a straightforward refund. This shows that despite the bad publicity and incidences of COVID-19 outbreaks on cruise ships, the industry still had loyal clients who could lift it up as soon as things normalised. The industry has also seen a surge in demand for the 2021 booking of cruises earmarked for all regions of the globe. This is a good sign of the resilience of the sector and its popularity with tourists. To show investor confidence in the future of the sector, during the COVID-19 low in April 2020, Saudi Arabia's Investment Fund purchased an 8.2% stake in Carnival. The investment amounted to $400 million and signalled investor confidence in the sector in the future (Ebelthite 2020) and also signals Saudi Arabia's endeavour to become a cruise ship destination.

Given that diseases can easily spread in the confined and crowded spaces of cruise ships, the industry needs to improve its state of medical preparedness so that it relies less on ground-based facilities in the case of emergencies and future outbreaks. It cannot be business as usual in the future. Cruise ships need to be better equipped to handle more severe medical conditions on board in the event that they are denied medical help from land, as was often the case during the COVID-19 pandemic. They also need to improve the sanitation procedures on cruise ships in order to enable them to mitigate the impacts of fast-spreading diseases like COVID-19. Guests need to be educated on the importance of good hygiene, with crew members assisting in making sure that their guests sanitise before visiting public areas on the ship. There is also a need for the industry to work on better evacuation and quarantine procedures in the event of future outbreaks.

6.6 Conclusion

The cruise ship industry has been heavily impacted by the COVID-19 pandemic. This saw the industry grinding to a halt and cancelling almost all cruises across the world and losing billions of dollars in potential revenue in the process. Before the

COVID-19 induced meltdown, the industry had witnessed sustained dynamic growth from the year 2009 and became one of the fastest-growing segments of the tourism sector. At its peak in 2019, the industry carried over 30 million passengers and created over 1.11 million jobs, which paid salaries over US$45.6 billion and was in total worth over US$134 billion. All these economic benefits were annihilated during the peak of the COVID-19 pandemic. The pandemic further witnessed stock values of most cruising companies dropping by as much as 80% within 2 months. This signalled the worst ever performance by these companies on stock markets.

Besides the economic costs, the COVID-19 pandemic also came at an unprecedented cost to the passengers and crews of cruise liners. They had to endure psychological trauma when their cruise liners were denied docking rights at various ports around the world, even when there were no coronavirus incidences on board. Many of the passengers were infected by the virus on the ships, and those who were free from the virus had to endure weeks of quarantine on the same ship as those who were ill. Some passengers were denied critical medical care, which was only available on land, when they were not allowed to disembark from the ships, thereby leading to preventable loss of life. Some passengers and crew members died. Such nightmare experiences on the part of the passengers and crews led to the loss of the good reputation the industry had gained over the years and to questions over their social licence to operate. The cruise ship industry is also being hit by lawsuits and criminal investigations due to incidences that occurred during the peak and early days of the COVID-19 pandemic. The cruise companies have been accused of putting money before the welfare of their passengers and crews. As such, there is a chance that biosafety laws of different countries were broken. With all the loss of potential revenue and no rescue package for them on the horizon, the industry faces an uncertain future.

However, there is light at the end of the tunnel, with some regular passengers expressing a willingness to return to the cruise ships once the virus is under control, with some even making future bookings during this period. Investors are also showing faith in the industry by investing funds into it during the COVID-19 pandemic. This confidence booster, coupled with forceful marketing and the offering of discounted tours, will go a long way towards helping the industry bounce back from the COVID-19 disaster. Finally, the United Nations (2020) strongly recommends learning from this crisis. For example, countries like Singapore, Vietnam and Taiwan learnt from their experience of SARS in 2003. This experience incentivised them to act promptly and got their citizens to cooperate with specific measures. This paid off in reducing the spread and incidences of COVID-19. In that spirit, the cruise ship industry needs to improve on its management of public health systems on board in order to reduce passenger vulnerability to infection, should outbreaks similar to COVID-19 recur.

References

Belke, A. (2009). Fiscal stimulus packages and uncertainty in times of crisis: Economic policy for open economies. *Economic Analysis and Policy, 39*(1), 25–46.

Bloom, J. (2020). Will we ever take cruise holidays again? *BBC*. Retrieved from www.bbcnews.com.news/amp/business-52182509. Accessed 9 Apr 2020.

Boseley, M. (2020). Criminal investigation launched into Ruby Princess cruise ship coronavirus disaster. *The Guardian*. Retrieved from www.theguadian.com/world/2020/apr/05/criminal-investigation-launched-into-ruby-princess-cruise-ship-coronavirus-disaster. Accessed 9 Apr 2020.

Brida, J. G., & Aguirre, S. Z. (2008). The impacts of the cruise industry on tourism destinations. Sustainable tourism as a factor of local development. Monza, Italy, 7–9/11/2008.

Business Research and Economic Advisors (BREA). (2018). *The global economic contribution of cruise tourism 2017*. Phillipsburg: BREA.

Caribbean Public Health Agency. (2017). *ZIKA: Impacts on tourism*. Jamaica Boulevard: CARPHA.

Centers for Disease Control and Prevention. (2014). *Surveillance for norovirus outbreaks*. Retrieved from http://www.cdc.gov/features/dsNorovirus/. Accessed 17 Apr 2020.

Centers for Disease Control and Prevention. (2020). *CDC announces modifications and extension of no sail order for all cruise ships*. Retrieved from www.cdc.gov/media/releases/2020/s0409-modifications-extension-no-sail-ships.html. Accessed 9 Apr 2020.

Chang, B. (2020). 21 cruise ships around the world have been affected by the coronavirus so far – Here is a list. *Business Insider US*. Retrieved from www.BusinessInsider.com/cruises-that-have-been-affected-by-coronavirus-2020-3%3famp. Accessed 9 Apr 2020.

Cho, S. Y., Kang, J. M., Ha, Y. E., Park, G. E., Lee, J. Y., Ko, J. H., et al. (2016). MERS-CoV outbreak following a single patient exposure in an emergency room in South Korea: An epidemiological outbreak study. *The Lancet, 388*, 994–1001. https://doi.org/10.1016/S0140-6736(16)30623-7.

Crockett, Z. (2020). The economics of cruise ships. *The Hustle*. Retrieved from https://thehustle.co/the-economics-of-cruise-ships/amp. Accessed 9 Apr 2020.

Cruise Lines International Association (CLIA). (2019). *2019 cruise trends and industry outlook*. Retrieved from https://cruising.org/news-and-research/press-room/2019/april/clia-reveals-growth. Accessed 14 Apr 2020.

Davies, A., & Butler, B. (2020). Ruby Princess: Battle begins to hold someone accountable for cruise ship coronavirus debacle. *The Guardian*. Retrieved from www.theguadian.com/business/2020/apr/10/ruby-princess-battle-begins-to-hold-someone-accountable-for-cruise-ship-coronavirus-debacle. Accessed 14 Apr 2020.

Djalante, R., Shawb, R., & DeWit, A. (2020). Building resilience against biological hazards and pandemics: COVID-19 and its implications for the Sendai Framework. *Progress in Disaster Science, 6*(2020), 100080. https://doi.org/10.1016/j.pdisas.2020.100080.

Ebbs, W. (2020). Cruise stocks may never recover from the Coronavirus. *CNN*. Retrieved from https://www.cnn.com/grand-princess-cruise-ship-stupidity-puts-everyone-in-danger. Accessed 9 Apr 2020.

Ebelthite, S. (2020). Saudi Arabia buys stake in cruise giant Carnival as it seeks to become cruise destination. *Cruise Arabia and Africa*. Retrieved from www.cruisearabiaonline.com/2020/04/07/saudi-arabia-buys-stake-in-cruise-giant-carnival-as-it-seeks-to-become-cruise-destination/amp. Accessed 9 Apr 2020.

European Centre for Disease Prevention and Control. (2010). *Expert forum on communicable disease outbreaks on cruise ships. Stockholm, 28–29 September 2009*. Stockholm: ECDC.

Ferson, M. J., & Ressler, K. A. (2005). Bound for Sydney town: Health surveillance on international cruise vessels visiting the Port of Sydney. *Medical Journal of Australia 182*(8), 391–394.

Forbes. (2020). *Norwegian and Royal Caribbean down 80%. Is either now a good value?* Retrieved from www.forbes.com/sites/greatspeculations/2020/04/06/norwegian-royal-caribbean-down-80-is-either-now-a-good-value-amp/. Accessed 14 Apr 2020.

Frittelli, J. (2020). *COVID-19 and the cruise ship industry*. Washington, DC: Congressional Research Service, IN11245·Version 2·New.

Gan, N., Thomas, N., & Culver, D. (2020). Over 1,700 frontline medics infected with coronavirus in China, presenting new crisis for the government. *CNN*. Retrieved from: https://www.edition.cnn.com/2020/02/13/asia/coronavirus-health-care-workers-infected-intl-hnk/index.html. Accessed 9 Apr 2020.

González, A. T. (2018). Venice: The problem of overtourism and the impact of cruises. *Journal of Regional Research, 42*, 35–51.

Hadjichristodoulou, C., Mouchtouri, V. A., Guglielmetti, P., Lemos, C. M., Nichols, G., Paux, T., Schlaich, C., Cornejo, M. D., Martinez, C. V., Dionisio, M., Rehmet, S., Jaremin, B., & Kremastinou, J. (2013). Actions for prevention and control of health threats related to maritime transport in European Union. *Travel Medicine and Infectious Disease, 11*(4), 238–242.

Holmes, R. (2020). African cities invest to boost cruise ship arrivals. *Skift*. Retrieved from www.skift.com/2020/01/02/african-cities-invest-to-boost-cruise-ship-arrivals/amp. Accessed 9 Apr 2020.

Hung, K. K. C., Mark, C. K. M., & Yeung, M. P. S. (2018). The role of the hotel industry in the response to emerging epidemics: A case study of SARS in 2003 and H1N1 swine flu in 2009 in Hong Kong. *Globalization and Health, 14*(1), 117. https://doi.org/10.1186/s12992-018-0438-6.

Inui, S., Fujikawa, A., Jitsu, M., Kunishima, N., Watanabe, S., Suzuki, Y., Umeda, S., & Uwabe, Y. (2020). Chest CT findings in cases from the cruise ship "Diamond Princess" with Coronavirus Disease 2019 (COVID-19). *Radiology: Cardiothoracic Imaging*, in press.

Jamal, T., & Budke, C. (2020). Tourism in a world with pandemics: Local-global responsibility and action. *Journal of Tourism Futures*. https://doi.org/10.1108/JTF-02-2020-0014.

Kelley, C. (2020). Centuries-old laws may shield the cruise industry from huge payouts in coronavirus suits. *USA Today*. Retrieved from www.ustoday.com/amp/511173002. Assessed 9 Apr 2020.

Koh, D. (2020). Occupational risks for COVID-19 infection. *Occupational Medicine, 70*, 3–5. https://doi.org/10.1093/occmed/kqaa036.

Kumar, A., Singh, N., & Neelu, J. A. (2020). Learning styles based adaptive intelligent tutoring. *International Journal of Cognitive Research in Science, Engineering and Education, 5*(2), 83–98.

Kuusi, M., Nuorti, J. P., Maunula, L., Minh Tran, N. N., Ratia, M., Karlsson, J., & von Bonsdorff, C. H. (2002). A prolonged outbreak of Norwalk-like calicivirus (NLV) gastroenteritis in a rehabilitation centre due to environmental contamination. *Epidemiology and Infection, 129*(1), 133–138.

McCormick, E., Greenfield, P., & Goñi, U. (2020) Revealed: 6000 passengers on cruise ships despite coronavirus crisis. *The Guardian*. Retrieved from https://www.theguardian.com/environment/2020/apr/09/revealed-6000-passengers-on-cruise-ships-at-sea-despite-coronavirus-crisis. Accessed 9 Apr 2020.

Micallef, J. V. (2018). The cruise industry's boom is primed to continue. *Forbes*. Retrieved from www.forbes.com/sites/joemicallef/2018/09/01/the-cruise-industrys-boom-is-primed-to-continue/amp/. Accessed 16 Apr 2020.

Miller, J. M., Tam, T. W. S., Maloney Fukuda, K., Cox, N., Hockin, J., Kertesz, D., Klimov, A., & Cetron, M. (2000). Cruise ships: High-risk passengers and the global spread of new influenza viruses. *Clinical Infectious Diseases, 31*, 433–438.

Mole, B. (2020). 175 now infected with coronavirus on cruise ship, including quarantine officer: Japan is now planning to test everyone aboard. *Arstechnica*. Retrieved from https://arstechnica.com/science/2020/02/175-now-infected-with-coronavirus-on-cruise-ship-including-quarantine-officer. Accessed 9 Apr 2020.

Munarriz, R. (2020). Will carnival, Royal Caribbean, and Norwegian cruise line all survive. *The Motley Fool*. Retrieved from www.themotleyfool.com/investing/2020/04/05/will-carnival-royal-caribbean-and-norwegian-cruise.aspx. Accessed 9 Apr 2020.

National Institute of Infectious Disease. (2020). *National Institute of infectious diseases. Field briefing: Diamond princess COVID-19 cases*. Tokyo: National Institute of Infectious Diseases.

Retrieved from https://www.niid.go.jp/niid/en/2019-ncov-e/9407-covid-dp-fe-01.html. Accessed 20 Apr 2020.

Perez, E. and Mckay, B. (2003). Swab the decks! Cruise lines guard against SARS threat. *Wall Street Journal*. Retrieved from www.wsj.com/articules/SB105164036784562700. Accessed 21 Apr 2020.

Peterson, B., Matousek, M., & Cain, A. (2020). Cruise lines told workers to carry on as normal as the coronavirus spread. Now, many crew members are infected or unemployed. *Business Insider*. Retrieved from www.businessinsider.sg/cruise-line-crew-workers-lose-salaries-benefits-over-coronavirus-2020-4/amp. Accessed 20 Apr 2020.

Pieters, J. (2020). Dutch cruise ship allowed to dock in Cambodia; health ministers gather to discuss coronavirus. *NL Times*. Retrieved from https://nltimes.nl/2020/02/13/dutch-cruise-ship-allowed-to-dock-in-cambodia-health-ministers-gather-to-discuss-coronavirus. Accessed 9 Apr 2020.

Reuters. (2020). Royal Caribbean sheds 26% of U.S. workforce as coronavirus hits travel. *Reuters*. Retrieved from www.reuters.com/articule/amp/idUSKCN21Y02S. Accessed 10 Apr 2020.

Sawano, T., Ozaki, A., Rodriguez-Morales, A. J., Tanimoto, T., & Sah, T. (2020). *Limiting spread of COVID-19 from cruise ships – Lessons to be learnt from Japan*. Oxford: Oxford University Press.

Singh, A. (2020). The U.S. stock market crash will bankrupt these 3 companies. *CCN*. Retrieved from www.ccn.com/the-us-stock-market-crash-will-bankrupt-these-3-companies/. Accessed 10 Apr 2020.

South African Tourism. (2009). *South Africa cruise tourism: Prospects, benefits and strategies* (Final Report November, 2009 Version 5.0, Pretoria).

Stock, D., Becken, S., & Davis, C. (2015). *Impact of Norovirus in the cruise ship industry* (Griffith Institute for tourism research report no 8). Queensland: Griffith University.

Swift, R. (2020). Global cruise ship fleets lose US$4 billion in value as coronavirus pandemic idles industry, halving the number of vessels as sea. *South China Morning Post*. Retrieved from www.amp.scmp.com/business/companies/article/3078347/global-cruise-ship-fleets-lose-us4-billion-in-value-coronavirus. Accessed 10 Apr 2020.

United Nations. (2020). *Shared responsibility, global solidarity: Responding to the socio-economic impacts of COVID-19*. New York: The United Nations Sustainable Development Group.

Vera, A. (2020). Major cruise lines suspending operations at US ports for 30 days over coronavirus pandemic. *CNN*. Retrieved from www.cnn.com/travel/amp/cruise-line-suspensions-coronavirus/index.html. Accessed 9 Apr 2020.

Webeck, E. (2020). Grand Princess Passengers sue cruise line for negligence over COVID-19 outbreak. *The Mercury News*. Retrieved from www.mecurynews.com/2020/04/09/grand-princess-passengers-sue-cruise-line-for-negligence-over-covid-19-outbreak/amp/. Accessed 10 Apr 2020.

Woodyard, C., & Hines, M. (2020). More the 90,000 cruise crewmembers left to battle coronavirus – At times without pay. *USA Today*. Retrieved from www.usatoday.com/amp/2959362001. Accessed 5 Apr 2020.

Chapter 7
Impact of COVID-19 on Global Car Rental Industry and Ride and Share Transport Services

Abstract The tourism sector has numerous intertwined subsectors, with the performance of each critical to the performance of the other. The car rental is critical to the success of the tourism industry, yet it remains under, and thus there is limited knowledge on how the previous pandemic impacted the car rentals industry. This study examines the impact of COVID-19 on the car rental as well as ride and share transport niches of the tourism industry. Making use of secondary and archival data, the study found that the two subsectors lost revenue amounting to billions of dollars in potential revenue during the COVID-19 pandemic. The pandemic resulted in company rating downgrades, market volatility and loss of share value for leading entities such as Hertz, Avis, Europcar, Uber and Lyft. However, the impact of the pandemic on ride and share transport was moderate compared to car rentals. The pandemic led to Hertz filing for bankruptcy, albeit after paying US$16 million in retention bonuses on the eve of bankruptcy to its executives. Losses led to various cost-containment measures, which saw a record number of employees being laid off and disposal of non-core assets. The chapter recommends a robust, coordinated health and safety certification and protocol for the industry to ensure the safety of passengers and employees to boost confidence in the sector as it emerges from the pandemic. Furthermore, a call is made to revisit employment conditions in this sector with the view to provide better protection of employees from the COVID-19 and other disaster-related market shocks. A revision of the current business model is imperative, given the vulnerability of the sector to shock to ensure disaster resilience. Impact assessment post the pandemic will be critical to get a full picture of the impact, particularly on small businesses that were not covered in this study.

Keywords Car rental · Ride and share · Tourism · Coronavirus · COVID-19 · Transport · Furloughing

G. Nhamo et al., *Counting the Cost of COVID-19 on the Global Tourism Industry*,
https://doi.org/10.1007/978-3-030-56231-1_7

7.1 Introduction

Emerging insights have shown how the tourism industry has been and will be, affected by COVID-19 pandemic together with other economic sectors (Dube et al. 2020). In a fast-changing environment, the preliminary evidence shows that the impact of measures put in place to control the spread of the pandemic, such as lockdowns, has had a devastating impact on the global economy. To this effect, the International Monetary Fund (IMF) has since announced that the world has entered into an economic recession (IMF 2020). The IMF paints a gloomy picture for the global economy for 2020 and beyond. The belief is that this recession will surpass, by far, the magnitude and the impact of the Great Depression of 1932, the impact of World War II, the impact of the 2007/2008 economic recession and the economic impact experienced during the severe acute respiratory syndrome (SARS) outbreak (Gössling et al. 2020).

At the beginning of May 2020, the United Nations World Tourism Organization (UNWTO) envisioned that global tourist arrivals were expected to drop by between 60% and 80% in 2020, threatening many livelihoods supported by this industry, including those from the car rental and related spaces. The UNWTO (2020) indicated that the pandemic was reversing the progress that had been made in advancing the achievement of Sustainable Development Goals (SDGs). The drop-in tourist arrivals cost the world close to $80 billion in export revenue in the first 3 months of the year 2020. Data from UNWTO shows that various global tourism regions have been severely affected by the pandemic, albeit variations in impact can be observed across various regions with the least impact noted in the Middle East and Africa (Fig. 7.1).

Figure 7.1 shows that in the first quarter, the worst affected region was Asia and the Pacific. This could be attributed to the fact that the region was the first to record cases of the coronavirus and consequently the first region to take preventative action by instituting travel restrictions and social distancing measures seeking to suppress the spread of the virus. Globally, the impact was also significant as more than 20% of tourist arrivals at most destinations and airports were wiped out by the impact of the pandemic. This ultimately had implications on industries within the tourism value chain. Zheng et al. (2020) note that the outbreak of the pandemic led to the stigmatisation of Chinese tourists as people saw them as carriers of the dreaded coronavirus. Hall et al. (2020) warn about the transformative impact of the socio-economic impact of pandemics on society, which can have far-reaching effects. This is an important observation and measures have to be taken by all tourism players to investigate how the pandemic will affect them. A look into previous disease outbreaks reveals that while studies were done to investigate the impact of previous disease outbreaks such as SARS on the tourism economy, very few had been undertaken on car rental companies, regardless of their interlinkage to many other tourism subsectors. This is regrettable given that the car rental industry is well-established, dating back to 1908 (Zhang et al. 2014.) Car rental companies do not only provide convenience to travellers but are also a significant segment of the tourism economy,

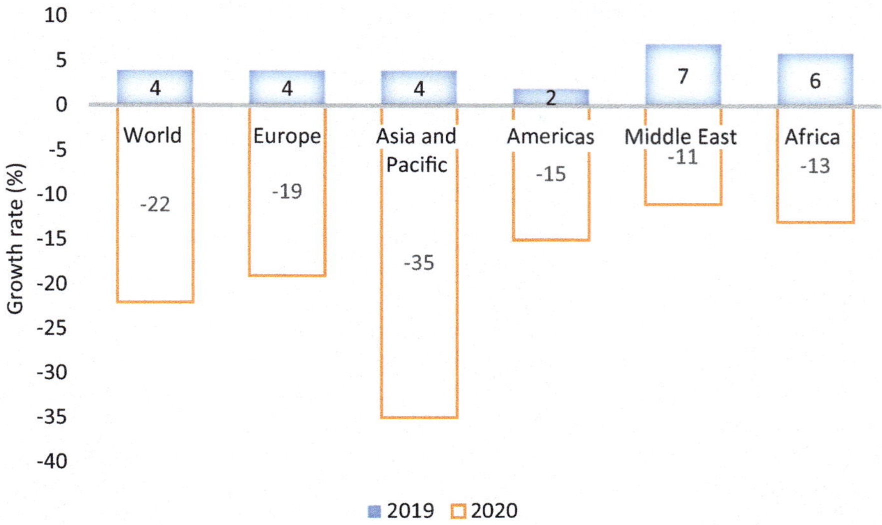

Fig. 7.1 Global tourism growth rate for first quarter 2019 compared to 2020 assessing the impact of COVID-19. (Source: Authors, data from UNWTO 2020)

which generates revenue in both local and international travel. The car rental industry's performance is also closely tied to gross domestic performance (GDP) and passenger arrivals and by proxy air traffic volumes (Russell 1999).

Given that COVID-19 adversely affected national GDPs of many tourism destinations, places and countries, this study investigates its impact on the global travel sector with a focus on car rental companies as well as share and ride services. The study provides the basis for potential policy intervention and stakeholder support. It will also provide a critical reference point for future disease outbreaks.

7.2 A Literature Survey

The transport sector is an essential sector of the tourism economy in that it offers convenience to the travelling public and enables the tourist to better explore their destinations. Airports across the world, towns and regions offer vehicle hire services that provide convenience to the travelling public. The car hire industry can be traced back to the early twentieth century (Yang et al. 2008). A review of the largest listed car rental companies shows that the sector has since grown into a multibillion-dollar industry, generating hundreds of billions into the global economy annually. Due to the high demand for rented cars, car rental companies contribute immensely to the car industry purchases in the USA and other countries. Purchases by car rental companies account for about a third of car purchases in the USA and about 15% of new car sales in Japan (ibid).

As observed by the Grand View Research (2019), the major markets for car rentals are in the USA, France, Germany, the UK, Japan, Korea, China and India. In 2018 the car rental market size reached $88.2 billion with a projected growth rate of 15% between 2019 and 2025. A report by Acumen Research and Consulting (2019) estimated a growth rate of 13.7% of the car rental market value between 2019 and 2026. This was likely to result in a value of $187.5 billion by 2026. In 2019 the market share for car rentals rose to $92.92 billion, with growth figures set at 10.7% between 2020 and 2026. By 2026, the car rental industry was expected to rake in $214.4 billion (Allied Market Research 2020). These figures justify why this tourism subsector needs more research now and in the future.

North America constitutes the most significant share market of this global portfolio. The sector's steady growth over the years can be traced to growth in tourism figures. The demand for car rental services has been huge at airport deports. The fastest-growing region of the car rental market, according to Acumen Research and Consulting (2019), is Asian. The most prominent industry players are the Avis Budget Group Inc., Enterprise Holdings Inc., Europcar Group, Car2go, Hertz Corp and Sixt, among others. The Grand View Research (2019) notes that from the top markets, airport car bookings resulted in $35.0 billion in 2018, and this was expected to continue growing in the coming years.

However, the advent of the Internet and increased Internet penetration presented a competitor for car rental services in the form of hailers and ride and share services (Puschmann and Alt 2016). This competition threatens the market segment of lower and middle-income earners and some leisure markets that find the ride and share facility much more affordable. Indeed, there are other traditional competitors, such as public transport and the increased penetration of more efficient rapid transport systems, particularly in urban areas. While literature is still emerging on the impact of the ride and share services, there is an acknowledgement that it will be disruptive to current modes of transport (McKenzie 2019), including car rentals. This is so given that there is a move towards the promotion of ride and sharing services, such as e-scooters, to deal with the challenges of air pollution and congestion in urban areas (Gössling 2020; Nishad et al. 2020). Shared transport, such as bikes, has been found to reduce the mileage on the usage of cars (Fishman et al. 2014) with health benefits too (Otero et al. 2018). Whatever competition emerges between the two transport subsectors, they remain in the tourism space and are both impacted by the COVID-19 pandemic.

To this end, the advent of Uber has been hailed as a game-changer that is disruptive in the tourism industry (Henama and Sifolo 2017). Giddy (2020) argues that the disruptive usage of Uber was beneficial to the generality as it has facilitated convenient mobility even in the global south. Kayi (2019) equally praises the role of Uber transport sharing facility in facilitating tourism in Kampala – Uganda's capital city. Rizk (2017) postulates that the Uber facility, besides facilitating movement, is a vital employment creator in Egypt. This brings to the fore the importance of this subsector.

No one ever envisioned that the biggest threat to car rental companies would emerge in 2020 from a pandemic. In the annual review of global risks by the World

Economic Forum (WEF), disease outbreaks ranked low in the 2020 Global Economic Risks Index and were classified as unlikely to occur. The WEF, however, bemoans the lack of investment and preparedness of the world in programmes dealing with infectious disease outbreaks. The WEF (2020:76) notes that:

> Considerable progress has been made since the Ebola epidemic in West Africa in 2014–2016, but health systems worldwide are still under-prepared for significant outbreaks of other emerging infectious diseases, such as SARS, Zika and MERS. A recent first-of-its-kind comprehensive assessment of health security and related capabilities across 195 countries found fundamental weaknesses around the world: no country is fully prepared to handle an epidemic or pandemic. Meanwhile, our collective vulnerability to the societal and economic impacts of infectious disease crises appears to be increasing.

Such a warning was telling. The warning came at a time when infectious diseases accounted for only 1.5% of global deaths in 2019. Infectious diseases were ranked number ten out of ten in terms of likely impact but were not in the top ten of risks likely to occur. Hall et al. (2020) are worried about this ignorance and/or neglect of the threat of pandemics by societal leaders. A rapid assessment report by Gössling et al. (2020) makes no mention of how car rental companies were going to be affected by the pandemic which could have affected the response of the world to the pandemic with significant socio-economic ramifications. It is in this light that this chapter seeks to investigate this seemingly forgotten part of the tourism industry in terms of how it was affected by the COVID-19 pandemic (Nepal 2020). The next section presents the methodology.

7.3 Materials and Methods

The chapter relied on archival and secondary data from authoritative sources to answer research questions and to draw findings and conclusions. To this end, quarterly financial reports, company advisories to shareholders, reports from popular newspaper articles, car hire companies and the UNWTO were used as secondary sources of data. There was also the utilisation of data from market research companies to determine stock movements of car rental companies as well as ride and share rental companies. However, data availability was the main determining factor of the cases that were examined further. The use of secondary data in transport and tourism research has been noted to be on the increase and has become an acceptable standard. Dube and Nhamo (2020) have utilised the methodology to investigate air manufacturing industries, while Gössling (2020) used a similar approach to investigate shared transport facilities across the world. The usage of secondary and archival data is also found to be convenient as it is inexpensive and allows geographic reach to data that might not be easily accessible to the researcher due to geographic or financial reasons. Data analysis followed the ethos of content and thematic analysis and usage of descriptive statistics.

7.4 Presentation of Results and Discussions

The study found that the COVID-19 outbreak affected both the ride and sharing facilities, as well as car rental companies. The pandemic management strategies, most notably am ban on both business and leisure inter-country as well as cross-border travel, threatened the viability of many companies, particularly the tourism industry. The closure of borders led to the cutting off of both international and domestic tourism as countries enforced lockdown and social distancing measures. As can be seen in Fig. 7.2, there was an increase in booking cancellations in between February and March. This period coincides with the period where infections were spiking which led to the disease being declared a pandemic on 11 March by the World Health Organization.

During this period, Italy had the highest booking cancellations as it was by then, together with other European countries, at the epicentre of new infections and deaths. The cancellations and increased border closures also had a significant impact on the capitalisation of the three most prominent global market leaders as of April 2020 (Table 7.1).

The following subsections will highlight how various car rental companies, followed by the ride and sharing companies, were impacted by COVID-19.

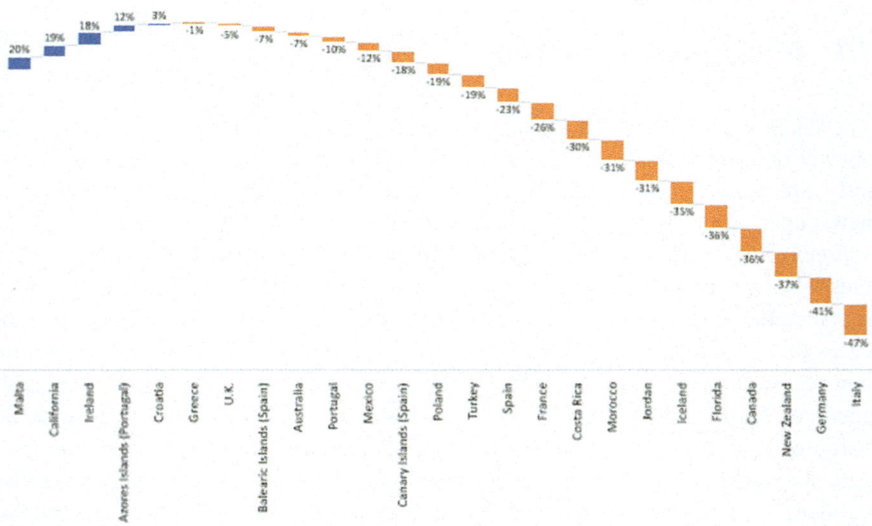

Fig. 7.2 Car booking cancellations due to COVID-19 from 23 February to 10 March 2020. (Source: Authors, data from Auto Rental News 2020a)

Table 7.1 Impact of COVID-19 on capitalisation of the three largest car rental companies

Parameter	Europcar (euros)	Hertz (US$)	Avis (US$)
Capitalisation 24 April 2020 (million)	244	528	909
% loss of share value	63%	83%	74%

Source: Authors, data from Global Fleet (2020)

7.4.1 *Impact of COVID-19 on AVIS Budget Group*

The AVIS Budget Group, which purchased car-sharing group Zipcar in 2013, is one of the largest car rental companies in the world. Avis was founded soon after World War II in 1946, while its biggest competitor Hertz was founded in 1918 (Goldstein 2020). Since then, the companies have grown to the extent of almost global presence. The coronavirus pandemic severely affected Avis Budget as car rental booking declined due to the pandemic-linked travel restrictions, looking at the company's performance from the last quarter of 2019 to the first 5 months of the year 2020. Figure 7.3 shows that during this period, the company recorded its best performance around February 2020. Post-February 2020, the company's stock prices started to fall, reaching a peak low at the end of March 2020. Some marginal gains were made, but these were not enough to offset earlier losses. The trajectory of the share price was a direct response to the company's position of not being able to generate revenue due to COVID-19 pandemic-linked travel restrictions.

The 2020 first quarter (Q1) financial report shows that the company suffered severe financial losses. The Avis Budget Group (2020) reports that although it generated revenue of $1.8 billion, this generated revenue was a 9% decrease compared with the same period the previous year (Fig. 7.4). As of 31 March 2020, the group had an approximate $1.6 billion in liquidity ($1.4 billion of available cash and cash equivalents, and $225 million of availability under its revolving credit facility). It had no meaningful corporate debt maturities that were immediately due, with significant maturities pencilled for 2023. As such, the company believed it had enough finance to pull through 2020 and into 2021. Nevertheless, the poor performance of the company in Q1 resulted in a rating downgrade. DBRS Morningstar downgraded Avis from BB (low) to B and Maintained Under Review – Negative for Avis Budget Group, Inc. (Avis Budget or the company), and its related subsidiary, Avis Budget Car Rental, LLC, including the company's long-term issuer (DBR Morning Star 2020). This bears testament to the liquidity pressure Avis was experiencing as a result of low business due to the pandemic. The "Under Review with Negative Implications" further highlighted the pressure, uncertainty and the possibility of a further downgrade if the company's liquidity significantly deteriorated. The downgrade rating has negative implications on borrowing costs for the company.

The recorded first quarter revenue losses exceeded the expectations of Wall Street, which had projected losses of $1.91 per share. At the peak of the COVID-19 pandemic, the group cancelled 80% of incoming rental vehicle orders in the USA for the remainder of the year. The company also went on to dispose of 35,000 cars

Fig. 7.3 Market performance of Avis Budget Group and the impact of COVID-19 on its trading. (Source: Market Screener 2020)

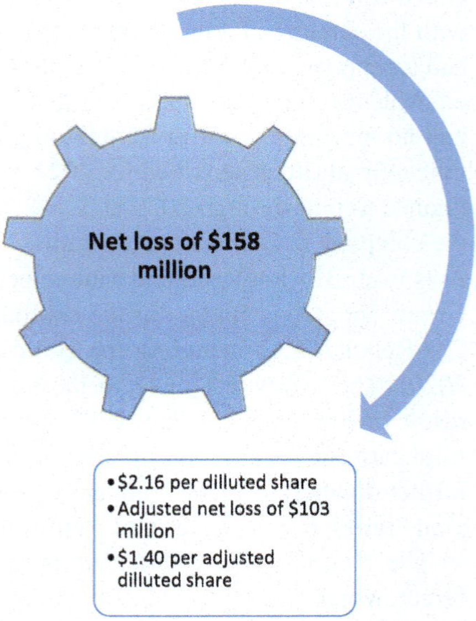

Fig. 7.4 First quarter revenue loss due to the impact of COVID-19. (Source: Authors, data from Avis Budget Group 2020a)

from its fleet. As a consequence, the company experienced an 8% revenue decrease compared to 2019. In April 2020 the company's earnings fell further by 80% on an annual comparison basis, with the expectation of a near similar picture by June 2020. The cash burn for April, May and June 2020 was projected to be $400 million, $250 million and $150 million, respectively (Avis Budget Group 2020).

This resulted in adjusted earnings before interest, taxes, depreciation and amortisation (EBITDA) loss of $87 million. In a bid to preserve liquidity, the company took measures that resulted in cost removal of about $2 billion. This was done by making cuts on expenditure, reducing non-essential capital assets, reducing fleet size and a reduction in the number of employees according to a shareholder's report. Given that, in the USA, the car rental industry was not covered under the federal bailout, the company trimmed its employee size by 70%, which translates to 21,000 employees (Auto Rental News 2020). The group indicated that it was determined to keep reducing its fleet to match demand to ensure a return to a positive balance.

7.4.2 *Impact of COVID-19 on Hertz Global Holdings*

As already indicated, Hertz Global Holdings, Inc., comprising Hertz, Thrifty, Dollar and Firefly, is one of the first car rental companies that started in 1918 and has evolved tremendously over time. Hertz has 9700 corporate and franchise rental locations in 150 countries (Goldstein 2020). Of the 9700 locations, 2983 are in the USA. The company has about 38,000 employees. The outbreak of the pandemic put a stop to ten consecutive quarters of year-over-year revenue growth and nine quarters of year-over-year adjusted corporate EBITDA improvement. The first 2 months of January and February 2020 witnessed a growth of between 6% and 8%, respectively, in the USA. In March 2020, Hertz started to feel the impact of city closures around the world as the company realised increased booking cancellations and a decline in forward bookings, which affected revenue earnings. Stocks for the company were the worst affected in the sector as can be seen in Fig. 7.5 falling from slightly over 25% in February to a record low level in June 2020. The stocks almost reached zero by that time.

Due to the threat that was posed by the pandemic, the company also took steps to protect itself by instituting measures outlined in Box 7.1.

From the Market Screener (2020a, b) report, the pandemic was expected to result in operating profit losses of around $506 million (Fig. 7.6). It is clear from the information that the pandemic will result in business losses for the global car rental company, which could affect the business and the supporting industries negatively such as the business suppliers in short to medium term. At the end of March 2020, Hertz had cash liquidity of about $1 billion with a moderate level of corporate debt due in 2020. The company was also looking into increasing its liquidity by accessing its surplus equity in its fleet-backed facilities. Before the announcement of its Q1 report, the company's rating was also downgraded from rating B (high) to BB (low) (DBRS Morningstar 2020). The downgrade was applied to Hertz Corporation,

Fig. 7.5 Hertz Global Holdings Inc. expert dynamic chart showing the impact of COVID-19 on sector market performance. (Source: Market Screener 2020a, b)

Box 7.1 Hertz Mitigation Efforts to COVID-19-Induced Reduction in Travel Demands

- Adjusted fleet levels to match demand.
- Cancelled new fleet orders and disposed of the excess fleet through multiple disposition channels before the shutdown of the used car market.
- Leveraged its multiple used car channels.
- Negotiated with suppliers to defer new fleet deliveries and modified placed orders.
- Furlough programme across North America field and USA corporate locations to match demand. A report in April 2020 indicated as many as 10,000 employees out of 29,000 were laid off.
- Senior leaders took significant pay cuts, with the CEO relinquishing her 100% base pay.
- Deferred capital expenditure projects and reduced discretionary spending.
- Renegotiated key contracts as well as reducing operating and overhead expenses.
- Took action to access surplus equity in its car rental fleet to increase liquidity position and lobbied for a government bailout.
- In April 2020 the company waived certain operating lease payments for its USA rental car fleet.
- In May 2020, the company entered into a forbearance and limited waivers with certain of its corporate lenders and holders of its asset-backed vehicle debt.

Source: Authors, based on Hertz (2020a, b)

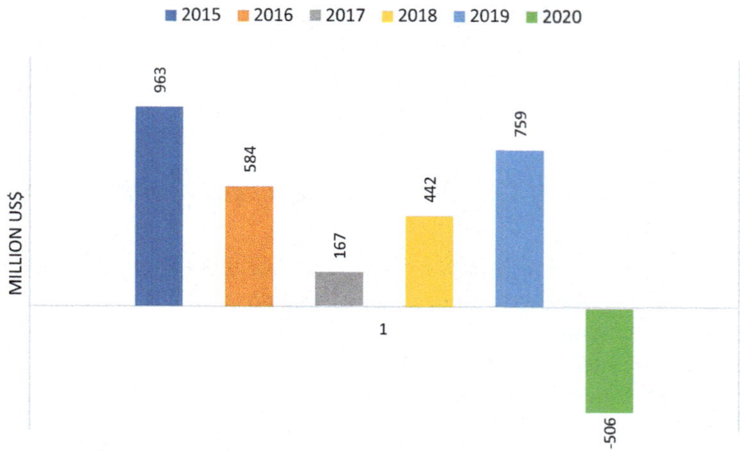

Fig. 7.6 Hertz annual operating profit from 2015 to 2019 and estimated operating profit of 2020. (Source: Authors, data from Market Screener)

including Hertz's long-term insurer. Besides, the company was put under review with adverse implications. The company also was at risk of a further downgrade should the liquidity situation drastically worsen as it was placed under negative review.

With about 201,364 vehicles, Hertz also reported that its finances were significantly affected by the COVID-19 pandemic. This resulted in consolidated revenue of \$1.9 billion (Hertz 2020b). During the same period, Hertz Global suffered a net loss of \$356 million and adjusted corporate EBITDA of -\$243 million. In as much as the company had recorded positive growth in the first 2 months, which resulted in 1% profits in February year-on-year, the company recorded a revenue decline of −12% year-over-year on a constant currency basis (Ibid).

A report by Goldstein (2020) indicates that after the severe impact of COVID-19 in March and April 2020, the company was looking into rescheduling its debt as it was, to some extent, unsustainable. The company had \$17 billion in debt, which translated to 966% debt-to-equity ratio. Ordinarily, the optimum debt-to-equity ratio (the amount of debt used to run a business) is 1:1.5 and not exceeding 2.0 depending on the industry. However, Gallo (2015), writing for *Harvard Business Review,* notes that "large manufacturing and stable publicly traded companies have ratios between 2 and 5. Any higher than 5 or 6 and investors start to get nervous". A high debt-to-equity ratio can adversely affect the company's borrowing from financiers as the ratio is often used to determine the approval or rejection of an application for business debt.

Hertz went on to file a chapter 11 bankruptcy after troubles in servicing its \$19 billion debt with its nearly 700,000 vehicles parked due to the impact of COVID-19. In a controversial development on the night of filing for bankruptcy, the company paid \$19 million retention bonus to its executives. The move left its strong 38,000 employees in limbo. Hertz, as one of the oldest companies, had survived even the great depression but owing to weak fiscal management was vulnerable to failure as many industry experts indicate that the company was already in trouble before the pandemic.

7.4.3 The Impact of COVID-19 on Europcar Mobility Group

Reports from the Europcar Mobility Group indicate that the company, while having recorded some robust growth in the first 2 months of 2020, experienced a drastically changed situation in March 2020. The Europcar Mobility Group (2020a) reported gains of 3.6% on a proforma basis. However, in March 2020 the company realised a − 34.6% drop on a performance basis. The Corporate EBITDA (IFRS 16) was -€64 m in Q1 2020 compared to -€14 m in Q1 2019. As a consequence, the company realised a net income of -€105 m (versus -€67 m in Q1 2019). The corporate operating cash flow declined by -€136 m, while the net corporate debt on 31 March 2020 was €1068 m. On the same date, the net corporate debt was €1068 million, while revenue declined by −10.1%. The cash flow position for Europcar was worsened by its acquisition of Fox Rent A Car in November 2019 (Business Wire 2020), which was aimed at accelerating the company's presence in the US market – the largest global market. Fox Rent A Car was one of the largest independent players in the US car rental market.

Faced with such financial challenges, the company went on a drive to ensure viability post the pandemic. The drive included cost reduction, cash preservation and liquidity-securing measures to contain about €850 m at the end of 2020 (Box 7.2). It also sought to bridge financial gaps by seeking €321 m new financing facilities of which €301 m was guaranteed by the French and Spanish governments. The financing line from the Spanish government for Europcar Mobility Group was meant for its two operating subsidiary companies in Spain and had a 3-year maturity. A summary of measures employed to save this company is outlined in Box 7.2.

Figure 7.7 shows that the pandemic had an impact on the market performance of Europcar Mobility Group. The company started to experience challenges at the end of February 2020, at which point there was a visible decline in the stock performance of the company. This decline persisted until the end of March 2020 when it reached its lowest levels and some slight gains that were made in April and May 2020.

7.4.4 Car Rentals and Ride and Sharing Funding Requests due to COVID-19

Operating vehicles during the pandemic calls for an improved standard of hygiene around cars and could affect businesses differently depending on the geography of the area. The economic recession triggered by the pandemic looks likely to reduce business revenue in both the short- and medium-term horizons. Indications are that

Box 7.2 Europcar COVID-19 Cost Containment, Reduction and Cash Preservation Measures

- Management Board pledged to reduce their base compensation by 25% from 1 April to 31 December 2020.
- Group's top managers (135 people) reduced their base compensation by 10% to 25% for a minimum of 3 months.
- Supervisory Board made a reduction of 25% of their annual remuneration.
- Renegotiated fixed fees and minimum fees that had been negotiated with airports and railways.
- Suspended temporary workers and subcontracted activities.
- Renegotiated commissions paid to brokers and travel agencies.
- It implemented partial unemployment for 80% of its workforce.
- Closed 88% of its stations.
- Negotiated for a postponement and reduction of rentals with station landlords.
- Got rid of external support and consultancy.
- Negotiated with its headquarters landlords.
- Made substantial reduction on its IT projects.
- Reduced all non-essential expenses and deferred expenses.
- Rationalised its real estate. Capital expenditure will be reduced up to −60% in 2020 compared to €76 m in 2019.
- Non-recurring items will be reduced by 60–70% for 2020 versus €56 m in 2019.
- Strict management of non-fleet working capital.
- Cancellation of the dividend initially proposed that amounted to €13 m.

Source: Authors, based on Europcar Mobility Group (2020b)

there was no immediate threat that could threaten the survival of these large corporates given their representations during their first quarter representations. This is regardless of the industry representative board, the American Car Rental Association (ACRA), advocating for funding in the three letters that were written to the federal government on 18 March, 27 March and 13 April 2020. The ACRA represents various companies that have a total of 160,000 employees and 2.2 million cars registered in the USA. In the first letter, the association sought relief and support to all stakeholders in the airport ecosystems, particularly car rentals.

In the other letter ACRA was acknowledging the benefits under the Payroll Protection Program (PPP) provisions in Title I of the Coronavirus Aid, Relief, and Economic Security Act (CARES Act) and the Main Street Program (MSP) established in Title IV of the CARES Act (ACRA 2020). However, the ACRA was requesting more to be done to assist the industry, such as the ones outlined in Box 7.3.

Box 7.3 Funding Request to the Federal Government by Car Rental Companies

- Increase funding under PPP for small businesses that remained vulnerable.
- Extension of the loan period covered under MSP and PPP in light of the continued impact of the pandemic.
- Extension of funding to cover other aspects of businesses beyond just salaries.
- The federal tax incentive to promote travel among Americans to induce business.
- Government insurance protection for affected businesses.
- An establishment of travel industry rainy day fund.
- Protection from a lawsuit from infections that can result from travels by tourists.

Source: Authors, based on ACRA (2020)

Fig. 7.7 Market performance of Europcar Mobility Group and the impact of COVID-19. (Source: Market Screener)

There are concerns that the car rental, as well as ride and share, businesses were unlikely to return to normal in the short term. This was likely to affect small players in particular and hurt the car manufacturing industry too due to depressed demand for new cars. The way issues around the COVID-19 pandemic were emerging and evolving at the time of this writing indicates a likelihood of job losses in the car rental and the larger motoring industry.

7.4.5 *The Impact of COVID-19 on the Ride and Share Transport*

The demand for social distancing also had an impact on the ride and share facilities for a large ride and share companies such as Uber, Lyft and Zipcar, among others. With the bulk of the countries under lockdown, borders closed and flying restricted (Fig. 7.8), the movement of transport and people was severely curtailed, which led to the decline of the ride and share businesses across the world.

Travel restrictions, therefore, led to the closure of ride and share businesses resulting in the suffering of employees due to low levels of demand and, in some cases, no business at all as businesses also feared for the safety of their employees. Table 7.2 shows how selected ride and share transport companies were affected and responded to the threat of total business collapse due to the threat of COVID-19. The companies profiled are some of the largest corporations with a presence (in most cases) across the world.

A look at the market performance of car ride and sharing facilities reveals that the uncertainties and anxiety caused by COVID-19 had an adverse impact on stock performance. This resulted in market volatility and a decline, which started the end of February and the beginning of March 2020, which was also marked by peak declines towards the end of March 2020. While the stock somewhat recovered, the recovery was not enough to offset earlier losses. Interestingly is that decline in the market performance of ride and share companies was better than that of car rental companies during the peak of the COVID-19 pandemic. This could be attributed to several factors, chief of which being that compared to car rental companies, the ride and share facilities are not severely dependent on the international tourism market, but also cater for local domestic travel. As a result, these firms could still generate revenue from their core activities.

The car ride and share facilities were not as severely impacted as car rentals in terms of usage and revenue. In the context of Uber Technologies, the company managed to ward off some fiscal pressure because of the demand that was witnessed in its Uber Eats segment and the launching of Uber Direct and Uber Connect to cater for retail deliveries. The Q1 financial report shows that Uber Eats and Uber Freight grew by more than 50% over the period, which offset declines in business in Uber rides declines. The demand for Uber Eats has to be understood in the context that the imposition of travel restrictions and lockdowns led to the closure of fast dining

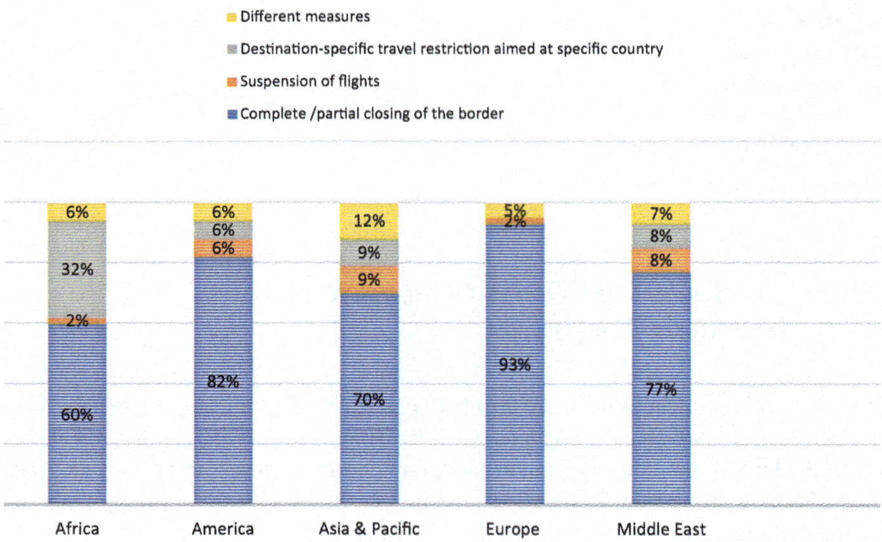

Fig. 7.8 Regional breakdown of travel restrictions as of 27 April 2020. (Source: Authors, data from UNWTO 2020)

sit-ins. This resulted in increased Uber Eats as in most cases; restaurants were allowed to do food deliveries. To this end, Uber benefitted sustaining the market performance and prevented its stocks from crashing. Some clients also used their services to seek medical attention and do shopping. The company also expanded over the period to more territories, although in some territories, closures were observed. Lyft launched Essential Deliveries in April 2020, a pilot initiative where organisations can request on-demand, contact-free delivery of meals, groceries and other necessities. Lyft Essential Deliveries positioned the company to compete with other Uber services, positioning Lyft to tap into that growing market even in the future (Figs. 7.9 and 7.10).

A study that was conducted by Car Gurus (2020) in the USA indicates that due to safety concerns, a significant number of travellers are considering switching from using public transport to private transport. A sizeable number of previous users of ride and share transport indicated that they intend to stop the usage of such transport facilities. As such, the ride and share revenue could be reduced in the foreseeable future owing to health and safety concerns by the customers, which will further hamper the sustainability of the businesses. This calls for proper tight fiscal management going forward.

Regardless of better performance, Lyft and Uber both shed a significant percentage of their employees. This contributed to the high level of unemployment that was recorded in the USA, which was only second to the Great Depression in the first week of April 2020. That witnessed a 14,7% unemployment rate, the level that was last seen post-World War II (Fig. 7.11). Projections by Trading Economics (2020) were that the unemployment figure was likely to reach 16.5% in May 2020, edging

Table 7.2 The impact of COVID-19 on the ride and share companies, Uber and Lyft

Uber impact and response to COVID-19	Lyft impact and response to COVID-19
• First quarter results show revenue of \$3.54 billion and a net loss of -\$2.9 billion • Trip volumes fell −5% from year-on-year. However, on the company's earnings call, it was indicated that the rides business was down around −80% in comparison to the same period in 2019 • Instituted a hiring freeze and reduced customer support and recruiting teams by 14% of its corporate employees (about 3700 full-time employee roles) • Folding JUMP e-bike and e-scooter business into Lime • Discontinued Uber Eats in the Czech Republic, Egypt, Honduras, Romania, Saudi Arabia, Uruguay and Ukraine, due to losses in those areas in Q1 2020 • Transferred Uber Eats operations to its subsidiary, Careem, in the United Arab Emirates • Sued by the state of California for allegedly misclassifying drivers as contractors, instead of employees, to avoid paying for health benefits and other worker protections • Loss of \$1.70 per share • Uber Eats was 50% up year-on-year as more people ordered food for delivery at home, and more restaurants signed up for deliveries • CEO waived his base salary of around \$1 million, for the rest of 2020	• May rides went down more than 70% year-on-year • Sued by the state of California for allegedly misclassifying drivers as contractors, instead of employees to avoid paying for health benefits and other worker protections • Termination of approximately 17% of its employees and furloughed approximately 300 other employees • Reduction in base salary for exempt employees for about 3 months, ranging from 10% for most non-hourly employees and up to 30% for the senior leadership team • Board members voluntarily agreed to forego 30% of their cash compensation for the second quarter of 2020 • Expected to incur approximately \$28 million to \$36 million of restructuring and related charges, mostly related to employee severance and benefit costs, exclusive of stock-based compensation related charges • Loss from operations was \$414.1 million in Q1 2020 • Realised a net loss of \$398.1 million in the first quarter of 2020

Source: Authors, based on Uber Investor (2020); CNBC (2020); Lyft Inc. (2020)

closer to the record high that was observed during the Great Depression in August 1932. The laying off of employees wantonly is questionable with regard to the extent, and the employment models employed by a ride and share institution are equally questionable with regard to their sustainability. This led to the State of California taking legal action to protect employees. Valuable lessons have to be learnt with regard to models of employment that leave employees vulnerable to shocks, exposing the general employee to poverty and destitution. With vehicles parked, there was very little support from parent companies to cushion employees from revenue losses.

7.5 Conclusion and Recommendations

The chapter examined the impact of COVID-19 on car rental as well as ride and share service companies. The chapter showed that the COVID-19 pandemic came at a considerable cost to the two sectors under investigation. The sectors lost several

Fig. 7.9 Market performance of Uber Technologies and the impact of COVID-19. (Source: Market Screener 2020a, b)

billions of dollars in potential revenue, mostly in the early months of the disease outbreak between March and April 2020. It emerged that the car rental industry was worse off than its competitors in the ride and share facilities. Companies such as Hertz despite having survived a number of recessions and depressions as one of the oldest car rental companies went on to file for USA chapter 11 bankruptcy at the height of the pandemic. The pandemic exposed companies that had weak liquidity and management systems to failure.

The pandemic also resulted in the loss of stock value of the companies with implications on shareholders and employees alike. Employees in the majority of cases were the worst affected as they were furloughed or laid off in cost containment measures that were instituted, which focused on protecting share value, survival and ensuring business viability. Although companies showed that they had challenges in most cases, they had enough financial liquidity to take them through to 2021. Unlike other sectors, the car rental market had the option of reducing expenses through asset reduction, with most big car rental companies showing that they had a stable fiscal position. The other strategy that was adopted by car rental companies was to negotiate with various suppliers and service providers in a bid to reduce expenses. In as much as there was government funding in some cases, jobs were still lost.

The car-sharing and ride facilities were not spared by the impact of the pandemic either. However, the impact was somewhat muted compared to car rental

Fig. 7.10 Market Performance of Lyft Inc. and the impact of COVID-19. (Source: Market Screener 2020a, b)

companies. This was due to the fact that there was a leeway to diversify from passenger services to offer other critical services such as food deliveries that supported the ailing restaurant industry from lockdowns, social distancing measures and drastically reduced travel that suffocated tourists and other patrons. There is evidence that some ride and share companies expanded their operations and increased offerings at the height of the pandemic, which allowed the businesses to derive revenue and reduced the level of losses. However, jobs and revenue were still lost, which led some to question the sustainability of the business of ride and share models. This scenario leads to one questioning the sustainability of jobs in the tourism industry, given the sensitivity of the sector to natural disasters, epidemics and pandemics.

Going forward, ensuring the comfort and safety of customers will become a critical aspect of the industry as a whole. There is a need for the sector to take severe and well-coordinated measures to ensure passenger safety amid not only the coronavirus and COVID-19 pandemic but also similar and other disasters in the future. Ensuring the cleanliness and sanitisation of vehicles will be the focus of many tourists. Education and training of drivers are crucial, and the development of safety protocols will be needed to protect the travellers and the drivers alike. Training the drivers on how to clean vehicles religiously and use of certified chemicals in fumigating and cleaning the vehicles should be a must to ensure passenger safety. Some companies had already instituted measures to instil a sense of customer safety. For

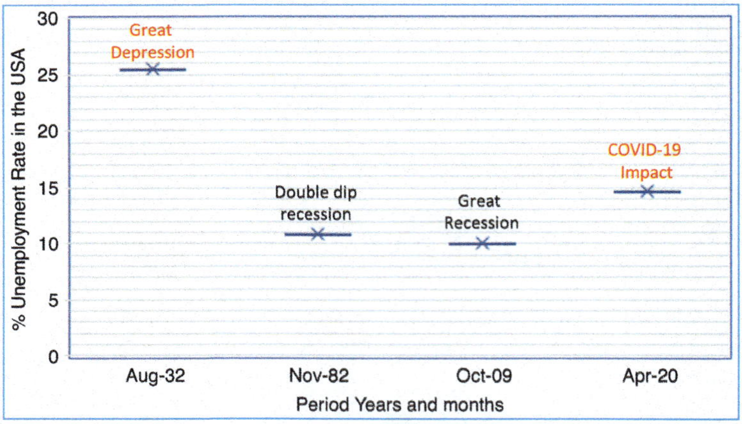

Fig. 7.11 Historical sources of monthly unemployment rates in the USA (1932 to 2020 COVID-19 times). (Source: Authors, data from The Washington Post 2020)

example, from its Q1 2020 report, Hertz indicated that it had started implementing stringent measures in line with the USA CDC guidelines to safeguard personnel and customers. In addition to following social distancing best practices at its locations, every vehicle was sealed and certified "Hertz Gold Standard Clean" after undergoing a 15-point cleaning and sanitisation process. This process follows the USA CDC guidelines and uses EPA-approved products. The study further recommends a post-pandemic impact assessment, particularly for the small companies in the sector, to see how they fared with a view to providing critical lessons for the future.

References

ACRA. (2020). *Implementation of CARES Act and Potential Additional Federal Legislation to Address Economic Distressed from COVID-19 Pandemic*. Retrieved May 11, 2020, from http://www.acraorg.com/wp-content/uploads/2020/04/acra-letter-to-congressional-leaders-on-next-stimulus-legislation-4-13-20.pdf

Acumen Research and Consulting . (2019). *Car rental market (By Booking Type: Online Access, Offline Access; By Application Type: Leisure/Tourism and Business; By Vehicle Type: Economy/Budget Cars, Luxury/Premium Car) - Global Industry Size, Share, Trends and Forecast 2019–2026*. Retrieved May 9, 2020, from https://www.acumenresearchandconsulting.com/car-rental-market

Allied Market Research . (2020). *Car rental market by application global opportunity analysis and industry forecast, 2020–2027*. Retrieved May 9, 2020, from https://www.alliedmarketresearch.com/car-rental-market

Auto Rental News . (2020). *Avis budget group sheds 70% of workforce*. Retrieved May 9, 2020, from https://www.autorentalnews.com/357515/avis-budget-group-sheds-70-of-workforce

Auto Rental News. (2020a). *Car rental cancelations rise as coronavirus cases increase*. Retrieved May 9, 2020, from https://www.autorentalnews.com/353242/car-rental-cancelations-increase-as-coronavirus-spreads

Avis Budget Group. (2020). *Avis budget group reports first quarter 2020 results and pandemic response actions*. Avis Budget Group. Retrieved May 9, 2020, from http://ir.avisbudgetgroup.com/static-files/244ef6bc-f027-40cf-a5c9-8674ab5054b4

Business Wire. (2020). *Europcar Mobility Group acquires Fox Rent A Car, to accelerate its growth momentum by a direct presence in the US, biggest worldwide market*. Retrieved May 10, 2020, from https://www.businesswire.com/news/home/20191106006193/en/Europcar-Mobility-Group-acquires-Fox-Rent-Car

Car Gurus. (2020). *U.S. COVID-19 sentiment study*. Retrieved May 12, 2020, from https://go.cargurus.com/rs/611-AVR-738/images/US-Covid19-Study.pdf

CNBC. (2020). *Uber shares shoot up after CEO says ride volume is increasing again after April bottom*. Retrieved May 11, 2020, from https://www.cnbc.com/2020/05/07/uber-uber-earnings-q1-2020.html

DBRS Morningstar. (2020). *Press release DBRS morningstar downgrades HTZ to B (high) and places it under review with negative implications*. Retrieved May 10, 2020, from https://www.dbrsmorningstar.com/research/359280

Dube, K., & Nhamo, G. (2020). Major Global Aircraft Manufacturers and Emerging Responses to the SDGs Agenda. In G. Nhamo, O. Odularu, & V. Mjimba (Eds.), *Scaling up SDGs Implementation Emerging Cases from State, Development and Private Sectors* (pp. 99–113). Cham: Springer.

Dube, K., Nhamo, G., & Chikodzi, D. (2020). COVID-19 cripples global restaurant and hospitality industry. *Current Issues in Tourism*, 1–4. https://doi.org/10.1080/13683500.2020.1773416.

Europcar Mobility Group. (2020a). *First quarter 2020 results*. Retrieved May 10, 2020, from https://finance.europcar-group.com/static-files/e04d9e91-678e-408c-8fbf-6b2fd3086f8c

Europcar Mobility Group. (2020b). *Europcar mobility group: First quarter 2020 results*. Retrieved May 10, 2020, from https://finance.europcar-group.com/news-releases/news-release-details/europcar-mobility-group-first-quarter-2020-results

Fishman, E., Washington, S., & Haworth, N. (2014). Bike share's impact on car use: Evidence from the United States, Great Britain, and Australia. *Transportation Research Part D: Transport and Environment, 31*, 13–20.

Gallo, A. (2015). *A refresher on Debt-to-Equity Ratio*. Retrieved May 10, 2020, from https://hbr.org/2015/07/a-refresher-on-debt-to-equity-ratio

Giddy, J. K. (2020). Travel patterns of Uber Users in South Africa. In J. M. Rogerson & G. Visser (Eds.), *New directions in South African tourism geographies* (pp. 113–127). Cham: Springer. https://doi.org/10.1007/978-3-030-29377-2_7.

Global Fleet . (2020). *How COVID-19 impacts car rental industry*. Retrieved May 9, 2020, from https://www.globalfleet.com/en/leasing-and-rental/smart-mobility/global/features/how-covid-19-impacts-car-rental-industry?a=PSE12&t%5B0%5D=Hertz&t%5B1%5D=Avis&t%5B2%5D=Sixt&t%5B3%5D=Europcar&curl=1

Goldstein, M. (2020). *Hertz car rental suffers massive layoffs, stock drop from COVID-19 pandemic*. Retrieved May 9, 2020, from https://www.forbes.com/sites/michaelgoldstein/2020/04/24/covid-19-creates-massive-layoffs-stock-drop-for-hertz-car-rental/amp/

Gössling, S. (2020). Integrating e-scooters in urban transportation: Problems, policies, and the prospect of system change. *Transportation Research Part D: Transport and Environment*, 79. https://doi.org/10.1016/j.trd.2020.102230.

Gössling, S., Scott, D., & Hall, C. (2020). Pandemics, tourism and global change: a rapid assessment of COVID-19. *Journal of Sustainable Tourism*. https://doi.org/10.1080/09669582.2020.1758708.

Grand View Research . (2019). *Car Rental Market Size, Share & Trends Analysis Report By Vehicle Type (Economy, Executive, Luxury), By Application (Local Usage, Airport Transport, Outstation), By Region, And Segment Forecasts, 2019–2025*. Retrieved May 9, 2020, from https://www.grandviewresearch.com/industry-analysis/car-rental-market

Hall, C. M., Scott, D., & Gössling, S. (2020). Pandemics, transformations and tourism: be careful what you wish for. *Tourism Geographies*, 1–22. https://doi.org/10.1080/14616688.2020.1759131.

Henama, U. S., & Sifolo, P. P. (2017). Uber: The south africa experience. *African Journal of Hospitality Tourism and Leisure, 6*(2), 1–10.

Hertz. (2020a). *Hertz global holdings taking proactive measures to address coronavirus impact on travel demand.* Retrieved May 10, 2020, from http://ir.hertz.com/2020-03-26-Hertz-Global-Holdings-Taking-Proactive-Measures-to-Address-Coronavirus-Impact-on-Travel-Demand

Hertz. (2020b). *Hertz globalholdings reports first quarter 2020 financial results* . Retrieved May 12, 2020, from http://ir.hertz.com/2020-05-11-Hertz-Global-Holdings-Reports-First-Quarter-2020-Financial-Results

International Monetary Fund -IMF. (2020). *Opening Remarks at a Press Briefing by Kristalina Georgieva following a Conference Call of the International Monetary and Financial Committee (IMFC).* International Monetary Fund. Retrieved April 9, 2020, from https://www.imf.org/en/News/Articles/2020/03/27/sp032720-opening-remarks-at-press-briefing-following-imfc-conference-call

Kayi, R. (2019). *Effectes of Uber transport to tourism development in Kampala city.* Makerere University: Doctoral dissertation, Makerere University.

Lyft Inc. (2020). *UNITED STATES SECURITIES AND EXCHANGE COMMISSION.* Retrieved May 11, 2020, from https://investor.lyft.com/static-files/3b79f72f-3160-4e9f-8f87-242158e2ca77

Market Screener . (2020a). *HERTZ GLOBAL HOLDINGS, INC. (HTZ).* Retrieved May 10, 2020, from https://www.marketscreener.com/HERTZ-GLOBAL-HOLDINGS-IN-29716824/charts/#.

Market Screener . (2020b). *Lyft, Inc Expert Dynamic Chart* . Retrieved May 11, 2020, from https://www.marketscreener.com/LYFT-INC-56481539/charts/

McKenzie, G. (2019). Spatiotemporal comparative analysis of scooter-share and bike-share usage patterns in Washington, DC. *Journal of Transport Geography, 78*, 19–28.

Nepal, S. K. (2020). Travel and tourism after COVID-19–business as usual or opportunity to reset? *Tourism Geographies*, 1–5. https://doi.org/10.1080/14616688.2020.1760926.

Nishad, D. S., Padhya, H. J., & Gandhi, Z. H. (2020). E-Scooter as an alternative non-motorized transport NMT for Urban Area: A case study for Surat City in Gujarat state in western India. *IJRAR-International Journal of Research and Analytical Reviews (IJRAR)7, 7*(1), 765–767.

Otero, I., Nieuwenhuijsen, M. J., & Rojas-Rueda, D. (2018). Health impacts of bike sharing systems in Europe. *Environment International, 115*, 387–394.

Puschmann, T., & Alt, R. (2016). Sharing economy. *Business & Information Systems Engineering, 58*(1), 93–99.

Rizk, N. (2017). A glimpse into the sharing economy An analysis of Uber driver-partners in Egypt. *SSRN*. Rizk, Nagla, A Glimpse into the Sharing Economy: An Analysis of Uber Driver-Partners in Egypt (February 22, https://doi.org/10.2139/ssrn.2946083.

Russell, P. (1999). Car rental in Europe. *Travel & Tourism Analyst, 1*, 1–28.

The Washington Post . (2020). *US unemployment rate soars to 14,7 percent the worst since the Great Depression era.* Retrieved May 9, 2020, from https://www.google.com/url?sa=t&rct=j&q=&esrc=s&source=web&cd=1&ved=2ahUKEwjUuM-W1afpAhWyURUIHWrMAtgQFjAAegQIBRAB&url=https%3A%2F%2Fwww.washingtonpost.com%2Fbusiness%2F2020%2F05%2F08%2Fapril-2020-jobs-report%2F&usg=AOvVaw0ke0QOj314Kcfg0aL10D5w

Trading Economics . (2020). *United states unemployment rate 1948–2020 Data 2021–2022 Forecast* . Retrieved May 11, 2020, from https://tradingeconomics.com/united-states/unemployment-rate

Uber Investor. (2020). *Uber announces results for first quarter 2020.* Retrieved May 11, 2020, from https://investor.uber.com/news-events/news/press-release-details/2020/Uber-Announces-Results-for-First-Quarter-2020/

UNWTO. (2020). *International tourist numbers could fall 60–80% in 2020, UNWTO reports*. Retrieved May 8, 2020, from https://www.unwto.org/news/covid-19-international-tourist-numbers-could-fall-60-80-in-2020
World Economic Forum. (2020). *The global risks report 2020*. World Economic Forum. Retrieved May 9, 2020, from http://www3.weforum.org/docs/WEF_Global_Risk_Report_2020.pdf
Yang, Y., Jin, W., & Hao, X. (2008). *Car rental logistics problem: A review of literature. 2*, pp. 2815–2819. Beijing, China: IEEE International Conference on Service Operations and Logistics, and Informatics. https://doi.org/10.1109/SOLI.2008.4683014.
Zhang, M., Xie, Y., Huang, L., & He, Z. (2014). Service quality evaluation of car rental industry in China. *International Journal of Quality & Reliability Management, 31*(1), 2012–0146.
Zheng, Y., Goh, E., & Wen, J. (2020). The effects of misleading media reports about COVID-19 on Chinese tourists' mental health: a perspective article. *Anatolia*, 1–4. https://doi.org/10.1080/13032917.2020.1747208.

Chapter 8
Impacts and Implications of COVID-19 on the Global Hotel Industry and Airbnb

Abstract While the tourism industry was on a steady growth path up to 2019, it remains sensitive and vulnerable to natural and political shocks. Such shocks have been on the increase in the recent past. Of concern has been the increase in diseases and pandemics which have become more prevalent since 2000 to 2020. As a result of the interlinkages in the tourism industry, disease outbreaks are often disruptive to global tourism economies. This chapter assesses the impact of COVID-19 on the hotel industry and shared accommodation in the form of Airbnb. Using critical document analysis and tracking hotel bookings from machine learning data for five top global online travel agencies and the American Hotel and Lodging Association, among other authoritative sources, the study found that COVID-19 has had a devastating impact on the sector. COVID-19 has led to massive booking cancellations and record low bookings with adverse effects on annual accommodation bookings for 2020 globally. This has, in turn, resulted in hotel closures and empty rooms for Airbnb properties. The slump in business has further led to record high hotel employee layoffs and/or furloughs, while Airbnb property owners battle to pay fixed expenses and mortgages on their properties. In addition, most global hotel chains have cancelled dividend payments and postponed capital expenditure to avoid bankruptcy and industry collapse. There are real fears that the impact of the pandemic, far exceeding previous disasters such as SARS, will reverse the gains and the contribution made towards the achievement of the Sustainable Development Goals. This study recommends the scaling up of several government and private sector relief funds and other measures aimed at assisting both the industry and its employees hardest hit by COVID-19. Particular attention should be given to small- to medium-size enterprises and establishments in developing countries.

Keywords COVID-19 · Tourism · Hotels · Airbnb · SDGs · Furloughed

G. Nhamo et al., *Counting the Cost of COVID-19 on the Global Tourism Industry*, https://doi.org/10.1007/978-3-030-56231-1_8

8.1 Introduction

Given the diversity of populations that intersect at hotels and other rented establishments in closed spaces, hotels are potential hotspots for disease spreading. A number of studies that have been conducted thus far have focused mainly on the spreading of food diseases through food contamination (Beauté et al. 2019; Hull-Jackson and Adesiyun 2019), environmental contamination (Kimura et al. 2011) and swimming pool contamination, with a few focusing on the spread of airborne diseases (Marks et al. 2000). Accordingly, hotels and other accommodation establishments within the hospitality industry are sensitive health risk areas. Disease exposure that emanates from hotels can result in criminal liability, and hotels have been found liable in various cases. Where liability has been proven, the implications on reputational damage could be dire; as such, the industry adopts stringent protocols, particularly when it comes to food safety (Gursoy 2019). Because diseases may emanate from hotels, the hotel sector is vulnerable to disease outbreaks, and, given the perishability of the tourism product, such threats may have severe implications. Disease outbreaks affect hotel room occupancy (Yang et al. 2020) with often devastating implications for the revenue and profits of hotels. Moreover, disturbances to mobility, which forms the basis of the tourism industry, can be detrimental to the hotel industry.

An inter-agency report produced by the United Nations paints a dire picture of the global economy, citing slow growth that extends from 2019 and which has been compounded by the outbreak of COVID-19, mainly in 2020 (United Nations 2020a). The report notes that the past couple of years have seen an increase in poverty levels across sub-Saharan Africa, a trend which will be perpetuated and worsened by COVID-19. The declaration of the coronavirus disease as a pandemic on 11 March 2020 sent shockwaves through the global market, triggering a wave of economic shockwaves that by far surpass those from the economic recession of 2008 (International Monetary Fund (IMF) 2020). The IMF later declared that with the impact of COVID-19, the world had entered a recession with marginal growth being expected for 2020. As a consequence, the world saw a massive capital outflow from emerging economies as the economic impact of the pandemic took hold. The United Nations (UN 2020b) notes that the pandemic was not only a health problem but also a socio-economic crisis which calls for global cooperation. The UN notes that, as a consequence, serious setbacks were expected to the achievement of all the Sustainable Development Goals (SDGs).

One of the sectors most affected by the demand for social distancing and avoidance of public gatherings was the tourism industry (UN 2020a). Previously Dube et al. (2020) touted tourism as the critical tool preferred to achieve several SDGs by many communities and countries across the world. Tourism creates and sustains jobs and creates entrepreneurship opportunities for millions of people around the world. The coronavirus affected both the supply and demand side of the economy, which crippled many industries, including the tourism industry. With several airlines and cruise ships grounded, the supply chain for the tourism industry, especially hotels and short-term accommodation rentals such as Airbnb, was severely disrupted. The International Air Transport Association (IATA) projected that as a result of the pandemic, about 65.5 million jobs in the tourism sector would be lost in 2020 (IATA 2020). The United Nations World Tourism Organization (UNWTO), on the other hand, postulated that the tourism industry was expecting to see a drastic slump in tourist arrivals of between 20 and 30% in 2020 (UNWTO 2020). This was projected to cost the industry $300 billion to $450 billion in international tourism receipts. This constitutes a third of the $1.5 trillion expected global losses under the worst-case scenario (UNWTO 2020).

Given that reality and the understanding that tourism was severely affected by the pandemic, there is a need to quantify the levels of the impact of coronavirus on various tourism sectors and subsectors at international, regional, national and local levels. This study investigated how the hotel and Airbnb arm of the tourism and hospitality industry was affected by COVID-19 pre and post its declaration as a pandemic by the World Health Organization (WHO).

8.2 Literature Review

There is a feeling that the world did not pay as much attention as it should have when the WHO flagged the coronavirus as a global threat in January 2020 (WHO 2020a). This could be attributed to several factors chief among them was a feeling that it would just end as an Asian Pacific problem. As such, attempts to institute quick intervention measures were ridiculed in a move that could have saved or slowed down the scourge. Little was known about the disease except by the Chinese, who had witnessed a first-hand account of the ravaging impact of the pandemic (Hjelmgaard 2020). The Chinese have been accused of hiding crucial information that could have reduced the impact of the disease; this has led to global political tension, particularly between the USA and China. The last two decades have seen an increase in endemic and pandemic disease outbreaks (Chung 2015), and these have had a significant impact on regional tourism economies. Some of these outbreaks have had a significant impact on global and regional tourism in the last two decades and include the severe acute respiratory syndrome (SARS) outbreak in 2003, the avian flu H5N1 threat in 2006 and the swine flu H1N1 pandemic in 2009–2010, Middle East respiratory syndrome (MERS) in 2012, Ebola virus disease (EVD) from 2014 to 2016 and now COVID-19 in 2020.

In 2009, Rittichainuwat and Chakraborty noted that there was a marked increase in the occurrence of diseases and other threats to tourism. The authors found that in as much as tourists understand the risk of diseases and other threats, they evaluated the risk before travelling to destinations and consequently take preventative or precautionary measures. Tourists are generally a risk-averse group of people who are sensitive to potential risks (Irvine and Anderson 2006; Tavitiyaman and Qu 2013). To this end, there is a need to consider tourism risk as a central element of tourism that takes into consideration, among other things, disease outbreaks (Perpiña, Camprubase outbreak 2019).

Quintal et al. (2010) reiterated the importance of attention to risk in the tourism industry, as it invokes feelings of anxiety, fear and insecurity among tourists. Such feelings could affect tourist appreciation of a destination and negatively affect their experience, hence the temptation to avoid or postpone visits in the light of a disaster, thus affecting hotel and Airbnb bookings. There are differences in attitude towards risk by different tourist typologies, which could affect arrivals during and after disease outbreaks. Wu et al. (2010) proposed a model for projecting the effect of disease outbreaks on hotel occupancy, with a view to buffering them from shocks. When tourists cancel visits, hotels suffer and revenue loss results. The fact that tourism is a highly perishable good makes it difficult at times to detect risk easily (Williams and Baláž 2013).

Jayawardena et al. (2008) noted that SARS pandemic had negative implications for the tourism and hospitality industry, with a global cost of $15 billion, and led to job losses in the hospitality and tourism industry. In Toronto, for example, job losses in tourism and hospitality contribute 75% of the jobs that were lost during the SARS outbreak. The SARS pandemic also resulted in significant conference cancellations and negatively affected tourism revenues, which may have been the main trigger for job losses at the iconic Niagara Falls (Jayawardena et al. 2008). The impact of the disease was felt far and wide, even though the pandemic had originated in China's Guangdong Province before emerging in neighbouring Hong Kong (Pine and McKercher 2004).

Hong Kong suffered immensely as a result of the SARS outbreak both in human and financial costs. According to Pine and McKercher (2004), the losses were mainly due to tourism capital investment postponement, dented destination image and travel cancellations and postponements, which had implications for hoteliers. As a consequence of the outbreak, tourist arrivals at Hong Kong International Airport slumped nearly 80%, declining to 565,000 from a peak of more than 2.8 million in May 2002. Massive job losses were experienced in the hotel, catering, transport, amusement and other recreational areas and hotel occupancy dropped 10–12% (Pine and McKercher 2004). Chen et al. (2007) found that the SARS pandemic led to hotel stock volatility in the Taiwanese tourism market, which led to losses in revenue by shareholders.

Rittichainuwat and Chakraborty (2009) argued that even though the WHO did not issue a warning to the world about bird flu (H5N1 virus), a number of people died and the outbreak had a damaging effect on the reputation of hotel and tourism industry in Thailand. Page et al. (2012) found that the outbreak of swine flu that coincided with the economic recession harmed the hotel and industry in the UK, as the tourism sector in that country reported drastic declines in bookings and arrivals from all its major tourism markets. Another country that was affected by swine flu was Brunei. According to Haque and Haque (2018), Brunei recorded a loss of nearly 30,000 (15%) tourists and B$15 million in 1 year as a result of swine flu.

The swine flu outbreak resulted in apprehensions about travelling for a number of tourists, which had an impact on several tourism destinations and markets, much to the detriment of the hotel industry (Lee et al. 2012). Hung et al. (2018) note that coronaviruses such as SARS and H1N1 swine flu place an increased burden of expenses on hotels as they raise the need for facilities to be sanitised. Other pandemics that have had a significant impact on the hotel industry include the Ebola virus. The outbreaks of the Ebola virus in North Africa affected hotel bookings across the continent (Novelli et al. 2018).

Given their exposure to crises such as pandemics, hotels have developed specific response measures as an adaptation mechanism for addressing continued external threats of disease. Israeli and Reichel (2003) highlighted 19 strategies that are often adopted by the industry to mitigate the impact of pandemics on hotel financial performance. These include cost-cutting and cost-containment measures which include offering massive discounts, making increased marketing efforts, cutting outsourced services, postponing capital expenditure and reducing the labour force. Such measures could have worked in regional disease outbreaks, but none of the measures has been tested for their effectiveness in the context where there is a global pandemic such as COVID-19. Regardless of the resilience that has been shown by the hotel and tourism industry to deal with the crisis, hotels have often found themselves either ill-prepared or unknowledgeable to deal with disease pandemics (Chien and Law 2003), as each disease outbreak has shown unique attributes which present challenges for each pandemic.

Yu and Aviso (2020) and also Dube et al. (2020) argued that on a global scale, COVID-19 has resulted in company closures, with a devastating impact on tourism sectors such as hotels and rented accommodation such as Airbnb. Preliminary studies have shown that China reported a hotel occupancy decline of 71% year-on-year in February 2020 (Yang et al. 2020). The authors further argued that travel restrictions were hurting economies. To this end, the United Nations (2020b) advocated for an undertaking of impact assessments that will translate into a better understanding of the pandemic's impact to assist in the response.

8.3 Research Methodology

This study examined and measured the impact of COVID-19 on hotels and Airbnbs. A multisource-critical document analysis (Karppinen and Moe 2019) was used to make meaning from a wide range of reports, notices and advisories by major hotel groups that have a footprint across the world. Authoritative sources such as national news media were interrogated to investigate the impact of COVID-19 on the hospitality industry. Borrowing from text data mining, the researchers used hotel group names and paired them with COVID-19 and coronavirus and searched Google to see all communications and news that were associated with the hotel with regard to the pandemic. A further perusal of company websites was made to ensure that the process was as exhaustive as possible. From there an automated Google Search algorithm populated all the stories and media coverage of the impact of COVID-19 on hotels and Airbnb. This ensured that sufficient attention was paid to all the reports that were emerging on the subject matter.

Data from HotelRunner Pulse was used to evaluate the impact of the disease being declared a pandemic on hotel bookings. Hotel Pulse integrates information and data from five major booking platforms, namely, Agoda, Expedia, Airbnb, Hotelbeds and Booking.com. HotelRunner dashboard data that were used were generated to provide HotelRunner Pulse's partners with a picture of industry performance. The dashboard uses the content based on anonymous booking data of more than 38,000 properties from 193 countries running on the HotelRunner platform (HotelRunner Pulse 2020). The data can be said to be representative of the global hotel industry. Data from the American Hotel and Lodging Association were also used to estimate the impact of COVID-19 on hotels and employment in the USA. Data on proportionate impacts of the US hotel industry based on historical performance during past shocks were measured on a 10% impact quartile. Data used for the Airbnb section came from AirDNA, which uses machine learning algorithms to track ten million listings in 80,000 markets globally on Airbnb, Vrbo and other booking sites. With millions of rooms to track globally, the use of machine learning to detect trends is encouraged (Al-Jarrah et al. 2015; Qiu et al. 2016). Qualitative data were analysed using the generic thematic and content analysis as espoused by Nowell et al. (2017), whereas quantitative data were analysed using Microsoft Excel ToolPak.

8.4 Presentation of Data and Discussion of Findings

The following sections present and discuss the main findings of the study. The first section presents the main findings, starting with showing general trends in booking and followed by a section that highlights results from selected global hotel groups. This is followed by a presentation on the impact of COVID-19 on the American hotel industry. The following section deals with the impact of the pandemic on rented accommodation and then makes a number of conclusions and recommendations.

8.4.1 General Trends in Bookings

The study found that there were three critical interventions that had significant negative effects on hotels and short-term rental accommodation establishments. These included closures of ports, national lockdowns and the active promotion of social distancing. With flight cancellations, there was no supply of tourists to accommodation establishments in many parts of the world, triggering a wave of hotel cancellations. Figure 8.1 shows three episodes pertaining to 3 weeks between March and April 2020 in relation to the confirmed gross book value. According to Expedia (2020:1):

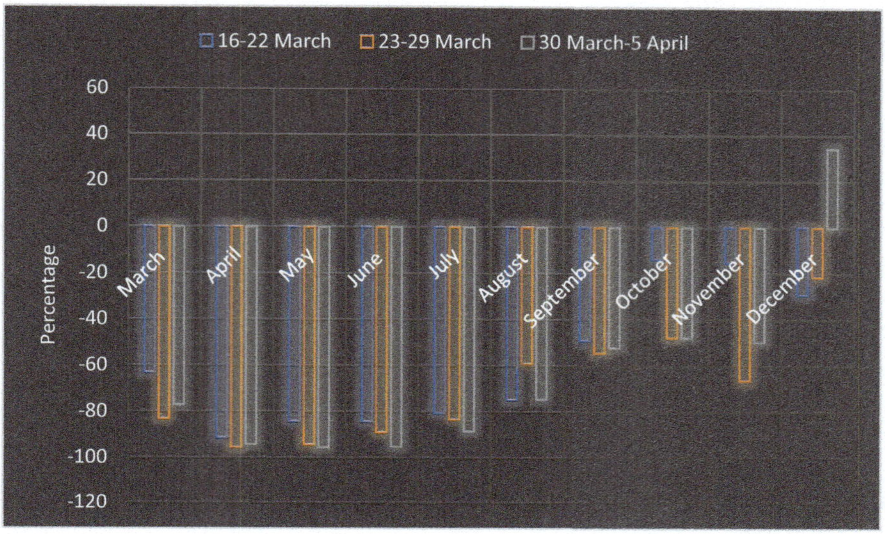

Fig. 8.1 Confirmed gross booking value: year-on-year growth in the same booking window. (Source: Authors, data from HotelRunner Pulse (2020))

> Gross value is the total retail value of transactions booked for both agency and merchant transactions, recorded at the time of booking. Bookings include the total price due for travel, including taxes, fees and other charges, and are generally reduced for cancellations and refunds.

It emerged that there was a significant decline in gross booking value due to cancellations between mid-March and early April 2020. Significant declines were observed, over 50% for the year 2020, save for the months of October and December 2020. Needless to say, the intermediate months post the announcement of COVID-19 as a pandemic has resulted in a considerable loss in revenue. There are declines reaching over 95% in April, May and June 2020. A significant decline in gross booking value is also seen in July 2020 although the decline starts to ease off a bit. This means that most travellers felt that the pandemic was going to have a severe impact on travel patterns in short to medium term. The pandemic will have a long-term effect, however, as it has negatively affected bookings for the whole of 2020 with prospects for affecting the early 2021 bookings.

The confirmed gross booking value for the period 30 March to 5 April 2020 looked promising, however, as it showed reduced percentages of decline and recorded a positive balance in December 2020. The glimmer of hope could be a testimony to the resilience of the tourism industry, as noted by Filimonau and De Coteau (2020). The positive outlook for December 2020 is encouraging as it could be a sign of market rebound as people look forward to the festive season.

There was also a consideration of forward bookings, taking two snapshots at two points in the third week of March 2020 and during the first week of April 2020 (Fig. 8.2a, b). From Fig. 8.2a, it emerges that in comparing 2019 and 2020, there

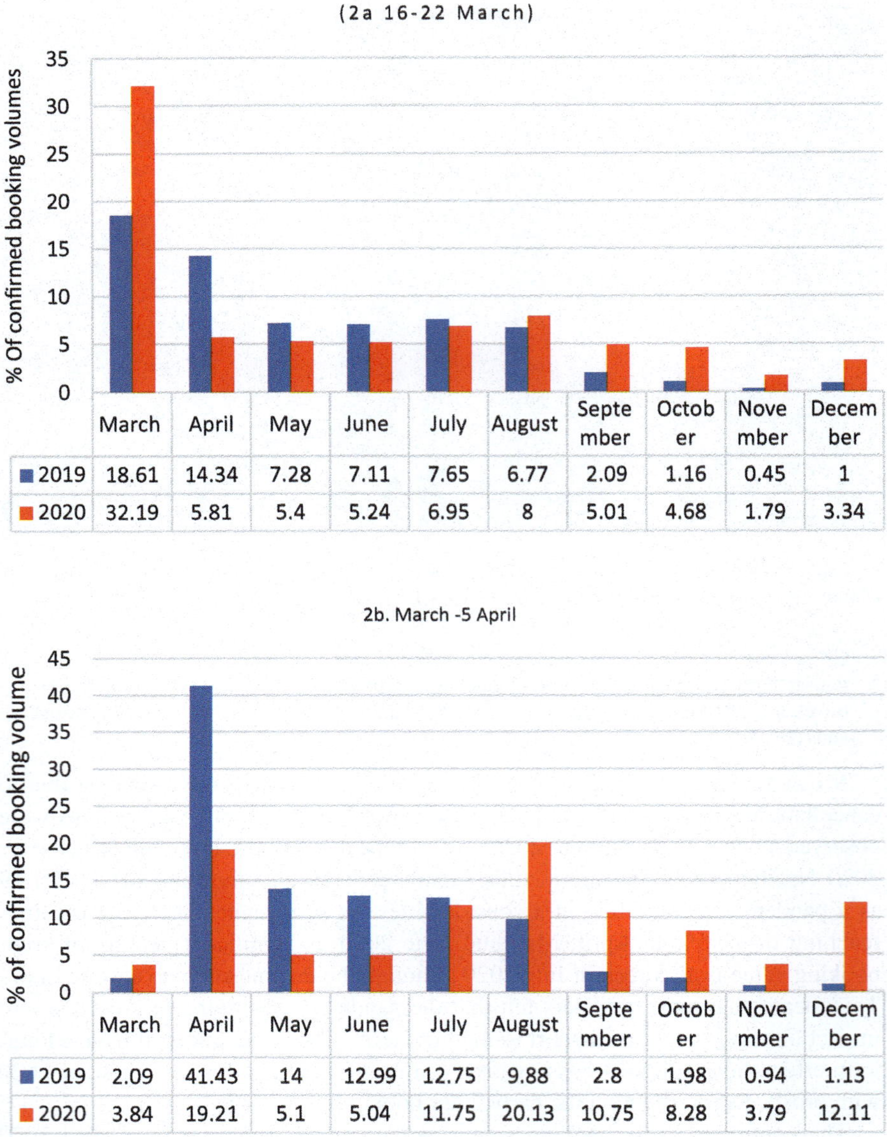

	March	April	May	June	July	August	September	October	November	December
2019	18.61	14.34	7.28	7.11	7.65	6.77	2.09	1.16	0.45	1
2020	32.19	5.81	5.4	5.24	6.95	8	5.01	4.68	1.79	3.34

	March	April	May	June	July	August	September	October	November	December
2019	2.09	41.43	14	12.99	12.75	9.88	2.8	1.98	0.94	1.13
2020	3.84	19.21	5.1	5.04	11.75	20.13	10.75	8.28	3.79	12.11

Fig. 8.2 (**a** and **b**) Forward-looking confirmed booking volume: monthly distribution of the confirmed booking volume in the same booking window. (Source: Authors, data from HotelRunner (2020))

was a higher booking volume year by year. The booking volumes were better for the third week of March 2020 compared to the same time in 2019. This means that the industry performed better than the previous year. The other months that showed healthier bookings in 2020 are August through to December 2020. For the two periods, however, April to June 2019 outperforms the same period in 2020, a situation that can be attributed to lower bookings as a result of cancellations caused by a coronavirus.

On a comparative basis, there is evidence that 2020 was set to outperform 2019 given stronger bookings in some months, including March, had it not been the impact of the pandemic that disrupted the trend. This can be attributed to the anticipated increase in growth pattern in the tourism industry, as seen on a year-to-year basis. However, the situation might change abruptly due to the announced recession caused by the COVID-19. It is crucial to note that in the absence of the pandemic, there is evidence that tourism would have done better in 2020 than last year (2019).

The booking cancellations between the 3 weeks focus of the study were also analysed. Wholesale cancellations of bookings were observed in the short term, with most of the bookings being cancelled in the third week of March 2020 and in April 2020. The upsurge in infections led to the tightening of border controls and the prohibition of public events and social gatherings. Record cancellations could have been triggered by the realisation that travel was not going to be feasible amid the raging pandemic. The long-term booking rate (July 2020 onwards) was not as affected, possibly because the tourists expected the disease to have died away in a few months. It could also be that people will be getting back into their routines, or it could be tourists adopting a wait-and-see attitude. However, in the third week of March 2020, there were record cancellations (Fig. 8.3). As discussed earlier, diseases instil fear and create anxiety among tourists (Tavitiyaman and Qu 2013). In this regard, the cancellations may be attributed to the fear people have about being infected or affected by the disease, hence cancelling or postponing holiday journeys.

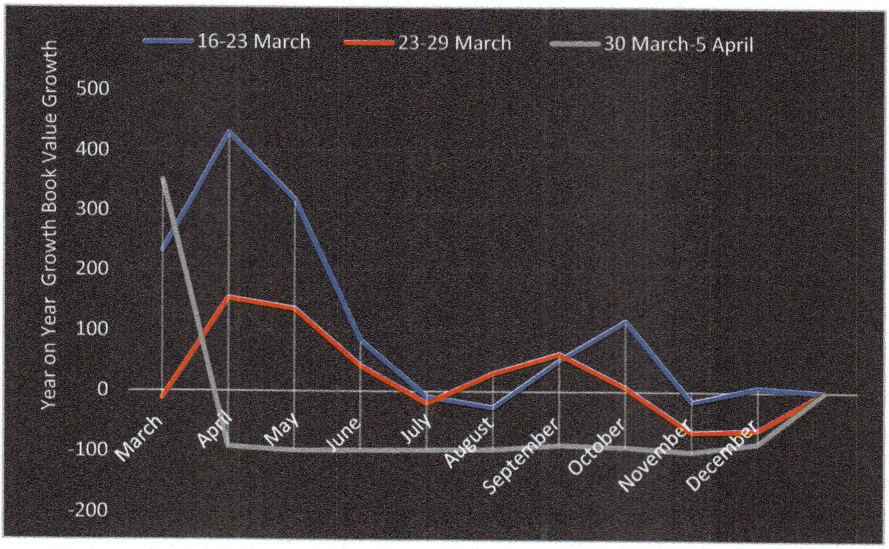

Fig. 8.3 Cancelled gross booking value: year-on-year growth in the same booking window. (Source: Authors, data from HotelRunner (2020))

On a comparable basis with 2019 in both snapshots taken for the weeks 16 to 23 March 2020 and 30 March 2020 to 5 April 2020 (Fig. 8.4a, b), a large number of booking cancellations are shown during March, April and May 2020, particularly during the third week of March 2020 (Fig. 8.4a). However, booking cancellations during the second episode under investigation for March and April were even higher (Fig. 8.4b).

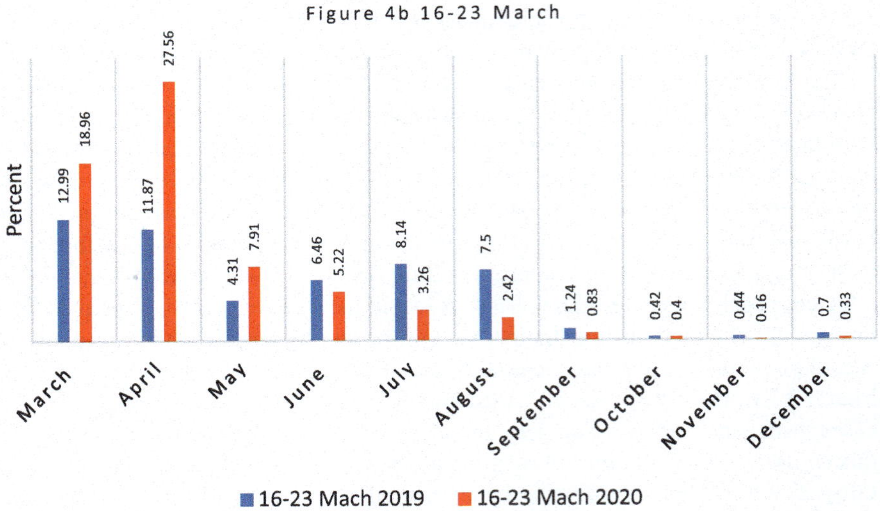

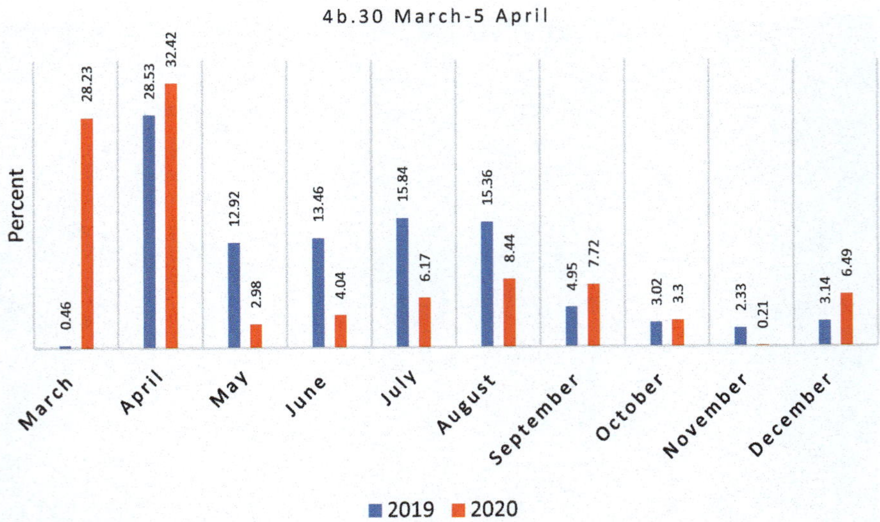

Fig. 8.4 (**a** and **b**) Forward-looking cancelled booking volume: monthly distribution of the cancelled booking volume in the same booking window. (Source: Authors, data from HotelRunner (2020))

Higher cancellations and lower bookings meant that hotels and lodges across the world were losing much-needed cash flow, given the uncertainties that were associated with the disease. The cancellations and weak confirmed bookings had a significant impact on the liquidity and cash flow of lodges and hotels in the short and long term. Such a scenario in the short term incapacitated the hotel industry which found itself in a forced liquidity crisis. Most hotels realised that they were bleeding cash, hence the need for cost-containment and cost-cutting measures to kick in, which meant freezing positions, job layoffs and furloughs in some instances. The next section outlines how selected group hotels and hotel organisations were affected and reacted to the problem.

8.4.2 Impact of COVID-19 on a Selected Group of Hotels

Table 8.1 summarises the effects of COVID-19 on the operations of selected hotel groups. Further investigation of the response strategies implemented by each hotel group was also done. It emerged that most hotels had their operations suspended and occupancy rates dwindled to single digits. This led to massive job furloughs across the sector, cancellation of dividend payouts and deferment of capital expenditure, among other things. Such cost-cutting and cost-containment measures have severe implications for the hotel industry and its stakeholders and the tourism industry as a whole, as accommodation performance is often used as a proxy for tourism industry performance.

One of the worst affected, and the largest, hotel market was the USA (United States of America). According to UNWTO (2020), the USA received 80 million visitors, with receipts worth $214 billion, in 2019. This made it the third most visited country and the top tourism revenue earner globally. The USA is one of the countries that was worst affected by COVID-19, recording 18,516 infections 1,003,974 and 57,730 deaths between 21 January and 30 April 2020 (WHO 2020a). A report by the American Hotel and Lodging Association (2020) shows that COVID-19 ravaged the hotel market with its impact far exceeding the September 11 bombing and the SARS pandemic on the USA hospitality market.

As a consequence of the coronavirus pandemic, the average occupancy rate for hotels in the USA dropped to an average of less than 25%. This resulted in between 2.8 and 3.5 million modelled job losses by the end of March 2020, with the expectation that up to 6.5 million jobs could be lost if hotel occupancy continues to slide (American Hotel and Lodging Association 2020). The distribution of job losses is displayed in Fig. 8.5. Two historical occupancy rate is 59% recorded after the September 11 bombing which resulted in the loss of 400,000 jobs and a 54% occupancy rate during the 2007–2009 economic recession that resulted in 470,000 job losses in the hospitality and tourism sector in the USA (American Hotel and Lodging Association 2020).

Significant job losses are expected in California, Florida, Texas and New York (Fig. 8.6). The results seem to tally with figures that emanate from the Statista,

Table 8.1 Impact of COVID-19 on selected and major global hotel chains

Tourism organisation	Impacts and responses to COVID-19
Accor (operates in 100 countries, with more than 4800 hotels and 280,000 employees worldwide)	Closed two-thirds (3200) of its hotels worldwide Cancelled 280 million euros ($304 million) in dividends Dedicated to allocating 25% of the planned dividend payout of €70 million to a fund dedicated to employees and "preserve" the remainder Introduced a new booking and modification policy that allowed tourists to flexibly manage their bookings or get credit notes and refunds 20-day plus booking discount of 25% Reduced the schedules of or furloughed 75% of its global head office teams
Marriott International	They furloughed about 174,000 employees at all levels and no payment for workers during the period, only health benefits (*Wall Street Journal* 2020) Put workers on unpaid leave
Hilton Hotel (has 6100 properties in 100 countries)	Closed 150 of its hotels in China, which represents 60% of its operations in China Announced suspension of operations on 26 March 2020 President and CEO had to forego salary for the remainder of the year The executive committee took a 50% pay cut for the period of the crisis Reduced schedules or be furloughed for up to 90 days and only be paid medical benefits Corporate team members who were not furloughed took a 20% pay cut Eliminated non-essential expenses and capital expenditures Suspended all share buybacks and payment of undeclared dividends (Hilton 2020)
The Trump Organization	Furloughed 1500 employees across the USA and Canada Several properties, including Trump National Doral Miami, Trump International Hotel Las Vegas and Trump International Hotel Waikiki closed temporarily 17 out of 24 properties in total closed, while others remained open with limited staff and services Hotels ran with restaurants closed, with golf clubs and clubhouses shut down 213 employees laid off at Trump's hotel in Vancouver, Canada, and 18 kept on, with 11 of those working reduced hours Some properties leasing state land had not paid the April rental and lease fees (Partlow et al. 2020)

(continued)

Table 8.1 (continued)

Tourism organisation	Impacts and responses to COVID-19
Intercontinental Hotel Group (5656 hotels across nearly 100 countries)	Reported 6% decline in business in January and February 2020 and 90% in China At the peak in China, 178 hotels were closed Further 60% decline in business in March 2020 Deferred renovations and imposed salary cuts for staff including board and executive level in March 2020 Cancelled S$150 million in dividends Cut capital expenditure by S$100 million for 2020 Cancelled fee waivers to existing and new bookings at all hotels globally for stays between 9 March 2020 and 30 April 2020 and later extended to 30 June 2020 Modified and/or reduced services in some hotels, e.g. reduction in restaurant and bar service, alternative guest room furnishings and/or amenities Suspended operations (InterContinental Hotels Group (IHG) 2020; Brandler 2020)
Indonesian Hotel and Restaurant Association (PHRI)	Shut down 698 hotels as of the beginning of April 2020 Open hotels were operating below 5% occupancy levels Several hotels in the country were providing delivery service of dishes from their restaurants
Tsogo Sun	Closed 36 hotels (i.e. 40% of group operations) Postponed capital expenditure and only kept up critical maintenance work Head office support functions operated on a limited staff-only basis to ensure good communication with guests, partners and suppliers The company noted that it was likely to fail to meet its debt obligations if poor occupancies continued beyond 3 to 6 months post-September 2020 Approached lenders to request waivers of covenants to ensure that the facilities to which the Group had access were maintained (Tsogo Sun Hotel Limited; Hospitality Fund Limited 2020)

which shows California, New York and Texas, among others, as leading in initial job loss claims in the past 3 weeks for each state (21 March–4 April 2020) (Roper 2020). Evidence from the Bureau of Labor Statistics (2020) suggests that over 60% of job losses in March 2020 were in the tourism and hospitality sector. It is important to note that the states with highest job losses were the most affected by COVID-19, of which New York has been one of the worst affected. For example, New York had 181,825 infection cases and 8650 deaths as of 11 April 2020 (University of Virginia 2020). As such, losses in employment have further increased the vulnerability of the respective communities with dire psychosocial and economic effects.

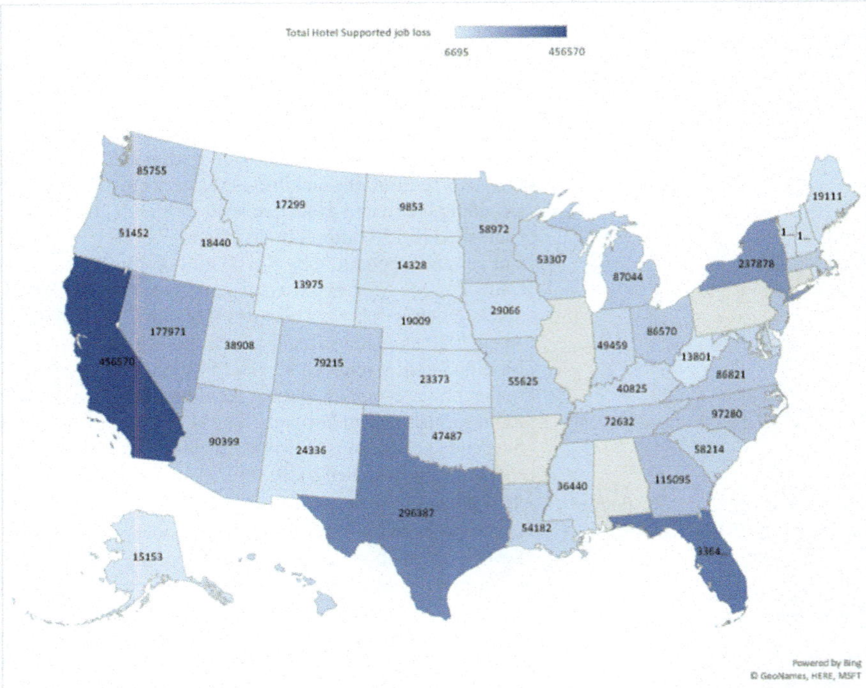

Fig. 8.5 Distribution of projected four million-plus job losses due to COVID-19 in various US state tourism markets as of the end of March 2020. (Source: Authors, data from American Hotel and Lodging Association (2020))

Figure 8.6 shows that since the beginning of the pandemic, American hotels have been bleeding cash and jobs. This has threatened to bring the hotel industry to a grounding halt if no intervention measures are taken to rescue the situation. The employees and small businesses are also vulnerable to the effects of the pandemic, with a sizeable number of companies at risk of collapse. If the pandemic persists, rescuing the tourism industry is going to be more costly, and it will take longer to ensure that the industry regains stability and starts generating new jobs.

It emerges from Fig. 8.6 and Table 8.1 that the hotel industry lost billions in potential revenue and resulted in job losses as never seen before. While it emerges that the industry chose in the main job furloughs as a cost-cutting measure, such a measure exacerbated the problems for hotel employees. While the employees had to deal with the fear and anxiety caused by the pandemic, they had to face a bleak and uncertain future. Baranik et al. (2019:381) note that furloughs are "characterised by decreased life satisfaction and increased work-family conflict and physical, cognitive, and emotional burnout even weeks after the shutdown ended". There is, therefore, a need for the hotel industry to come with some sort of safety net for its employees to shield them from such ever-imminent shocks to the hotel and tourism industry.

Fig. 8.6 Economic impact of COVID-19 on USA hotel economy. (Source: Authors, data from American Hotel and Lodging Association (2020))

The other concern is for small to medium enterprises which are vulnerable to the shock of the pandemic. Small to medium enterprises are often without the financial muscle to deal with shocks of such magnitude. Most such enterprises are outcompeted in the race for resources resulting in closures and permanent job losses. The tourism industry has a significant number of small to medium enterprises, and there is a need for these businesses to be protected from failure to protect jobs and the industry. It is appalling to note that despite the US government handout in place, take-up of the rescue package which was supposed to rescue the tourism sector was not embraced with open arms given the conditions that were attached it. The CARES Act imposed a condition that the beneficiaries of the government funding were supposed to use 75% of the fund on the wage bill, something that was strongly protested by the industry through their representative organisation American Hotels and Lodging Association (AHLA) in a letter that was written to Congress, requesting changes to the CARES Act.

The AHLA argued that the bill costs were quite minimal as compared to other fixed costs such as property tax, insurance, utility costs, bank loan payments, utilities and franchise fees to the chain brands. The hoteliers noted that in some instances, the Act also restricted hotels from laying off employees, something that was not acceptable to the industry as they pointed out that rehiring about 10% of employees would not make sense, as there was simply no work for employees and if the business could not service other fixed expenses, they were still going to shut down (*The Washington Post* 2020). The workers association fought back, noting that it was unfair for the industry to ask the government to divert public funds to pay to rich capitalists on Wall Street when workers were laid off. There is a feeling that in as much as it was raising some valid arguments, the main thrust was to protect the investor and the business and not necessarily the employee.

8.4.3 Impact of COVID-19 on Hotel Sector Shareholders and Value Chain

The coronavirus did not only affect the tourism industry; the impact was felt among all its stakeholders throughout the sector's value chain. The deferment and postponement of the payment of dividends adversely affected shareholders. Beyond the shareholders, suppliers and support services, the hoteliers were adversely affected by the pandemic through business losses. The shareholders are likely to fork out more money to remain afloat with the most affected being small and medium enterprises that are likely to go bankrupt in the not too distant future.

A number of hotel groups indicated that they had deferred capital projects. This meant a decline in tourism investments. A decline in investment would adversely affect support workers and the construction industry the most which may further worsen the recession and will likely have a knock-on effect on employment in that sector. In some instances, the hospitality industry has contacts with farmers and other fresh produce producers for the provision of food. This sector has also suffered due to the closure of hotels and the decline in demand, with implications for the global food market industry and probably its employees, which might see dwindling returns for farmers. Besides the food suppliers, the other daily consumables that are used in the hotel industry have also been adversely affected by the declining demand.

The hotel industry has some fixed obligations, which can be pretty substantial (Rushmore and O'Neill 2015). Such obligations need to be serviced regardless of whether the hotel is trading or not. Given the robust growth in the tourism industry which grew at a rate of 3.5% against a global economic growth rate of 2.5% (UNWTO 2020), many hoteliers were involved in the expansion and/or acquisition projects and thus are likely to find it challenging to continue with such projects. The indications from Tsogo Sun, a hotel operator in South Africa, are that it might be a challenge for some hoteliers to meet their debt obligations over a prolonged period of low hotel occupancies.

There were already indications from the US market that some enterprises had already defaulted on the payment of rentals and leases as of April 2020, ranging into millions of dollars (Schoening and Shapiro 2020). Towns and municipalities were likely to be affected, especially those that rely on tourism business revenue for their rates and taxes. The financial collapse was likely to result in businesses defaulting on their rates and taxes. Small municipalities such as the town of Victoria Falls in Zimbabwe have been bankrupted as a result of the pandemic and are most likely to struggle to raise adequate cash for their employees and social services due to a decline in revenue. South African hotels and lodges in Cape Town which were just emerging from one of the worst droughts to have affected the Western Cape have also been badly affected by the pandemic as they do not have enough cash reserves to rescue the situation. Following the wildfires in the Australia, where most resort towns battled and encountered severe losses, the tourism businesses had gone from bad to worse in terms of meeting its financial obligations, as areas in Victoria and New South Wales have yet to recover from the devastating impact of the fires that ended earlier in 2020.

8.4.4 *The Impact of COVID-19 on Airbnb*

The pandemic also had severe ramifications for vacation rentals such as Airbnb. With the COVID-19 affecting millions, the vacation rental and home-sharing companies were also as severely disrupted as their hotel counterparts as a result of travel cancellations. In particular, the blanket cancellations suffered by Airbnb left the owners vulnerable, with no income to pay for bonds and other associated rates. This caused a flurry of online complaints, petitions and lawsuits leading the company to set up a $250 million facility in order to compensate hosts for up to 25% of their lost income (Schaal 2020). An additional $10 million bailout was sought for bailing out superhosts. The cancellations came at a high cost to the company as it was billed for public listing some time in 2020. In 2019, the market value for Airbnb was set at $31 billion up from nothing 11 years back (Sherwood 2020). However, due to the pandemic and poor management of cancellations and refunds, the company plunged to almost half the value. The company was valued at 18 billion dollars as of April 2020 (Eaglesham and Grind 2020). In light of the depreciated value, it remains to be seen whether the company will proceed with its initial planned public offering.

Given the disgruntlement, there is evidence that some property owners took their property off short-term rentals, which might signal that property owners are going to put those properties up for long-term rentals. Figure 8.7 shows that all the sampled countries have reported a decline in the number of active properties. This has affected the annual returns and profitability of Airbnb. For property, the immediate future is bleak with the volume of cancellations very high and future bookings very low. There is, therefore, a need for property owners outside the USA to look for ways of paying the fixed expenses on their properties which might mean months of losses. The status quo brought into question the sustainability of business models adopted by companies such as Uber and Airbnb, which transfers all the risk to the service provider, while they take all the profits and no substantive risks. A better and more equitable and workable model has to be found to address such disparities.

8.5 Conclusion and Recommendations

The study investigated the impact of COVID-19 on the global hotel and Airbnb tourism sector. It emerged that the COVID-19 pandemic has had a devastating short- to medium-term impact. The pandemic has led to a spike in booking

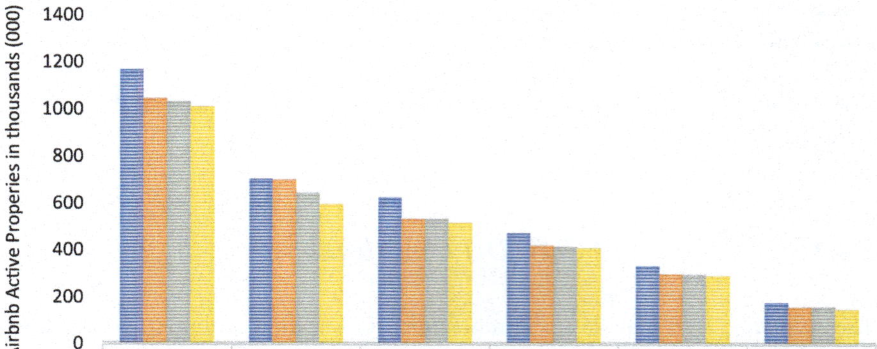

	United States	China	France	Italy	Spain	Germany
Dec-19	1169	705	628	475	336	180
Jan-20	1047	702	537	424	301	162
Feb-20	1033	645	536	416	298	163
Mar-20	1011	600	520	412	295	151

Fig. 8.7 Airbnb active properties. (Source: Authors, data from AirDNA (2020))

cancellations as witnessed on five platforms from which the data were collected, namely, Agoda, Airbnb, Booking.com, Expedia and Hotelbeds. A spike in cancellations and a decline in bookings has left many hotels vulnerable to the impact of financial losses. The pandemic also came at a higher cost to Airbnb, which has lost close to 50% of the company's market value. Cancellations for Airbnb have left property owners vulnerable to fixed expenses of property ownership such as rates, taxes and loan repayments. Faced with imminent business collapse, hotel organisations globally have resorted to massive job layoffs and furloughs with adverse impacts on employees as most hotels temporarily closed shop. As the chapter was being finalised, there was a real danger than most hotels would not be able to meet their financial obligations such as loan repayments and servicing of rates. Most global hotel groups deferred or cancelled the payment of dividends to shareholders and suspended capital expenditure. Initial assessment showed that for the tourism industry alone, hundreds of billions of dollars in potential revenue was lost. The figure could translate into trillions of dollars if one considers the knock-on effect of hotel shutdowns on the entire tourism value chain.

In light of the enormous liquidity challenges that were being faced by the tourism industry, especially hotels and Airbnb, there was a need for state and private sector intervention in the form of stimulus packages. Such stimulus packages would assist the industry to get back on track as soon as the disaster is over, and therefore a full chapter is dedicated to this in this book. It is anticipated that the first revival will come from the domestic tourism markets. The promotion of domestic tourism would offer much-needed relief to the industry as states were likely to open up to the domestic tourism market before they open up to the international markets.

Incentivising the domestic tourism market could offer the needed relief sooner than the international tourism market.

Governments could assist in waiving some of the taxes and charges as part of the relief package for the sector. Assisting employees with unemployment benefits could go a long way in relieving the burden from laid-off employees in the interim, but long-lasting solutions have to be found to assist laid-off employees. Fortunately, some of these interventions were already in place, as the chapter was being finalised. The tourism industry also includes many small to medium enterprises that needed special stimulus packages if they are to remain viable as they are more vulnerable to the impact of the pandemic.

In the absence of medication for the coronavirus, there was a need for the industry to be allowed to also open with new health and safety protocols that ensure the safety of guests and staff. A new deep cleaning mechanism was needed to ensure that the disease does not spread in accommodation establishments. In ensuring safety and health, it was imperative to ensure the existence and grafting of social distancing in hotels and hotel restaurants. Innovative technology such as contactless check-in and payments was a must to instil a sense of safety among guests. As such there was a need for the industry to come with health and certification scheme that shows the adherence of hotel and other accommodation to global health and safety that takes into consideration the new normal under COVID-19. Lastly, the chapter recommends further regional, national and local studies to ascertain the extent of the impact and industry response to COVID-19. Such work would assist the industry to better prepare for future pandemics when they occur.

References

Al-Jarrah, O. Y., Yoo, P. D., Muhaidat, S., Karagiannidis, G. K., & Taha, K. (2015). Efficient machine learning for big data: A review. *Big Data Research, 2*(3), 87–93.

American Hotel and Lodging Association. (2020). *Covid-19'S impact on the hotel industry.* American Hotel and Lodging Association. Retrieved April 12, 2020, from https://www.ahla.com/covid-19s-impact-hotel-industry.

Baranik, L. E., Cheung, J. H., Sinclair, R. R., & Lance, C. E. (2019). What happens when employees are furloughed? A resource loss perspective. *Journal of Career Development, 46*(4), 381–394.

Beauté, J., Sandin, S., de Jong, B., Hallström, L. P., Robesyn, E., Giesecke, J., & Sparén, P. (2019). Factors associated with Legionnaires' disease recurrence in hotel and holiday rental accommodation sites. *Eurosurveillance, 24*(20).

Bereau of Labor Statistics. (2020). *The employment situation — March 2020.* US Department of Labor. Retrieved April 13, 2020, from https://www.bls.gov/news.release/pdf/empsit.pdf.

Brandler, H. (2020). *IHG chief: "Demand for hotels is currently at the lowest levels we've ever seen".* BuisnessTraveller. Retrieved April 12, 2020, from https://www.businesstraveller.com/business-travel/2020/03/20/ihg-chief-demand-for-hotels-is-currently-at-the-lowest-levels-weve-ever-seen/.

Chen, M. H., Jang, S. S., & Kim, W. G. (2007). The impact of the SARS outbreak on Taiwanese hotel stock performance: an event-study approach. *International Journal of Hospitality Management, 26*(1), 200–212.

Chien, G. C., & Law, R. (2003). The impact of the Severe Acute Respiratory Syndrome on hotels: a case study of Hong Kong. *International Journal of Hospitality Management, 22*(3), 327–332.

Chung, L. H. (2015). Impact of pandemic control over airport economics: Reconciling public health with airport business through a streamlined approach in pandemic control. *Journal of Air Transport Management, 44*, 42–53.

Dube, K., Nhamo, G., & Chikodzi, D. (2020). COVID-19 cripples global restaurant and hospitality industry. *Current Issues in Tourism*, 1–4. https://doi.org/10.1080/13683500.2020.1773416.

Eaglesham, J., & Grind, K. (2020). *Airbnb Paying More Than 10% Interest on $1 Billion Financing Announced Monday Airbnb's new investors get free warrants based on company valuation of $18 billion.* The Wall Street Journal. Retrieved April 13, 2020, from https://www.wsj.com/articles/airbnb-paying-more-than-10-interest-on-1-billion-financing-announced-monday-11586297484.

Expedia. (2020). *Expedia, Inc. Financial results release glossary of business and financial definitions.* Retrieved April 11, 2020, from https://ir.expediagroup.com/static-files/a601b412-713f-4a4c-8fe3-8d7266cf7ddc.

Filimonau, V., & De Coteau, D. (2020). Tourism resilience in the context of integrated destination and disaster management (DM2). *International Journal of Tourism Research, 22*(2), 202–222.

Gursoy, D. (2019). Foodborne illnesses: An overview of hospitality operations liability. *Journal of Hospitality, 1*(1), 41–49.

Haque, T. H., & Haque, M. O. (2018). The swine flu and its impacts on tourism in Brunei. *Journal of Hospitality and Tourism Management, 36*, 92–101.

Hilton. (2020). *Hilton Corporate Response to COVID-19.* Hilton Hotel. Retrieved April 12, 2020, from https://newsroom.hilton.com/corporate/news/hilton-corporate-response-to-covid19.

Hjelmgaard, K. (2020). *Fact check: President Donald Trump vs. the World Health Organization.* USA Today. Retrieved April 14, 2020, from https://www.usatoday.com/story/news/factcheck/2020/04/11/coronavirus-fact-check-donald-trump-vs-world-health-organization/5128799002/

HotelRunner Pulse. (2020). *HotelRunner Pulse: Accommodation industry insights.* HotelRunner Pulse. Retrieved from https://hotelrunner.com/en/pulse

Hull-Jackson, C., & Adesiyun, A. A. (2019). Foodborne disease outbreaks in Barbados (1998–2009): a 12-year review. *The Journal of Infection in Developing Countries, 13*(01), 1–10.

Hung, K. K., Mark, C. K., Yeung, M. P., Chan, E. Y., & Graham, C. A. (2018). The role of the hotel industry in the response to emerging epidemics: a case study of SARS in 2003 and H1N1 swine flu in 2009 in Hong Kong. *Globalization and Health, 14*(1), 117.

InterContinental Hotels Group (IHG). (2020). *Our travel advisory policy.* InterContinental Hotels Group (IHG). Retrieved April 12, 2020, from https://www.ihg.com/content/gb/en/customer-care/travel-advisory

International Air Transport Association-IATA. (2020). *25 Million Jobs at Risk with Airline Shutdown.* IATA. Retrieved April 7, 2020, from https://www.iata.org/en/pressroom/pr/2020-04-07-02/.

International Monetary Fund-IMF. (2020). *Opening Remarks at a Press Briefing by Kristalina Georgieva following a Conference Call of the International Monetary and Financial Committee (IMFC).* International Monetary Fund. Retrieved April 9, 2020, from https://www.imf.org/en/News/Articles/2020/03/27/sp032720-opening-remarks-at-press-briefing-following-imfc-conference-call.

Irvine, W., & Anderson, A. R. (2006). The effect of disaster on peripheral tourism places and the disaffection of prospective visitors. In Y. Mansfeld & A. Pizam (Eds.), *Tourism, security & safety: From theory to practice* (pp. 169–186).

Israeli, A. A., & Reichel, A. (2003). Hospitality crisis management practices: the Israeli case. *International Journal of Hospitality Management, 22*(4), 353–372.

Jayawardena, C., Tew, P. J., Lu, Z., Tolomiczenko, G., & Gellatly, J. (2008). SARS: lessons in strategic planning for hoteliers and destination marketers. *International Journal of Contemporary Hospitality Management, 20*(3), 332–346.

Karppinen, K., & Moe, H. (2019). Texts as data I: Document analysis. In H. Van den Bulck, M. Puppis, K. Donders, & L. Van Audenhove (Eds.), *The Palgrave handbook of methods for media policy research* (pp. 249–262). Cham: Palgrave Macmillan.

Kimura, H., Nagano, K., Kimura, N., Shimizu, M., Ueno, Y., Morikane, K., & Okabe, N. (2011). A norovirus outbreak associated with environmental contamination at a hotel. *Epidemiology & Infection, 139*(2), 317–325.

Lee, C. K., Song, H. J., Bendle, L. J., Kim, M. J., & Han, H. (2012). The impact of non-pharmaceutical interventions for 2009 H1N1 influenza on travel intentions: A model of goal-directed behavior. *Tourism Management, 33*(1), 89–99.

Marks, P. J., Vipond, I. B., Carlisle, D., Deakin, D., Fey, R. E., & Caul, E. O. (2000). Evidence for airborne transmission of Norwalk-like virus (NLV) in a hotel restaurant. *Epidemiology & Infection, 124*(3), 481–487.

Novelli, M., Burgess, L. G., Jones, A., & Ritchie, B. W. (2018). No Ebola… still doomed'–The Ebola-induced tourism crisis. *Annals of Tourism Research, 70*, 76–87.

Nowell, L. S., Norris, J. M., White, D. E., & Moules, N. J. (2017). Thematic analysis: Striving to meet the trustworthiness criteria. *International Journal of Qualitative Methods, 16*(1). https://doi.org/10.1177/1609406917733847.

Page, S., Song, H., & Wu, D. C. (2012). Assessing the impacts of the global economic crisis and swine flu on inbound tourism demand in the United Kingdom. *Journal of Travel Research, 51*(2), 142–153.

Partlow, J., O'Connell, J., & Fahrenthold, D. (2020). *Trump Organization has laid off about 1,500 employees as pandemic hits business*. The Washington Post. Retrieved April 12, 2020, from https://www.washingtonpost.com/politics/trump-organization-has-laid-off-about-1500-employees-as-pandemic-spreads/2020/04/03/c413e42e-75d6-11ea-ae50-7148009252e3_story.html

Perpiña, L., Camprubí, R., & Prats, L. (2019). Destination image versus risk perception. *Journal of Hospitality & Tourism Research, 43*(1), 3–19.

Pine, R., & McKercher, B. (2004). The impact of SARS on Hong Kong's tourism industry. *International Journal of Contemporary Hospitality Management, 16*(2), 139–143.

Qiu, J., Wu, Q., Ding, G., Xu, Y., & Feng, S. (2016). A survey of machine learning for big data processing. *EURASIP Journal on Advances in Signal Processing, 1*, 67.

Quintal, V. A., Lee, J. A., & Soutar, G. N. (2010). Risk, uncertainty and the theory of planned behavior: A tourism example. *Tourism Management, 31*, 797–805.

Rittichainuwat, B. N., & Chakraborty, G. (2009). Perceived travel risks regarding terrorism and disease: The case of Thailand. *Tourism Management, 30*(3), 410–418.

Roper, W. (2020). *Unemployment.* Statista. Retrieved April 12, 2020, from https://www.statista.com/chart/21389/jobless-claims-by-state-coronavirus/?utm_source=Statista+Global&utm_campaign=4e326e7746-All_TheWeekAhead_weekly_COM_PM_CW16_2020&utm_medium=email&utm_term=0_afecd219f5-4e326e7746-307759193.

Rushmore, S., Jr., & O'Neill, J. W. (2015). Updated benchmarks for projecting fixed and variable components of hotel financial performance. *Cornell Hospitality Quarterly, 56*(1), 17–28.

Schaal, D. (2020). *Airbnb CEO apologizes to hosts with a $260 million relief package.* Skift. Retrieved April 13, 2020, from https://skift.com/2020/03/30/airbnb-ceo-apologizes-to-hosts-with-a-260-million-relief-package/amp/

Schoening, E., & Shapiro, M. (2020). *70 percent of U.S. Hotel employees furloughed or laid off due to COVID-19; Hotels losing $3.5 billion weekly.* Meetings Group. Retrieved April 14, 2020, from https://www.northstarmeetingsgroup.com/News/Hotels-and-Resorts/Coronavirus-Update-Hotel-Resort-Casinos-Closed-Economic-Impact

Sherwood, H. (2020). How Airbnb took over the world. *The Guardian*. Retrieved April 13, 2020, from https://www.theguardian.com/technology/2019/may/05/airbnb-homelessness-renting-housing-accommodation-social-policy-cities-travel-leisure

Tavitiyaman, P., & Qu, H. (2013). Destination image and behavior intention of travelers to Thailand: The moderating effect of perceived risk. *Journal of Travel & Tourism Marketing, 30*, 169–185.

The Washington Post. (2020). *Hotel industry asks Congress for changes to Cares Act.* The Washington Post. Retrieved April 16, 2020, from https://www.washingtonpost.com/context/hotel-industry-asks-congress-for-changes-to-cares-act/1dcc5e4f-35a9-44c6-928f-86a4933f196b/?itid=lk_inline_manual_11

Tsogo Sun Hotel Limited; Hospitality Fund Limited. (2020). *Joint voluntary announcement – response to the impact of covid-19.* Tsogo Sun Hotel Limited. Retrieved April 12, 2020, from https://www.satsa.com/wp-content/uploads/SENS_20200320_S428424.pdf.

United Nations. (2020a). *Shared Responsibility, Global Solidarity: Responding to the socio-economic impacts of COVID-19.* United Nations Retrieved April 2, 2020, from https://www.un.org/sites/un2.un.org/files/sg_report_socio-economic_impact_of_covid19.pdf.

United Nations. (2020b). *Policy brief: The impact of COVID-19 on women.* United Nations. Retrieved April 11, 2020, from https://www.un.org/sites/un2.un.org/files/policy_brief_on_covid_impact_on_women_9_apr_2020_updated.pdf.

University of Virginia. (2020). *COVID-19 surveillance dashboard.* University of Virginia Biocomplexity Institute. Retrieved April 12, 2020, from https://nssac.bii.virginia.edu/covid-19/dashboard/

UNWTO. (2020). *International tourism highlights 2019 edition.* Madrid: UNWTO. Retrieved April 12, 2020, from https://www.e-unwto.org/doi/pdf/10.18111/9789284421152.

Williams, A. M., & Baláž, V. (2013). Tourism, risk tolerance and competences: Travel organization and tourism hazards. *Tourism Management, 35*, 209–221.

World Health Organization. (2020a). *Global- United Sates.* World Health Organization. Retrieved April 12, 2020, from https://who.sprinklr.com/region/amro/country/us.

World Health Organization. (2020b). *WHO Director-General's opening remarks at the media briefing on COVID-19 – 11 March 2020.* World Health Organization. Retrieved April 13, 2020, from https://www.who.int/dg/speeches/detail/who-director-general-s-opening-remarks-at-the-media-briefing-on-covid-19%2D%2D-11-march-2020

World Health Organization. (2020c). *Novel coronavirus(2019-nCoV) situation report – 10.* World Health Organization. Retrieved March 29, 2020, from https://www.who.int/emergencies/diseases/novel-coronavirus-2019/situation-reports

Wu, E. H., Law, R., & Jiang, B. (2010). The impact of infectious diseases on hotel occupancy rate based on independent component analysis. *International Journal of Hospitality Management, 29*(4), 751–753.

Yang, Y., Zhang, H., & Chen, X. (2020). Coronavirus pandemic and tourism: Dynamic stochastic general equilibrium modeling of infectious disease outbreak. *Annals of Tourism Research.* https://doi.org/10.1016/j.annals.2020.102913.

Yu, K. D., & Aviso, K. B. (2020). Modelling the economic impact and ripple effects of disease outbreaks. *Process Integration and Optimization for Sustainability*, 1–4. https://doi.org/10.1007/s41660-020-00113-y.

Chapter 9
Restaurants and COVID-19: A Focus on Sustainability and Recovery Pathways

Abstract The hospitality industry, specifically the restaurant sector, is one of the leading global sources of employment, and it generates considerable revenue. However, the industry is susceptible to disasters that destroy infrastructure and affect human movement, notably the outbreak of pandemics and epidemics. This study investigated the early impacts of COVID-19 on the global hospitality's restaurant sector. The study made of data from the OpenTable database and the National Restaurant Association of America. It emerged from the study that most restaurants were shut down for sit-in meals as governments increasingly promulgated regulations for social distancing and lockdowns. This adversely affected fine dining and family restaurants, pubs and taverns. The fast-food outlets were equally affected with businesses mostly operating at less than 20% capacity. This led to substantial financial losses and direct and indirect jobs losses thus bringing many restaurants into unprecedented liquidity challenges. At the peak of the pandemic impact in the USA, millions of restaurant employees lost their jobs. The study recommends a raft of tailor-made measures to assist the restaurant businesses and employees during and after the crisis so that both emerge out of it stronger and resilient to withstand such possible future events. Among such measures are tax rebates for employees and business; improved health and hygiene measures; the provision of grants, loans and debt relief interventions; decreased interests on loans; and other innovative measures to ensure business viability post the pandemic.

Keywords SDG 8 · SDG 12 · COVID-19 · Jobs · Restaurants · Hospitality · Tourism · USA

9.1 Introduction

Kim et al. (2005) and Tse et al. (2006) argue that restaurants, as well as hotels, airlines, travel agencies and resorts, are usually at the receiving end of catastrophes both natural and man-made. The tourism and hospitality sector is vulnerable to human calamities and extreme weather events calamities such as hurricanes, tropical cyclones, tornados, tsunamis, financial crisis and wars. The past decade was been characterised by weather extreme events at the instigation climate variability.

G. Nhamo et al., *Counting the Cost of COVID-19 on the Global Tourism Industry*,
https://doi.org/10.1007/978-3-030-56231-1_9

The year 2019 while it started with news headlines of the fires in South West Australia, these were nothing compared to the global impact of coronavirus that crippled many sectors of the tourism economy (Dube et al. 2020; Gössling et al. 2020).

COVID-19 pandemic presented a challenge to the world's quest to attain the 2030 Agenda for Sustainable Development (AfSD), with its interlocked 17 sustainable development goals (SDGs). While the pandemic affected all goals, two goals and their relation to the tourism industry, namely, SDG 8 and SDG 12, are the focus of this chapter. SDG 8 stipulates the desire to "(promote) sustained, inclusive and sustainable economic growth, full and productive employment and decent work for all" (United Nations 2015:14). Critical to this chapter is that Target 8.9 of this goal encourages the world to "devise and implement policies to promote sustainable tourism that creates jobs and promotes local culture and products." Complementarily, Target 12.b focuses on developing and implementing "tools to monitor sustainable development impacts for sustainable tourism that creates jobs and promotes local culture and products" (United Nations 2015, pp. 20, 23).

Kim et al. (2020) investigated the influence of infectious disease epidemic outbreaks on the financial performance of the restaurant sector. Their work focused on four epidemic disease outbreaks: swine flu (2009), *Salmonella infantis* infection (2011 and 2014), the avian flu (2014) and the bovine spongiform encephalopathy (BSE) (2015). Their findings confirm the negative effect of epidemic disease outbreaks on the restaurant industry. Pandemic and epidemics can be damaging to three organisation characteristics, namely, organisations' value, namely, brand reliability, advertising effects and service types. It is highly likely that COVID-19 will have a similar impact. This carries dire consequences for the global economy because of the contribution of restaurants to economic growth and development through employment opportunities as well as related product and services inputs. Interestingly, the employment opportunities in the sector often provide a crucial entry point to the labour market for many youths.

This chapter documented the impacts of COVID-19-induced panic and fear (CIPF), real or otherwise, and its implications on restaurants' unemployment numbers and recovery pathways globally. Drawing from both the literature and emerging data, valuable lessons were presented for the restaurant sector, especially to operation that had and were/are still experiencing the hurricane-like impacts from the COVID-19 pandemic. This chapter is set out to investigate the impact of COVID-19 on global restaurant industry meltdown.

9.2 A Literature Survey

Lee and Warner (2005a) bemoaned the consequences of the SARS epidemic on China's employment. In most circumstances, the impact of pandemics affects employees differently across the tourism and hospitality sector. According to Lee and Warner (2006), migrants were often among the hardest-hit group in Guangzhou and Shangai. They were laid off after the government ordered the closure of the entertainment business as part of places where there may be substantial public gatherings. An estimated eight million workers, out of about 100 million, employed in the restaurant and related sectors, returned to their rural areas. The same authors also focused on Singapore's service sector, which included restaurants. They found that, in 2002, the hotels and restaurants sector employed 125,300 people (Lee and Warner 2006), and, like many tourism-dependent economies in Asia, the sector contributed up to 10% of the gross domestic product (GDP). During the SARS epidemic, revenue, at selected restaurants, fell by as much as 50%, resulting in the hotel and restaurant sector experiencing significant job cuts losing 5800 jobs out of the 12,100 in the sector (Ibid). The SARS outbreak also impacted on social activities in Hong Kong, where people became afraid to venture out into crowded public places and shopping malls (Tse et al. 2006). As a result, restaurant patronage significantly declined.

The impacts of SARS in these countries and region were not only financial. The impact was also psychosocial, and some of the matters from that era remain real under the COVID-19 scenarios. McKercher (2003) documented the concept of SARS-induced panic (SIP) in Hong Kong. In his view, although none of the over six million residents of Hong Kong had succumbed to SARS, the SIP potentially contributed to the loss of between 50,000 and 100,000 jobs in the local hotel, tourism, hospitality and retail sectors. The losses were largely attributed to perception and fear of reports from the region. Lee and Warner (2005b) found that 25 restaurants in Hong Kong closed within the first 2 weeks of April 2003, resulting in 1600 restaurant staff losing their jobs. Wilder-Smith (2006) highlighted the severe customer drop in Chinese cuisine restaurants in Guangdong, Hong Kong and Chinatowns in North America, with some as high as a 90% decrease. From McKercher's (2003) assessment, the SIP was a hysteria surrounding SARS that induced irrational behaviour in the population, and this may not be ruled out in the face of COVID-19. The CIPF implies that the impacts of the pandemic would be far less than the fear it injects and provokes in the global citizenry. Rumour spreading, misinformation, fearmongering and ignorance all played a part in consumer panic during SARS, and it was mainly the tourism sector that was highly vulnerable to SIP. This resulted in an up to an 80% decline in tourist arrivals, in affected areas and their vicinity. Breda (2004) indicates that Hong Kong's travel and tourism sector was predicted to have lost about UDS$3.6 billion and 41,700 jobs, with the restaurant association predicting that 5000 of its members will be shutting their doors (McKercher 2003).

After a crisis like SARS, Tse et al. (2006) encourage restaurants to start by assessing the nature and extent of the damage. During the SARS outbreak in Hong Kong, the primary damage was sudden cash flow problems, which led to job losses. Hence, an appropriate response measure will be to aim at maintaining adequate liquidity for the business. To this end, "operating costs must be minimised, and low-cost loans must be solicited", with revenue enhancement plans put in place to generate more revenue and cash inflow to the business (Tse et al. 2006: 7). Some of the cost-cutting measures included temporary closures of restaurants, shortened business hours, the redeployment of staff and promoting travel packages (Lee and

Warner 2005b). This witnessed the Peninsula's restaurants regaining their clientele and starting to fill again. Further cost reductions implemented included reducing advertising and promotion, negotiation with suppliers to lower costs for foodstuff, talking to landlords to reduce rentals and having staff pay cuts and placing staff on unpaid leave (Tse et al. 2006).

Additional measures focused at lobbying government for financial support through relief packages that include interest-free loans and negotiating with banks for suspension of loan payments for a certain period of time. Such negotiations should be led by organised labour movements, chambers of commerce and/or restaurant and tourism industry associations. This paid off in Hong Kong, as ultimately, the Hong Kong General Chamber of Commerce announced that it succeeded in getting an HKD11.8 billion relief packages for the restaurant sector impacted by SARS. Included in the relief package were tax rebates, lower rent for shops in public shopping malls and reduced water and sewage charges for restaurants (Lee and Warner 2005b).

There were other onside innovations from Hong Kong restaurants. These included improving food quality and service; restaurants offered discounts and new forms of promotions to entice the old and potentially new patrons (Wilder-Smith 2006); and anti-SARS menus emerged to align with strong government health promotion messages. The menus highlighted the potential to boost customers' immune system and were based on Chinese herbal medicine (Tse et al. 2006). This approach was well-received by the people of Hong Kong, who traditionally have lots of Chinese medical herbs in their diets. Take-away and door-to-door deliveries were enhanced to cater to customers that were still afraid of getting out to the restaurants.

Given that SARS transition included airborne water molecules, restaurants had to deploy clear and adequate sanitation (including sanitising) interventions on-premises, to lower the perceived risk of possible contamination. Many restaurant-goers will have this fear after COVID-19. Hence, restaurants in Hong Kong had to advertise their hygiene policies and highlight measures that they have put in place to protect patrons (Tse et al. 2006). Cleanliness became a selling point (Lo et al. 2006). Box 9.1 summarises some of the measures introduced by Hong Kong restaurants.

While restaurants in Hong Kong had to develop new disaster risk reduction and management plans, one restaurant group in the USA, the Waffle House Restaurants (Ergun et al. 2010), introduced proactive measures to deal with hurricanes. As preparation begins before the hurricane season, over time, the group drew lessons and documented them for future hurricane seasons. In a way, the COVID-19 pandemic will present lots of lessons for the future in the restaurant sector, with the view to minimise job losses and bouncing back to normal operations quicker and smarter. Many countries in Africa are experiencing lockdowns for the first time in their histories.

Drawing on experiences in Indonesia, Rindrasih et al. (2019) discuss the performance of the tourism industry in response to disaster events. After analysing multi-year disasters, from 1998 to 2016, the findings revealed that disasters affect the

Box 9.1 Restaurant Measures after SARS 2003

- All staff were instructed to wear protective surgical (face) masks.
- Patrons were provided with disposable disinfectant tissue packs upon their arrival at the restaurants.
- A "hygiene ambassador" was employed to greet customers and offer anti-bacterial wipes.
- Regular restaurants disinfection, with all eating utensils, kept in a sterilised compartment or immersed in hot water to eradicate the virus.
- Dishes were accompanied with a "common" pair of chopsticks that everyone used to transfer food to his/her bowl instead of using the pair for eating.
- Restaurant buffets were temporarily stopped.
- New interventions would be put in place as more information is received in the mainstream.
- New crisis management plans had to be put in place, and these were not only for SARS but also for other potential crisis, as highlighted earlier.
- The team approach to crisis management was improved, and this remains key in all disaster situations.

Source: Tse et al. (2006, pp. 9-10); Lo et al. (2006, pp. 73-74)

tourism industry differently. The level of impact depends on, for instance, the scale of the destruction, location and nature of the disasters – whether human- or nature-induced. However, whatever the nature of the disaster, there is a need for coherent recovery policy frameworks, driven by all key stakeholders, including government, the industry concerned and organised labour. The media too should take responsibility by portraying destinations positively, as tourism is a susceptible sector in terms of perceptions (Dube and Nhamo 2020a).

The next section presents the methodological framework informing this work.

9.3 Methods and Materials

COVID-19 impacts on various types of restaurants, which includes family dining, casual dining, fine dining, quick service, fast casual, coffee and snack shops and bars and taverns, were explored. Secondary and archival data formed the bulk of the data used in this study. The authors took advantage of the publicly available data to cut financial research costs. Part of the rich data sources used included the OpenTable platform. This is an online restaurant reservation company is related to Booking.com an online platform. The OpenTable platform hosts over 60,000 restaurants across the world. The research sample targeted states or cities with more than 50 restaurants and focused on seated dinners at restaurants. The comparison was at the

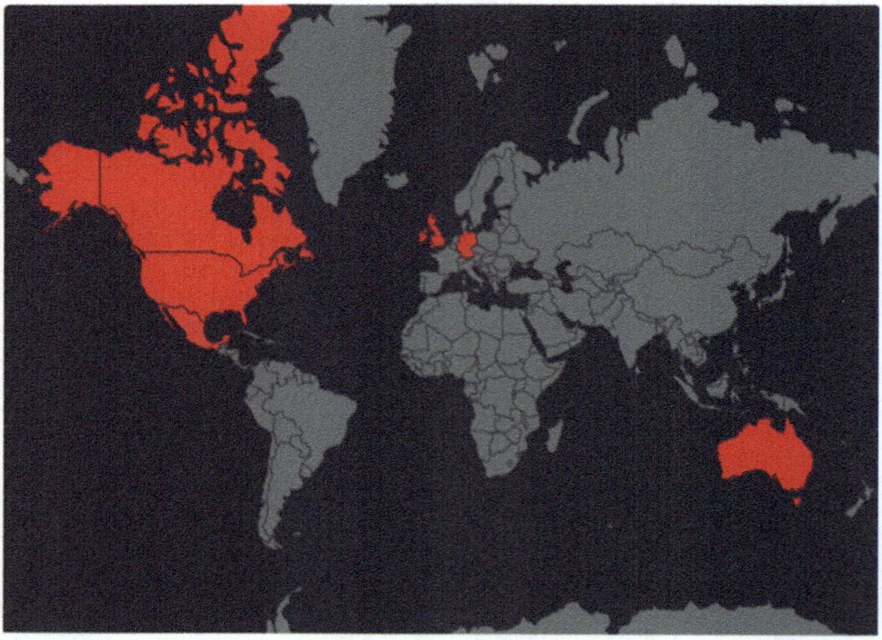

Fig. 9.1 Coverage of sampled restaurants across the world. (Source: OpenTable 2020)

annual, weekly and, where possible, even at the daily levels of restaurant patronage. The data only included states or cities with more than 50 restaurants. The sample selection criteria meant that most restaurants informing this study were from Australia, the USA, Canada, Germany, Ireland, Mexico and the UK (Fig. 9.1).

Given that the USA has arguably the highest number of restaurants, there was a need to examine the trends in the country using both archival and secondary data from the National Restaurants Association (NRA). The NRA is the official representative for all restaurants across the 51 states in the USA, including Hawaii.

The baseline data that were used for OpenTable was for 18 February to 28 March for 2019, which was cross compared with data for the same period in 2020. Other secondary and archival data used were from authoritative sources such as government agencies and mainstream media outlets. Content and thematic analysis were used to analyse qualitative data as advised by Vaismoradi et al. (2013), while the Microsoft Excel Toolpak was used to analyse quantitative data.

9.4 Presentation Data and Discussion of Key Findings

Like any other part of the tourism industry, the restaurant industry is sensitive, fragile too, and quickly responds to external shocks (Dube et al. 2020). While cases of COVID-19 infections first appeared in December 2019, the disease gained prominence in early 2020, when the World Health Organisation (WHO) received notification from China, on 31 December 2019. With 7818 confirmed cases in 19 countries (Fig. 9.2) and 170 deaths in China, the outbreak was declared a public health emergency of international concern on 30 January 2020 and a pandemic on 11 March 2020 (WHO 2020a). Thereafter number of mortality and morbidity soared (Fig. 9.3) leading to broad concerns, and the related alarm bells have not ceased since then persisting to the time when this chapter was being finalised.

In as much as there was a lag of 18 days between the available data and the declaration of COVID-19 as a global emergency, in most countries the number of restaurant bookings in mid-February 2020 was similar to those in 2019 (Fig. 9.4). Most notable is that despite the 30 January WHO declaration, some restaurants continued operating as usual without taking precautions, and some were even witnessing growth relative to the previous years. Germany, for example, recorded better business than in previous years up to 1 March 2020. Despite all this, by the beginning of March 2020, it was clear to most people that the disease was a severe challenge, with WHO calling on a raft of measures to help curb and contain it. This call culminated in the declaration of a global pandemic on 11 March 2020, as mentioned earlier.

One of the critical measures heavily promoted by the WHO and many governments to fight the human-to-human transmission of COVID-19 transmission was social distancing. By the time the disease was declared a pandemic, most people were already heeding calls to for social distancing. Complementing this call was avoiding areas where large groups of people could socially interact as this was perceived to be a high-risk practice where the disease could be easily spread. Post declaration of the diseases as a pandemic, there was a sharp decline in restaurant

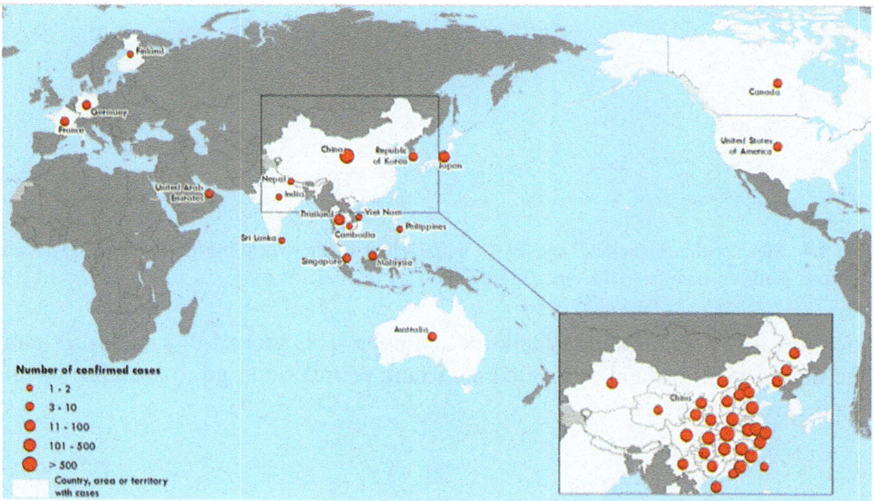

Fig. 9.2 Distribution of COVID-19 cases as of 30 January 2020 (Public Health Emergency of International Concern). (Source: World Health Organization 2020a)

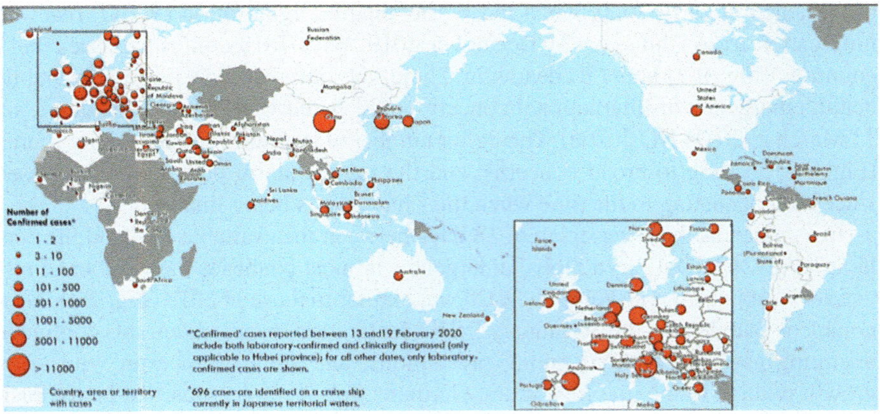

Fig. 9.3 Distribution of confirmed COVID-19 cases as of March 112,020 (Pandemic Declaration). (Source: World Health Organization 2020b)

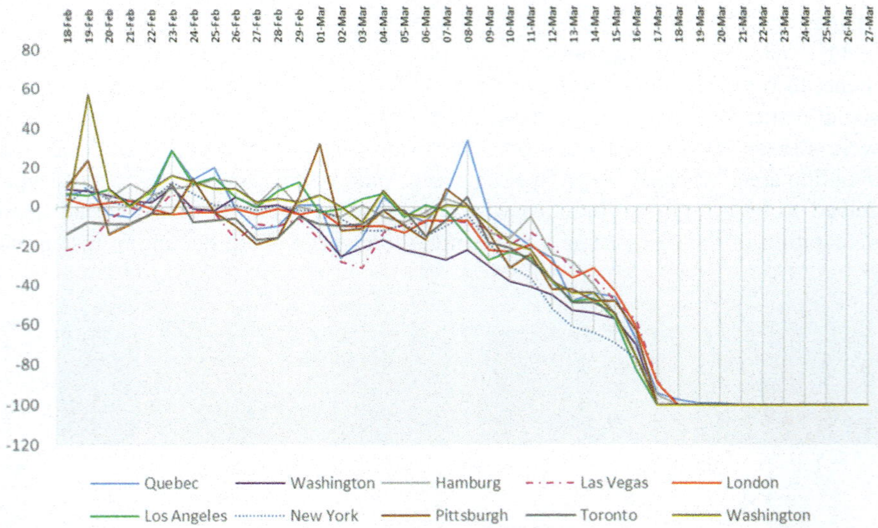

Fig. 9.4 State of the restaurant industry in major cities of the world during the COVID-19 era. (Source: Authors, data from the OpenTable 2020

bookings for sit-in restaurant facilities and pubs. This ground to a standstill many fine dining facilities and other eateries where people often go to socialise having meals across the world.

9.4.1 *Impact of COVID-19 on the Global Restaurant Chain*

According to data from OpenTable, the sit-in facilities experienced a total shutdown marked by 100% decline in sit-in bookings and registrations at restaurants, pubs and inns. In as much as the year started on a high note for most restaurants in major cities across the world, the situation drastically changed in early March where restaurants in various towns across the world started to record business way below the business; they recorded in 2019. Post the 10 of March, some business in some cities were more resilient, namely, London, Hamburg, Las Vegas and Washington. Despite some measure of resilience amid declining business by 22 March 2020, almost all restaurants in major cities were recording business levels of between 2 and 0% as compared to 2019 figures in most cities. Figures 9.4 show how that the restaurant industry was utterly decimated by the COVID-19 pandemic in the month of March 2020 in various cities across the world.

Arguably, the USA has one of the largest restaurant industries globally, and coincidentally, the country became the global epicentre for COVID-19 infections and deaths. As of the 29 April 2020, the USA had 981,246 confirmed COVID-19 cases and 55,258 deaths (Centers for Disease Control and Prevention 2020). This was significant in the context of 3,117,880 global confirmed cases and 217,212 deaths (Johns Hopkins University and Medicine 2020). Despite this, pandemic managing restrictions within the country were not as stringent as in many other countries with notably lower infections and deaths.

9.4.2 *Impact of COVID-19 on USA Restaurant Industry*

It emerged that before COVID-19, the USA had about 64,000 restaurants (Fig. 9.5) which employed about 15 million workers (Fig. 9.6) with a sales revenue of about US$714 billion annually (Fig. 9.7). This shows that the restaurant industry is quite prominent in the USA with significant revenue. The states which benefit a lot from the restaurants business include California, New York, Florida and Texas. Coincidentally these states were among the worst affected by COVID-19.

The study revealed that the occurrence of COVID-19 had a devastating impact on the tourism-linked and dependent restaurant industry that was performing well before the pandemic. Prior to the pandemic, the industry was showing signs of growth and creating jobs but then started shedding jobs as from March 2020. Facing with a grim future, the restaurant industry representative association wrote to the Speaker of House of Representatives on 20 April 2020 highlighting the challenges in the sector and thus requested for government assistance in a number of areas (Association National Restaurant 2020a, 2020b).

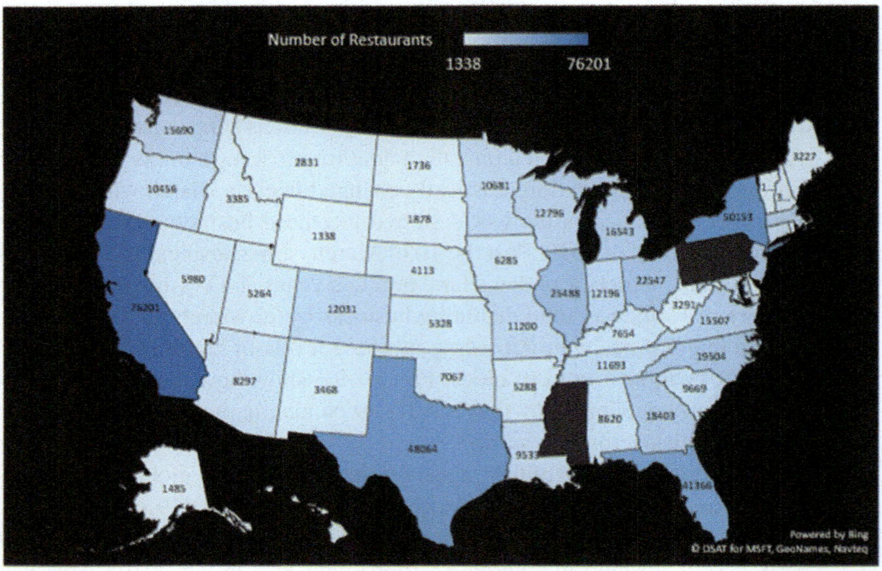

Fig. 9.5 Distribution and number of restaurants across the 51 USA states before COVID-19 (n = 63,6154). (Source: Authors, data from the National Restaurant Association 2020a, 2020b)

An interesting issue is the rapidity of job losses that relate to and indicate the delicacy of the cash flow position of the sector. Figure 9.8 shows a downturn in industry job performance of industry jobs. Employment figures in the industry show that the restaurant sector had been creating jobs up to February 2020, where 50,600 jobs were created in 1 month alone. In March 2020, the rapid job losses wiped off all the jobs that were created in 1 year.

Figure 9.9 indicates that all the seven restaurants types reported that business had slumped in comparison to 2019. Data were gathered through a survey conducted by the NRA of America among 6500 restaurant operators across the 51 states between 10 and 16 April 2020. Over 95% of the respondents indicated that business had declined. The worst affected restaurants types were fine dining, followed by bars and taverns. These findings confirm earlier findings that sit-ins almost evaporated in the USA due to COVID-19. Fine dining restaurants, family dining, casual dining and bars and taverns ordinarily promote close personal contact a favourable condition for the spread of COVID-19. Most governments across the world closed down restaurants' operations. However, quick service and fast casual were not as severely affected, reporting a decline of 57% and 64%, respectively. This could have been attributed to the fact that these could offer home delivery, drive-throughs and take-aways. This approach limits contact between community members. More important is in the USA, take-away and drive-through services were allowed to operate. This somewhat mitigated the level of job losses in the industry and has implications for the industry in cases of future pandemics or disasters that affect the restaurant

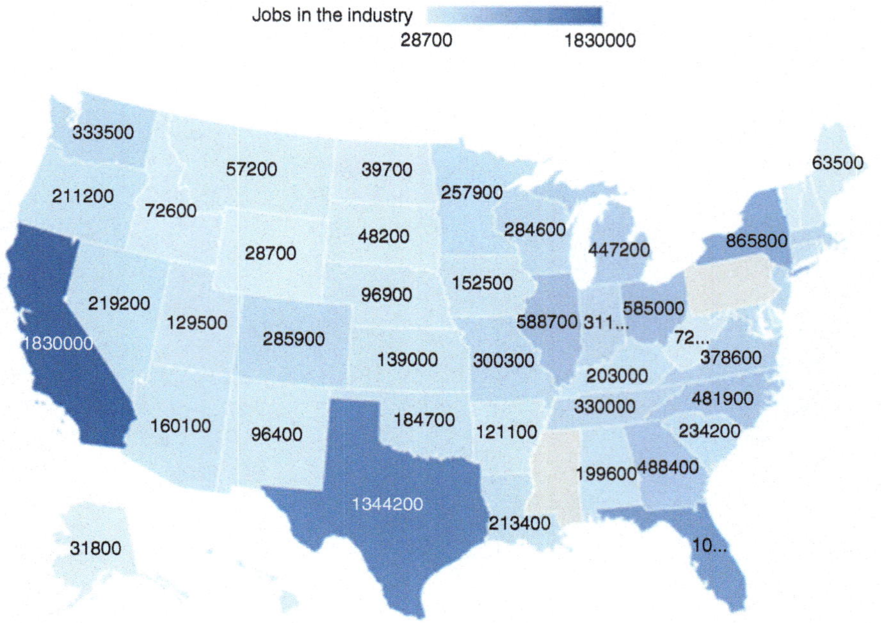

Fig. 9.6 Jobs in the restaurant industry before COVID-19 (n = 15, 109, 900). (Source: Authors, data from the National Restaurant Association 2020a, 2020b)

industry. The popularity of fast-food restaurants has been growing in several countries across the world including in China (Zhang 2018), Zimbabwe (Shumba and Zindiye 2018), various countries (Khan et al. 2018), South Africa (Thompson et al. 2018) and indeed the USA (Zagorsky and Smith 2017). However, countries such as Zimbabwe and South Africa did not permit the operation of fast-food such services during the early periods of their lockdowns.

The slump in business resulted in jobs layoffs and furloughs. The highest number of staff reduction was experienced in the fine dining, bars and taverns as well as family dining where operations went down to less than 20% of the staff (Fig. 9.10). The quick service and fast casual dining places had somewhat moderated job losses as they could operate with about 60% of their staff. The option for takeaways provided a much-needed buffer for that sector.

By the end of March 2020, data from the NRA shows that in the USA, job losses had been extended to three million. The industry recorded $25 billion losses across its members, representing a 50% decline in anticipated revenue. As of April 20, a total of more than eight million restaurant employees had been either furloughed or laid off, and $80 billion was lost as sales revenue, and 40% of the restaurants were already shut (National Restaurant Association 2020b). Figure 9.11 outlines some of the highlights from that report.

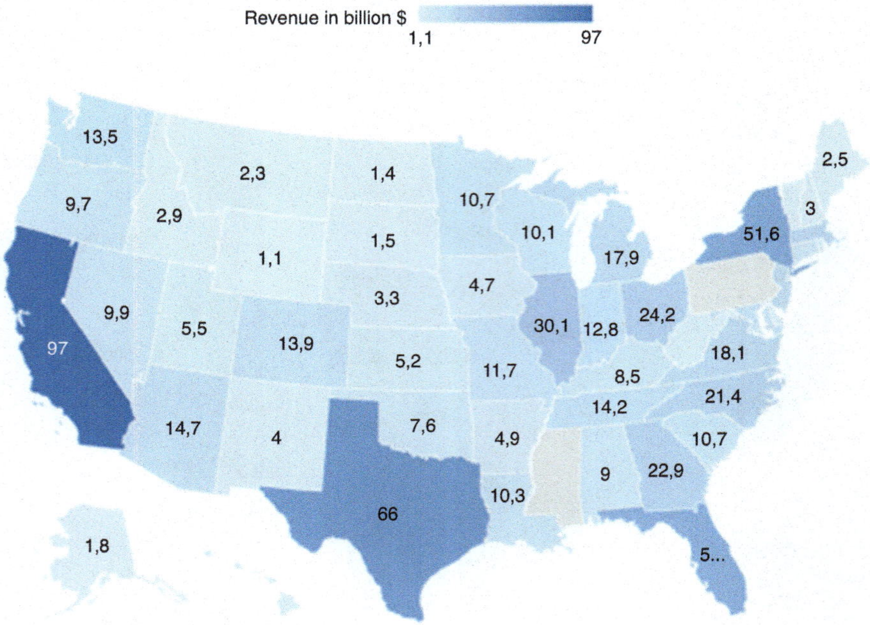

Fig. 9.7 Estimated revenue from restaurant sales by the state before COVID-19 ($714,3 billion). (Source: Authors, data from the National Restaurant Association 2020a, 2020b)

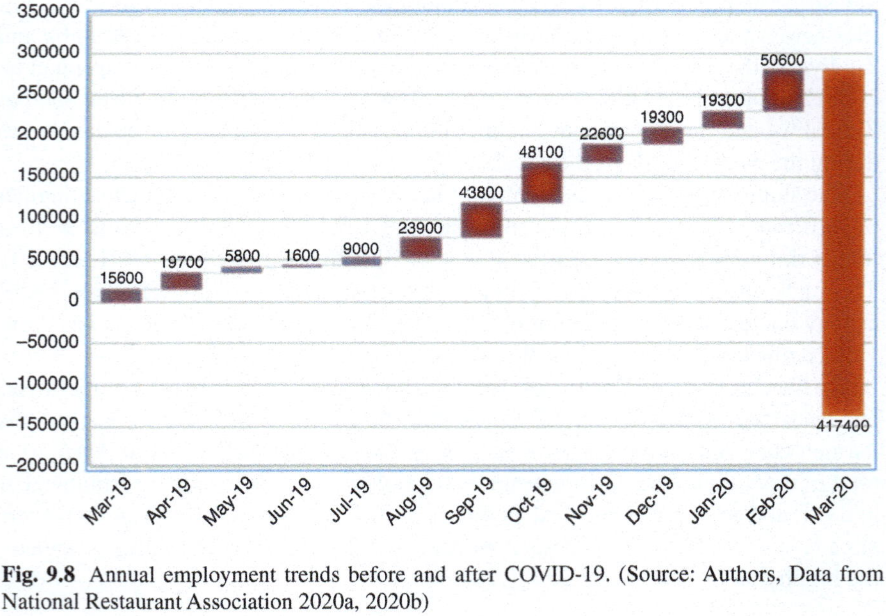

Fig. 9.8 Annual employment trends before and after COVID-19. (Source: Authors, Data from National Restaurant Association 2020a, 2020b)

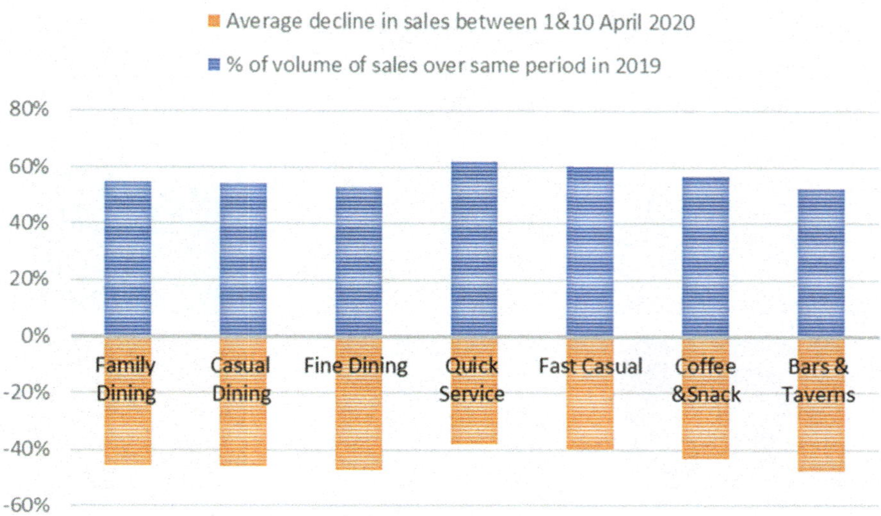

Fig. 9.9 Business trends due to COVID-19 on various types of restaurants in the USA. (Source: Authors, data from the National Restaurant Association 2020a, 2020b)

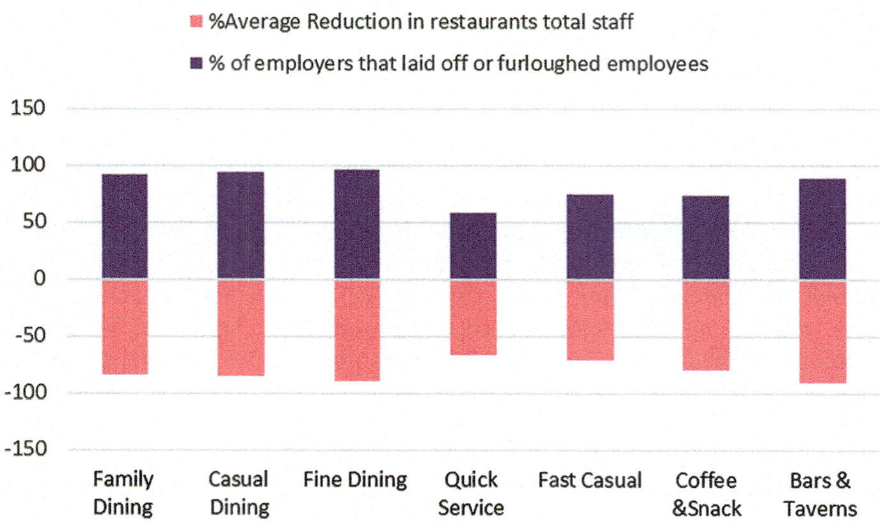

Fig. 9.10 Distribution of furloughed and laid-off workers by type of restaurant due to COVID-19 impacts. (Source: Authors, data from the National Restaurant Association 2020a, 2020b)

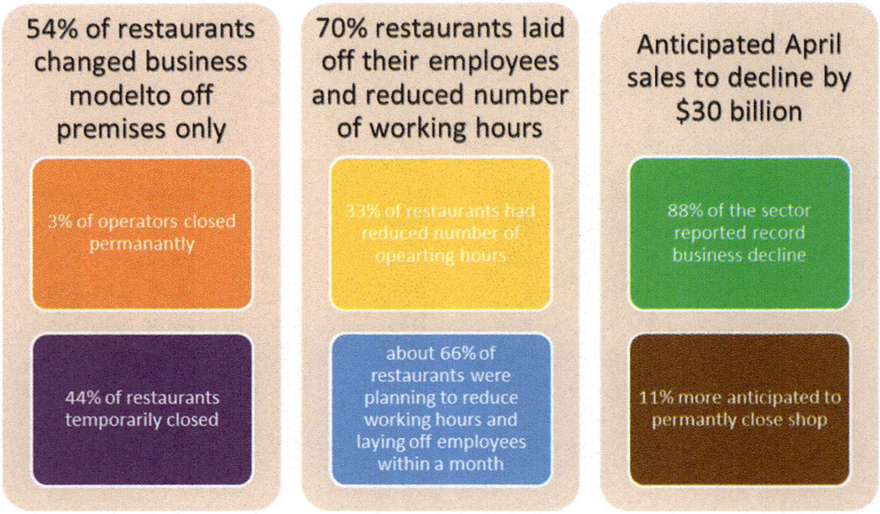

Fig. 9.11 Impact of COVID-19 on USA restaurant revenue, employment and sustainability end of March 2020. (Source: Authors, data from the National Restaurant Association 2020)

9.4.3 COVID-19 Impacts on Restaurants Sustainability and Recovery Pathways

As stated earlier, the April 2020 NRA letter to House of Representatives indicated that the restaurant sector was in dire straits and was anticipating unprecedented job and revenue loss levels. The letter indicated that at least eight million restaurants employees were either furloughed or laid off since the beginning of the COVID-19 pandemic. The NRA further revised the loss in revenue, predicting it to go up to $ 50 billion by the end of April 2020. At least 40% of the restaurants had closed down with some having no prospects of reopening again. As a consequent the industry representative launched a bid for a $240 billion to cater for immediate expenses between March and June 2020 (Association National Restaurant 2020). The Association further highlighted that given the unique nature of the restaurant industry and the tourism industry in general, they were bidding for a unique structured relief programme to alleviate challenges that the sector was facing. At that point most small industry players were at higher risk of collapsing.

While the tourism industry is often touted as a vehicle for employment creation and most recently as a tool to achieve sustainability (Dube 2020), the sector's sensitivity to natural disasters worrying. The sensitivity of the restaurant and generally tourism industry led many to question the sustainability of the sector. It emerged from the study that the industry took less than a month to reach rock bottom as many companies indicated incapacity to pay their franchise fees and bills let alone pay salaries to their employees regardless of years of growth in some countries. The pandemic threatened the viability of many restaurant businesses across the world as most were either closed or operating below sustainable levels to allow them to break even.

This led many restaurants to lay off workers at historical scales never seen in history. As a consequence of industry disruptions by COVID-19, the Private Sector Job Quality Index (JQI) in the USA estimated that about 10.8 million employees working in bars and restaurants lost their jobs (JQI 2020). With the industry at a standstill globally, other industry employees in the hospitality sector supply chain were also faced with the same predicament as the supply chain and the demand was severely disrupted.

Most affected people from the COVID-19 were already vulnerable employees, who faced temporary and permanent job layoffs. The move by the American government to provide a US$1200 once-off payment to employees affected by the coronavirus and USS$2 trillion stimulus package set aside to improve unemployment benefits and to offer several tax cuts, grants, loans and relief that also benefited the tourism sector was commendable. While the industry was pleading for government assistance, it may be critical for the tourism industry reflect on its relations with most of its employees and in turn assist even beyond the COVID-19 pandemic. Some governments, including South Africa, were assisting employees whose income were cut or had lost their jobs through a reduction in their pay-as-you-earn (PAYE) liability or by giving them tax breaks (News24 Team 2020).

In as much as the USA injected a substantial amount of money to help the tourism economy, it was sad to note that the sector still lost a considerable number of jobs with the restaurant industry shedding millions of jobs between March and April 2020. This could have been attributed to several factors. One of the factors is that the USA funding was closely tied to the Pay Cheque Protection Programme, which was primarily tailor-made to keep workers at work even if there was no work to be done. However, employees often found it challenging to keep industry doors open as other expenses such as franchise costs, rates, debt interest and other such fixed business costs were more burdensome than the burden of paying employees their salaries and benefits. One other aspect about the Pay Cheque Protection Programme was that it was for payment of salaries and benefits of up to 8 weeks starting in April with the repayment due after 6 months (period around December 2020) and within that period the employer could not reduce employee salaries or reduce the number of workers. With a lot of uncertainties, most businesses found it to be unpalatable for a good cause as it would ultimately bankrupt employees. The package was designed with voting in mind with the aim of keeping as many people as possible at all cost to benefit the incumbent President Donald Trump and his party in power as presidential elections were in early November 2020. The National Restaurant Association (NRA) then launched a bid that outlined industry requirements for April to June with a bid for a $240 billion industry relief package. The sector suggested several things to be changed to make them attractive (Box 9.2).

From Box 9.2, it emerges that the industry was uncertain about the future. However, one thing that was certain is that the COVID-19 epidemic transformed the restaurant industry, and the disruptions were likely to be long term. And most likely change in the industry was a need for measures for social distancing, particularly in

Box 9.2: Making the Restaurant Sector Attractive Again

- The $240 billion RFIRF compensates restaurants for 1) government-ordered closures; 2) reopening "start-up" capital to meet massive challenges of reorienting the restaurant space, restocking inventory and supplying the workplace with new safety equipment; and 3) rehiring and retraining the workforce.
- To allow restaurant entities that had experienced a reduction in sales revenue of 25% or more due to coronavirus to apply to the account, which would be administered by the US Department of the Treasury.
- Allow restaurants to select their loan period after government restrictions end, revise loan forgiveness requirements and restore the 10-year loan-term written in the CARES Act.
- Allow federal support to help restaurants modify physical facilities to accommodate continued social distancing, enhance sanitisation and employee education and expand the use of personal protection equipment and disposable products when employees interact with both customers and each other.
- Increase funding for economic injury disaster loans (EIDLs):
- Restore $50 billion in needed appropriations for restaurants in economic distress, including those that need to access a second EIDL.

Source: Authors, based on National Restaurant Association 2020a, b:1

the absence of a vaccine and treatment. This meant redesigning spaces for business if sit-ins were to be entertained and was imposing an additional fiscal burden on the business.

The demand for new seating arrangement meant a reduction in the number of people that can dine at any given time resulting in a reduction of capacity and equally business. The industry was also likely to be left with no option but to adopt new hygiene measures among employees and patrons. The demand for hygiene also meant increased operating costs in the purchasing of sanitisers in safety and health training in some respects as the industry begins to open. Some sectors of society were also demanding contactless experience hence the need to invest in QR machines and contactless card machines and replacement of menu boards with electronic boards. In that regard there was need for fresh injection of cash from either government or shareholders.

The industry had to find innovative solutions to raise capital to ensure liquidity in difficult times. This could be done in several ways through both private and public sector support in the quest to find alternative revenues for the restaurant sector. Initiatives such as a pay later voucher system or starting online branded cooking lessons remained good positives although such measures were largely inadequate to meet the cash demands of the sector that was losing more revenue than it was making. In many countries, where COVID-19 containment measures led to restaurants closures such as in South Africa, well-coordinated systems could have been created to allow food deliveries or pick-up systems as it was likely to take long before people could sit and dine. By the time such measures were finally allowed a lot of potential revenue was already lost. The World Bank recommended similar measures to be undertaken to rescue the tourism industry to ensure its continued contribution to global and local economies (Freund, 2020).

Additional efforts that can assist the industry may include a well-coordinated knowledge exchange process among industry players through sector organisations. This would facilitate the sharing of knowledge on sector funding opportunities and how to plan and cement ties for destination marketing post the pandemic. Government and financiers could play a critical role in providing rescue solutions for the suffocating small-to-medium enterprises (SMEs) by means of debt holiday breaks and reduced interests on borrowings. These measures were already in place in Rwanda and South Africa (Money Marketing 2020; National Bank of Rwanda 2020). The same agreement could also be forged between industry players and suppliers where feasible to provide the needed fiscal space ensuring as many businesses as possible are rescued. Where establishments were leasing and renting, and based on merit, there was a need to provide rental holidays allowing the industry to come out of the pandemic with minimal revenue and job losses. This worked during the SARS pandemic in 2003 (Lee and Warner 2005a, b). The government could also support troubled tourism businesses with tax breaks or reduced taxes, for example, personal income tax, business rates, value-added tax and other licensing expenses that can be deferred so that business has enough liquidity as they strive to remain afloat.

9.5 Conclusion and Recommendations

The study evaluated the impact of COVID-19 on the hospitality industry, with a focus on the restaurants sector. It emerged that due to the pandemic, most of the restaurants, just like the rest of the tourism industry, were brought to a halt. Business dropped to near zero in most countries across the world and cities. Pandemic management measures threatened the survival of most restaurants, some to the point of failure to recover post the pandemic. Of concern were the related job losses mainly affecting the chronically vulnerable societies, largely women and youth who make up the bulk of restaurant employees. The majority of the restaurant workers are usually underpaid and often have their salaries subsidised by tips from clients. This exacerbates the vulnerability of this group.

Given the challenges faced by tourism employees, the chapter recommends measures that could ease pressure on both businesses and employees. The measures include having several concessions aimed at cushioning businesses from collapse. Among other things, the study recommends the enhancement of already emerging tax rebates, loan repayment holidays, government grants, preferential loans and a reduction on VAT and personal income tax to ensure the survival of the industry post the pandemic. To cushion employees, there was a need to pay out leave days, where possible, and have government institute additional and more aggressive PAYE relief and unemployment grants to cover employees whose salaries were cut or wholly laid off. Central banks could continue cutting repo rates as appropriate to near zero and/or even negative until there was no more room for it.

The hospitality industry and the restaurant sector may have changed forever, with virtual and remote spaces being the new normal. To this end, there is a need for the industry to increase the pace of automation and embrace the possibility of increased online dealings. The pandemic made customers feel uneasy and fearful of touching menu boards and also to make physical contact with the payment machines. To that end, we recommend innovative alternative solutions such as the use of electronic display screens and QR codes to limit contact with physical infrastructure. This would assist in bringing back confidence to dinners as it was likely to take time for customers to recover from the trauma of COVID-19. Perceptions of the possibility of getting infected in a restaurant are likely to persist for a while. Other innovations could include riding on the already existing use of drones and robots in the restaurant space. Drones could be used for deliveries. This space would also include the rapid deployment of food deliveries, i.e. Uber Eats, in meeting customer and business demands. However, there is a caveat: a massive roll-out of the technology may lead to technology-induced joblessness.

Given that the study was done during the pandemic, there will be a need for follow-up studies. Such studies could be done at the company, regional, national, continental and international levels. This will assist in building industry and broad knowledge and experience base for use in future events that are most likely to occur than not.

References

Association, National Restaurant. (2020). *Letter to house senate leaders*. Washington DC: National Restaurant Association. Retrieved April 28, 2020, from https://www.restaurant.org/Downloads/PDFs/business/COVID19-Letter-to-House-Senate-Leaders.pdf

Breda, Z. (2004). The impact of severe acute respiratory syndrome (SARS) on China's tourism sector. *Tourism Research Journal, 1*(2), 5–14.

Centers for Disease Control and Prevention. (2020). *Coronavirus Disease 2019 (COVID-19)*. Centers for Disease Control and Prevention. Retrieved April 29, 2020, from https://www.cdc.gov/coronavirus/2019-ncov/cases-updates/cases-in-us.html

Dube, K. (2020). Tourism and sustainable development goals in the African context. *International Journal of Economics and Finance Studies, 12*(1), 88–102.

Dube, K., & Nhamo, G. (2020a). Vulnerability of nature-based tourism to climate variability and change: Case of Kariba resort town, Zimbabwe. *Journal of Outdoor Recreation and Tourism, 29*, 5–15. https://doi.org/10.1016/j.envdev.2018.12.006.

Dube, K., Nhamo, G., & Chikodzi, D. (2020). COVID-19 cripples global restaurant and hospitality industry. *Current Issues in Tourism*, 1–4. https://doi.org/10.1080/13683500.2020.1773416.

Ergun, Ö., Stamm, J., Keskinocak, P., & Swann, J. (2010). Waffle House restaurants hurricane response: A case study. *International Journal of Production Economics, 126*(1), 111–120. https://doi.org/10.1016/j.ijpe.2009.08.018.

Freund, C. (2020). *We can't travel, but we can take measures to preserve jobs in the tourism industry.* World Bank Blogs. Retrieved March 30, 2020, from https://blogs.worldbank.org/voices/we-cant-travel-we-can-take-measures-preserve-jobs-tourism-industry

Gössling, S., Scott, D., & Hall, C. (2020). Pandemics, tourism and global change: A rapid assessment of COVID-19. *Journal of Sustainable Tourism.* https://doi.org/10.1080/09669582.2020.1758708.

Job Quality Index (JQI). (2020). *Statement from the U.S. Private sector Job Quality Index ("JQI") team on vulnerabilities of jobs in certain sectors to the Covid-19 economic shutdown.* The U.S. Private Sector Job Quality Index (JQI). Retrieved March 30, 2020, from https://www.jobqualityindex.com/

Johns Hopkins University and Medicine. (2020). *COVID-19 dashboard by the center for systems science and engineering (CSSE) at Johns Hopkins University (JHU).* Johns Hopkins University. Retrieved April 29, 2020, from https://coronavirus.jhu.edu/map.html

Khan, M. A., Khan, M. M., & Gadelrab, R. M. (2018). Sodium content in fast foods: Assessment of menu items in selected countries. *Journal of Foodservice Business Research, 21*(5), 553–569.

Kim, S., Chun, H., & Lee, H. (2005). The effects of SARS on the Korean hotel industry and measures to overcome the crisis: A case study of six Korean five-star hotels. *Asia Pacific Journal of Tourism Research, 10*(4), 369–377.

Kim, J., Kim, J., Lee, S., & Tang, L. (2020). Effects of epidemic disease outbreaks on financial performance of restaurants: Event study method approach. *Journal of Hospitality and Tourism Management, 43*, 32–41.

Lee, G., & Warner, M. (2005a). The consequences of the SARS epidemic for China's employment and human resources. *The Journal of Comparative Asian Development, 4*(1), 9–28.

Lee, G., & Warner, M. (2005b). Epidemics, labour markets and unemployment: The impact of SARS on human resource management in the Hong Kong service sector. *The International Journal of Human Resource Management, 16*(5), 752–771.

Lee, G., & Warner, M. (2006). The impact of the SARS epidemic on the service sector in Singapore. Human resources, labour markets and unemployment. *Asia Pacific Business Review, 12*(4), 507–527.

Lo, A., Cheung, C., & Law, R. (2006). The survival of hotels during disaster: A case study of Hong Kong in 2003. *Asia Pacific Journal of Tourism Research, 11*(1), 65–80.

McKercher, B. (2003). SIP (Sars induced panic) a greater threat to tourism than SARS (severe acute respiratory syndrome). *e-Review of Tourism Research, 1*, 1.

National Restaurant Association. (2020a). *Restaurant industry blueprint for recovery.* National Restaurant Association. Retrieved April 29, 2020, from https://www.restaurant.org/Manage-My-Restaurant/Business-Operations/preparedness/Restaurant-Industry-Blueprint-for-Recovery

National Restaurant Association. (2020b). *COVID-19 update.* National Restaurant Association. Retrieved April 28, 2020, from https://restaurant.org/Downloads/PDFs/business/COVID19-Infographic.pdf

OpenTable. (2020). *The state of the restaurant industry.* Retrieved March 29, 2020, from https://www.opentable.com/state-of-industry

Rindrasih, E., Witte, P., Spit, T., & Zoomers, A. (2019). Tourism and disasters: Impact of disaster. *Scientific Research, 12*(2), 93–115. https://doi.org/10.4236/jssm.2019.122006.

Shumba, K., & Zindiye, S. (2018). Success factors of franchise entrepreneurs operating in a volatile business environment: A case of the fast food industry Harare, Zimbabwe. *The Social Sciences, 13*(4), 908–991.

Thompson, C., Ponsford, R., Lewis, D., & Cummins, S. (2018). Fast-food, everyday life and health: A qualitative study of 'chicken shops' in East London. *Appetite, 128*, 7–13.

Tse, A., So, S., & Sin, L. (2006). Crisis management and recovery: How restaurants in Hong Kong responded to SARS. *Hospital Management, 25*(2006), 3–11.

Vaismoradi, M., Turunen, H., & Bondas, T. (2013). Content analysis and thematic analysis: Implications for conducting a qualitative descriptive study. *Nursing & Health Sciences, 15*(3), 398–405.

Wilder-Smith, A. (2006). The severe acute respiratory syndrome: Impact on travel and tourism. *Travel Medicine and Infectious Disease, 4*(2), 53–60.

World Health Organization. (2020a). *Novel Coronavirus(2019-nCoV) situation report - 10.* World Health Organization. Retrieved March 29, 2020, from https://www.who.int/emergencies/diseases/novel-coronavirus-2019/situation-reports

World Health Organization. (2020b). *Coronavirus disease 2019 (COVID-19) situation report – 51.* World Health Organization . Retrieved March 29, 2020, from https://www.who.int/emergencies/diseases/novel-coronavirus-2019/situation-reports

Zagorsky, J. L., & Smith, P. K. (2017). The association between socioeconomic status and adult fast-food consumption in the US. *Economics & Human Biology, 27*, 12–25.

Zhang, X. (2018). Spatial patterns and social/cultural implications of Japanese fast food chains in China. *Asian Geographer, 35*(1), 15–34.

Chapter 10
Impact of COVID-19 on the Global Sporting Industry and Related Tourism

Abstract Infectious diseases present an omnipresent threat to the health and safety of the global community. In 2020, the COVID-19 pandemic that began in China in 2019 swept around the world imposing unprecedented impacts on all the sectors of the world economy. The multibillion-dollar sports tourism industry was hit hard. This chapter unravels the impacts of COVID-19 on sporting events and, by default, sporting tourism. Such information is key for the industry to develop resilience against future pandemics and other natural disasters. Document and critical discourse analysis was used in gathering and analysing data. The findings show that most sporting events were either cancelled or postponed during the peak of the pandemic because of COVID-19 containment measures across the world, such as travel bans and port closures. Many sports governing bodies were also bankrupted and left in financial distress. Individual athletes lost income as their salaries were reduced or completely cut. Sponsors, betting firms, broadcasters and others in the sports value chain also incurred huge losses during this period. Among the key events affected were the 2020 Olympics, football leagues, golf, cricket, marathons, rugby and tennis. As such, there was a need for bailout packages and seeking alternatives for the sports industry to recover from the shock. This chapter recommends post-COVID-19 reviews of the sports industry and putting in place alternative future intervention measures for similar pandemics and other disasters.

Keywords COVID-19 · Sports tourism · Coronavirus · Football · Olympics · Comrades Marathon

10.1 Introduction and Background

With a basic reproduction rate of 2.2 (every 1 person infects 2.2. others), of relevance to the COVID-19 outbreak was its timing and the potential for super-spreader events. COVID-19 originated in Wuhan, which is a city of 11 million people, and

G. Nhamo et al., *Counting the Cost of COVID-19 on the Global Tourism Industry*, https://doi.org/10.1007/978-3-030-56231-1_10

coincided with the world's largest mass population movement, namely, the Chinese Spring Festival (Ebrahim and Memish 2020). Chinese nationals come from all over the world to attend this festival – hence the heightened risk of global spread. Mass gatherings have been observed to be fertile grounds for the super-spreading of infectious diseases (News24 2020a). Participation in sports training, sports competitions and spectating have been noted to be possible mass transmission pathways of infectious diseases. The transmission is facilitated by the overcrowded conditions and sharing sporting equipment and practice surfaces (Raymond et al. 2004).

The sports industry has been one of the fastest growing industries in the world. In North America, for example, the industry reached the US$67.3 billion value mark in 2016, with the figures projected to reach US$78.5 billion by 2021 (PricewaterhouseCoopers 2017). This means that sports tourism is fast becoming an important driver of economic development. Within the sports industry, spectators, players, sponsors, broadcasters, betting networks and the host community derive important economic value from sporting events (Choa et al. 2019). Sports tourism can be defined as "sport-based travel away from the home environment for a limited time, where the sport is characterised by unique rule sets, competition related to physical prowess and play" (Hinch and Higham 2011: 21–22). Sport tourism events are conceptualised as a hybrid between participation in sport by the tourist as a recreational or high-performance athlete and recreational spectator activities. Sports tourism is therefore the use of sport as a touristic endeavour (Hinch and Ito 2018). In Britain, for example, as of 2015 nearly 800,000 overseas visitors each year travel to watch football and visit famous football grounds (Macgowan 2015). Despite their importance to the local economy, sports events and tourism are very sensitive to both natural and human-induced disasters, and especially health-related pandemics, given the overcrowded conditions that dominate the industry's activities.

The world was put on hold by the emerging 2019 coronavirus pandemic, which surpassed the combined toll of the 2003 severe acute respiratory syndrome (SARS) and Middle East respiratory syndrome (MERS) outbreaks in terms of deaths (WHO 2020). With the public health burden of COVID-19, ever-increasing, urgent and unprecedented decisions to contain the outbreak had to be made the world over (Poon and Peiris 2020; Dube et al. 2020). Sporting events and sport tourism are known to boost local and national economies, because they bring diversified revenue channels to the host communities. Due to the outbreak of COVID-19, sporting events and the chain of activities that support these events were seriously impacted. There was an almost global shutdown on sporting events and the related sports tourism.

This chapter investigates the impacts of the COVID-19 pandemic on sporting events and, by default, sports-related tourism. The purpose is to document experiences with a view to learning and building the resilience of the sporting industry against future shocks.

10.2 Literature Survey

The potential damage posed by the start and spread of emergent infectious diseases has been illustrated recently by the H1N1 (2009), SARS, MERS, Ebola and Zika outbreaks. There is considerable evidence showing that such epidemics are likely to become more frequent and more fatal, unless drastic action is taken to mitigate their spread on a global scale (Massaro et al. 2018). In 2020 the world witnessed the emergence of a new, viral, zoonotic pathogen (SARS-CoV2) causing an outbreak of the coronavirus disease 2019 (COVID-19) (Rodriguez-Morales et al. 2020). This led the World Health Organization (WHO) to declare a Public Health Emergency of International Concern. COVID-19 originated in Wuhan, Hubei province, China, in 2019. The epidemic then extended to all provinces of the country and then to neighbouring countries, such as Vietnam, Russia, Nepal and other closely connected countries such as Thailand, Singapore, Malaysia, the Philippines, South Korea and Japan. Within 2 months, COVID-19 had affected every continent of the world (Nishiura et al. 2020), with the epicentre shifting from China to Europe (especially Italy and Spain) and then to the USA (WHO 2020).

The COVID-19 pandemic, as expected, disrupted the health status and economies of almost all the countries across the globe. Zumla et al. (2015) observed that in the past there was wishful anticipation by the political and scientific communities that coronaviruses like MERS-CoV and SARS-CoV would disappear with time. However, this has not happened and cases of especially MERS have continued to be reported from the Middle East throughout the years, since its discovery in 2012 (WHO 2015). This is because there is a large reservoir of these coronaviruses in animals such as bats, camels and pangolins, in addition to there being no specific treatment or vaccine. With the interconnected global village we live in, the potential risk of a global spread of an infectious disease like COVID-19 is ever-present (Rodriguez-Morales et al. 2020).

The rapid expansion of modern transportation and the improved mobility of people and goods has led to the fast spread of emerging communicable diseases. Local transportation is key in sustaining epidemics at the national level, while international air travel may facilitate cross-border and continental spread (Biscayart et al. 2020). On a global scale, passenger air travel is known to play a critical role in the spread of infectious disease (WHO 2020), and this is a common mode of transportation in the sports industry and sports tourism. In the city of Wuhan, where COVID-19 began, the local airport, Tianhe International, is a hub for major Chinese airlines. Although it is mostly a domestic airport, code-sharing with several European and North American airlines allows airlines to fly, with a single stop, to the major capitals and main cities around the world in a few hours. In 2018 alone, about 24.5 million passengers arrived and departed at this airport (Wuhan Airport 2020). An estimated five million people left Wuhan weeks before the travel ban and lockdown took effect on 23 January 2020 (Chen et al. 2020), and one would assume these included sports personalities and sports tourists. This fact alone accounts for a realistic possibility of the global dispersal of the causative agent from the source (Biscayart et al. 2020).

Due to air transit, the SARS outbreak that initially emerged in Guangdong, China, in November 2002, quickly spread to 25 countries as far away as Canada by the end of March 2003 (Hui-Ju et al. 2015). Another example is the biggest Ebola outbreak, which occurred in West Africa in 2014, when the virus quickly spread and resulted in enormous infections in countries such as Guinea, Liberia, Sierra Leone and Nigeria. The virus also spread to the USA through infected patients who travelled from West Africa (Von Drehle 2014). COVID-19 was declared a pandemic on 11 March 2020, when it had affected countries on all continents (WHO 2020). The reported case numbers (over 800,000 at the time of writing) were most likely underestimates, given the shortages or unavailability of test kits in many developing countries, the 2.2 basic reproduction value (R_0) of the virus and evidence of viral shedding from asymptomatic infected people (Ebrahim and Memish 2020). In addition to other strategies of reducing the spread of COVID-19, efforts to reduce crowds and mass events have reached unprecedented levels across the world (Ahmeda and Memish 2020; Dube et al. 2020).

Previous outbreaks of coronaviruses such as SARS-CoV had a negative impact on sporting activities at all levels, from the recreational to professional levels. This is because there is always concern about the possibility of infection among athletes and audiences in competitive sports. Most sporting activities involve close contact and usually bring together large crowds from different backgrounds and, sometimes, from different countries. This creates situations that may enhance the risk of a contagious disease spreading locally and even globally (Nishiura et al. 2020). Not surprisingly, after the SARS outbreak, many international events (including both contact and non-contact sports) which were originally scheduled for the first half of 2003 were either cancelled or postponed, thus impacting the sports tourism industry.

Sports tourism occurs when individuals or groups of people travel with sport as the prime motivation for the journey, despite other variations in the activities of the participants. Sports events may stimulate local economic development, for example, in 2010, the New York City Marathon boosted the city's economy by US$340 million (Choa et al. 2019). In 2013, the sports industry in the USA produced 456,000 jobs with an average salary of US$ 39,000. These jobs included athletes, referees, coaches and agents (Depta 2015). In the European Union (EU), the sports industry accounts for 2.12% of the gross domestic product (GDP), amounting to €279.7 billion annually. The industry also employs 2.72% of the total EU workforce, which translates to about 5.7 million people (EU 2018).

Over the last number of decades, many countries and cities have been bidding to host mega sporting events such as the Olympics and the World Cups (rugby, football, cricket, etc.). This is done with the expectation that hosting such events will bring a cocktail of potential benefits to the host countries and cities (Wan and Song 2019) – especially lavishly spending sports tourists. Other sports competitions also bring travellers from all over the world, because there are cricket, athletics and rugby participants and enthusiasts across the world. The chosen host cities are popularly visited throughout the event, even by non-sports personalities who may just be accompanying participants and friends. Occasionally, post-event tourism is positively impacted by the staging of such events, with Barcelona, which hosted the 1992 Olympics, being a good example. Kim et al. (2020) highlighted three main benefits of hosting mega sporting events as infrastructure development, image promotion and economic growth. For example, Beijing's economic growth by 0.8% from 2005 to 2008 was in part attributable to its hosting the 2008 Beijing Summer Olympics (Wan and Song 2019). Hosting sporting events can therefore significantly influence social and economic aspects of the host country, as well as the physical environment. Regrettably, this industry is also very sensitive to disasters of any magnitude.

Not just single-event competitions but also multi-event games were affected by the SARS outbreak. For example, Raymond et al. (2004) pointed out that athletes from Hong Kong and other SARS-affected countries were originally banned from attending the 2003 Special Olympics World Summer Games in Ireland because of contingency measures taken to tackle the SARS problem. It was particularly rare for governments, rather than sports bodies, to limit participation in a sporting event, as happened during the period of the SARS outbreak. Due to the SARS outbreak, the fourth FIFA Women's World Cup was postponed and later moved from China to the USA. The world track-cycling championships, which were scheduled for July in 2002 in Shenzhen, China, were also cancelled because of SARS. This put into jeopardy the economic benefits associated with hosting these events and imposed losses on the organisers and their sponsors.

There are also other necessities during pandemic outbreaks. The organisers of sporting events, in times of disease outbreaks, need to be equipped with sufficient and correct knowledge to effectively address any infection from the perspective of prevention, early recognition and timely control of spread. Appropriate hygiene and etiquette measures have to be taken during sports training and during the actual matches (So et al. 2003). This is because participation in sports training and competition may act as a pathway for possible transmission of a disease (Nishiura et al. 2020). Sharing sporting equipment and practice surfaces can be another potential means of transmission. SARS-infected bodily fluids and respiratory secretions can contaminate sporting equipment and practice surfaces. Close personal contact during training and competition in contact sports like football, basketball and netball can further transmit the contagion from person to person (Raymond et al. 2004).

Raymond et al. (2004) considered crowded changing rooms with dense steam from showers as possible causes of infection due to aerosolised viral particles that can be transmitted from any viral carrier to the entire team. There is also a high probability of viral transmission because steam enhances the floating time of the virus-carrying droplets. Frequent travel to sports destinations may further increase the probability of viral transmission in sports settings. Presence in a relatively crowded aircraft or bus with poor ventilation can be risky if an infected person is travelling with the team. Rothe et al. (2020) and Tian et al. (2020) in their study of coronaviruses highlighted that the chances of infection on an aircraft when infected persons travel during the asymptomatic phase of illness were there, but very low. Therefore, it was advisable to take precautions to prevent possible transmission during flights.

Since there are no specific drugs or vaccines available for COVID-19, the risks associated with infection remain very high. Health systems are overburdened everywhere. Hence targeted and non-coercive community interventions put in place to deal with COVID-19 had an impact on sporting activities and sports tourism. These measures included the cancellation of ad hoc events and the suspension of events with super-spreader potential such as sports crowds. Respiratory infections have been noted to be the most commonly transmitted at such events (WHO 2020). Even when the R_0 is low, the crowd density during mass gatherings predisposes to high rates of transmission (Ebrahim et al. 2020). Other community-based measures taken to control the spread of COVID-19 that had implications for sports were the use of social distancing measures to reduce direct and close contact between people in the community and travel restrictions to affected areas, including reduced flights and public transport and route restrictions (Ahmeda and Memish 2020). Voluntary home quarantines when a team member gets into contact with an infected person(s) were also instituted, as this has been noted to reduce stress on the emergency healthcare system. Although family clusters of infections may occur, the numbers of affected people are usually far lower than in institutional settings (Ebrahim et al. 2020). These measures, when followed religiously, may help delay the exponential spread of an outbreak until drugs and vaccines become available, but they also destroy the sports tourism industry.

To control the COVID-19 outbreak in Japan, Sugishita et al. (2020) observed that sports and entertainment events were cancelled on government advisory for 2 weeks from 26 February to 11 March 2020. This was designated in Japan as a voluntary event cancellation. At the same time, it was advised that small business and private meetings be cancelled voluntarily. Epidemiological models showed that these measures were able to reduce COVID-19 infections by up to 35%. The peak number of cases was reduced by about a third compared to what it would have been without the adoption of these measures.

Most of the literature describes how previous coronavirus outbreaks occurred, the areas and places that were impacted and the interventions that were undertaken to try and contain the outbreaks. There is very limited literature that takes an in-depth look into how exactly sporting events at different levels were impacted and what measures could be taken to be resilient to future shocks caused by the virus, hence the need to explore how the industry was impacted by the COVID-19 pandemic. The next section presents the methodological underpinnings for this work.

10.3 Materials and Methods

This chapter analyses case studies of the impacts of the COVID-19 pandemic on sporting events – from the global to continental and national levels. Documented in the study are the impacts of COVID-19 on the staging of the Tokyo 2020 Olympic Games, cricket, golf, tennis, football, the Comrades Marathon and rugby championships. Critical document analysis of the games provided the main source of data. Critical document analysis is a mostly qualitative data collection method in which documented information is interpreted by the researcher to produce meaning around the study objectives. The content analysis gives rise to emerging themes, much like the themes that emerge from focus group discussions and in-depth interviews. Selected locations and countries with severely impacted major sporting events are shown in Fig. 10.1.

The documents used in the study were mainly public records and reports from local and international organisations, personal documents such as first-person accounts and experiences, incident reports, electronic media and newspapers. Critical document analysis is fast becoming a popular method of research and has been used in several similar studies, for example, the one by Nhamo and Mjimba (2020).

10.4 Findings and Discussions

It can be argued that the COVID-19 coronavirus pandemic "decimated" the sporting industry, with all forms of sports, at all levels, impacted negatively. Most sporting events and fixtures were either suspended or cancelled altogether,

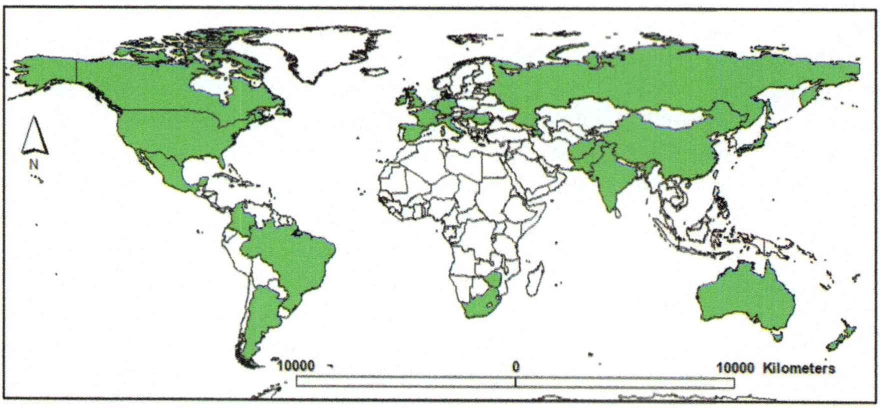

Fig. 10.1 Selected locations and host countries of severely impacted sporting events. (Source: Authors)

on a scale that was unprecedented in the modern history of sports. The impacts were felt from the Tokyo 2020 Olympics, through the cricketing world, the major professional football leagues and basketball, to rugby championships, marathons and other related events. The details will be presented in the following subsections.

10.4.1 Controversies and Ultimate Postponement of the Tokyo 2020 Olympics

The Tokyo 2020 Olympic Games was undoubtedly the largest mass sports event of 2020 that was affected by the COVID-19 pandemic. Approximately 204 countries and regions take part in the Olympic Games, with 164 countries participating in the Paralympics (Nakamura et al. 2018). The host city, Tokyo, was expecting to receive about 20 million visitors that were to be attended to by over 70,000 volunteers from the games and 8000 from the city. Some 11,090 Olympic athletes and 4400 Paralympic athletes were going to participate, with 14 million food dishes expected to be delivered to participants (Ingle 2020). These conditions alone, before factoring in the COVID-19 pandemic, presented a major challenge for the organisers (Nishiura et al. 2020). As the pandemic continued to spread the world over and all sporting events were being cancelled as a containment measure, the inevitable occurred on 24 March 2020: the hard, but rational decision to postpone the Tokyo 2020 Olympics was made, after a discussion between the International Olympic Committee (IOC) president and the Japanese Prime Minister. This was after "strong viewpoints" had been raised by several National Olympic Committees that postponement was the only course of action to take, for the health and wellbeing of athletes. Other countries like Canada and Australia had even indicated that they would boycott the games if they were to be staged in the midst of the COVID-19 pandemic (Canadian Olympic Committee 2020). They argued that nothing was more important than the health and safety of athletes and the world community.

The decision to postpone the games due to the COVID-19 pandemic, rational as it seemed, was prolonged. Scott et al. (2020) referred to the delayed decision to postpone the Tokyo 2020 Olympics as emanating from what they called a "game of chicken" between the International Olympic Committee (IOC) and the Tokyo 2020 organisers. This was due to the commercial and legal ramifications that could follow such a decision. In such situations, who it is that effectively postponed the games would be crucial for litigation purposes, because that could open them to potential allegations of breach of contract. However, given the impossibility of hosting the games during the period of the COVID-19 pandemic, the IOC had the contractual right to cancel the games on safety grounds and was protected from any claims for damages by the host city in such an event. On the other hand, as highlighted by Scott et al. (2020), the host city contract did not refer to postponements, and hence the IOC rather preferred Japan to make the ultimate decision. Boycotts by certain key members of the IOC such as Canada and Australia gave the impetus to the IOC and local organisers to postpone the games on the basis of incomplete games. This meant that it was no longer possible to proceed with the planned schedule, and it also provided some protection from possible legal and insurance claims from the multiple commercial entities that had contracts tied to the event, such as broadcasters and sponsors.

The postponement of the Olympic Games due to the COVID-19 pandemic is considered unprecedented and unquantifiable. This is because it was the first time this had occurred in the 124-year modern history of the event and also because of the complexity of seeking a solution to issues that the IOC and Organisers had no control over (Dang and Hals 2020). Furthermore, regardless of how understandable the basis of the postponement was, there were some in Japan who felt that the decision amounted to an embarrassment for the hosts (Scott et al. 2020). This is due to the fact that no other Olympics had ever been rescheduled before, but only cancelled during World War I and II periods (Ingle 2020). The Olympic organisers also highlighted that postponing the Games, which had a budget of US$12.6 billion, came with its own set of challenges. In fact, addressing the unique situation required close coordination with all partners, such as the Japanese government, the Tokyo Metropolitan Government, broadcasters, Tokyo 2020 marketing partners, suppliers and contractors. Numerous practicalities arose, such as that the hotel and other bookings that had been made previously would not necessarily be available during the rescheduled Games. However, despite the severe financial toll on the sponsors and sports tourists, the major corporate sponsors of the Tokyo 2020 Olympics stood by the IOC after the Games were postponed. Still, the corporate sponsors were not likely to get the expected return on billions of dollars committed to their agreements (Dang and Hal 2020). There were further questions on whether the Athletes' Village and other key venues would still be available given pre-agreed deals with private occupants and the need to find other tenants. Another key question was if Japan could really afford the costs of keeping these venues empty for another year, without them being viewed as white elephants or ivory towers.

Another issue that the IOC had to worry about was the prospect of having to bail out some of the international sports federations that depend on financial handouts and had not been having games due to the effects of COVID-19. Some of these federations had insurance to mitigate the negative impacts of the COVID-19 shock, but most faced a worrying future in the aftermath of COVID-19. During the delay in postponing the Olympic Games, several athletes expressed grave concerns about the need to prepare for the Games as normal, during the COVID-19 pandemic. They argued that this was putting them and their families at risk of contracting the virus and therefore pushed for a postponement. They also noted that for them to fittingly prepare for the Games, their health and safety needed to be considered first (Scott et al. 2020). By the time the Games were postponed, only about 57% of the athletes set to take part in the Tokyo Olympics had qualified. The rest still had to attend qualifying tournaments that were supposed to have taken place around the world, but had been cancelled because of the pandemic (Reddy 2020; Murakami and Grohmann 2020). Those that had qualified still faced the challenge of adequate preparation, since most gyms, stadiums and swimming pools were shut down across the world, as most countries went under lockdown to try and contain the COVID-19 pandemic (Murakami and Grohmann 2020).

Upon the postponement of the Tokyo 2020 Olympics, no specific dates were mentioned, except that the Games had been rescheduled to no later than the summer of 2021. There was no clear consensus on new dates for the Games, mainly because each sport discipline has its own calendar to consider and because of the continued uncertainty about when the pandemic may be under control. For example, a year's delay of the Games would clash with the World Athletics Championships, already scheduled to run from 6 to 15 August 2021 in Oregon. Moving them to October would make them clash with the European football season and major US sports events – hence the need for an all-stakeholder approach in determining the new dates (Murakami and Grohmann 2020). The Games were then later rescheduled from 23 July 2021 to 8 August 2021 and would maintain the name "Tokyo 2020", despite taking place in 2021.

From an athlete's perspective, the postponement of the Olympic Games has also brought with it scores of unknowns. This is so, for one, because the older athletes who were at the tail-end of their careers may not be able to compete in 2021. Having waited for 4 years to compete in their final Olympics, it is uncertain whether they would be willing to finance another year of training and also put themselves through the gruelling training, given their advanced age and the fact that some had already announced their retirement soon after the 2020 Olympics. For sports with an age limit at the Olympics, such as football, which is played by the under-23 age group, will the players who are over 23 years in 2021, but were eligible to play in 2020, be allowed to compete? On the other end of the spectrum, will athletes who were underage to compete in 2020 be allowed to take part in the 2021 Games, given the fact that the Games will still retain the name Tokyo 2020? Another question is if the qualifying tournaments of the postponed Tokyo 2020 Olympics will continue to be held in 2020, or will they now be held within a few months of the rescheduled Games? Of critical importance is, what will happen to those who had already qualified for the Games – will they keep their spots, or will they have to qualify again, considering that there are always young and new competitors who may emerge during the year? Many would surely advocate to have a clean slate and start the qualification process all over. Shelnin (2020) argues that for finely tuned Olympic athletes, a year's postponement changes everything against the backdrop of so many variables and the uncertainty surrounding the build-up to the Games and the COVID-19 pandemic. Moreover, by April 2020 certain experts, like Kentaro Iwata, a professor of infectious diseases at Kobe university, expressed doubts that the Games could even be held in 2021, because a year-long delay might not be sufficient (News24 2020b).

The lessons from the postponement show that there is a need to press the refresh button on the structure of the Games. The costs of hosting the Olympics are becoming unbearable. There is a lot of pressure on the host cities and organisers, who end up making too many promises to the local communities. There is a need to rationalise and scale down the Games to manageable levels. The Games need to be sustainable in both the socio-economic and environmental dimensions, with the athletes' representative bodies having more say in how the Games are run, in order to guarantee their health, safety and interests. Who knows, COVID-19 could have brought an opportunity to downsize this event too.

10.4.2 Disruption in Football Events

Football, like most sports, was also greatly affected by COVID-19. Inasmuch as all leagues were affected, the smallest and less-sponsored ones took a bigger knock than the bigger and well-sponsored ones. In English football, the lockdown on sports due to the COVID-19 pandemic made teams outside the lucrative Premier League face unprecedented periods of financial uncertainty. Stone (2020) pointed out that for all football in England below the three divisions that make up the National League, the leagues were made to stop in March 2020, in the middle of the pandemic, before the scheduled end of the season – with all the results being expunged. This meant that there was no promotion or relegation in these leagues. The same conditions applied to the women's game below the Women's Super League and Championship level. The football body made this decision in the light of the financial impact on football clubs and the uncertainty brought about by the COVID-19 pandemic. There were also concerns for the safety and welfare of clubs, players, staff, officials, volunteers and supporters during this unprecedented period. The decision meant that the planned restructuring of the non-league football system at the lower levels, which was planned for 2020, had to be suspended until the 2021/2022 season. Even though some of the teams in these leagues had already won enough points to gain promotion to the next level of the game even before the start of the COVID-19-disrupted season, these were still cancelled. In addition to the big let-down for these clubs, the decision had a huge financial impact on them, since they had invested heavily in that success, and the loss of promotion also affected some of their revenue assumptions for the coming season. However, with a quarter of the fixtures left to play when football was cancelled, the association saw it as being unfair to award promotion or relegate teams when there were so many matches left to play.

During the suspension, the Football League agreed on a £50 million relief package for teams – mostly those in the Championship League, League One and League Two. The money was based on an early payment of bonuses, television rights and interest-free loans to assist financially stricken clubs. Without match-day revenue, some (even rich) clubs had already started making the tough decision of laying off some of their staff and asking their players to accept wage cuts of as high as 70%. This was because the rescue package was only enough to support the clubs for 4 weeks and far from a complete answer to the financial problems that they were facing because of the pandemic. Some clubs started implementing wage deferral arrangements, which involved putting caps on salaries until the situation has normalised. As Stone (2020) observed, the extension of the football season into the time period it was supposed to be over had financial implications for clubs that were changing kit manufacturers and sponsors. The deals, together with the massive marketing that comes with them, were seriously impacted by the delays in completing the season, and millions in potential revenue were lost. At the time of writing, all football was still suspended, and the prospects of abandoning the leagues in the hardest-hit countries were very high. With hundreds of millions of dollars at risk due to the pandemic, there was consensus on the need to devise ways to reach the finishing line. Taking the English Premier League (EPL), for example, not finishing the season would have an impact on broadcasting to the amount of about £760 million. The appetite to finish major football leagues even in the middle of a global pandemic led some players to feel as if they were being used and forced to play in dangerous conditions. The players argued that it was only after some players got sick that serious action was taken to stop football during the COVID-19 pandemic. Other interested community members argued that football was existing in a "moral vacuum" and did not deserve any support, especially from the tax payer's money. This, they argued, was justified because the clubs make enough money which they do not use wisely, causing them to suffer financially within the first month of a global pandemic.

At the time, there was still some uncertainty surrounding the spread of the coronavirus, and some of the football matches that were played in Europe when the COVID-19 pandemic was starting to run amok have turned out to be super-spreading events. Doyle (2020) observes that the European match between Atalanta from Italy and Valencia from Spain is now viewed as "partita zero" or game zero and a "biological bomb" that exploded in Italy. In the middle of the COVID-19 crisis, the winning team's (Atalanta) 40,000 strong fans who had travelled from Bergamo set up wild celebrations, hugging and shaking hands with one another. Although there were several major triggers and catalysts for the diffusion of the virus in Italy, it is widely believed that during the euphoric celebrations that night, the fans unintentionally played a part in the spread of COVID-19, which devastated that part of Italy during the pandemic. Since it was a key match, most key informants noted that any fan of the playing teams would not miss the match if they had tickets, even if they felt a slight fever. In the aftermath, 35% of Valencia's staff and players tested positive for COVID-19. The area around Bergamo, where Atalanta football club comes from, went into lockdown just 4 days after the game, as the number of confirmed cases of the COVID-19 virus rose rapidly. The game now goes down in memory as the night of joy which created a tragedy (Doyle 2020).

The 2020 European Football Championship (known as Euro 2020) was one of the high-profile victims of the COVID-19 outbreak. Euro 2020, which coincided with the 60th anniversary of the Union of European Football Associations (UEFA), was supposed to be played for the first time across 12 European countries, instead of the traditional one country hosting or two countries co-hosting. The move was meant to celebrate the 60-year milestone of UEFA and was also necessitated by challenges in finding committed hosts with 12 satisfactory football stadia, following the expansion of the tournament from 16 to 24 teams, after the Euro 2016 hosted by France (Ludvigsen 2019). The outbreak of the COVID-19 pandemic in most of these host countries in early 2020 raised concerns regarding its potential impact on players, staff and visitors to the tournament. As the impacts of the COVID-19 pandemic on football grew, a number of domestic and UEFA competition matches were first played in empty stadia, and eventually all matches of Europe's major leagues were suspended. Ultimately, Euro 2020 was postponed to 2021. The rationale behind the shift was to allow for pressure to be reduced on the public services in the affected countries while also providing space in the calendar for domestic European leagues that had been suspended to complete their seasons, once the pandemic is under control.

In South African football, it was business unusual after a national state of disaster was declared on 15 March 2019 because of COVID-19, and all mass gatherings, including sports, were suspended. While the Premier Soccer League (PSL) players were still being paid by the clubs during the football shutdown period, the match officials were left to bear the brunt of the pandemic from a salary point of view. This was because they are not permanently employed by the South African Football Association and are remunerated as and when they work or according to the number of matches at which they officiate (Makhaya 2020). At a more localised level, fitness trainers in South Africa also highlighted that the lack of sports and the restrictions on movement during the COVID-19 pandemic had a negative impact on their way of life. As Magasela (2020) observed, fitness gymnasiums had to be shut down to mitigate against the spread of the virus: business suffered seriously because people were first scared to go to gyms and as restrictions tightened, the people were banned from going to fitness clubs. Job losses were inevitable as the COVID-19 pandemic persisted.

Sports people with sponsors also acknowledged that due to the universal shutdown of businesses during the pandemic, sponsors were facing their own challenges in getting value for their money from them, which could only happen if they were working.

Ultimately, the postponement and/or cancellation of all of these sporting events had an effect on sports tourism and the associated benefits for the cities and countries hosting them.

10.4.3 COVID-19 and Cricket Events

Cricket is another sport that was hit hard by the COVID-19 outbreak globally. The cricket-playing nations are shown in Fig. 10.2. Of special note was the postponement of the high-profile Indian Premier League (IPL). The decision to suspend the

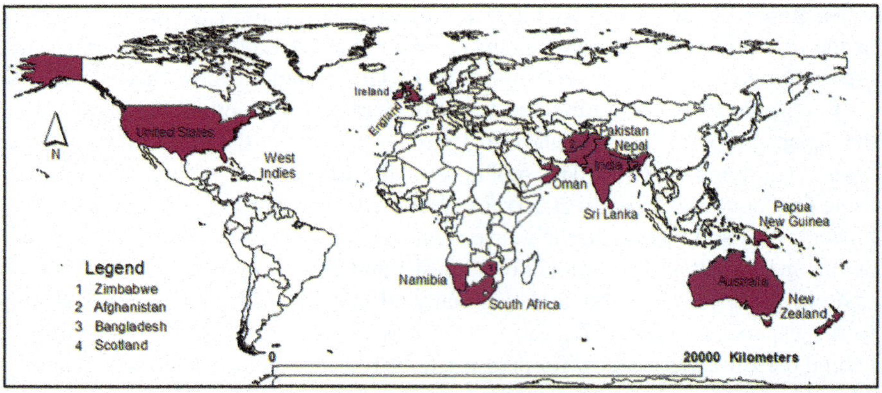

Fig. 10.2 International Cricket Council One-Day International playing nations. (Source: Authors)

multibillion-dollar IPL was taken as a precaution to stop the spread of COVID-19 as India's 1.3 billion people were placed under lockdown for 21 days to try and contain the killer virus. The IPL cricket tournament draws the world's top players from all the major cricket-playing nations. The postponement came as India also cancelled its two remaining one-day cricket internationals against South Africa because of the pandemic (Chakraborty 2020). In fact, the touring South Africa team had to return home without playing.

COVID-19 also led to the halving of the winner's prize of the 2020 IPL and put a dent in its US$6.8 billion brand value. The postponement or eventual cancelling of the tournament meant further losses for the Board of Control for Cricket in India (BCCI), which gets around US$532 million annually from broadcasters and its central pool of sponsors. The Chinese smartphone maker Vivo, which sponsored the tournament to the value of 219 million rupees, stood to suffer a big loss if the postponed tournament were eventually cancelled. The BCCI and IPL franchises also pay 20% of a player's annual fee to his home board, which stands to lose that income if the postponed tournament were eventually cancelled.

Cricket Australia is also at risk of losing close to US$174 million if the coronavirus outbreak derails their high-profile home test series against India in October 2020 (Chakraborty 2020). The English and Welsh Cricket Board in turn announced that they would be reducing the salaries of their national team players but also offered a rescue package for the cricket clubs that were in financial distress because of the COVID-19 outbreak. They observed that players also needed to bear the financial consequences of the meltdown caused by the coronavirus.

10.4.4 Impacts on Rugby Championships

As with the other sports, the cancellation of rugby fixtures deprived teams of much-needed ticket sales and broadcasting money. Some of the rugby tournaments that were affected by the coronavirus outbreak were the Six Nations, the World Rugby Sevens Series, the Olympics, the European Cups, the Guinness Pro14, Super Rugby and The Rugby Championship. Among the most impacted rugby competitions were the Super Rugby and Rugby Championship tournaments (Fig. 10.3). These involve teams from Argentina, Australia, Japan, New Zealand and South Africa. In New Zealand, each of the five Super Rugby teams had to be given a grant of US$149,000 by New Zealand Rugby to assist them to survive for up to 3 months after their funds had dried up under the COVID-19 lockdown (Mulvenney 2020). The lockdown on sporting activities had left the teams with players to pay, but no revenue streams. The decision to fund these teams was mainly aimed at stopping them from collapsing and also to protect the core capabilities of the clubs, so that they would be ready to continue where they left off once the COVID-19 pandemic was under control. Players in these teams also played their part by taking salary cuts (Mulvenney 2020).

In South Africa, the teams in the Super Rugby league implemented a remote training programme to keep players active even during the lockdown periods. This included players doing daily wellness and temperature reports to make sure that they were not infected. The plan of the clubs was to be match-fit within 3 weeks of the end of the COVID-19 pandemic. However, huge challenges come with practising as an individual indoors, when players had been used to working, training and travelling together as a team. In South Africa, rugby and football were the greatest losers sportswise, during the COVID-19 outbreak. In 2018, South Africa Rugby earned R714 million in broadcast rights, sponsorships of R388 million and R100 million in gate revenue (Ray 2020). All this revenue, and potentially more, stood to

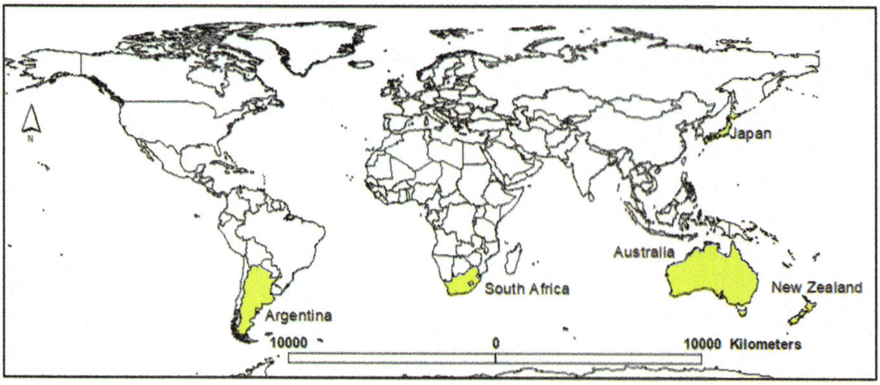

Fig. 10.3 Super Rugby and The Rugby Championship nations. (Source: Authors)

be lost because of the COVID-19 pandemic. Technically, the rugby board was in breach of its contracts with broadcasters for not supplying content during the lockdown and also for not giving its sponsors value for money. Whereas South African rugby teams were financially precariously sound during the first 21 days of the COVID-19 lockdown, cost-cutting measures would be a necessary reality further down the line if the pandemic continued unabated.

Due to the COVID-19 pandemic, Rugby Australia was forced to lay off staff in a bid to survive the shock of the huge losses they had incurred when lucrative games were cancelled. Three-quarters of staff lost their jobs, and even the CEO of Rugby Australia had to take a 50% salary cut. This came in the wake of a loss of up to $74 million in potential revenue when their test matches against Ireland and Fiji were cancelled due to COVID-19. Extreme measures therefore had to be taken to ensure cost reduction for the sport to remain financially viable, so that when the global crisis ended, the board would be able to rebuild the sport (AFP 2020).

10.4.5 Implications for Major Golf Tournaments

There are four major golf championships with several events. These include The Masters Tournament, the PGA Championship, the US Open and The British Open. The countries that host these championships and tournaments are shown in Fig. 10.4. Based on the locations of the host countries, the golf scene presents tourists (both players and fans) with opportunities to travel around the world. On 12 March 2020, the PGA Tour made a statement concerning The Players Championships and COVID-19 (PGA Tour 2020a). In the statement, they

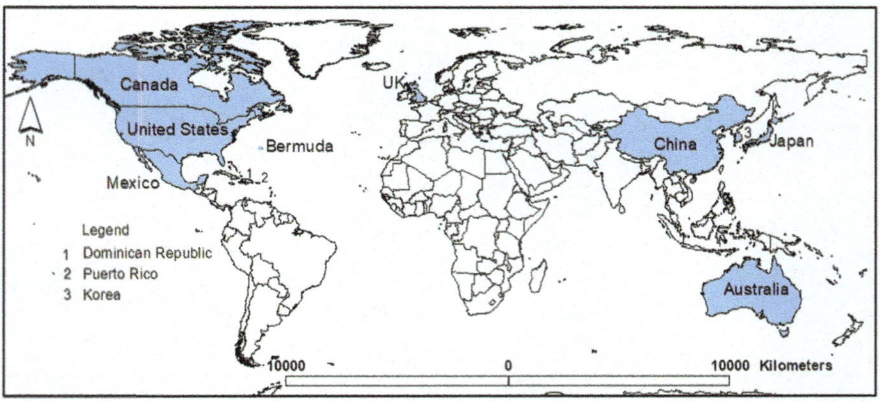

Fig. 10.4 Major golf tournaments' host nations. (Source: Authors)

indicated that they were aware of the challenges emerging from the spreading pandemic, with the situation rapidly changing. However, the decision was made to continue with The Players Championships, although it would be reviewed constantly. Fans who no longer wished to attend were free to request refunds through laid-down procedures. The tournament went ahead for day 2, but with no fans. Eventually, the tournament was cancelled (PGA Tour 2020b). In the same statement announcing the cancellation there was further bad news, as all upcoming PGA Tour events were cancelled.

On 13 March 2020, an announcement was made that The Masters was postponed (The Associated Press 2020). The Masters was due to take place from 9 to 12 April 2020 at the Augusta National Golf Club. This cancellation also impacted the Augusta National Women's Amateur and the Drive, Chip and Putt National Finals. However, no new dates were announced for The Masters, with the decision coming only 12 hours after the PGA Tour commissioner Jay Monahan announced the cancellation of four PGA-related tournaments. These cancellations included three rounds of The Players Championship discussed earlier. It emerged that The Masters had historically finished in April each year since 1935, with the only exception being cancellations from 1943 to 1945 because of World War II. The picture of the total number of events that were eventually cancelled or postponed per major tour is presented in Fig. 10.5.

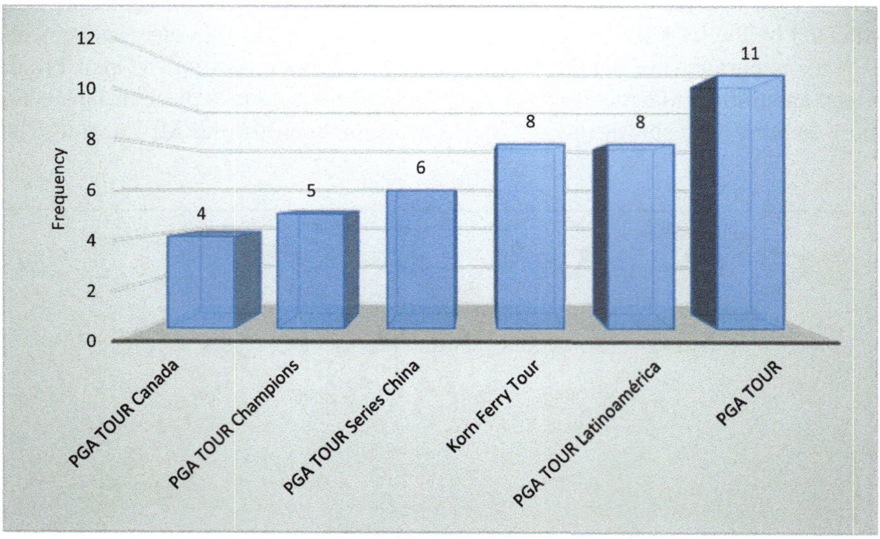

Fig. 10.5 Total number of events cancelled or postponed per major tour. (Source: Authors, data from PGA Tours (2020c))

On 17 March 2020, more bad news came in. The PGA Championship scheduled for 11 to 17 May 2020 at TPC Harding Park in San Francisco was also cancelled (PGA Tour 2020c). In the statement, the PGA Tour commissioner indicated that the health of the players remained the number one priority. Other major events cancelled included the RBC Heritage (13–19 April 2020), Zurich Classic of New Orleans (20–26 April 2020), Wells Fargo Championship (27 April-3 May 2020) and the AT&T Byron Nelson (4–10 May 2020).

A number of players took to social media to react to the cancellations and postponements. On 16 March Tiger Woods wrote, "There are a lot more important things in life than a golf tournament right now. We need to be safe, smart and do what is best for ourselves, our loved ones and our community" (PGA Tour 2020d). On 20 March 2020 Webb Simpson wrote, "So proud of the @WMPhoenixOpen. The Thunderbirds are donating $1 million to charities in need in the Phoenix area … What the Thunderbirds do each year is incredible, and I'm glad to be a part of it". Lastly, Ian Poulter wrote, "Hi guys I know it's difficult but if we all listen and distance ourselves and not gather in groups we can stop the spread. I'm seeing so many videos of selfish, disrespectful people who think this is a joke. If you don't listen this will spiral out of control even more" (ibid.).

10.4.6 Impacts on Major Tennis Events in 2020

There were five major tennis championships scheduled for 2020, and some were affected by the COVID-19 outbreak, with negative impacts on tourist movements. The main tournaments included Wimbledon, the US Open, Australian Open, French Open and Summer Games Tennis Tournament that was to take place in Japan. The host countries are shown in Fig. 10.6. The Main Board of the All England Club

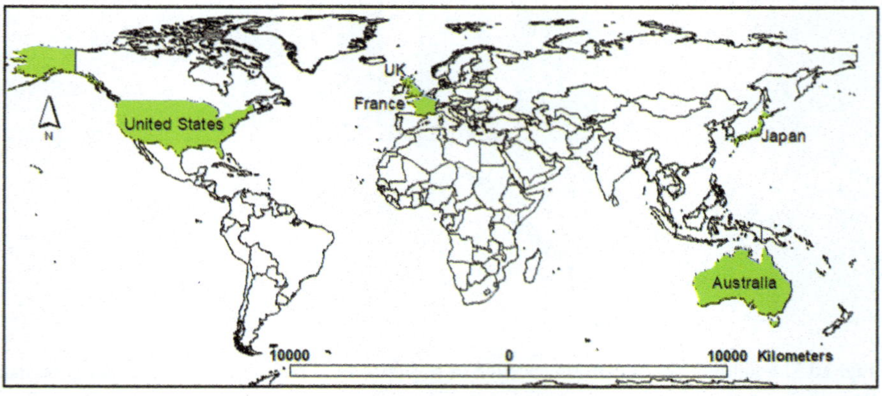

Fig. 10.6 Countries hosting major tennis events. (Source: Authors)

(AELTC) and the Committee of Management of The Championships announced the cancellation of Wimbledon's 134th Championships on 1 April 2020. The tournament was shifted to 28 June to 11 July 2021 (AELTC 2020). The main reason was to ensure that the multitudes of tourists that usually grace the occasion do not get infected with the SARS-CoV-2 virus which causes COVID-19. This was also to comply with the UK government directives for combatting COVID-19.

The Wimbledon cancellation also covered the entirety of the ATP/WTA European grass-court swing. This included the ATP events in 's-Hertogenbosch, Stuttgart, London-Queen's, Halle, Mallorca, Eastbourne as well as WTA events in 's-Hertogenbosch, Nottingham, Birmingham, Berlin, Eastbourne and Bad Homburg (Tennis.com 2020). The US Open was not affected at the time of the announcement. However, events were to be continuously monitored. At the time of completing this chapter, the USA was the new global COVID-19 epicentre, both in terms of infections and deaths. As in the case of the PGA Tour, the last time Wimbledon was not played was in 1945, during World War II. Serena Williams reacted with shock (Tignor 2020). Wimbledon is the oldest tennis tournament globally and attracts huge numbers of tourists and visitors over its duration of 13 days. For example, the following visitor figures were reported: for 2017 (473,372), 2018 (473,169) and 2019 (500,397) (Lange 2020). This is why it made sense for the organisers to cancel the tournament due to COVID-19 – many were going to be in real danger of contracting the disease.

10.4.7 Other Major Sporting Events Impacted by COVID-19

In South Africa, the Comrades Marathon was one of the most prominent victims of the COVID-19 pandemic. The marathon is the largest and oldest ultramarathon in the world and is held annually between the cities of Durban and Pietermaritzburg. The event attracts participants from all over the world and has been growing in popularity over the years. Figure 10.7 shows the number of capped participants in the marathon from the years 2011 to 2020. In 2020, when it was cancelled, it was expected to attract 27,500 participants, compared to the 19,591 it attracted in 2011 (Comrades 2020).

Some of the other key sporting events that were postponed because of the COVID-19 pandemic were the Dubai World Cup and several USA events. The Dubai World Cup race, which is held at the Meydan Racecourse, is one of the world's richest horse races, with a total prize money of US$35 m, including US$12 m in the feature race (ESPN 2020). In the USA, major professional sports leagues such as Major League Baseball, the National Basketball Association, Major League Soccer and the National Hockey League suspended the playing of games as the virus spread rapidly across the country. Major marathons such as the Boston, London, Paris, Great Wall and Barcelona marathons were either cancelled or postponed over COVID-19 concerns. One of the first major events to see a shift in schedule was the 2020 Tokyo Marathon, at the beginning of March 2020. The

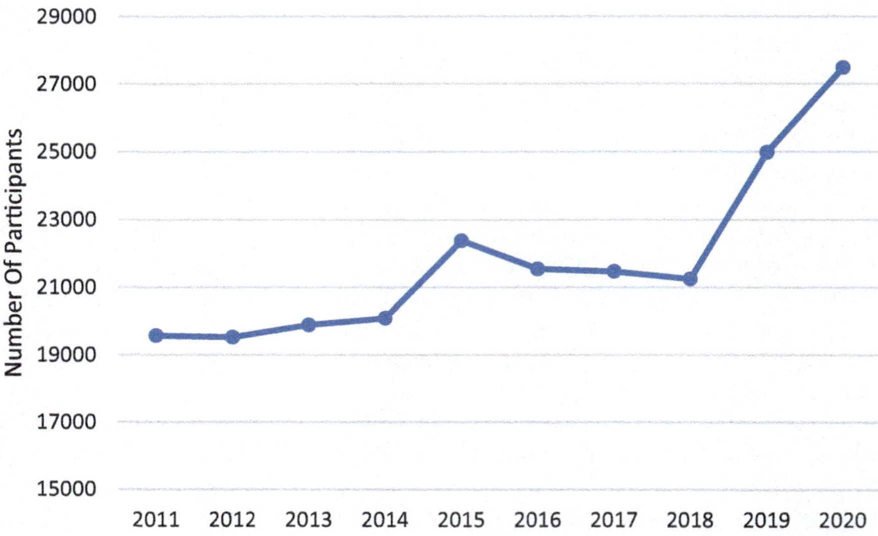

Fig. 10.7 Comrades Marathon capped entries (2011–2020). (Source: Authors, based on Comrades (2020))

marathon was only held for the elite and non-elite runners given entry to the 2021 race. As the pandemic became widespread and the uncertainty of its containment grew, sports organisers proactively started to reschedule or cancel events set to take place in the first half of the year 2020 (ESPN 2020).

As Reuters (2020) observed, the anti-doping system was facing challenges with most countries going into lockdown and closing their borders to contain the spread of COVID-19. The World Anti-Doping Agency (WADA), in giving guidance on taking samples from athletes during the COVID-19 pandemic, emphasised that the collection officials must be healthy, have no symptoms related to the virus and must be provided with protective medical equipment. Where movement was restricted, WADA advised to concentrate on targeted athletes from high-risk sports and disciplines. However, another challenge was that most accredited laboratories had suspended their normal operations due to the pandemic.

Owing to the suspension of professional sporting activities worldwide, the US$450 billion global betting industry also suffered a big setback. As soon as there was a global shutdown on sporting – especially horse racing – most of the prominent betting companies issued profit warnings and most started having financial hardships (Smith 2020). In East Africa, gambling sales went down by as much as 99% following the devastation of global sports in the wake of the COVID-19 pandemic. The suspension of the world's leading football leagues left many gamblers in East Africa with very little to bet on. In addition, all gambling shops were closed in an attempt to control the pandemic, and only as few as 30% of the gamblers could gamble online. As a result, this industry, with an annual turnover of US$42 million in Kenya, Uganda and Tanzania, was almost decimated during the pandemic (Karoney and Edwards 2020).

Such was the unbearable negative impact of COVID-19 across several global sporting disciplines and the supporting sports tourism industry.

10.5 Conclusions

The study revealed that sports events and sports tourism were negatively affected by the COVID-19 pandemic, as measures to control its spread took preference over sports all over the world. From the banning of air travel and closure of airports to lockdowns in many countries, the odds were against sporting events and their stakeholders. The fact that sport comes secondary to the wellbeing of communities is unquestionable. However, we cannot ignore the millions of dollars that have been lost by major sport governing bodies, individual teams and athletes and their dependants due to the pandemic. The sports industry was both decimated and bankrupted. Most sporting events around the world, ranging from individual events to regular league matches and mega sports events lined up for the first half of 2020, were cancelled and/or postponed. The most salient victim was the 2020 Tokyo Olympics. The cancellation of sporting activities led to losses for the individual teams, broadcasters and sponsors, leaving most of them in financial distress. From an athlete's perspective, the COVID-19 pandemic resulted in the loss of income, as most sporting clubs significantly cut salaries, with some reducing salaries by as much as 70%. The athletes also faced a major battle to remain fit during the period of the COVID-19 pandemic, given the fact that they had no access to training grounds and centres. The most vulnerable members of the sporting industry were confronted with job losses and an indefinite suspension of income. As a sensitive tourist industry, COVID-19 also impacted heavily on the sports tourism sector, leading to a loss of income in the entire value chain, including hotels, restaurants and food production. Undoubtedly, plans needed to be put in place to reduce the impact of COVID-19 on sportsmen, administrators and other stakeholders. It is clear from what happened during the COVID-19 pandemic that athletes need to speak with one voice to protect themselves against the commercial interests of their employers. The employers seemed to have been out of touch with the gravity of the pandemic during its initial phases. They continued with sporting events even when it was no longer safe to do so, in some cases facilitating the spread of the disease and also risking the health and wellbeing of the athletes and spectators. Only after several athletes had contracted the disease was action taken. With the industry having hit rock bottom, an urgent rescue package needs to be designed and delivered by governments and interested stakeholders to help it survive the post-pandemic period. In future, it will be imperative for the industry to put in place a fund to help them survive disaster periods as that imposed on it by the COVID-19 pandemic.

References

AELTC (Main Board of the All England Lawn Tennis and Croquet Club). (2020). *Cancellation of the championships 2020*. Retrieved from https://www.wimbledon.com/en_GB/news/articles/2020-0401/cancellation_of_the_ championships_2020.html. Accessed 4 April 2020.

Agence France-Presse (AFP). (2020). Rugby Australia forced to lay off staff in bid to survive. *Business Day*. Retrieved from www.businesslive.co.za/amp/bd/sport/rugby/2020-03-31-rugby-australia-forced-to-lay-off-staff-in-bid-to-survive/. Accessed 5 April 2020.

Ahmeda, Q. A., & Memish, Z. A. (2020). The cancellation of mass gatherings (MGs)? Decision making in the time of COVID-19. *Travel Medicine and Infectious Disease*. https://doi.org/10.1016/j.tmaid.2020.101631.

Biscayart, C., Angeleri, P., Lloveras, S., Chaves, T. S. S., Schlagenhauf, P., & Rodríguez-Morales, A. J. (2020). The next big threat to global health? 2019 Novel Coronavirus (2019-nCoV): What advice can we give to travellers? – Interim recommendations January 2020, from the Latin-American Society for Travel Medicine (SLAMVI). *Travel Medicine and Infectious Disease, 33*. https://doi.org/10.1016/j.tmaid.2020.101567.

Canadian Olympic Committee. (2020). *Team Canada will not send athletes to Games in summer 2020 due to COVID-19 risks*. Retrieved from https://olympic.ca/press/team-canada-will-not-send-athletes-to-games-in-summer-2020-due-to-covid-19-risks/. Assessed 23 March 2020.

Chakraborty, A. (2020). *Cricket Boards brace for losses in the time of Coronavirus*. Retrieved from www.iol.co.za/sport/cricket/international/cricket-boards-brace-for-losses-in-the-time-of-coronavirus-45272764.

Chen, Y., Liu, Q., & Guo, D. (2020). Emerging coronaviruses: genome structure, replication, and pathogenesis. *Journal of Medical Virology*. https://doi.org/10.1002/jmv.25681.

Choa, H., Joo, D., & Chic, G. C. (2019). Examining nostalgia in sport tourism: The case of US college football fans. *Tourism Management Perspectives, 29*, 97–104.

Comrades. (2020) *Comrades Marathon*. Retrieved from https://www.comrades.com/index.php/home/home-nav-a/history.

Dang, S., & Hals, S. (2020). In coronavirus-hit world, sponsors to stand by delayed Olympics. *TimesLive*. Retrieved from www.timeslive.co.za/amp/sport/2020-03-25-in-coronavirus-hit-world-sponsors-to-stand-by-delayed-olympics/. Accessed 30 March 2020.

Depta, L. (2015). 12 ways sports make a positive impact. *Bleacher Report*. Retrieved from www.bleacherreport.com/amp/2347988-12-ways-sports-make-a-positive-impact.amp.html. Accessed 20 March 2020.

Doyle, M. (2020). How Covid-19 turned Atalanta's dream night into Bergamo's worst nightmare. *Goal*. Retrieved from https://www.goal.com/en-za/news/how-covid-19-turned-atalantas-dream-night-into-bergamos/mkdmvlojfyi11plf44dk67eul. Accessed 10 April 2020.

Dube, K., Nhamo, G., & Chikodzi, D. (2020). COVID-19 cripples global restaurant and hospitality industry. *Current Issues in Tourism*. https://doi.org/10.1080/13683500.2020.1773416.

Ebrahim, S. H., & Memish, Z. A. (2020). COVID-19: Preparing for superspreader potential among Umrah pilgrims to Saudi Arabia. *Lancet, 395*. https://doi.org/10.1016/S0140-6736(20)30466-9.

Ebrahim, S. H., Ahmed, Q. A., Gozzer, E., Schlagenhauf, P., & Memish, Z. A. (2020). Covid-19 and community mitigation strategies in a pandemic, BMJ 368:m1066. doi: https://doi.org/10.1136/bmj.m1066

ESPN. (2020). List of sporting events cancelled because of the Coronavirus. *ESPN*. Retrieved from www.espn.com/olympics/story/_/id/28824781/list-sporting-events-canceled-coronavirus%3fplatform=amp. Accessed 30 March 2020.

European Union. (2018). *New study on the economic impact of sport by the European Commission*. Brussels: House of Sport (EPSI).

Grohmann, K. (2020). Olympics-Tokyo 2020 to be rescheduled to no later than summer 2021. *Reuters Sport*. Retrieved from www.reuters.com/articule/amp/idUSKBN21B1ZP. Accessed 28 March 2020.

Hinch, T., & Higham, J. (2011). *Sport tourism development*. Bristol: Channel View.
Hinch, T., & Ito, E. (2018). Sustainable sport tourism in Japan. *Tourism Planning and Development, 15*(1), 96–101. https://doi.org/10.1080/21568316.2017.1313773.
Hui-Ju, H., Hong-ling, W., Chuan-Min, Z., Fang-Fang, C., Li-Mei, L., Jian-wei, L., & Xue-Jie, Y. (2015). Bats as reservoirs of severe emerging infectious diseases. *Virus Research, 205*, 1–6.
Ingle, S. (2020). IOC insists Tokyo Olympics will go ahead despite coronavirus. *The Guardian*. Retrieved from https://www.theguardian.com/sport/2020/feb/14/ioc-insists-tokyo-olympics-will-goahead-despite-coronavirus.
Karoney, C., & Edwards, P. (2020). Coronavirus: East African gambling sales down 99%. *BBC*. Retrieved from www.bbc.co.uk/sport/amp/africa/52037039. Accessed 30 March 2020).
Kim, Y., Kim, J., Cho, S., & Yim, K. (2020). The impact of mega sporting events on host country's destination images: The cases of the 2014 Sochi Winter Olympics and 2015 Beijing IAAF World Championships. *Sport Marketing Quarterly, 28*(3), 148–164.
Lange, D. (2020). *Wimbledon: Aggregate attendance 2017–2019*. Retrieved from https://www.statista.com/statistics/957232/wimbledon-championships-aggregate-attendance/. Accessed 4 April 2020.
Ludvigsen, J. A. (2019). "Continent-wide" sports spectacles: The "multiple host format" of Euro 2020 and United 2026 and its implications. *Journal of Convention & Event Tourism, 20*(2), 163–181. https://doi.org/10.1080/15470148.2019.1589609.
Macgowan, A. (2015). Overseas football fans visiting Britain now at 800,000. *BBC Sport*. Retrieved from http://www.bbc.com/sport/football/34197370.
Magasela, B. (2020). Personal trainers feel the pinch caused by covid-19. *Sowetan Live*. Retrieved from www.sowetanlive.co.za/amp/sport/boxing/2020-03-23-personal-trainers-feel-the-pinch-caused-by-covid-19/. Accessed 25 March 2020.
Makhaya, E. (2020). *Coronavirus: The impact on PSL and South African football*. Goal.com.
Massaro, E., Ganin, A., Perra, N., Linkov, I., & Vespignani, A. (2018). Resilience management during large-scale epidemic outbreaks. *Scientific Reports, 8*, 1859. https://doi.org/10.1038/s41598-018-19706-2.
Mulvenney, N. (2020) New Zealand Super Rugby teams get financial lifeline. *Business Day*. Retrieved from www.businesslive.co.za/amp/bd/sport/rugby/2020-04-01-new-zealand-super-rugby-teams-get-financial-lifeline/. Accessed 10 April 2020.
Murakami, S., & Grohmann, K. (2020). Relief for athletes, blow for Japan as 2020 Olympics postponed. *Reuters*. Retrieved from www.sowetanlive.co.za/amp/news/world/2020-03-25-relief-for-athletes-blow-for-Japan-as-2020-olympics-postponed. Accessed 28 March 2020.
Nakamura, S., Wada, K., Yanagisawa, N., & Smith, D. R. (2018). Health risks and precautions for visitors to the Tokyo 2020 Olympic and Paralympic Games. *Travel Medicine and Infectious Disease, 22*, 3–7.
News24. (2020a). More than 2.6 billion worldwide told to observe lockdowns. Retrieved from https://m.news24.com/SouthAfrica/News/coronavirus-all-the-latest-news-about-covid-19-in-south-africa-and-the-world-20200312. Accessed 30 March 2020.
News24. (2020b). Japan coronavirus expert 'pessimistic' Olympics can be held in 2021. Retrieved from https://m.sport24.co.za/OtherSport/Olympics2020/japan-coronavirus-expert-pessimistic-olympics-can-be-held-in-2021-20200420?isapp=true. Accessed 20 April 2020.
Nhamo, G., & Mjimba, V. (Eds.). (2020). Sustainable Development Goals and institutions of higher education. *Sustainable Development Goals Series*. https://doi.org/10.1007/978-3-030-26157-3_1.
Nishiura, H., Sah, R., & Rodriguez-Morales, A. J. (2020, February). The COVID-19 outbreak and implications for the Tokyo 2020 Summer Olympic Games. *Travel Medicine and Infectious Disease*. https://doi.org/10.1016/j.tmaid.2020.101604.
PGA Tour. (2020a). *PGA tour statement on the players championship and coronavirus*. https://www.pgatour.com/company/2020/03/12/pga-tour-statement-the-players-championship-coronavirus.html. Accessed 4 April 2020.

PGA Tour. (2020b). *PGA TOUR statement regarding cancellation ofThe Players Championship and upcoming PGA TOUR events.* Retrieved from https://www.pgatour.com/company/2020/03/12/pga-tour-statement-regarding-cancellation-of-the-players-championship-and-upcoming-pga-tour-events.html. Accessed 4 April 2020.

PGA Tour. (2020c). *PGA Championship announces postponement.* https://www.pgatour.com/news/2020/03/17/pga-championship-harding-park-postponed-pga-tour.html. Accessed 4 April 2020.

PGA Tour. (2020d). *Players take to social media during suspension of TOUR season.* https://www.pgatour.com/news/2020/03/20/tiger-woods-brooks-koepka-phil-mickelson-social-media-pga-tour-season-suspension-coronavirus.html. Accessed 4 April 2020.

Poon, L. L. M., & Peiris, M. (2020, February 27). Emergence of a novel human coronavirus threatening human health. *Nature Medicine.* https://doi.org/10.1038/s41591-020-0796-5. [Epub ahead of print].

PricewaterhouseCoopers. (2017). *PwC sports outlook – at the gate and beyond: Outlook for the sports market in North America through 2021.* Retrieved from https://www.pwc.com/us/en/industries/entertainment-media/publications/sports-Outlook-north-america.html. Accessed 20 March 2020.

Ray, C. (2020). Future of SA's big three sports is precarious in a COVID-19 world. *Daily Maverick.* Retrieved from www.dailymaverick.za/articule/2020-03-23-future-of-sas-big-three-sports-is-precarious-in-a-covid-19-world/amp. Accessed 10 April 2020.

Raymond, C. H., Ko, J., Yuan, Y. W. Y., Lam, J. J., & Louie, L. (2004). Severe Acute Respiratory Syndrome and sport: facts and fallacies. *Sports Medicine, 34*(15), 1023–1033.

Reddy, L. (2020). Coronavirus: Tokyo 2020 Olympic organisers respond to frustrated athletes. *BBC Sport.* London. Retrieved from www.bbc.co.uk/sport/amp/olympics/51943083. Accessed 20 March 2020.

Rodriguez-Morales, A. J., Bonilla-Aldana, D. K., Balbin-Ramon, G. J., Paniz-Mondolfi, A., Rabaan, A., & Sah, R. (2020). History is repeating itself, a probable zoonotic spillover as a cause of an epidemic: the case of 2019 novel Coronavirus. *Infez Med (Le Infezioni in Medicina), 28*, 3–5.

Rothe, C., Schunk, M., & Sothmann, P. (2020). Transmission of 2019-nCoV infection from an asymptomatic contact in Germany. *The New England Journal of Medicine, 382*, 970–971. https://doi.org/10.1056/NEJMc2001468320035551.

Scott, L., Roan, D., Capstick, A. (2020) Tokyo 2020: Why is Olympic decision taking so long? *BBC Sport.* London. Retrieved from www.bbc.com/sport/olympics/52007137. Accessed 23 March 2020.

Shelnin, D. (2020). For finely tuned Olympic athletes, a one year postponement changes everything. *The Washington Post.* Retrieved from www.washingtonpost.com/sports/2020/03/24-olympics-postponed-athlete-impact/?outputType=amp. Accessed 24 March 2020.

Smith, G. (2020). Life without sports – betting firms begin to tally up their coronavirus losses. *Fortune.* Retrieved from www.fortune.com/2020/03/21/betting-firms-tally-coronavirus-losses/amp/. Accessed 30 May 2020.

So, L. K. Y., Lau, A. C. W., & Yam, L. Y. C. (2003). Development of a standard treatment protocol for severe acute respiratory syndrome. *Lancet, 361*(9366), 1615.

Stone, S. (2020). Coronavirus: what next for football as cracks begin to appear. *BBC.* London. Retrieved from www.bbc.co.uk/sport/amp/football/52063085. Accessed 28 March 2020.

Sugishita, Y., Kurita, J., Sugawara, T., & Ohkusa, Y. (2020). Preliminary evaluation of voluntary event cancellation as a countermeasure against the COVID-19 outbreak in Japan as of 11 March, 2020. *medRxiv.* https://doi.org/10.1101/2020.03.12.20035220.

Tennis.com. (2020). *Wimbledon cancelled.* Retrieved from https://www.tennis.com/pro-game/2020/04/2020-wimbledon-cancelled-all-england-club-grass-major-covid-19-calendar/88193/. Accessed 4 April 2020.

The Associated Press. (2020). *Augusta National Golf Club announces postponement of the Masters.* Retrieved from https://www.pgatour.com/news/2020/03/13/augusta-national-golf-club-announces-postponement-of-the-masters-tournament.html. Accessed 4 April 2020.

Tian, H., Li, Y., & Liu, Y. (2020). Early evaluation of Wuhan city travel restrictions in response to the 2019 novel corona virus outbreak. *MedRxiv*. https://doi.org/10.1101/2020.01.30.20019844.

Tignor, S. (2020). *Shocked: Wimbledon's cancellation isn't tragic, but it still hits hard.* Retrieved from https://www.tennis.com/pro-game/2020/04/shooked-wimbledons-cancellation-isnt-tragedy-we-can-be-bumped/88230/. Accessed 4 April 2020.

Von Drehle, D. (2014). Now arriving, the deadly Ebola virus lands in America. *Time, 184*(14), 34–37.

Wan, S. K., & Song, H. (2019). Economic Impact Assessment of Mega-Events In The United Kingdom And Brazil. *Journal of Hospitality & Tourism Research, 43*(7), 1044–1067. https://doi.org/10.1177/1096348019851838.

World Health Organization. (2015). *Clinical management of severe acute respiratory infection when Middle East respiratory syndrome coronavirus (MERS-CoV) infection is suspected.* Interim guidance. Updated 2 July 2015. Retrieved from http://www.who.int/csr/disease/ coronavirus_infections /case-managementipc/en/. Accessed 20 March 2020.

World Health Organization. (2020). *Novel coronavirus (2019-nCoV)-situation report-7-27 January 2020.* https://www.who.int/docs/default-source/coronaviruse/situation-reports/20200127-sitrep-7-2019–ncov.pdf?sfvrsn=98ef79f5_2020. Accessed 20 March 2020.

Wuhan Airport. (2020). *Wuhan airport 2020.* https://www.wuhan-airport.com/. Accessed 20 March 2020.

Zumla, A., Azhar, E. I., Arabi, Y., Alotaibi, B., Rao, M., McCloskey, B., Petersen, E., & Maeurer, M. (2015). Hostdirected therapies for improving poor treatment outcomes associated with the Middle East respiratory syndrome coronavirus infections. *International Journal of Infectious Diseases*. pii:S1201–9712(15)00215–5.

Chapter 11
Impact of COVID-19 on Global Religious Tourism and Pilgrimages

Abstract The past three decades have witnessed an increase in zoonotic infections leading to severe effects on global public health. These emerge and re-emerge when animal infections breach species barriers to infect human beings. At the same time, the world has witnessed an increase in the popularity and commodification of religious tourism and pilgrimages. During religious festivals, there is always the risk of infectious diseases being spread, which may result in these festivals being postponed and/or cancelled. This chapter examines the effects of the COVID-19 pandemic on religious tourism and pilgrimages around the world, especially the Hajj and the Umrah, both hosted in Saudi Arabia. Document analyses supplemented with fine-grain analysis were the preferred methods for collecting and analysing data. The research observed that, globally, iconic religious events had to be cancelled, postponed or significantly scaled down, resulting in reduced tourist arrivals. It also emerged that decisions that were made to cancel, postpone or significantly scale down such festivals or pilgrimages may have saved the world from aggravated circumstances causing a spike in COVID-19 numbers. The chapter recommends that aggressive marketing for religious and pilgrimage tourism be embarked on once COVID-19 has been contained and people are comfortable travelling again. Where possible, some events could be rescheduled, although this will not be possible with religious holidays like Easter and Ramadan. The chapter also acknowledges the interventions made by several governments to mitigate the impacts COVID-19, which could be repeated in future if still effective for similar pandemics.

Keywords COVID-19 · Pilgrims · Religious tourism · Moria · Hajj · Ramadan · Umrah

11.1 Introduction and Background

Tala and Padurean (2008) maintain that religious tourism (also referred to as spiritual tourism) is gaining popularity and is on an upward trend. Religious tourism associated with pilgrimage is potentially the oldest form of tourism, dating back thousands of years (Wilson 2014). The religious tourism sector comprises a large

G. Nhamo et al., *Counting the Cost of COVID-19 on the Global Tourism Industry*,
https://doi.org/10.1007/978-3-030-56231-1_11

number of stakeholders which include governments, religious communities, international organisations, host communities, local authorities, tour operators, transport companies, local tourism service providers and academia as well as other private sector stakeholders such as the air travel and cruise ship industry (ibid). Yakunin et al. (2016) have created a typology for religious tourism, including the following: the celebration of religious rites, self-improvement and affirmation of the spiritual state, raising of personal status in the religious community, worship of the holy place and secular motives. Consequently, several sites globally attract religious tourists for one reason or another.

As early as 2012, the United Nations World Tourism Organization (UNWTO) identified the Asia Pacific region as the world's religious tourism epicentre. In 2012, the UNWTO estimated that about 600 million national and international religious and spiritual voyages took place across the world to attend different religious events planned throughout the year by different religions (UNWTO 2012). Of these, about 40% took place in Europe and more than 50% in Asia. Apart from Saudi Arabia's Hajj and Umrah pilgrimages, where religious tourism has been prioritised after oil wealth (Sherbini et al. 2016), Rot et al. (2014) identified the shrine of Mary of Bistrica as one of the key religious attractions in Croatia, although its promotion as a destination remained low at that time. Twenty-two such religious tourism shrines were available in the country. From Greece, Samos and Thessaloniki have been identified as the most suitable sites for religious tourism, given their rich religious and historical monument (Balomenou et al. 2015; Zouni and Digkas 2019). In South Africa, more than a million pilgrims attend the Easter weekend gathering of the Zion Christian Church in the Limpopo province of South Africa (Saayman et al. 2014; Nyikana 2017). With all these religious tourism attractions in place, their maximum utilisation is always disturbed by epidemic and pandemic outbreaks like all other niches of the tourism sector.

Shinde (2015) defines religious tourism as visits to religious places, with the intention of satisfying religious and recreational needs. One of the world's largest gatherings of religious tourists occurs annually during the Hajj pilgrimage to Mecca in Saudi Arabia. This pilgrimage attracts more than two million Muslims from all over the world for a month-long event that is typified by intense rituals and conditions of overcrowding (Ebrahim and Memish 2020). For Christians, millions visit the Vatican City, which is the home of the Pope and the centre of the Roman Catholic Church. Millions also visit Israel, especially Jerusalem and Bethlehem, during key religious periods such as Easter and Christmas. Infectious diseases that require person-to-person transmission such as coronaviruses and, specifically, COVID-19 are known to be amplified during mass gatherings such as these pilgrimages and have the potential to increase their spread.

Over the past 30 years, several viruses with high effects on public health have emerged after cross-species passage from animal reservoirs to human populations. The Ebola virus and two different coronaviruses causing severe acute respiratory syndrome (SARS-CoV) and the Middle East respiratory syndrome (MERS-CoV) diseases are typical examples. These biological agents are new to humanity, have high epidemic potential but are also prone to disappear if early detection and interventions are implemented (Al-Tawfiq 2020). Other viruses have expanded their geographical area of existence from the original ecological niche to new places and continents, as recently shown by the extensive outbreaks of the Zika virus (Ippolito and Rezza 2017).

Towards the end of 2019, a new virus called the 2019-novel coronavirus (2019-nCoV) (COVID-19) emerged at a seafood wet market in Wuhan, China (Peng et al. 2020). The virus gradually spread regionally to countries like Thailand, Japan, South Korea, Singapore, Vietnam and Taiwan and then overseas to countries including Italy, Spain, the UK, France, the USA and South Africa (Nishiura et al. 2020). At the time of writing this chapter, COVID-19 had reached every continent of the world and, on 11 March 2020, was declared a pandemic with incidences of over a million globally (WHO 2020a). Coronaviruses are zoonotic because they originate in wildlife. Emergence occurs at the human-animal interface when animal infections breach species barriers to infect human beings. Historically, the emergence of these coronaviruses has been associated with global challenges in public health and has been a matter of concern for both internal and international travellers all over the world (Gallego et al. 2014), including those from the religious and pilgrimage sections. The United Nations has identified the COVID-19 pandemic as the biggest challenge the world has faced since the World War II (United Nations 2020) because the measures imposed to fight the virus have dire socio-economic effects that are unparalleled in the recent past. The new coronavirus disease attacked societies at their core, claiming lives, as well as people's health and their livelihoods in the process.

Mass gatherings are events attended by enough people to strain the planning and response resources of the host community, state or nation. The decision to host a mass gathering is usually made well in advance by the stakeholders involved such that effective prior planning is executed. Effective prior planning is of key importance because of the need to prevent and respond to communicable disease from religious tourists (WHO 2020b). The crowding of people has a significant impact on public health, exceeding the generally recognised limits of public health. Even when the existing health status of a host community and other support services are adequate to deal with the regular disease burden affecting the normal population, the influx of large numbers of people caused by mass gatherings can place a severe strain on such services (Ahmed and Memish 2020). Effectively, this also compromises the ability to detect a developing problem and carry out an effective response (Ebrahim et al. 2020).

Religion is a key component of the human race and dictates society's way of life. It can be a source of enormous individual comfort and community resilience in times of crisis. The COVID-19 pandemic has led to unprecedented levels of control and restrictions on mass gatherings in order to try and contain the contagion (Dube et al. 2020). This has led to either adjustments, cancellations or postponement of religious festivals and events, with severe negative effects on religious tourism of different magnitudes. This chapter seeks to determine the effect that the COVID-19 pandemic has had on religious tourism and pilgrimage around the world. Further, the chapter seeks to document the methods used by different religions and religious units to try and mitigate the shock created by the pandemic on their way of worshipping. This will help the different religions to be prepared for future outbreaks, given the worrying trend of increases in the emergence of new and re-emerging communicable diseases with high mortality rates such as COVID-19.

11.2 Literature Survey

Carbone et al. (2016) highlight that the pilgrimage phenomenon has continued over the centuries. Tourists arrive in Europe by various means of transport from across the world and walk through one of the traditional ways to Santiago linking up France and Spain. However, other destinations, including Italy and Portugal, have been added, as religious tourism continuous to undergo transformation (Rejman et al. 2016). This transformation meant that travel agencies have to package religious tours with luxuries including hotel stays, spas and wellness facilities. Hotel growth has been witnessed in Poland due to Pope John Paul II's numerous pilgrimages there. Furthermore, growth in hotels and other supporting infrastructure has been massive in Mecca and Medina in Saudi Arabia. This transformation led Ladki and Mazeh (2017) to investigate the cost of Mecca among three groups of religious tourists that involved (1) Indonesia and India, (2) Lebanon and Tunisia and (3) Dubai and Qatar. The three groups were selected based on social class, namely, the most populated and less affluent Islamic countries, the less affluent Arab states with moderate income and the richest Arab Islamic countries. The finding was that countries with wealthy economies charge much higher prices for Hajj packages than those with moderate economies. So, the Hajj remains intertwined with business too.

The historical emergence of coronaviruses has always been associated with a heavy toll on public health systems and serious disruptions to the socio-economic fabric of the affected places. The previous serious coronavirus outbreaks were the 2002/2003 SARS and the 2012 MERS epidemics (Al-Tawfiq et al. 2014). In December 2019, the world saw the emergence of yet another coronavirus named COVID-19. A significant knowledge gap existed in the COVID-19 outbreak, especially when looking at the ratio of mild and asymptomatic cases to fatal cases (surveillance pyramid), which were unknown since it was new. This situation was observed to have seriously hampered the assessment of the epidemic and complicated the response to the outbreak (Al-Tawfiq and Gautret 2019; Peng et al. 2020). In the end, mass gatherings, especially religious festivals, needed to be controlled in order to suppress the spread of the new disease.

The Kingdom of Saudi Arabia has taken a leading role in the management of communicable diseases during mass gatherings. With guidance from the WHO, the country established the Saudi Global Centre for Mass Gatherings Medicine (GCMGM). The Centre studies human gatherings, coordinates international health measures and offers its experience on the patterns and lifestyles of human gatherings in order to ensure collective coordination and the provision of necessary health services. These services include epidemiological surveillance, infection control, planning for disaster prevention, preservation of people's health and many other health-associated topics. All this is done in order to promote tourism and the associated benefits from such movements. The Centre has brought together global academic and public health institutions with complementary expertise to gather and translate the most appropriate public health policy evidence for use by countries that host, or are planning to host, mass gathering events (Memish et al. 2014).

Planning and preparing public health systems and services for managing mass gathering is a complex procedure that involves advanced risk assessment and system enhancement. These are critical to identifying potential public health risks and the so-called Black Sky hazards, as well as preventing, minimising and responding to public health emergencies (WHO 2020b). If the gathering draws visitors from different nations, regions and cultures, as do many of the religions ceremonies, the potential for increased risk of importation of infectious diseases creates additional challenges. The potential benefit of the GCMGM became evident when there was an outbreak of MERS in September 2012 (Memish et al. 2014).

Travelling for religious observance has been in existence since antiquity and was the first form of tourism (Handriana et al. 2019). Religious tourism occurs from the instant individuals begin a trip influenced by faith. Religion and pilgrimage tourism is referred to by the UNWTO (1995) as all travel away from the usual environment for religious purposes. However, this excludes travel for professional purposes like a pastor travelling for work. Mintel (2012) broadens the definition of religious tourism to include travel for the purposes of touring places, routes and festivals of spiritual importance. It further covers journeys where the foremost purpose of the visit is participation in activities that have religious significance such as conferences, conventions, ceremonies, retreats and camping. Handriana et al. (2019) highlight the motives of religious tourism as being to increase levels of faith; remind oneself that death is a necessary end; remember and honour the struggle of the saints in that religion; bring children with the hope that they can also increase their levels of faith; obtain peace of mind and get closer to God and obtain his blessings. Table 11.1 shows some of the popular religious tourism sites in the world and the average number of visitors to special events.

It is estimated that about 300–330 million tourists visit the world's key religious sites annually, with approximately 600 million national and international religious expeditions also realised. However, these numbers could be underestimates because they have been cited by the UNWTO and many other agencies for the best part of a decade. This is despite every indicator of volume suggesting active growth in the sector. It is also estimated that religious tourism is an $18 billion global industry, with North America alone having a significant share of the value, where one in four North American travellers is interested in "spiritual vacations" (Griffin and Raj 2017). This value and benefits of religious tourism stand to be lost when global pandemics such as COVID-19 occur and measures are put in place to control them.

Religious tourism gatherings can also be spontaneous and hosted simultaneously on several locations within a country, across regions or globally, for example, the funeral of a global or prominent religious leader and/or an ordination of a Pope. An estimated 10 million visitors from over 184 countries travel to Saudi Arabia each year to participate in the Hajj pilgrimage, the mini-pilgrimage Umrah or during the

Table 11.1 Popular religious tourism sites in the world and average visitor during special events

Country	Nature of Gathering	Place	Visitors/year (million)
India	Kumbh Mela (Hindu)	Ganges, River Godavari	100
	Shrine to Ayyappan (Hindu)	Sabarimala	30
	Harmandir Sahib/Darbar Sahib (Golden Temple) (Sikh)	Punjab	13
Mexico	Basilica of Our Lady of Guadalupe (Christian)	Mexico City	20
Iran	Qom (Muslim)	Qom	20
Saudi Arabia	Hajj (Muslim)	Makkah	10
Iraq	Arba'een (Muslim) – visit shrine of Imam Hussain	Karbala	15
China	Nanputuo Temple (Buddhist)	Xiamen	10
	Taishan Temple (Daoist)	Shandong	0.5
Philippines	The annual feast of the Black Nazarene (Christians)	Manila	8
France	Sanctuary of Our Lady Lourdes (Christian)	Lourdes	8
Israel	Western Wall (Jewish/Christian)	Jerusalem	8
Brazil	Basilica of the National Shrine of Our Lady of Aparecida (Christian)	Aparecida	6.6
Poland	Jasna Gora monastery (Christian)	Czestochowa	5
Portugal	Fatima (Christian)	Cova da Iria	5
South Africa	The Zion Christian Church (Christian)	Moria	5
Nigeria	Qadiriyya shrine	Kano	3
UK	Canterbury Cathedral – St Thomas Becket (Christian)	Canterbury	1.5

Source: Authors, based on Griffin and Raj (2017: vii), ARC (2014)

month of Ramadan. The vast majority of these pilgrims come from developing countries with compromised public health systems (Ahmed and Memish 2020). In the context of COVID-19, which has had a severe effect on older people, it is of interest to note that close to 43% of the pilgrims to the Hajj are 56 years of age or older. In addition, two-thirds of pilgrims are from countries with suboptimal disease surveillance systems or travel health counselling services (Ebrahim and Memish 2020). A further 50% usually have pre-existing chronic diseases, and respiratory infections are the most commonly observed illnesses among pilgrims. With constrained social distancing among pilgrims as they engage in religious rituals, continued engagements during the COVID-19 outbreak would have amplified their risk of acquiring and transmitting respiratory diseases.

However, during the MERS outbreak, no cases were recorded among Hajj pilgrims (Waldron and Doherty 2015; Soliman et al. 2015). During the SARS outbreak and its aftermath in 2002, Saudi Arabia banned the entry of people who had visited any of the six countries where local SARS transmission had occurred. These countries included China, Hong Kong, Taiwan, Singapore, Vietnam and Canada. Besides, the Saudi trade and diplomatic missions in China, Singapore, Hong Kong and the Philippines were instructed to stop issuing Umrah visas to pilgrims. Consequently, no SARS was imported into the Kingdom of Saudi Arabia, and no case of SARS was detected by airport screening in any pilgrims visiting the country. This was most likely the result of the immediate ban on any incoming visitors from SARS-affected countries with the ban only being lifted in July 2003, following positive reports from the WHO that no new cases had been reported for the previous 20 days (Azhar et al. 2016).

Given the intense and overcrowded conditions that dominate the pilgrims in Saudi Arabia, disease outbreaks have always been of much concern. Although no outbreaks have occurred in the recent past, in 1821 and 1865, cholera outbreaks killed an estimated 20,000 and 15,000 pilgrims, respectively (Zumla et al. 2017). These went on to spread all over the world. Given the risk of a major outbreak, during the pilgrimage season, a mega semi-permanent and highly mobile healthcare system is activated for the Hajj, that is, the months leading from Ramadan to after the Hajj. This provides on-site acute medical care, including acute care for critical illness at all the Hajj sites. Given the enormous effort of the temporary but massive and sophisticated healthcare system for Hajj, many thousands of healthcare workers also travel to the Hajj sites, thereby becoming religious tourists too (Ahmed and Memish 2020). Sometimes they are hired from outside the country, as they seek to serve the "guests of God".

11.3 Materials and Methods

The chapter sought to document the impact of COVID-19 on religious tourism. As such, the chapter uses case studies from all over the globe, particularly those that are very popular with pilgrims and religious tourists. Figure 11.1 shows the countries that have sites which are popular with pilgrims and religious tourists that were used in the study. This chapter mainly used document and critical discourse analysis as a method for data gathering. Document analysis was done systematically by reviewing and evaluating documents from both printed and electronic sources (Bowen 2009). The documents analysed in the study included editorials, scientific papers, professional reports and newspaper and magazine articles. As in other analytical methods in qualitative research, document analysis requires that data be examined and interpreted in order to extract themes and meaning, gain understanding and develop empirical knowledge from them.

The fine-grain analysis then followed the completion and cleaning of the collected document content data. This was performed by making multiple readings of

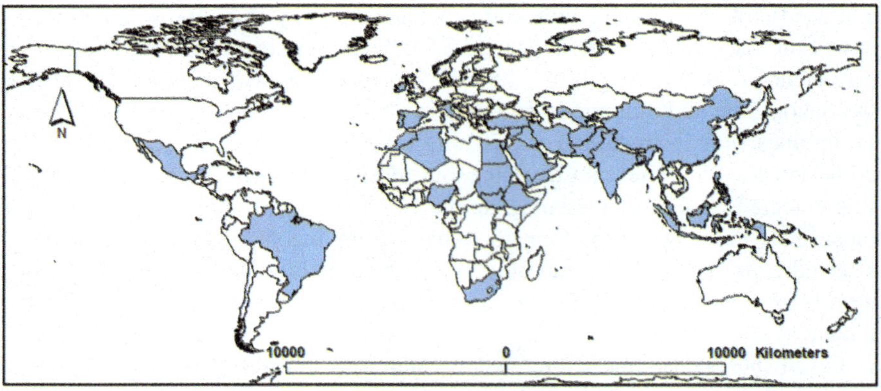

Fig. 11.1 Countries with pilgrims and religious tourism sites used in the study. (Source: Authors)

the collected data, analysing what was present in and/or missing from each. This was followed by the grouping of the data into several broad themes and examining divergences and convergences within and between different documents to address the research question. The themes were then itemised, and content analysis was used to extract meaning from the documents. At this point, a finer examination of themes that may have been previously unnoticed was performed. Finally, a tailor-made narrative was constructed to answer the research objective.

Based on Griffin and Raj (2017), religious tourism products include traditional pilgrimage, religious tourism, church tourism, religious events, missionary tourism, retreats, faith-based cruises and religious routes. The authors classify sacred religious sites. These include single nodal features, archaeological sites, burial sites, detached temples and/or shrines, whole towns, shrine and/or temple complexes, sacred mountains, pilgrimage foci and secular pilgrimage. The next section presents the data and discusses the key findings.

11.4 Presentation of Data and Discussion of Findings

In most countries of the world, religious tourism and pilgrimage were negatively affected by the COVID-19 outbreak. Many religious and pilgrimage events had to be scaled down initially and eventually postponed and/or cancelled altogether. This was in response to the pandemic as it ravaged many parts of the world.

11.4.1 A Focus on Saudi Arabia's Hajj and Umrah Pilgrimages

Religious tourism and pilgrimage come at a scale and magnitude that are difficult to match anywhere in the world as in Saudi Arabia. Two major pilgrim gatherings were on the horizon when the outbreak of COVID-19 occurred. These were the Umrah and the Hajj. In 2017, more than 2.4 million Muslims visited Mecca for Hajj, with some 1.8 million out of this total coming from outside Saudi Arabia (Cochrane 2018). The numbers are restricted as there are about 1.8 billion Muslins globally (Bensaid 2019). Figure 11.2 shows the main source countries for Hajj.

The pilgrimage starts on the 8th and lasts until the 12th of the last month of the Islamic lunar calendar (Dhu al-Hijjah). Hajj further coincides with the Eid al-Adha holiday that comes immediately after Ramadan, which is the month of fasting in the Moslem faith. The tourists (so-called hajjis) often arrive several days in advance and stay for a week or more. The Hajj is estimated to generate up to $8 billion and comes second only to the country's big oil industry in terms of revenue generation for the country (Cochrane 2018). In addition, Mecca also hosts Umrah, which is a nonobligatory pilgrimage that takes place throughout the entire year, adding to the already huge arrivals of Hajj. In 2017, eight million pilgrim tourists visited Mecca, generating $4 billion, bringing the total contributed to the GDP of Saudi Arabia by

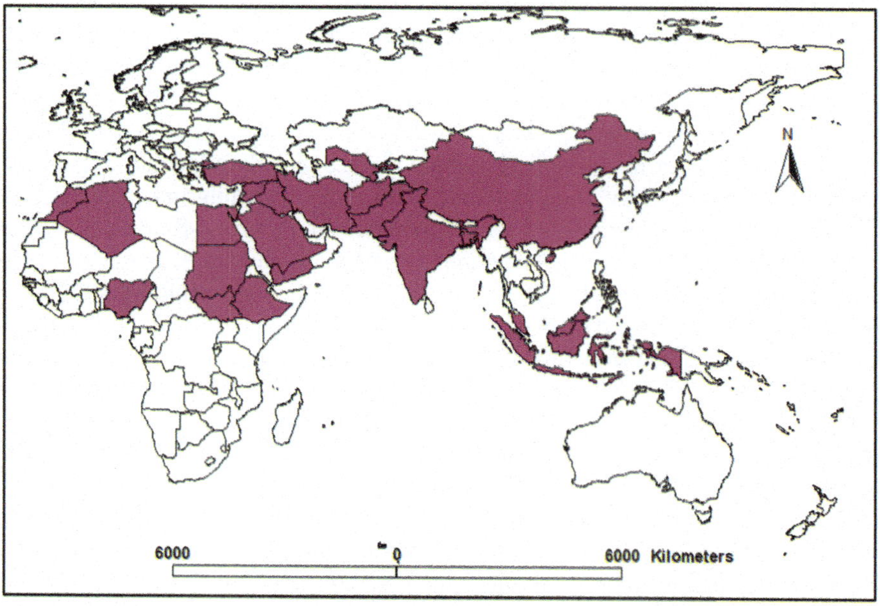

Fig. 11.2 Major Hajj source countries. (Source: Authors)

the two pilgrimages to $12 billion (*Times of India* 2018). With this, an estimated 25–30% of the private sector income in Mecca and Medina comes from the pilgrims. Overall, the pilgrimage sector contributed about 20% of non-oil GDP and 7% of the total GDP (ibid).

The economics of Hajj and Umrah has resulted in Saudi Arabia's Government's Vision 2030 strategy, instituting a massive expansion plan to accommodate +/–30 million visitors for Umrah (Daye 2019). Investments in Mecca and Medina are financed through a sovereign wealth fund, the PIF that had about $230 billion in assets in 2017. The massive investments in Mecca have transformed it into a modern entity with new hotels that include a $3.5 billion Abraj Kudai hotel complex with twelve 45-storey towers. The towers added 10,000 bedrooms with further plans to add 80 000 hotel rooms near the Prophet's Mosque in Medina (Cochrane 2018). Other infrastructure investments included a railway line, metro and parking for 18,000 buses for pilgrims' transportation. In 2018, up to 466 flights took Indian pilgrims to Mecca (*Times of India* 2018) igniting the airlines, travel agents and related tourism sector. An estimated 54 million pilgrims had attended the Hajj in the past 25 years up to 2018. Figure 11.3 summarises Hajj 2019 arrivals in Mecca as per the data provided by the Saudi Arabia General Authority for Statistics (SAGAS). Overall, about 2.49 million pilgrims made it to Mecca. Other numbers for the 2018 Hajj are shown in Table 11.2.

The packages for Hajj differ by country. For example, in the UK, they average £4750, with Umrah sitting at £1050 (Cochrane 2018). These figures bring about £100 to £200 profitability per pilgrim for the operators. Bensaid (2019) gives another range of $800 to $7000 for the trip, depending on the package. This makes it a big tourism industry. However, to manage the numbers during the construction phases of Mecca, especially up to 2017, Saudi Arabia instituted the Hajj quota system. The Hajj quotas permit 1000 pilgrims per million of the total Muslim population in each sending country, although there is some flexibility. However, 75% of non-Saudi Hajj pilgrims come from eight countries including Indonesia, Pakistan, Bangladesh, Iran, Egypt, Nigeria, Turkey and India. Hajj pilgrimage funds have been set up in several countries to assist poorer Muslims and also for those that must wait for many years for a visa, sometimes as long as 30 plus years. In Malaysia, the Tabung Haji (the Pilgrims Fund) established in 1963 had assets of $18.3 billion in 2017 (Cochrane 2018). There were also huge Hajj funds in Indonesian trusts, including Sharia Financial Institutions and BPKH (Hajj Financial Management Agency/Ministry of Religious Affairs) (Jumali 2018). The savings stood at RPH 98.34 trillion in 2018. Looking into the future, both the Hajj and Umrah were predicted to generate up to $150 billion in the 5 years 2018–2022 (Thaqafi 2017), resulting in about 100,000 permanent Hajj-related jobs.

With little sign of containment of the COVID-19, Ahmed and Memish (2020) observed that Saudi Arabia had to take difficult and unprecedented action in the interests of public health, global safety and protecting its citizens. They argued that Islam mandates that the sanctity of human life be placed above all other rights, including the right of God to demand worship by humankind. With this in mind, the cancellation of religious gatherings, including the Hajj 2020, was well

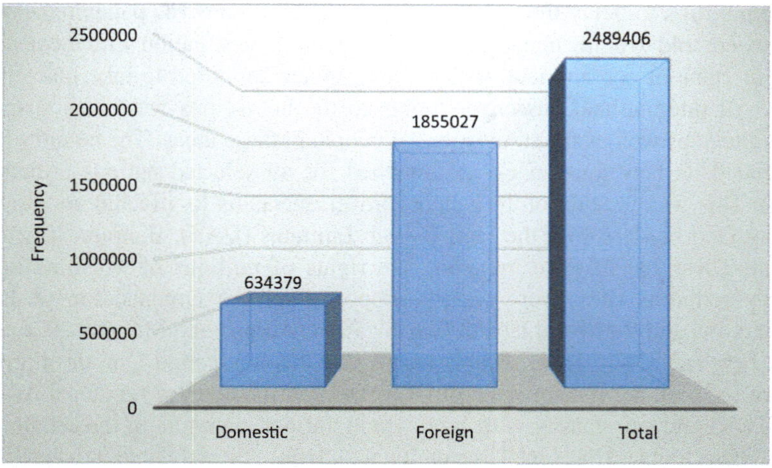

Fig. 11.3 Hajj 2019 pilgrims. (Source: Authors, data from SAGAS (2019: 10))

Table 11.2 Other Hajj numbers for 2018

Item	Numbers
Pilgrims that took part	2 million
International and domestic flights operated	14,000
Buses used to ferry pilgrims	21,000
Civil defence employees	18,000
Copies of the Quran distributed	8 million

Source: Authors, based on the Times of India (2018)

communicated in advance and was very much in line with Islamic ideals. The reasoning was that cancelling it contributed immensely to the safety of humanity. More important is that had the authorities not cancelled the event and allowed it to proceed, then the ramifications would have been global, with even worse reputational damage in the eyes of the touring pilgrims. Given the nature of COVID-19, where many individuals remain asymptomatic, this scenario presented a major risk for spreading the disease if persons were allowed to move freely into shrines during the gatherings (Al-Tawfiq 2020).

On the other hand, the host, Saudi Arabia, was faced with the extremely delicate balance of welcoming religious pilgrims for both Hajj and Umrah in 2020, many of whom will have waited a lifetime to enact their religious rites. Many of the tourists would also have invested lifetime savings to make the journey. Two unprecedented and courageous decisions had to be made. Hence, on 27 February 2020, Saudi Arabia suspended the year-round Umrah pilgrimage in an attempt to slow the spread of the virus. It also suspended visas for Muslims seeking to visit the holy cities of Mecca and Medina (Ahmed and Memish 2020). On 31 March 2020, Saudi Arabia asked pilgrims to delay their Hajj plans amid the coronavirus pandemic. The pilgrims were told to delay making Hajj contracts until the situation with the pandemic showed signs of containment, with the pilgrimage supposed to take place in July 2020. All international passenger flights to the holy cities were also suspended indefinitely, as well as free movement of people in these areas. The country further suspended its new tourism e-visa launched for 49 selected nations (Abuhjeeleh 2019). This was in addition to banning travel of Saudis to affected countries and closing land borders with the United Arab Emirates (UAE), Bahrain, Kuwait and Jordan. These bans, while impeding the rights of millions of Muslims to fulfil Islamic religious rites, were widely supported by the Organisation of Islamic Cooperation and the World Health Organization (Ahmed and Memish 2020).

In February 2020, Saudi Arabia suspended the year-round Umrah pilgrimage, and, in the end, the transmission of COVID-19 in the country remained relatively low. This contrasts sharply with Iran, which did not intervene in the religious pilgrimage in Qom and has seen large regional outbreaks directly linked to the pilgrimage. The holy city of Qom, which receives up to 20 million pilgrims annually from the whole world, became the starting point of the COVID-19 pandemic in Iran. The religious leaders had vehemently and successfully resisted quarantine and social distancing, even initially downplaying the severity of the virus; hence it spread rapidly across the country. The COVID-19 pandemic, therefore, exposed the social and religious rifts that have been growing in the country and usually come to light during the time of crisis (Sadrzadeh 2020). This was so when religious leaders ignored the advice from health officials to cancel their events in the initial stages of the pandemic. In the end, Qom became a hotspot for the coronavirus outbreak in Iran with 18,407 infections and 1284 fatalities (Ziabari 2020). Furthermore, temporary closure of the holy sites was agreed by the religious authorities only after almost a month of debate on whether such a decision could be authorised.

However, after the closure announcement, groups of firebrands and angry protesters smashed the closed gates of the holy sites trying to forcefully gain access and demanded the complexes be reopened. This led some to call the incident the epitome of sacred ignorance (Sadrzadeh 2020). The decision to allow pilgrims by Iran also had global ramifications: one of the first COVID-19 cases in New Zealand came from a family that had recently taken the pilgrimage to Qom, three of the first cases in Canada had links to the religious event in Qom, and all 33 initial cases in Iraq had links to this event, as well as the first confirmed case in New York City which later became a hotspot of its own. In addition, India evacuated hundreds of Muslim pilgrims from the affected areas, many of whom tested positive to COVID-19 on arrival (Blum 2020). However, the Iranian example shows that early interventions to stop large religious gatherings can help to contain an outbreak rather than just issuing advisories to social distance.

Although it was rather early to assess the extent of the impact of the intervention on tourism and Saudi Arabia's economy, the suspension of pilgrims for both Hajj and Umrah in 2020 led to potential revenue loss for the country. The intervention also negatively affected revenues for the entire tourism value chain from airlines, to travel agents, hotels, restaurants and local guides that earn a living from these mega pilgrimage events. However, this loss was not unique to Saudi Arabia as the world over, governments closed borders and also embarked on lockdowns. This meant that even if Saudi Arabia had remained open to the business of the Hajj and Umrah, the arrivals from outbound tourists would still have gone down drastically.

11.4.2 Other Religious and Pilgrimage Cancellations

Religious tourism is big business for many other host countries and a significant source of income. In India, "Hajjonomics" (Hajj linked economics) propels private tour operators as they try to make the most of their share of India's Hajj market. Some tour operators confirmed that as much as 20% of their business came from religious tourism (Sanghera 2020). However, they faced some handicaps in trying to scale up their market for Hajj tourists as the Indian Government allowed each operator a maximum of only 150 packages per season, and the host puts a cap on the number of pilgrims each country is allowed to send per year. Unfortunately, in the year 2020, when the Saudi Government increased India's allocation to a total of 200,000 from 170,000, there was the COVID-19 pandemic. Hence the pandemic constituted a loss of potential revenue for the tour operators given India's 140 million-plus Muslim population. Packages from the tour operators start from US$4220 per pilgrim. These packages offer perks such as conveniently located hotels, comfortable transport options and flexibility with dates. In the process, the operators had found ways to leverage the Hajj pilgrimage to sell add-on holidays such as visits to nearby countries like Turkey (Sanghera 2020).

In Iraq, the majority of pilgrims to Karbala agreed in 2020 to perform the ritual remotely in order to curb the spread of COVID-19 (Ibrahim and al-Salam 2020). With public gatherings including religious visits banned in the country, many in Iraq opted to perform the visits to Karbala's shrines remotely using free telematic services like telephones, live streaming and television channels dedicated to the pilgrimage. Many pilgrims noted that the visit meant a lot to them and the fact that they could not visit the shrine for their annual pilgrimage was very painful, but they agreed with the decision. However, some observed that remote visits would not feel the same as physical presence, and one would definitely feel the void of not being in Karbala.

In South Africa, Limpopo province is host to one of Africa's largest pilgrimage by the ZCC, with an estimated membership of 20 million across Southern Africa alone (Saayman et al. 2014). At its headquarters called Zion City Moria, the church receives close to 5 million pilgrims from all over South Africa, Zimbabwe, Zambia, Mozambique, Botswana, Malawi, the UK, the USA, Australia and many more. These pilgrims gather twice a year, during Easter and for the September festival (Ebrahim 2020). The 2020 Easter gathering coincided with the COVID-19 outbreak and a declaration of a state of national disaster in South Africa. The declaration also came with strict measures to control large gatherings, which were banned, and a lockdown that prevented people from leaving their homes. National borders were also closed. Gatherings of 100 people were initially allowed, but this was scaled down to almost zero by the lockdown which went on for a full month, including the Easter period. So serious was the lockdown that it came in two phases of 3 weeks with an additional extension of 3 weeks. Even the president of the country, Cyril Ramaphosa, had to give a virtual Easter message. All these measures implicitly meant that the ZCC Easter pilgrimage could not take place legally given that the pilgrims would have to come from other parts of South Africa and the world and also given that some of the countries from which the pilgrims would travel were also at the epicentre of the pandemic. Even if South African borders were opened (of which they were not), its neighbours which are major source countries for the ZCC pilgrimage also proclaimed lockdowns.

On Friday 13 March 2020, the ZCC agreed to postpone its Easter pilgrimage indefinitely and promised new dates as soon as the pandemic had been brought under control. Although a new date was promised, the meaning would not be the same as it would no longer be over Easter. During Easter, many Christians globally remember the life of Jesus Christ. The unprecedented decision was made after the church leadership had met both local and national government officials to discuss the impact of the disease and how the church could help in mitigating its effects. The ZCC further called a meeting with its District Property Task Team in which it sought further professional advice from the members of the church who were health professionals. Never in the history of the ZCC since its formation in 1910 had the church postponed or cancelled its Easter pilgrimage (Kekena 2020). The ZCC did not want to risk people's lives or do anything that would escalate the spread of the pandemic and called on its members to pray for the pandemic and work together with civil society, religious organisations and the government to limit the spread of COVID-19 (Karrim and Raborife 2020).

Figure 11.4 shows the countries that most of the ZCC pilgrims come from. The decision to postpone the pilgrimage may have been a contributory factor to the lower than anticipated spread of the COVID-19 pandemic in South Africa, although this had a telling negative impact on the tourism industry in Limpopo province and the entire regional economic system driven by air travel, hotels, restaurants and small-scale enterprises that thrive on selling ZCC-associated merchandise and other artefacts. This was unlike in other countries, for example, Indonesia, where Muhtada (2020) observed that one of the difficult challenges in the efforts to combat the spread of COVID-19 was the counterproductive attitude exhibited by some religious communities. This came as some religious groups still intended to hold gatherings involving many people despite the government ban on these. This was also combined with government indecisiveness in dealing with religions that defied the threat of COVID-19 spread during such ceremonies, possibly thinking of their economic impacts, especially on the tourism sector. Indonesia is one country that is heavily dependent on tourism. The defiance was also largely attributed to religious understandings that were fatalistic and deterministic which simply implies that everything happens by God's will and has been well arranged by Him. Furthermore, this understanding in Indonesia was complicated by political sentiments spurred by the wounds of the 2019 elections, the outcome of which some groups still had not

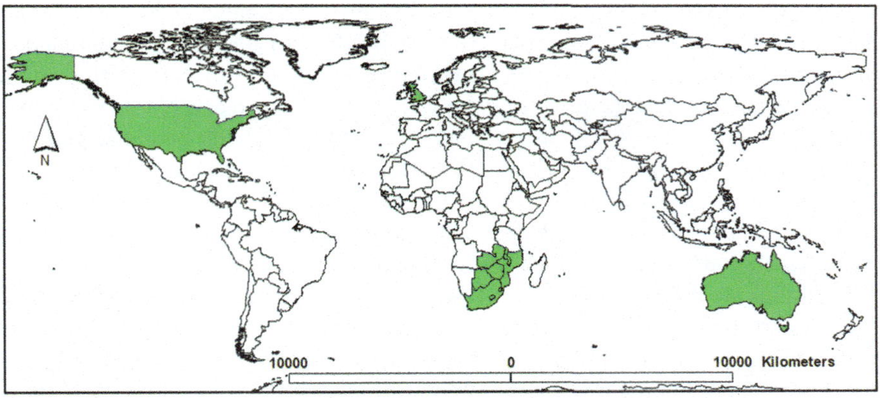

Fig. 11.4 Countries that ZCC pilgrims come from. (Source: Authors)

accepted and which found extra reasons to defy government's call to stop large gatherings.

Although public gatherings in most parts of the world were banned to curb the spread of the coronavirus, including masses in Italy, the Vatican was not spared. The Vatican decreed to its churches the world over that the Easter holy week celebrations for the entire Roman Catholic Church would be held for the first time without the participation of congregants. The move was seen as unprecedented in the modern times of the church. The holy week services that begin with Palm Sunday leading up to Easter are the second most important days in the Roman Catholic liturgical calendar after Christmas for the world's 1.3 billion members of the church (Pullella 2020). Other events of the holy week held without the participation of the public including the Holy Thursday, Mass of the Lord's Supper, Good Friday, the liturgy of the Lord's passion, way to the cross, Easter vigil mass and finally the Easter morning mass. The Church noted that the solemnity of Easter could not be transferred or postponed; hence bishops and priests had to celebrate the rites of the holy week but without the presence of the people. The church noted that the decree to prevent the public from attending was intended to stop the contagion, not the prayer. The church encouraged priests to inform the faithful of the times of the liturgical celebrations so that they could also prayerfully unite in their homes. They encouraged the use of live telematic broadcasts such as the Internet and television and not pre-recorded ones during these celebrations (Vatican 2020). The traditional decoration of St Peter's Square (the world's largest church) with olive trees during the Palm Sunday and the traditional tens of thousands of flowers used to decorate the papal altar and the rest of the square flown from the Netherlands were not possible during the COVID-19 outbreak. As a result, the world's biggest church was forced to change a tradition that had been followed for centuries in order to try and minimise the spread of the virus.

On 17 March every year, the Irish celebrate the St Patrick's Day, which is a religious and cultural event. The day marks the death of St Patrick, the patron saint of

Ireland, who is credited with bringing Christianity to the country. The day has been celebrated by the Irish for the past 1000 years. Dublin hosts the largest parade, attracting an estimated 500,000 people including religious tourists from all over the world. Parades are also held in several towns in both Ireland and Northern Ireland. Around the globe, the day is also celebrated, with 2019 having over 400 places in 50 countries marking the occasion (Mc Donald 2020). However, in 2020, the parades and events associated with the day were cancelled due to the COVID-19 outbreak. Bishops and priests celebrated the St Patrick Masses via webcam and parish radio stations with most highlighting that the strength of culture was seen not when things are going well but when crisis strikes. They also appealed to the people to listen to the experts and respect measures proposed to combat the spread of COVID-19. They went on to pray for the selfless health workers and medical scientists who were searching for a vaccine and better treatment for the virus (Mc Donald 2020). The festivities and the tourism associated with this event have been noted to have a big economic footprint in the places where it is celebrated, with US$5.3 billion being spent on the event in 2017 in the USA alone (Ogg 2017).

11.4.3 The Consequences of COVID-19 "Disobedience"

A chronicle of some religious events that were allowed to take place amid the COVID-19 pandemic shows that these ended up creating a biological time bomb not only for the countries involved but for the whole world. Therefore, these events need to be managed differently should the world find itself in such a predicament again. In India, for example, there was outrage over a Muslim congregation called Tablighi Jamaat, which sparked a massive cluster of COVID-19 cases. More than 647 cases across the country were traced to the religious event that had been held in Delhi earlier in March 2020, with officials saying the pilgrims had not followed social distancing as advised (Jha and Dixit 2020). As cases directly linked to the event continued to rise, there was also a rise in Islamophobia and vilification of Muslims, raising tensions in a country severely polarised on religious grounds. Health officials claimed a definite spike in COVID-19 cases in the whole country after this gathering. So serious were sentiments towards religious gatherings in India during the pandemic that some went to the extent of calling them "corona terrorism". In the end, panic gripped much of India as authorities launched a search to identify those who had attended the congregation, with as many as 25,000 Jamaat members and their contacts quarantined across nearly 15 Indian states. Such an incident presents a future real risk to religious tourism. In another case, Indian authorities in the state of Punjab had to quarantine around 40,000 people from 20 villages following a COVID-19 outbreak linked to one person – a religious preacher who had ignored advice to self-quarantine after returning from a trip to Europe. He attended a Sikh religious festival, a 6-day event that attracts over 60,000 people. A week after his death, 19 of his relatives tested positive, and up to 550 people were traced to have had direct contact with him (Frias 2020).

In South Africa, a church that hosted a religious gathering in Free State province which lasted for 6 days in March 2020 led to over 67 infections. The congregants were unwittingly exposed to five international guests from France, Israel and the USA who later tested positive for the virus. With the pandemic taking a strong foothold in South Africa during that period, the event led to the tracing of all the people who had come into direct contact with the infected persons to test and quarantine them. In the end, they identified 1259 people and 67 tested positive, including the church leaders. The COVID-19 infected guests of the church had not presented any symptoms on their arrival in South Africa when they were screened (Hosken and Wicks 2020).

In South Korea, the Shincheonji Church of Jesus was almost singlehandedly responsible for the COVID-19 outbreak in that country. The church's doctrine exposed its members to the coronavirus in the form of practices like not accepting illness as a valid reason to miss a service, taking a rollcall on attendance using special swipe cards and making strong follow-ups on absenteeism, praying in close proximity and not being allowed to wear anything on your face during a church service including glasses and masks as this was deemed disrespectful to God even when the COVID-19 pandemic was going on (Rashid 2020). Such cultish practices combined with overcrowding during the long hours of church services created the most conducive environment for the spread of the virus.

When the Korean Government traced the COVID-19 outbreak to the church, it was very difficult for them to track down everyone who had been exposed due to the church's inward-looking doctrine and suspicion of outsiders. For example, most of the 230,000 plus church members do not answer calls from non-members, so the authorities had to persuade close to 200 of the church members to call their fellow members to ask them about the symptoms of the virus. This was after the church's leaders deliberately withheld information about its membership. In the end, by the 9th of March, South Korea had 7400 COVID-19 cases with almost 63% of these traced to the Church. It was widely believed that a church congregant known as patient number 31 by the Korean Centres for Disease Control and Prevention super-spread the virus to many people during their church services (Rashid 2020). The result was the leader of the church being investigated over some of the coronavirus deaths and potentially facing charges of criminal negligence. Such glaring examples serve as a lesson to prevent or cancel gatherings when the world is in the middle of an epidemic such as that of COVID-19.

The religion to which individuals are affiliated seems to have been a contributory factor to the spread and infection by COVID-19 in the initial phase of the pandemic. The Orthodox Jewish or Haredi communities emerged as a classical case. In Israel, they make up about 12% of the population but accounted for as much as 60% of COVID-19 infections as of 1 April 2020 (Estrin 2020). In the UK they make up less than 0.03% of the population, but as of 31 March 2020, 2.5% of the total tally of fatalities were Jews. In France, Israel, New York and the UK, most places with high incidences of COVID-19 coincided with the Orthodox Jewish neighbourhoods (Liphshiz 2020). The burning question is why Haredi communities had such sharp spikes in COVID-19 cases.

With nothing conclusive, some of the reasons were because the Orthodox Jewish live a closely knit communal lifestyle, and many initially refused to follow social distancing advisories. For weeks they ignored government bans on large weddings, neglected to close schools and held prayer services even when close to a quarter of Israel's cases were contracted in synagogues. Their teachings are also associated with touching the same surface and the same siddurim or prayer book, which makes them vulnerable to infection if one single member is infected. In Israel, at some point, a popular rabbi told thousands of his followers to defy the health ministry's advice on social distancing and mass gatherings, but as the coronavirus outbreak took a heavy toll on the community with many experiencing severe outbreaks and religious leaders falling ill, things started to change (Staff 2020). After discussions with senior government officials in Israel and the USA, senior rabbis started to urge their followers to obey the government's stay home order. Many in the Haredi communities only follow orders from rabbis, not health officials. Some noted the timing of the Jewish holiday Purim which fell on 9 March 2020 as having contributed to the spread of COVID-19 within the Jewish community as it coincided with the outbreak of the coronavirus in Europe. Festivals such as Purim bring even more people together than usual, in the process hastening the spread of the virus. Also noted in the UK was that almost 25% of Jewish adults attend synagogues every week compared to 10% for Christian adults. All this made them more vulnerable as large gatherings were conducive for the virus to multiply (Liphshiz 2020).

Ultimately, deliberate or forced, cancellation of religious gatherings and pilgrimages had a negative impact on the tourism industry. The negative effects were felt across the entire tourism value chain from air travel, airport companies, travel agencies, local tour guides, hotels, B&Bs, backpackers, restaurants and other stakeholders.

11.5 Conclusions

It may be concluded from the chapter that religious activities and, by default, religious tourism were negatively affected by the COVID-19 pandemic. Almost all the major pilgrimages that were planned for the first half of 2020 for all the regions were either scaled down, cancelled or postponed on the advisory of public health experts. Religious activities affected included the Hajj and Umrah, Easter festivities for the Roman Catholic Church, the Moria pilgrimage for the ZCC church in South Africa and St Patrick's Day. Those who also made losses from these cancellations include service providers at both the host and receiving places who make a living from the commodification of such religious events. This includes travel agencies, hotels, transport and restaurants.

Religion during the COVID-19 outbreak was also used as a source of community resilience. Wherever religious leaders accepted advice from medical experts, there was, in many instances, a harmonious relationship with the government as given by the South African example. This helped the incidence and spread of the virus to remain suppressed in the country. However, religion was also blamed for spiking the incidence of COVID-19 and making the already bad situation worse. This was mainly in countries and regions where the religious leaders either ignored or defied expert advice to practise social distancing and cancel mass religious gatherings, and in most cases, their relationship with the governments was less cordial.

As a result, it is recommended that in the future, religious gatherings that draw huge numbers of people from different regions or countries should be cancelled during periods of pandemics of an international nature such as COVID-19. Equally important is the need to engage with religious leaders in terms of the role that they can play as leaders to build community resilience to the threat without unilaterally impeding of their right to worship. The use of social influencers and religious and community leaders has been shown to bear good results in the fight against COVID-19 in many places; hence scaling it up to other places affected is strongly recommended. Although global preparedness for emerging infections with pandemic potential has improved by building and learning from past experiences, the focus has mainly been on the action when the infection has emerged not on prevention. It is recommended that more emphasis needs to be placed on prevention which is less costly than taking action when there is already an outbreak, which is more expensive to control in the long term. Adopting this one health approach could substantially improve the situation and reduce the threat from future incidents.

References

Abuhjeeleh, M. (2019). Rethinking tourism in Saudi Arabia: Royal vision 2030 perspective. *African Journal of Hospitality, Tourism and Leisure, 8*(5), 1–16.

Ahmed, Q. A., & Memish, Z. A. (2020). The cancellation of mass gatherings (MGs)? Decision making in the time of COVID-19. *Travel Medicine and Infectious Disease*. https://doi.org/10.1016/j.tmaid.2020.101631.

Alliance of Religions and Conservation (ARC). (2014). *Pilgrim numbers*. Accessed from http://www.arcworld.org/projects.asp?projectID=500. Accessed Mar 2020.

Al-Tawfiq, J. A. (2020). Asymptomatic coronavirus infection: MERS-CoV and SARS-CoV-2 (COVID-19). *Travel Medicine and Infectious Disease, 27*, 101608. https://doi.org/10.1016/j.tmaid.2020.101608.

Al-Tawfiq, J. A., & Gautret, P. (2019). Asymptomatic Middle East Respiratory Syndrome Coronavirus (MERS-CoV) infection: Extent and implications for infection control: A systematic review. *Travel Medicine and Infectious Disease, 27*, 27–32.

Al-Tawfiq, J. A., Zumla, A., & Memish, Z. A. (2014). Travel implications of emerging coronaviruses: SARS and MERS-CoV. *Travel Medicine and Infectious Disease, 12*, 422–428. https://doi.org/10.1016/j.tmaid.2014.06.007.

Azhar, E. I., Lanini, S., Ippolito, G., & Zumla, A. (2016). The Middle East Respiratory Syndrome Coronavirus: A continuing risk to global health security. *Advances in Experimental Medicine and Biology-Advances in Microbiology, Infectious Diseases and Public Health*, Springer, https://doi.org/10.1007/5584_2016_133.

Balomenou, C., Poulaki, P., & Lagos, D. (2015, 25–28 August). Religious tourism in Greece and regional development: The case of Samos Island. In *55th Congress of the European Regional Science Association: "World renaissance: Changing roles for people and places"* Lisbon, Portugal, European Regional Science Association (ERSA), Louvain-la-Neuve.

Bensaid, A. (2019). *Is Saudi Arabia unfairly profiting from its holy sites?* Accessed from https://www.trtworld.com/magazine/is-saudi-arabia-unfairly-profiting-from-its-holy-sites-28899. Accessed 11 Apr 2020.

Blum, N. (2020). How Iran became a global vector of infection for COVID-19. *Tablet*. Available at: www.tabletmag.com/amp/sections/israel-middle-east/articules/iran-coronavirus-covid-19. Accessed 26 Mar 2020.

Bowen, G. A. (2009). Document analysis as a qualitative research method. *Qualitative Research Journal, 9*(2), 27–40. https://doi.org/10.3316/QRJ0902027.

Carbone, F., Corinto, G., & Malek, A. (2016). New trends of pilgrimage: Religion and tourism, authenticity and innovation, Development and Intercultural Dialogue: Notes from the diary of a Pilgrim of Santiago. *AIMS Geosciences, 2*(2), 152–165. https://doi.org/10.3934/geosci.2016.2.152.

Cochrane, P. (2018). *The economics of the Hajj*. Retrieved from https://www.accaglobal.com/an/en/member/member/accounting-business/2018/07/insights/economics-hajj.html. Accessed 11 Apr 2020.

Daye, A. S. (2019). Rising tourism in Saudi Arabia: Implications for real estate investment. *Cornell Real Estate Review, 17*, 98–105.

Dube, K., Nhamo, G., & Chikodzi, D. (2020). COVID-19 cripples global restaurant and hospitality industry. *Current Issues in Tourism*. https://doi.org/10.1080/13683500.2020.1773416.

Ebrahim, S. (2020). *Moria: The largest Christian gathering in SA over Easter*. Available at: www.thedailyvox.co.za/moria-the-largest-christian-gathering-in-sa-over-easter-shaazia-ebrahim. Accessed 26 Mar 2020.

Ebrahim, S. H., & Memish, Z. A. (2020). COVID-19: Preparing for superspreader potential among Umrah pilgrims to Saudi Arabia. *Lancet, 395*. https://doi.org/10.1016/S0140-6736(20)30466-9.

Ebrahim, S. H., Ahmed, Q. A., Gozzer, E., Schlagenhauf, P., & Memish, Z. A. (2020). Covid-19 and community mitigation strategies in a Pandemic. *BMJ, 368*, 1066. https://doi.org/10.1136/bmj.m1066.

Estrin, D. (2020). Israel faces challenges in fighting coronavirus in ultra-orthodox communities. *NPR*. Available at: www.npr.org/2020/04/01/825260941/israel-faces-challenges-in-fighting-coronavirus-in-ultra-orthodox-communities. Accessed 25 Mar 2020.

Frias, L. (2020). Officials say a Covid positive Indian who refused to isolate left 40,000 in quarantine. *Business Insider South Africa*. Available at: www.businessinsider.co.za/coronavirus-super-spreader-linked-to-quarantine-of-40000-india-2020-3. Accessed 26 Mar 2020.

Gallego, V., Berberian, G., Lloveras, S., Verbanaz, S., Chaves, T. S., & Orduna, T. (2014). The 2014 FIFA World Cup: Communicable disease risks and advice for visitors to Brazil – A review from the Latin American Society for Travel Medicine (SLAMVI). *Travel Medicine and Infectious Disease, 12*, 208–218.

Griffin, K., & Raj, R. (2017). The importance of religious tourism and pilgrimage: Reflecting on definitions, motives and data. *International Journal of Religious Tourism and Pilgrimage, 5*(3). https://doi.org/10.21427/D7242Z.

Handriana, T., Yulianti, P., & Kurniawati, M. (2019). Exploration of pilgrimage tourism in Indonesia. *Journal of Islamic Marketing*. https://doi.org/10.1108/JIMA-10-2018-0188.

Hosken, G., & Wicks, J. (2020). Free State officials still tracking 286 people after COVID-19 church outbreak. *TimeLive*. Retrieved from www.timeslive.co.za/amp/news/south-africa/2020-03-30-free-state-officials-still-tracking-286-people-after-covid-19-church-outbreak/. Accessed 26 Mar 2020.

Ibrahim, A., & al-Salam, A. (2020). Remote pilgrimage: Iraqis mark Imam al-Mahdi's birth amid COVID19. *Aljazeera*. Retrieved from www.aljazeera.com/amp/news/2020/04/remote-pilgrimage-iraqis-mark-imam-al-mahdi-birth-covid19-200408095252041.html. Accessed 26 Mar 2020.

Ippolito, G., & Rezza, G. (2017). Preface – Emerging viruses: From early detection to intervention. *Advances in Experimental Medicine and Biology – Advances in Microbiology, Infectious Diseases and Public Health*. https://doi.org/10.1007/5584_2016_133.

Jha, N., & Dixit, P. (2020). A cluster of coronavirus cases can be traced back to a single mosque and now 200 million Muslims are being vilified. *BuzFeed News*. Retrieved from www.buzzfeednews.com/amphtml/nishitajha/coronavirus-india-muslims-tablighi-jamaat. Accessed 31 Mar 2020.

Jumali, E. (2018). Management of hajj funds in Indonesia. *Journal of Legal, Ethical and Regulatory Issues, 21*(3), 1–9.

Karrim, A., & Raborife, M. (2020). Trips to Moria on hold until further notice over the coronavirus pandemic, ZCC tells congregants. *News24*. Retrieved from www.news24.com/southafrica/news/trips-to-moria-on-hold-until-further-notice-over-coronavirus-pandemic-zcc-tells-congregants-20200317. Accessed 30 Mar 2020.

Kekena, S. (2020). 'Concerned' ZCC postpones Easter pilgrimage as coronavirus cases rise to 85. *SABC*. Available at: www.sabcnews.com/sabcnews/concerned-zcc-postpones-easter-pilgrimage-as-coronavirus-cases-rise-to-85/. Accessed 16 Mar 2020.

Ladki, S. M., & Mazeh, R. A. (2017). Comparative pricing analysis of Mecca's religious tourism. *International Journal of Religious Tourism and Pilgrimage, 5*(1), 20–28. https://doi.org/10.21427/D76Q7Z.

Liphshiz, C. (2020). Jews are vastly overrepresented in coronavirus deaths in the UK: Here are some theories why. *Jewish Telegraphic Agency*. Available at: www.jta.org/2020/03/31/global/jews-are-vastly-overrepresented-in-coronavirus-deaths-in-the-uk-here-are-some-theories-why/amp. Accessed 20 Mar 2020.

Mc Donald, S. (2020). With St Patrick's day canceled by Covid-19, Irish bishops offer prayers. America. *The Jesuit Review*. www.americamagazine.org/politics-society/2020/03/17/st-patricks-day-canceled-covid-19-irish-bishops-offer-prayers. Accessed 26 Mar 2020.

Memish, Z. A., Zumla, A., McCloskey, B., Heymann, D., Al Rabeeah, A. A., Barbeschi, M., & Horton, R. (2014). Mass gatherings medicine: International cooperation and progress. *The Lancet, 383*. https://doi.org/10.1016/S0140-6736(14)60225-7.

Mintel. (2012). *Religious and Pilgrimage Tourism – International – February 2012*. London: Mintel International Group Ltd.

Muhtada, D. (2020). Religion and COVID-19 mitigation. *The Jakarta Post*. Retrieved from www.theJakartapost.com/academia/2020/03/26/religion-and-covid-19-mitigation.html. Accessed 31 Mar 2020.

Nishiura, H., Sah, R., & Rodriguez-Morales A. J. (2020, February). The COVID-19 outbreak and implications for the Tokyo 2020 Summer Olympic Games. *Travel Medicine and Infectious Disease*. https://doi.org/10.1016/j.tmaid.2020.101604.

Nyikana, S. (2017). Religious tourism in South Africa: Preliminary analysis of a major festival in Limpopo. *African Journal of Hospitality, Tourism and Leisure, 6*(1), 1–8.

Ogg, J. C. (2017). St Patrick's Day in 2017 to set record for economic impact. *27/7 Wall Street. Newsletter.* New York.

Peng, X., Xu, X., Li, Y., Cheng, L., Zhou, X., & Ren, B. (2020). Transmission routes of 2019-nCoV and controls in dental practice. *International Journal of Oral Science, 12*(9). https://doi.org/10.1038/s41368-020-0075-9.

Pullella, P. (2020). Pope in dramatic visit to empty Rome to pray for end if virus. *Reuters*. Retrieved from www.reuters.com/articule/amp/idUSKBN212094. Accessed 26 Mar 2020.

Rashid, R. (2020). Being called a cult is one thing, being blamed for an epidemic is quite another. *New York Times*. Retrieved from www.nytimes.com/2020/03/09/opinion/coronavirus-south-korea-church.amp. Accessed 26 Mar 2020.

Rejman, K., Maziarz, P., Kwiatkowski, C. A., & Haliniarz, M. (2016). Religious tourism as a tourism product. *World Scientific News, 57*, 562–575.

Rot, E., Mikinac, K., & Bogdan, B. (2014). General characteristics of religious tourism in Croatia. *UTMS Journal of Economics, 5*(1), 79–87.

Saayman, A., Saayman, M., & Gyekye, A. (2014). Perspectives on the regional economic value of a pilgrimage. *International Journal of Tourism Research, 16*, 407–414.

Sadrzadeh, A. (2020). Faith and the coronavirus. *Qantara*. Retrieved from https://en.qantara.de/content/covid-19-epidemic-in-iran-faith-and-the-coronavirus. Accessed 24 Mar 2020.

SAGAS (Saudi Arabia General Authority for Statistics). (2019). *Hajj Statistics 2019 – 1440*. Riyadh: Saudi Arabia General Authority for Statistics.

Sanghera, T. (2020). Hajjonomics: The business of getting India's pilgrims to Mecca. *Aljazeera*. Retrieved from www.aljazeera.com/ajiimpact/hajjonomics-business-indias-pilgrims-mecca-190808202539949. Accessed 24 Mar 2020.

Sherbini, A., Aziz, Y. A., & Sidin, S. M. (2016). Income diversification for future stable economy in Saudi Arabia: An overview of the tourism industry. *International Journal of Economics, Commerce and Management, IV*, (11), 173–189.

Shinde, K. A. (2015). Religious tourism and religious tolerance: Insights from pilgrimage sites in India. *Tourism Review, 70*(3), 179–196.

Soliman, T., Cook, A. R., & Coker, R. J. (2015). Pilgrims and MERS-CoV: What's the risk? *Emerging Themes in Epidemiology, 12*, 3.

Staff, T. (2020). As virus hit tight-knit Haredi communities, rabbis among the first to fall ill. *Times of Israel*. Available at: www.timesmesofisrael.com/as-virus-hit-tight-knit-haredi-communities-rabbis-among-the-first-to-fall-ill/. Accessed 24 Mar 2020.

Tala, M. L., & Padurean, A. M. (2008). Dimensions of religious tourism. *Amfiteatru Economic*, November Special Issue:242–253.

Thaqafi, T. A. (2017). Hajj revenues poised to exceed $150 bn by 2022: Experts. Retrieved from https://www.arabnews.com/node/1151751/saudi-arabia. Accessed 11 Apr 2020.

Times of India. (2018). Why Haj will soon be the new oil for Saudi Arabia. Retrieved from https://timesofindia.indiatimes.com/world/middle-east/why-hajj-will-soon-be-the-new-oil-for-saudi-arabia/articleshow/65469627.cms. Accessed 11 Apr 2020.

United Nations. (2020). *Shared responsibility, global solidarity: Responding to the socio-economic impacts of COVID-19*. New York: The United Nations Sustainable Development Group.

United Nations World Tourism Organisation (UNWTO). (1995). *Concepts, definitions and classifications for tourism statistics*. Madrid: UNWTO.

United Nations World Tourism Organisation (UNWTO). (2012). *Religious tourism in Asia and the Pacific*. Geneva: UNWTO.

Vatican. (2020). COVID-19 Decree regarding the celebration of the Easter Triduum. *Vatican News*. Retrieved from www.vaticannews.va/en/vatican-city/news/2020-03/covid-19-decree-regarding-easter-triduum-liturgical-celebration.html. Accessed 26 Mar 2020.

Waldron, G., & Doherty, L. (2015). Low public health risk of MERS-CoV in people returning from the Hajj. *BMJ, 351*, h5543.

Wilson, J. A. J. (2014). The halal phenomenon: An extension or a new paradigm? *Social Business, 4*(3), 255–271.

World Health Organization (WHO). (2020a). *Coronavirus disease 2019 (COVID-19) situation report – 51*. Retrieved from https://www.who.int/docs/default-source/coronaviruse/situation-reports/20200225-sitrep-36-covid-19.pdf?sfvrsn=2791b4e0_2. Accessed 11 Mar 2020.

World Health Organization (WHO). (2020b). *Key planning recommendations for mass gatherings in the context of the current COVID-19 outbreak. WHO/2019-nCoV/POE mass gathering/2020.1*. Geneva: WHO.

Yakunin, V. N., Yamashev, V. M., Anuchin, O. I., Adaevskaya, T. I., & Ovsyannikova, N. V. (2016). Religious tourism and pilgrimage in Russia: Cultural-historical analysis. *The Turkish Online Journal of Design, Art and Communication*, 2595–2603. https://doi.org/10.7456/1060NVSE/064.

Ziabari, K. (2020). 'Sacred ignorance': COVID-19 reveals Iran split. *Asia Times*. Available at: www.asiatimes.com/2020/03/sacred-ignorance-covid-19-reviews-iran-split/. Accessed 26 Mar 2020.

Zouni, G., & Digkas, D. (2019). Marketing suggestions for multireligious tourism development: The case of Thessaloniki. *Journal of Tourism, Heritage & Services Marketing, 5*(2), 36–42.

Zumla, A., McCloskey, B., Endericks, T., Azhar, E. I., & Petersen, E. (2017). Challenges of cholera at the 2017 Hajj pilgrimage. *The Lancet, 17*(9), 895–897.

Chapter 12
Implications of COVID-19 on Gaming, Leisure and Entertainment Industry

Abstract The global entertainment sector is a multibillion-dollar economy whose size surpasses several countries' annual domestic gross product. Regardless of the significance of this sector, it is often an under-researched branch of the tourism industry. For example, there is barely a study on how the previous epidemics and pandemics affected this critical sector. This study, therefore, investigates the impact of the coronavirus (COVID-19) on entertainment (music, television and special events). Using archival and secondary data from authoritative sources, the study showed that the impact of the pandemic was devastating on the sector, leading to multibillion-dollar losses for global entertainment companies. Global entertainment companies such as Walt Disney World and Box Office suffered massive losses, which forced several companies to furlough employees and executives in the battle for survival and to protect capital. The study also showed that the pandemic disrupted both the demand side and the supply side of business in the entertainment sector, which resulted in cancellations of festivals and film production at a time of surging demand for entertainment. Given the nature of the pandemic, 2020 proved to be a challenging year for many entertainment companies – a situation likely to play out for the next 2–3 years as companies battle to recover. The study recommends a post-pandemic assessment of each segment of the sector to allow it to build a disaster preparedness and management system, which does not seem to be in place at the moment.

Keywords Entertainment sector · COVID-19 · Film segment · Tourism · Box Office · Netflix

12.1 Introduction

One of the often-overlooked segments of the tourism industry is the gaming, leisure and entertainment sector, which is an economic contributor generating vast sums of revenue annually across the world. This sector is varied and diverse and, like the

G. Nhamo et al., *Counting the Cost of COVID-19 on the Global Tourism Industry*,
https://doi.org/10.1007/978-3-030-56231-1_12

broader tourism industry, creates many jobs for skilled and unskilled personnel. The entertainment sector is comprehensive, which often makes it difficult to define (Rodríguez-Ferrándiz, 2014). Hennig-Thurau and Houston (2019:41) define entertainment as "any market offering whose main purpose is to offer pleasure to consumers, versus offering primarily functional utility". At most, the entertainment sector provides a platform for individuals to showcase their talent through art performance, which can also be showcased through film and other recreational activities.

The advancement of society, which has seen several countries moving towards being intermediate and advanced states, has raised the demand for leisure and entertainment (Aratuo and Etienne 2019). The sector is crucial in that it enables people to rest and relax, with the demand such greater in countries with an ageing population. The entertainment sector nonetheless caters to people of all ages, young and old (Yan 2020; Glaser 2019). The young and middle-aged are often attracted to events such as festivals, musical shows and theme parks that draw large crowds of enthusiastic fans.

Due to the advent technologies in recent years, the entertainment sector has evolved to offering advanced cinemas where blockbuster movies are screened. The advent of 3D a few years ago and augmented reality led to the expansion, for example, of the film segment (Knapp 2015). There is a contested view that developments of various technologies resulted in increased revenue collection by the cinema segment, with evidence suggesting that the increase could have been driven by ticket price increases (Bestvina Bukvić et al. 2017). Nonetheless, the entertainment sector has witnessed tremendous growth, becoming one of the fastest growing economic sectors globally. In addition, the advent of compact discs and the Internet opened up whole new segments of online sports, gaming, streaming and video on demand. Online gaming, for example, has allowed people to engage in gaming as a source of livelihood away from the sports field (Hennig-Thurau and Houston 2019).

It is imperative to note that the entertainment sector has led to the development of a global niche market that is crucial in shaping both the social and cultural values of society as we know it and its evolution going forward. The sector has enabled the development of artistic values in some communities, which in turn is an anchor of the sector. Angelini and Castellani (2019) argue that artistic value goes beyond economic value, as it is much more vibrant and encompasses culture.

Throsby (2003:280), however, argues that art is made up of several values, including the aesthetic value that refers to "properties of beauty, harmony, form, and other aesthetic characteristics of the work". Several other authors provide a critique of what they believe art represents. Hernando and Campo (2017) provide a multidisciplinary understanding of art, which includes borrowing from economics, marketing, sociology, psychology and a historical perspective of art. Smith (2008) brings the idea of art as having a productive value in that it has inspirational properties, be it inspiration from music or visual performance. As such, art cannot be limited to

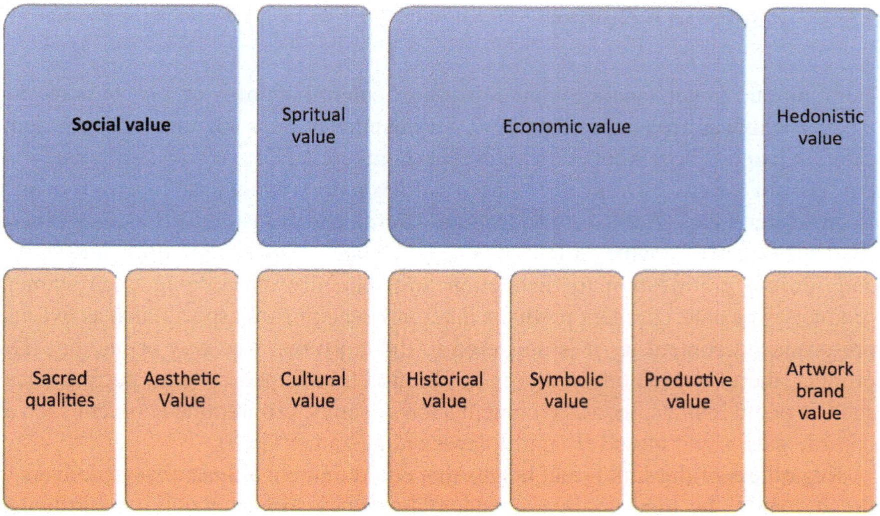

Fig. 12.1 General pillars of art and entertainment

consumptive value alone (i.e. just for enjoyment). Of interest also are the assertions of Klamer (2008) that over and above everything, art has an economic value driven from charging prices. Art also has a social value in that it creates interpersonal relationships in communities and groups of people. Figure 12.1 is a summary of the value of art and entertainment for society.

The outbreak of COVID-19 disrupted the social interactions between people due to the demand for social distancing to limit the spread of the disease. Various countries banned social gatherings, and, even in the absence of such bans, the "stay home now" campaign and #traveltommorrow campaign at the behest of the World Tourism Organization (UNWTO) were instrumental in motivating people to stay at home (UNWTO 2020). In addition, there were fears of contracting the disease, which led people to limit travelling as this could risk infection. These measures came at a considerable cost to the leisure and entertainment sectors. In other settings, early signs indicated the negative impact of COVID-19 on Box Office movie revenue (Kim 2020).

This chapter, therefore, concerns the impact of travel bans, social distancing, lockdowns and other COVID-19 restrictive measures on the global entertainment sector. The study is imperative as it will not only assist in documenting the impact of the pandemic on the entertainment sector but also has the potential to inform future policy and practice. It is in this vein that the United Nations called for sector economic impact studies (UN 2020), with emerging evidence that COVID-19 was also likely to negatively impact the entertainment sector (Roggeveen and Sethuraman 2020).

12.2 Literature Review

Defining and understanding what constitutes entertainment is crucial in assessing what or who was affected by the COVID-19 pandemic. As such, this section is dedicated to mapping out what constitutes entertainment and the sector as a segment of the tourism industry. Drake (1919:666) defines entertainment as “a certain quality in our experience”. According to Hennig-Thurau and Houston (2019), this experience encompasses a wide variety of consumer states. These range from amusement and sensual gratification to distraction and mentally challenging entertainment products. The entertainment products inherently encompass aspects such as live and pre-produced content, as it is delivered to the consumer one way or another. The product, therefore, includes – but is not limited to – filmed content, performances such as poem recitals, recorded content such as songs, programmed content such as console games and massively multiplayer online games (Ibid).

Regardless of the smiles and laughs that entertainment almost always delivers to the audience, the industry is a considerable money-spinner involving substantial financial cash outlays for production. In this way, the sector has the potential to create instant fortunes. According to Hennig-Thurau and Houston (2019), the entertainment sector has made fortunes and, in turn, imparted fortunes to its wide array of stakeholders. Given the social value of entertainment, Vogel (2014) notes that several people across the world spend hours on entertainment which includes television, gaming, amusement parks, musical shows, casinos and performing arts. The citizens of the USA spend 160 billion person-hours per year on entertainment (Ibid).

Regardless of the sector’s financial strength, like every other segment of the tourism industry, it is vulnerable to economic downturns and other shocks. Given that some of the large entertainment companies are listed on various stock exchanges, the economic downturn between 2009 and 2011 led to a decline in stock values of entertainment companies (Huy 2018). The entertainment sector is also known for its contribution to the spread of diseases, which saw several artists getting infected with HIV/AIDS, and this had an adverse impact on the sector’s economy. Xia and Yang (2005) report that some female entertainment workers in Shanghai often participated in practices that increased their vulnerability to HIV and AIDS, including indulging in unprotected sex. The entertainment sector has indirectly been blamed for sexually priming its consumers, exposing them to the risk of promiscuity. This makes the audience vulnerable to various sexual diseases, exposes them to the objectification of women and promotes gender violence (Yao et al. 2010).

Roggeveen and Sethuraman (2020) highlight that there were fears that COVID-19 would have a transforming effect on the retail culture, which could see people, for example, using home gyms and online television services. These engagements were likely to have a damaging impact on revenue for movie houses. Koku (1995) indicates that regardless of the economic contribution of this fast-growing sector, the sector is mostly under-researched. There are fears, therefore, that while studies might be conducted to ascertain the impact of COVID-19 on other closely related tourism sectors such as the transport and hospitality sectors, an assessment of how the coronavirus impacted the entertainment sector would likely be omitted again.

While there are already some preliminary assessment studies on the impact of COVID-19 on the tourism industry covering sectors such as hotels, restaurants and aviation (Gössling et al. 2020; Hoque et al. 2020) and on tourists' mental health (Zheng et al. 2020), there is limited evidence of studies done on the impact of COVID-19 on several other sectors (Fernandes 2020; IMF 2020).

12.3 Research Design

In a bid to unpack the impacts of COVID-19, multiple cases of how the coronavirus affected the entertainment sector were investigated (including the performing arts, film and television production, gaming and festivals segments). The primary data collection mechanism was an intensive Internet search of archival material and secondary data from reputable news reports. The sources included professional blogs, company websites and social media handles of concerned media and entertainment companies. Data from Google Trends were also explored. This methodology was followed as it provided convenience and real-time information required to populate the research. Given the global coverage of the study, it was found to be challenging to go to every company profiled in the research.

Mellon (2014) argues that regardless of reliability issues, online searches offer numerous advantages such as being cost-effective, data sets are readily available, and at times data are frequently generated. McLaren and Shanbhogue (2011) support the use of Internet search engines in economic studies, pointing out that the increasing use of the Internet has created a useful source of information which can be used for statistical analysis of trends, and a number of studies have already been conducted using Internet search generated data. The use of social media data in tourism studies has also grown, with researchers such as Dube and Nhamo (2019, 2020a) using it to analyse tourist trends. In the current study, content and thematic analysis were used to analyse the generated data, and the Microsoft Excel Toolpak was used to analyse the numeric data and generate graphics.

12.4 Results and Discussion

The different segments of the entertainment sector were affected differently, with some being negatively affected and other sectors making unexpected record gains. The following subsections concerns how each segment was affected, with a focus on employment and potential revenue losses.

12.4.1 Impact of COVID-19 on Art and Performance

The study found that a number of scheduled events had been cancelled across the world, with most of the events being postponed from the data obtained from IndieWire. IndieWire is a reputable digital media coverage publication for film, television and digital news. It also does reviews and interviews for passionate fans, who diarised the impact of COVID-19 on the entertainment sector daily since early March 2020 when the impact of the pandemic on the sector became evident. It was found that about 36 major international festivals had been postponed due to the pandemic (Table 12.1). Most of the events had been postponed, while about 24 international festivals had been cancelled altogether. The organisers of the postponed events shifted them to the last quarter of 2020, from October onwards. The postponement of such events came at a high cost to the organisers and those hosting events.

Festivals contribute millions and billions to host countries and communities and offer multiple benefits to societies. Due to the large crowds drawn by international film festivals, these often benefit other segments of the tourism industry such as accommodation, restaurants, cruise ships, airlines and road transport. International events temporarily raise demand for additional labour providing access to short- and long-term employment as companies try to cater to demands for goods and services in the host community. Indirect revenue is also generated through media houses, which are used as channels for marketing and publicity, generating revenue from advertisement charges. There is a growing consensus that filming is a growing component of the tourism industry and helps in marketing destinations, too (Park et al. 2011; Yolal et al. 2015). Filming further assists in transforming the urban culture of the hosts or destinations (Yolal et al. 2016). Large crowds of people checking in at locations and reviewing the events often create interest in other people to visit the area or future events. The cancellations, therefore, had a considerable impact on host communities, especially in areas where such an event is the iconic feature of the destination in forming part of the seasonal tourism calendar. The cancellations, therefore, had a significant socio-economic impact on the affected destinations across the world.

Besides the tourism and cultural benefits of festivals, Acciari (2019) argues that film festivals provide an essential interactive platform that enables community members and visitors to create new business linkages. With hundreds of filmmakers attending each film festival, the events provide important platforms where filmmakers link with commissioning editors. Commissioning editors can also use the space for hunting new content and getting new releases before competitors. This is done with a view to providing ultimate entertainment to viewers, which can benefit viewership numbers and attract advertisers and associated revenue.

Independent cinemas also use film festivals as an opportunity to present their content to new audiences, while film production houses market their content and test screenings. Test screenings are essential in that they afford film producers' space to interact and get feedback from the audience, which ultimately assist in

Table 12.1 List of international festivals that were disrupted and cancelled in 2020 due to COVID-19

Postponed Events	Cancelled Events
1. 2020 Rock & Roll Hall of Fame Induction Ceremony and Induction Week festivities initially scheduled for 2 May 2020	1. Annecy Animation Film Festival 60th anniversary in Annecy, France, 15–20 June 2020 (12,000 attendees; tickets cost full price €6.50)
2. Montréal Just for Laughs Festival moved from 15–26 July 2020 to 29 September–11 October 2020	2. Munich Film Festival 25 June–4 July 2020
3. 2020 Frameline LGBTQ + Film Festival scheduled for 18–28 June 2020 in San Francisco delayed until the fall of 2020	3. 2020 Cannes Lions Festival planned for 22 June–26 June 2020
4. Mammoth Lakes Film Festival rescheduled from 27 February–2 March 2020 to 16–20 September 2020	4. Governors Ball 2020 music festival scheduled for 5–7 June 2020
5. Inside Out anniversary festival moved from 21–31 May to 1–11 October 2020 (film festival for LGBTQ North America)	5. Provincetown International Film Festival scheduled for 17–21 June 2020
6. 24th American Black Film Festival 2020 scheduled for 17–21 June in Miami Beach moved to 21–25 October 2020	6. 46th Seattle International Film Festival scheduled for 14 May–7 June 2020
7. Havana Film Festival New York originally billed for 28 April–5 May 2020 moved to the end of 2020	7. 2020 Glastonbury Music Festival scheduled for 24–28 June (200,000 guests)
8. Cannes Film Festival scheduled for 12–23 May 2020; new dates being considered for the end of June 2020	8. 2020 Uranium Film Festival in Rio de Janeiro scheduled for 21–31 May 2020
9. 2020 Bonnaroo Music and Arts Festival in Manchester, Tennessee, scheduled for 11–14 June 2020 at Great Stage Park moved to 24–27 September 2020	9. 2020 BFI Flare: London LGBTQ+ Film Festival scheduled for 18–29 March 2020
10. 2020 Sarasota Film Festival penned for 1–5 April 2020	10. 63rd Annual San Francisco International Film Festival scheduled for 8–12 April 2020
11. 2020 Edinburgh Film Festival postponed from 17–28 June 2020 to 7–31 August 2020	11. 2020 River Run International Film Festival scheduled for 26 March–5 April 2020
12. Netflix Is A Joke Fest week-long comedy festival The Los Angeles event planned for 27 April 2020	12. 42nd edition of France's Cinéma du Réel
13. 2020 Women + Film Festival planned for 13–22 March 2020 moved to late 2020	13. 22nd Annual Roger Ebert's Film Festival scheduled for 15–18 April 2020 in Champaign, Illinois
14. 2020 Sonoma International Film Festival scheduled to take place on 25 March 2020	14. 10th Luxembourg City Film Festival planned for 13–15 March 2020
15. Newport Beach Film Festival planned for 30 April moved to 6–13 August 2020 in Newport Beach, California	15. Sun Valley Film Festival planned for 18–22 March 2020

(continued)

Table 12.1 (continued)

Postponed Events	Cancelled Events
16. Theatrical release for Christophe Honoré's "On a Magical Night" postponed by Strand Releasing	16. New York International Children's Film Festival final weekend scheduled for 14–15 March 2020
17. Chicago Media Project's Doc10 Film Festival postponed	17. Miami Film Festival was cancelled in the middle as it was taking place from 6 and 15 March 2020
18. 2020 Ambulante Documentary Film Festival, travelling throughout Mexico in 8 states at more than 130 locations from 19 to 28 March 2020 postponed indefinitely	18. Hollywood's TCM Classic Film Festival scheduled for 16–19 April 2020
19. Chicago Critics Film Festival scheduled for 1–7 May 2020 at Music Box Theatre postponed indefinitely	19. 2020 Full Frame Documentary Film Festival scheduled for 2–5 April 2020
20. 2020 Canada Film Festival scheduled for 24–28 March 2020 postponed to 19–21 November 2020	20. France's television festival (Series Mania), which attracts 80,000 attendees and 3000 industry professionals annually, planned for 20–28 March 2020
21. 58th Ann Arbor Film Festival planned for 24–29 March 2020 postponed	21. Environmental Film Festival in Washington, DC scheduled for 12–22 March 2020
22. 39th Istanbul Film Festival scheduled for 10–21 April 2020 postponed	22. SXSW 2020, with its parallel film, music, interactive and education festivals/ conferences, scheduled for 13–22 March 2020
23. The Indian Film Festival of Los Angeles initially planned for 1–5 April 2020 postponed.	23. 6th edition of Qumra which was to be hosted by the Doha Film Institute initially planned for 20–25 March 2020
24. St Patrick's Film Festival London scheduled for 13–15 March 2020 postponed	24. 44th Annual Entertainment Symposium planned for 20–21 March 2020
25. 49th New Directors/New Films scheduled for 25 March–5 April 2020 postponed	
26. 46th Chaplin Award scheduled for 27 April 2020 postponed	
27. 2020 Seattle Jewish Film Festival planned for 19 March–5 April 2020 postponed	
28. PaleyFest LA 2020 set for 13–22 March 2020 at the Dolby Theatre in Hollywood postponed	
29. 2020 Montclair Film Festival scheduled for 1–10 May 2020 postponed	
30. 2020 Beverly Hills Film Festival was planned for 1–5 April 2020	
31. The Cinequest Film & Creativity Festival scheduled for 3 March 2020 in San José, California, rescheduled to 16–30 August 2020	

(continued)

Table 12.1 (continued)

Postponed Events	Cancelled Events
32. The Bentonville Film Festival scheduled for 29 April 2020 postponed to 2 May 2020	
33. 2020 Thessaloniki Documentary Festival in Greece set for 5–15 March postponed	
34. The inaugural Red Sea Festival in Saudi Arabia scheduled for 12–22 March postponed	
35. HBO postponed the all-star live benefit "Night of Too Many Stars: America Unites for Autism Programs" from 18 April 2020 to a later date	

Source: Authors; data from event websites

bettering the offering to the audience. Film festivals are a crucial platform for artists to market their talents and win awards that assist in boosting profiles, while directors use the platforms to scout for new talent (Benson 2020). Given the global nature of the festivals, they further act as platforms for celebrating diversity. As such, this learning and networking platform is crucial in shaping and developing the multibillion entertainment sector. Hence, the cancellations would indeed be a sad loss to film stakeholders.

12.4.2 Impact of COVID-19 on Movie Production and Releases

Each time when there is a movie release, millions of fans are waiting to be the first in the queue. Film enthusiasts usually wait for months, if not years, to be the first to watch the release of their favourite soapie or movie. This often results in a flurry of customers, particularly moving to movie cinemas or turning to their stream channels. This means that the movie houses make much revenue, especially when a renowned movie series is released. The push back and postponement of several movies that were due for release in 2020 negatively impacted movie houses' revenue collection base. About 54 movies and/or series were suspended, had production cancelled or had their release pushed back either to 2021 or 2022 (Table 12.2).

The cancellation and/or suspension of some of the movies that were in production was forced upon production houses when some of the actors contracted the COVID-19 during shooting. This forced the production houses and artists into quarantine or to take other precautionary measures given the contagious nature of the disease, resulting in forced shutdowns of productions. Shutting down production in session is very expensive, especially if it is being shot outside the territory of the movie house as this involves permit fees (which are charged daily), venue bookings and hotel bookings for the crew and artists. Shutting down production has additional overheads from returning all the crew, actors and equipment to their areas, often by

Table 12.2 Impact of COVID-19 on selected popular production houses in 2020

Production House	Movie/Series Affected
HBO	1. Halted the "Friends" reunion production
	2. Halted production of "Third Day" originally scheduled to debut 11 May 2020
	3. Stopped production of Season 3 of "Succession" and Season 3 of "Barry"
	4. "Undoing" drama miniseries shifted from May 2020 to last quarter of 2020
	5. Halted production of Season 2 of the comedy series "The Righteous Gemstones"
	6. Delayed start of production of Season 2 of "Euphoria"
	7. Cancelled the New York City premiere of the documentary "After Truth: Disinformation and The Cost of Fake News"
Warner Bros	8. Moved Matt Reeves' "The Batman" back by 4 months; film now opening 1 October 2021
	9. Postponed DC film "The Flash" from 1 August 2020 to 3 June 2022
	10. "Shazam 2" moved from 1 April 2020 to 4 November 2022
	11. Moved "The Sopranos" prequel "The Many Saints of Newark" from 25 September 2020 to 21 March 2021
	12. Postponed release of "Wonder Woman 1984" from 5 June to 14 August 2020
	13. The musical adaptation of "In the Heights" postponed indefinitely from 26 June 2020
	14. "Scoob" initially set for 15 May 2020 postponed indefinitely
	15. "Malignant" postponed from 14 August 2020
	16. Production on Baz Luhrmann's Elvis Presley biographical drama suspended indefinitely after one co-star and partner tested positive for coronavirus
	17. Stopped production on "The Matrix 4", which had moved from San Francisco (USA) to Berlin (Germany)
	18. Postponed production of "Fantastic Beasts"
	19. Postponed production of "The Batman"
	20. WarnerMedia and Xandr cancelled "Upfront" presentations for their networks, which include CNN, TBS and TNT
	21. Cancelled production of "Riverdale"
Universal Pictures	22. Moved Judd Apatow's "The King of Staten Island" straight to video on demand (VOD), premiering 12 June 2020
	23. Moved the "Candyman" sequel from 12 June to 25 September 2020
	24. Rescheduled "Minions: The Rise of Gru" from 2 July 2021 to 22 December 2021
	25. "Wicked" musical adaptation postponed from Christmas 2021
	26. Halted filming of "Jurassic World: Dominion"
	27. Suspended production of the forthcoming boxing biopic "Flint Strong"
	28. Delayed start on Billy Eichner's rom-com "Bros"
	29. Closed Universal Studios Hollywood theme park
	30. Postponed release of the ninth "Fast and Furious" to 2021
	31. Delayed the production of "Little America" Season 2 for Apple TV+
	32. Delayed production of the "Russian Doll" Season 2 for Netflix
	33. Delayed production of "Rutherford Falls" for Peacock

(continued)

Table 12.2 (continued)

Production House	Movie/Series Affected
	34. Postponed the release of "No Time to Die" to 12 November 2020 in the United Kingdom and 25 November 2020 in the USA; cancelled the film's China premiere – the movie was scheduled to open on 8 April 2020
	35. Cancelled the release of "1917" and Searchlight's "Jojo Rabbit," in China
	36. Cancelled the release of "Dolittle"
Sony	37. Reordered its slate, moving the untitled "Spider-Man"
	28. "Far from Home" sequel postponed from 16 July to 5 November 2021
	39. Moved the release of "Spider-Verse" from 8 April to 7 October 2022
	40. "Hotel Transylvania 4" now opening on 6 August 2021 instead of 22 December 2021
	41. Moved back the release of "Ghostbusters: Afterlife" to 5 March 2021
	42. Moved the release of "Morbius" to 19 March 2021
	43. Moved the release of "Uncharted" to 8 October 2021
	44. "Peter Rabbit 2: The Runaway" postponed to 15 January 2021
	45. Release of "Greyhound" and another untitled Sony-Marvel movie planned for October 2021 delayed indefinitely
	46. Release of "Fatherhood" moved closer to premiere on 23 October 2020
	47. Postponed indefinitely production of "Uncharted", the video-game adaptation, set to start in Berlin
	48. The delayed release of "The Climb" scheduled for 20 March 2020
	49. Cancelled the live-action musical "Cinderella"
	50. Suspended production of "The Nightingale and "The Wheel of Time" fantasy series
	51. "Release of "Peter Rabbit 2 in the USA postponed from 3 April to 7 August 2020
	52. "Wheel of Fortune" and "Jeopardy" announced filming episodes without a live studio audience indefinitely
Lionsgate	53. "John Wick: Chapter 4" postponed from May 2021 to 27 May 2022.
	54. Cancelled release of "Spiral: From the Book of Saw Run" (scheduled for 8 May 2020) and "Antebellum" (scheduled for 24 April 2020)

Source: Authors; data from IndieWire (2020)

flight. Therefore, the shutting down of movies in production came at a cost to movie production houses. Furthermore, there were fears among industry players that cancellations and delayed release of movies and other productions could trigger massive piracy that could cost production houses and Box Office.

Given the demand for social distancing, which had resulted in the closure of several movie houses in critical markets such as in the USA and China, cinemas and Box Office lost a vast amount in potential revenue. China (with 70,000 theatres), which had been on course to replace the USA with its 41,172 theatres (National Association of Theatre Owners 2019; as the lead movie revenue earner, was the first epicentre of the COVID-19 pandemic. This meant an extensive shutdown of theatres in response to social distancing measures and partial lockdowns in Hubei

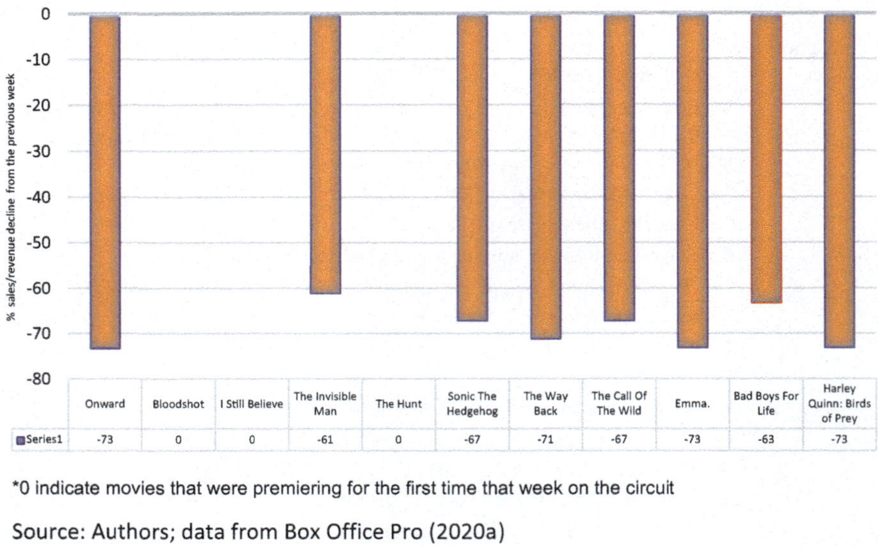

Fig. 12.2 Movie performance on the Box Office circuit in the week of 7 March 2020
*0 indicate movies that were premiering for the first time that week on the circuit

Province, where Wuhan is located. A preliminary study by Kim (2020) found that Box Office was losing revenue in China due to social distancing measures.

The USA and several other countries suffered a similar fate, resulting in shutdown of theatres. As a result of the increased closure of theatres, Box Office suffered its worst revenue loss in more than 25 years (Box Office Pro 2020a). On 15 March 2020, a few days after the pronouncement of COVID-19 as a pandemic, Box Office reported a weekly revenue of US$53.6 million (the lowest since September 2000 when it recorded US$49 million). In the USA, a sample of movies premiering in more than 4000 theatres showed a severe decline in terms of sales revenue (Fig. 12.2). Consequently, the revenue performance of the movies on the circuit did not reach their maximum potential, resulting in losses for both Box Office and production houses. This forced Box Office to suspend some releases that were due for that period, particularly in China. As a result, Box Office announced an adjusted annual revenue decline of −6.7% over the same period in 2019, and it later emerged that this was only the beginning of what was to come in 2020 (Box Office Pro 2020a).

Box Office went on to record one of the worst quarters in terms of revenue, marked by a −51% decline (equivalent to US$1.71 billion) (Figs. 12.3 and 12.4). One, therefore, may speculate that revenue for Box Office for 2020 was likely to decline by between a minimum of 60% to a maximum of 75%, given the prevailing market conditions with most blockbuster movies shut down. This was exacerbated by production challenges and the deferment of some blockbusters' release from 2020 to 2021 and 2022.

Although the ban on theatre was due to be lifted around July 2020, as envisaged by Box Office's calendar, there was very little content available – an aspect that

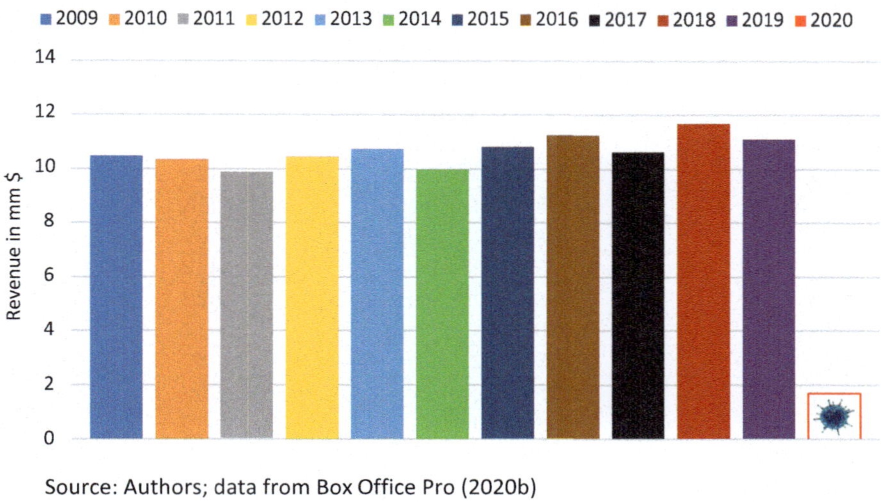

Fig. 12.3 Box Office annual revenue from 2009 to 2019 and the first quarter of 2020 impacted by COVID-19

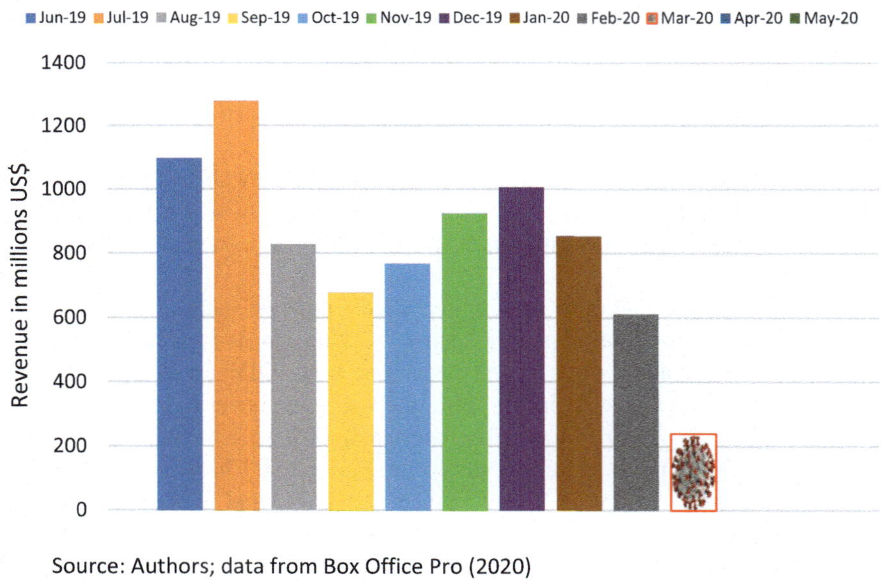

Fig. 12.4 Month on month revenue and the impact of COVID-19 on Box Office

would discourage customers from risking COVID-19 infection to go to a cinema. Box Office almost admitted this, pointing out that there was a risk in putting out a number of offerings because there would likely be a slow uptake after the lockdowns (Box Office Pro 2020a). Safety-conscious tourists are unlikely to flock to theatres even if they do open due to fear of contracting the disease. There is still a

Table 12.3 Share price movement of selected entertainment companies due to the impact of COVID-19

Entertainment company	Stock price (6 April 2020)	Share price movement
Apple	US$1.01 trillion	−13%
Amazon	US$963.18 billion	+1%
Google	US$781.04 billion	−12%
Facebook	US$456.41 billion	−11%
AT&T	US$203.28 billion	−23%
Disney	US$173.38 billion	−16%
Comcast	US$160.29 billion	−12%
Netflix	US$160.42 billion	−1%
Sony	US$75.22 billion	−5%
Fox	US$14 billion	−20%
Viacom CBS	US$8.72 billion	−35%
Lionsgate	US$1.14 billion	−22%

Source: Authors; data from Katz (2020)

risk that the pandemic could altogether change people's appetite for cinemas, with them preferring to stream services going forward. Trends seem to suggest that cinemas were already losing popularity due to several factors, including costs and interruptions from phone usage during screening.

Month on month, April 2020 was likely one of the worst for Box Office as the revenue figures show a 75% decline compared to April 2019 (Box Office Pro 2020b).

The industry's disruption impacts were not limited to Box Office. They extended to other sectors and stakeholders in the entire value chain. A look at the share price for selected entertainment companies is telling of what happened as the lockdowns intensified (Table 12.3).

Table 12.3 shows that only Amazon recorded a positive share price movement, while Netflix recorded a marginal loss of −1%. This can be attributed to the fact that the two companies offer online streaming and VOD services. As theatres closed due to the lockdowns, people were likely to switch over to VOD, streaming and online gaming. As a result, Netflix reported aggressive growth as more people subscribed to the online streaming platform (Fig. 12.5). The countries that had national lockdowns (namely, Spain, France and Italy) had a more pronounced growth rate compared to the countries that had strict and fewer restrictions on social distancing. Regardless of the growth, Netflix was anxious about new content in the long run as it feared a drag of the pandemic would adversely affect new productions (Netflix 2020). Due to delayed production, Netflix reported that it had incurred US$218 million in incremental content costs. This was, however, offset by a record increase in subscribers which saw the company gaining about 16 million subscribers, pushing the total number to 182.9 million paid streaming subscribers globally. As a result, Netflix had a bit more cash to spend and invest even in third parties.

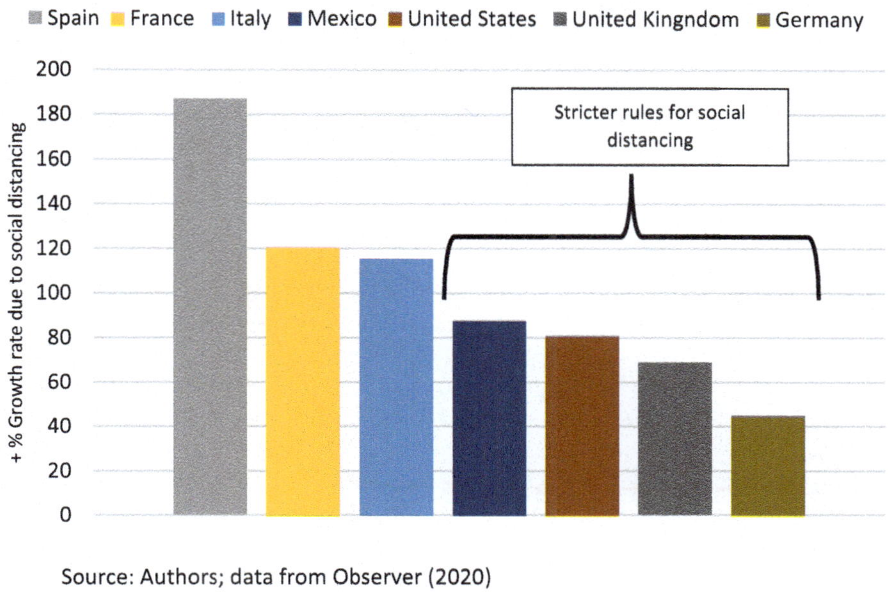

Fig. 12.5 Impact of the demand for social distancing due to COVID-19 on Netflix

Due to industry disruptions, the first quarter of 2020 was characterised by massive job layoffs. A look at Google Trends shows that the spike in the pandemic was matched by a devastating loss of jobs, with employees being furloughed or laid off altogether in various economic sectors. When analysed, the two terms show a similar trend. The peak of furloughs occurred at the end of March 2020, resulting in the spike on Google Trends, with the highest score 100 (Fig. 12.6).

The entertainment sector was not spared the furloughs and job losses that took place due to COVID-19. Table 12.4 highlights some of the cases that affected the sector. As a cost-containment measure, most large corporations furloughed their staff (with some companies furloughing as high as 95%) as the pandemic gave rise to uncertainties and led to disturbances in production. COVID-19 seriously affected production and supply, leading to losses in revenue – an aspect that undermined business in the entertainment sector. The impact on employees was noted as dire because there was very little economic or psychosocial protection for the sector. Instead, the focus was on protecting capital to the detriment of employees. This increased the gap between the rich and the poor, with poor employees in the entertainment sector who survived from cheque to cheque left vulnerable. Dube and Nhamo (2020b) highlight the vulnerability of employees in the tourism industry. An analysis of annual reports from various tourism companies shows that most companies had collected substantial profits in the previous years, and the shareholders were somehow cautioned from dividends received in 2019 (UNIC 2020).

Besides furloughing employees, some companies disinvested in the recruitment of new talent and independent contractors. This adversely affected independent

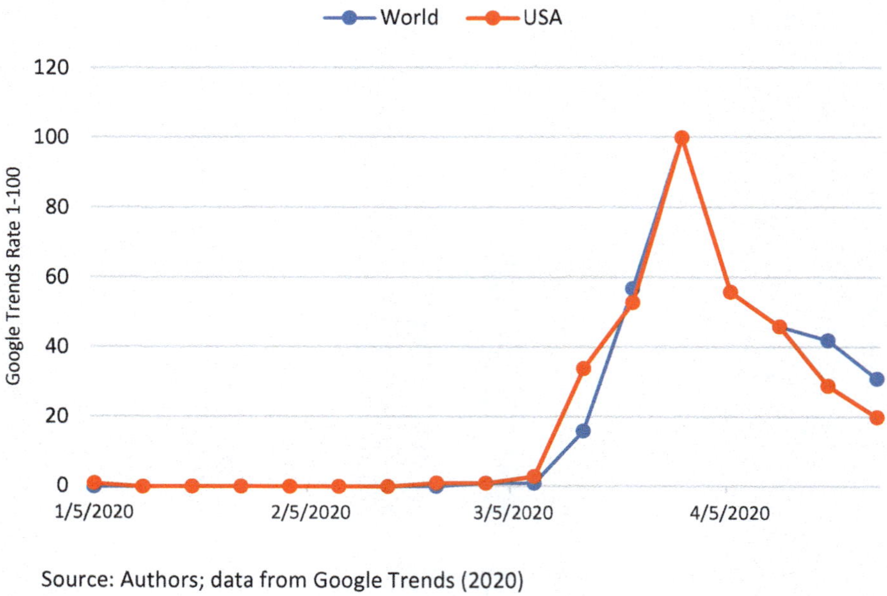

Fig. 12.6 Global and USA trends in job furloughs due to COVID-19 (January–April 2020)

content producers and their employees, with a ripple effect on the supply chain. The long-term implications of such interventions are yet to be quantified, particularly in terms of the human and social costs of the pandemic. Other organisations also deferred capital investments to ensure that they manage liquidity. Very few companies (such as Netflix, given their vantage point) spent some money investing in new content and talent. Netflix saw significant growth in revenue alongside technology companies offering entertainment, VOD and streaming services. Amazon, Apple, Alphabet (parent company of Google), Microsoft and Facebook all recorded huge gains, as shown in Fig. 12.7. It is believed that the demand for communication and entertainment was the most significant driver for growth in revenue.

While there were vociferous calls in other sectors of the tourism industry for state aid and funding to avert the crisis, the demand for support for the entertainment sector was by and large muted. The European Union (EU), however, was one of the regions that had explicit support for the entertainment sector. The European Parliament's resolution of 17 April 2020 on EU-coordinated action to combat COVID-19 and its consequences acknowledges the impact of the pandemic on the entertainment sector. The parliament called for solidarity and support for cinemas, theatres, music venues, freelance and self-employed workers and creative professionals whose income had been cut off due to the closure of the sector and stoppage of ticket sales triggering suffering among many entertainment workers and artists (European Parliament 2020). The resolution opened up opportunities for the entertainment sector to access funding in the form of grants, loans, EU guarantees, investment facilities and other financial initiatives.

Table 12.4 Impact of COVID-19 on employment patterns in the entertainment sector

Name of Company	Number of Employees Affected
Walt Disney World	• About 43,000 employees affected • Halted salaries for more than 100,000 employees, which led to savings of about US$500 million a month across its theme parks and hotels in Europe and the USA from 19 April 2020
AMC Theatres	• Furloughed 600 corporate staff, including the CEO • Furloughed 26,000 of the employees working at locations around the country
Caesars Entertainment	• Furloughed 90% of staff, including corporate employees
Cirque du Soleil	• Laid off 95% of employees to remain with only 259 staff to plan and sell tickets for future tours
New York's Metropolitan Opera	• Laid off all of its union employees for the duration of the pandemic
Annapurna Pictures	• Laid off the CFO and other film executives, two television executives and two assistants, bringing the total to seven (which represent 10% of the company's staff)
World Wrestling Entertainment (WWE)	• Reduced executive pay and furloughed staff • Cut down on talent expenses • Cancelled third-party staffing and consulting arrangements • Deferred spending on new headquarters in Stamford (Connecticut) for at least 6 months
Cinemark	• Furloughed 50% of staff • Laid off 17,500 hourly workers • Furloughed 50% of corporate employees at its Plano (Texas) headquarters at 20% of salary but with full benefits • Slashed salaries of remaining employees by 50%
AMC Entertainment	• Furloughed the CEO and all corporate employees, and shuttered its 1000 locations around the world
The general state of gaming and entertainment in the USA	• Nearly 662,000 casino gaming and resort employees (98% of the total casino workforce in the USA) prevented from working, risking nearly US$74 billion in total wages • Closed 443 (95%) of the country's commercial casinos and 83% of the USA's 524 tribal casinos
Netflix	• Pledged US$100 million to out-of-work production crews and cast members • Provided US$15 million to third parties and non-profits providing emergency relief to out-of-work crew and cast • Pledged US$1 million of that to the SAG-AFTRA COVID-19 disaster fund, the motion picture and television fund and the actors fund emergency assistance in the USA; and another US$1 million between the AFC and foundation des artistes

Source: Authors; Borden (2020); Geller & de León (2020)

12.4.3 *Impact of COVID-19 on the Music Segment*

The music segment provides a livelihood to several music artists and several stakeholders. Most musicians depend on revenue from live performance for the bulk of revenue collection. The pandemic had a severe impact on live performance due to

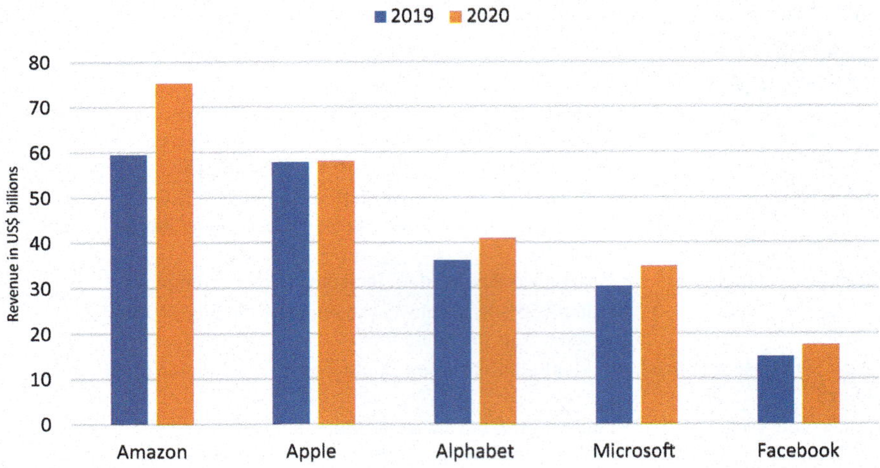

Source: Authors; data from Statista (2020)

Fig. 12.7 Impact of COVID-19 on the revenue collection of technological entertainment companies for the first quarter of 2020

the closure of many performance arenas. According to records from Billboard's (an American entertainment media brand owned by the Billboard-Hollywood Reporter Media Group) coronavirus diary between 24 January and 6 May, 2020, 301 live shows were cancelled, postponed or adversely disturbed because of social distancing measures (Billboard 2020).

In January alone, about 11 shows were cancelled in China and Hong Kong after the closure of the music venues. The first closure was the Mercedes-Benz Arena as the government moved to ensure public safety. On 26 January, the cancellation of the My Love Andy Lau World Tour Hong Kong 2020 that was scheduled to start in February affected 100,000 fans according to a Billboard report. This started a wave of cancellations across the world. Other iconic artists who were affected were Snoop Dogg's UK & Ireland I Wanna Thank Me Tour, Kelly Clarkson's Las Vegas residency at the Zappos Theatre in Planet Hollywood that was pushed to 2021, Lionel Richie cancelled his 2020 European tour, Bon Jovi's cancelled 2020 North American tour, Shania Twain cancelled her Las Vegas residency performances and the Pussycat Dolls postponed their UK and Ireland tour. Many other musical festivals, concerts and award shows were cancelled, postponed or presented online. This deprived musicians and their bands of significant amounts of potential revenue. Musical venues were also negatively affected.

12.4.4 Entertainment Sector Support to Mitigate the Impact of COVID-19

The cinema representative board, the International Union of Cinemas (UNIC), highlighted several efforts and guidance on how the cinema industry, in particular, could move forward. Most importantly, it compiled a list of governments from its 38 member states offering relief to the entertainment sector (UNIC 2020). The UNIC report shows that the sector was receiving various types of assistance ranging from financial support to support in kind. The interventions were predominantly premised on four pillars, as shown in Fig. 12.8.

It is under these principles that several state parties offered support to the cinema and other segments of the entertainment sector. Germany's federal agencies set aside an aid package equivalent to €15 million for the film and media segments (UNIC 2020). In addition, the Federal Film Fund (FFA) deferred reimbursements of loans and payment of levies in a bid to provide relief and liquidity eroded by COVID-19. The Polish government, working with the support of its Ministry of Culture and National Heritage, was crafting support measures to tap into the €22 billion announced by the government and had already exempted the entertainment sector from paying royalties (UNIC 2020). The Swedish government set aside €46 million for businesses in the cultural sector that had lost revenue due to the impact of social distancing measures precipitated by

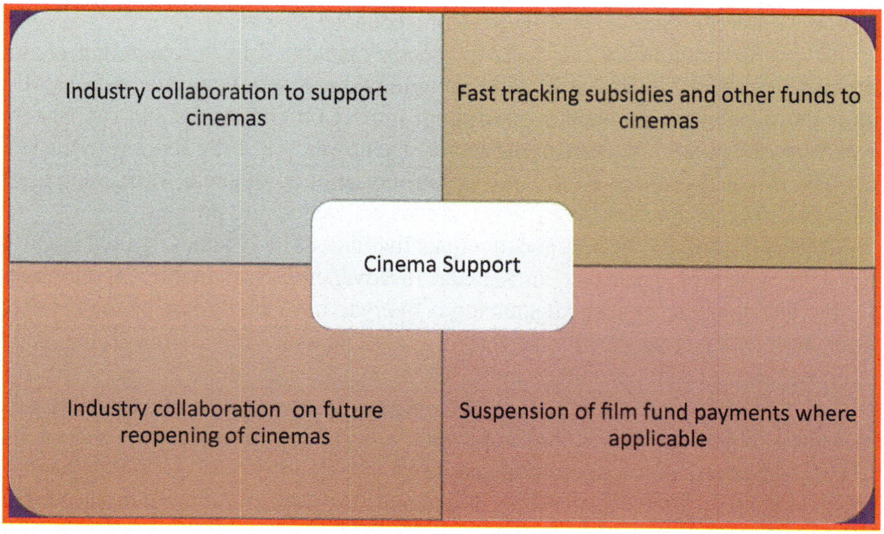

Source: Authors; data from UNIC (2020)

Fig. 12.8 Principles underpinning cinema support in Europe

COVID-19 (Ibid). The Swiss government had set aside CHF280 million to assist entertainment companies that had lost revenue due to either cancellation or postponement of events. The government committed to pay about 50% of the revenue loss (Ibid).

12.5 Conclusion and Recommendations

The study concerned the impact of COVID-19 on the entertainment sector. It was clear that the pandemic had a devastating impact on the entertainment sector, resulting in the disruption of the sector and setting it on unchartered territory. The pandemic led to the loss of production time as well as the cancellation of productions, films and music festivals, which had a devastating impact on the livelihood of artists and entertainment companies. Most entertainment companies' share values were significantly eroded, and only a few were spared. The pandemic altered the supply and demand sides of the entertainment sector. Global giants in entertainment such as Box Office lost billions in potential revenue due to postponements and the shutting down of studios, which could affect the company's revenue collection adversely for the next 2 to 3 years. The entertainment sector found itself compromised in terms of providing new content amid the postponement of production across the world. This came at a time when the demand for entertainment increased during lockdown periods when most people were at home. As a result, VOD and streaming companies performed better financially than other companies in the entertainment sector. Netflix and Showmax showed unprecedented growth.

As a cost-containment measure, the sector responded by rescheduling capital investments as well as laying off and furloughing employees to ensure sustainability after the pandemic. This had a devastating impact on employees and freelancers, who form the bulk of workers in the sector. There was very little support to the sector, given that the sector is not well organised and coordinated. Most companies chose to protect capital at the expense of workers, deepening poverty.

COVID-19 will likely continue to impact livelihoods and increase global inequality. This will push back the gains made in advancing Sustainable Development Goals. The demand for social distancing is likely to have a long-lasting impact even as economies open up despite the fear of infection, as most activities in the entertainment sector do not promote social distancing. As such, there is a need for support to be extended to this fragile sector. There is also a need for the sector to take measures to protect artists and employees who are at the receiving end of the COVID-19 crisis. Better protection should be afforded them by instituting policies and providing insurance against loss of revenue. The impact of the pandemic should be documented on a case-by-case basis to allow for disaster and risk management in the sector.

References

Acciari, M. (2019). The permanency of film festivals: Archiving the changing India. *Studies in South Asian Film & Media, 10*(1), 41–57.

Angelini, F., & Castellani, M. (2019). Cultural and economic value: A critical review. *Journal of Cultural Economics, 43*(2), 173–188.

Aratuo, D. N., & Etienne, X. L. (2019). Industry level analysis of tourism-economic growth in the United States. *Tourism Management, 70*, 333–340.

Benson, C. (2020). Out in the regions: Queer film festivals, community-building and the cultivation of creative talent in regional Victoria. In A. Campbell, M. Duffy, & B. Edmondson (Eds.), *Located research* (pp. 351–368). Singapore: Palgrave Macmillan.

Bestvina Bukvić, I., Borić Cvenić, M., & Buljubašić, I. (2017). Comparative analysis of the impact factors influencing the cinema and theatre attendance and their business performance. *Media, Culture and Public Relations, 8*(1), 56–71.

Billboard. (2020). *CORONAVIRUS Here are all the major music events cancelled due to coronavirus (updating)*. Billboard. Retrieved 7 May 2020, from https://www.billboard.com/articles/business/touring/9323647/concerts-canceled-coronavirus-list

Borden, T. (2020). *The coronavirus outbreak has triggered unprecedented mass layoffs and furloughs. Here are the major companies that have announced they are downsizing their workforces*. Business Insider. Retrieved 3 May 2020, from https://www.businessinsider.com/coronavirus-layoffs-furloughs-hospitality-service-travel-unemployment-2020?IR=T

Box Office Pro. (2020a). *Weekend Box Office report: COVID-19 hits the domestic Box Office*. Box Office Pro. Retrieved 4 May 2020, from https://www.boxofficepro.com/weekend-box-office-report-covid-19-hits-the-domestic-box-office/

Box Office Pro. (2020b). *Pulse report April*. Box Office Pro. Retrieved 4 May 2020, from https://pulse.boxofficepro.com/month

Drake, D. (1919). Is pleasure objective? *The Journal of Philosophy, Psychology and Scientific Methods, 16*, 665–668.

Dube, K., & Nhamo, G. (2019). Climate change and potential impacts on tourism: Evidence from the Zimbabwean side of the Victoria Falls. *Environment, Development and Sustainability, 21*, 2025–2041. https://doi.org/10.1007/s10668-018-0118-y.

Dube, K., & Nhamo, G. (2020a). Tourist perceptions and attitudes regarding the impacts of climate change on Victoria Falls. *Bulletin of Geography. Socio-economic Series, 47*, 27–44.

Dube, K., & Nhamo, G. (2020b). Sustainable development goals localisation in the tourism sector: Lessons from Grootbos private nature reserve, South Africa. *International Journal of Economics and Finance Studies, 12*(1), 88–102. https://doi.org/10.34109/ijefs.202012106.

European Parliament. (2020). *EU coordinated action to combat the COVID-19 pandemic and its consequences*. European Parliament. Retrieved 5 May 2020, from https://www.europarl.europa.eu/doceo/document/TA-9-2020-0054_EN.pdf

Fernandes, N. (2020). Economic effects of coronavirus outbreak (COVID-19) on the world economy. SSRN. Retrieved 5 May 2020, from https://ssrn.com/abstract=3557504 or https://doi.org/10.2139/ssrn.3557504

Geller, J. & de León, R. (2020). *Here's what every major company is doing about the coronavirus pandemic*. CNBC. Retrieved 3 May 2020, from https://www.cnbc.com/2020/03/13/workforce-wire-coronavirus-heres-what-every-major-company-is-doing-about-the-pandemic.html

Glaser, S. P. (2019). *When old becomes new again: The emergence of classic TV broadcasting*. Philadelphia: Drexel University.

Google Trends. (2020). *Google trends*. Google Inc. Retrieved 5 May 2020, from https://trends.google.com/trends/explore?q=furlough

Gössling, S., Scott, D., & Hall, C. (2020). Pandemics, tourism and global change: A rapid assessment of COVID-19. *Journal of Sustainable Tourism, 24*(4), 527–548. https://doi.org/10.1080/09669582.2020.1758708.

Hennig-Thurau, T., & Houston, M. B. (2019). Creating value, making money: Essential business models for entertainment products. In T. Hennig-Thurau & M. Houston (Eds.), *Entertainment science: Data analytics and practical theory for movies, games, books and music* (pp. 151–231). New York: Springer.

Hernando, E., & Campo, S. (2017). An artist's perceived value: Development of a measurement scale. *International Journal of Arts Management, 19*(3), 33–47.

Hoque, A., Shikha, F. A., Hasanat, M. W., Arif, I., & Hamid, A. B. (2020). The effect of coronavirus (COVID-19) in the tourism industry in China. *Asian Journal of Multidisciplinary Studies, 3*(1), 52–58.

Huy, D. T. (2018). The risk level of Vietnam listed hotel, tourism and entertainment industry firms after the global crisis 2009–2011. *Kwartalnik Naukowy Uczelni Vistula, 3*(57), 136–149.

IMF. (2020). *Opening remarks at a press briefing by Kristalina Georgieva following a conference call of the International Monetary Financial Committee (IMFC).* International Monetary Fund. Retrieved 9 April 2020, from https://www.imf.org/en/News/Articles/2020/03/27/sp032720-opening-remarks-at-press-briefing-following-imfc-conference-call

IndieWire. (2020). *Coronavirus cancellations: Every film, TV show, and event affected by the outbreak.* IndieWire. Retrieved 4 May 2020, from https://www.indiewire.com/2020/04/coronavirus-cancellations-hollywood-entertainment-covid19-1202215596/amp/Katz, B. (2020). *2020 will likely be the worst Box Office year in over 25 years.* Observer. Retrieved 4 May 2020, from https://observer.com/2020/04/coronavirus-box-office-covid-19-movie-effect/

Kim, I. K. (2020). The impact of social distancing on box office revenue: Evidence from the COVID-19 pandemic. *SSRN*. https://doi.org/10.2139/ssrn.3578822.

Klamer, A. (2008). The lives of cultural goods. In J. Amariglio, J. Childers, & S. Cullenberg (Eds.), *Sublime economy on the intersection of art and economics* (pp. 250–272). London: Routledge.

Knapp, A. K. T. (2015). Does 3D make sense for Hollywood? The economic implications of adding a third dimension to hedonic media products. *Journal of Media Economics, 28*(2), 100–118.

Koku, P. S. (1995). On bizarreness in the entertainment industry. *Journal of Services Marketing, 9*(4), 19–30.

McLaren, N., & Shanbhogue, R. (2011). Using internet search data as economic indicators. *Bank of England Quarterly Bulletin, Q2*, 134–140.

Mellon, J. (2014). Internet search data and issue salience: The properties of Google trends as a measure of issue salience. *Journal of Elections, Public Opinion & Parties, 24*(1), 45–72.

National Association of Theatre Owners. (2019). *Number of U.S. movie screens.* National Association of Theatre Owners. Retrieved 8 May 2020, from https://www.natoonline.org/data/us-movie-screens/

Netflix. (2020). *Letter to shareholders.* Netflix. Retrieved 7 May 2020, from https://www.netflixinvestor.com/investor-news-and-events/investor-events/event-details/2020/Netflix-First-Quarter-2020-Earnings-Interview/default.aspx

Observer. (2020). *Netflix stock rebounds as streaming traffic hits new high on AT&T networks.* Observer. Retrieved 4 May 2020, from https://observer.com/2020/03/netflix-stock-rebounding-coronavirus-impact-ratings/

Park, J., Lee, G., & Park, M. (2011). Service quality dimensions perceived by film festival visitors. *Event Management, 15*, 49–61.

Rodríguez-Ferrándiz, R. (2014). Culture industries in a postindustrial age: Entertainment, leisure, creativity, design. *Critical Studies in Media Communication, 31*(4), 327–341.

Roggeveen, A. L., & Sethuraman, R. (2020). How the COVID pandemic may change the world of retailing. *Journal of Retailing*. https://doi.org/10.1016/j.jretai.2020.04.002.

Smith, T. (2008). Creating value between cultures: Contemporary Australian aboriginal art. In M. Hutter & D. Throsby (Eds.), *Beyond price* (pp. 23–40). New York: Cambridge University Press.

Statista. (2020). *COVID-19 ECONOMIC IMPACT Tech giants shrug off COVID-19 crisis in Q1.* Statista. Retrieved 6 May 2020, from https://www.statista.com/chart/21584/gafam-revenue-growth/?utm_source=Statista+Global&utm_campaign=463a285823-All_

InfographTicker_daily_COM_PM_KW19_2020_Mo_COPY&utm_medium=email&utm_term=0_afecd219f5-463a285823-307759193
Throsby, D. (2003). Determining the value of cultural goods: How much (or how little) does contingent valuation tell us? *Journal of Cultural Economics, 27*(3–4), 275–285.
UN. (2020). *Financing for sustainable development report 2020 Inter-agency Task Force on Financing for Development.* United Nations. Retrieved 9 April 2020, from https://developmentfinance.un.org/sites/developmentfinance.un.org/files/FSDR_2020.pdf
UNIC. (2020). *UNIC RESEARCH The impact of the coronavirus outbreak on the European cinema industry.* UNIC. Retrieved 5 May 2020, from https://www.unic-cinemas.org/fileadmin/user_upload/Publications/Public_-_UNIC_research_-_Coronavirus_impact_on_the_cinema_industry_v22.pdfUNWTO. (2020). *STAY HOME TODAY, #TRAVELTOMORROW.* World Tourism Organization. Retrieved 1 May, 2020, from https://www.unwto.org/news/stay-home-today-traveltomorrow
Vogel, H. L. (2014). *Entertainment industry economics: A guide for financial analysis.* Cambridge: Cambridge University Press.
Xia, G., & Yang, X. (2005). Risky sexual behavior among female entertainment workers in China: Implications for HIV/STD prevention intervention. *AIDS Education & Prevention, 17*(2), 143–156.
Yan, L. (2020). Research on the creative style of DreamWorks' animated film script. 한국엔터테인먼트산업학회논문지, *14*(1), 97–106.
Yao, M. Z., Mahood, C., & Linz, D. (2010). Sexual priming, gender stereotyping, and likelihood to sexually harass: Examining the cognitive effects of playing a sexually-explicit video game. *Sex Roles, 62*(1–2), 77–88.
Yolal, M., Rus, R. V., Cosma, S., & Gursoy, D. (2015). A pilot study on spectators' motivations and their socio-economic perceptions of a film festival. *Journal of Convention & Event Tourism, 16*(3), 253–271.
Yolal, M., Gursoy, D., Uysal, M., Kim, H. L., & Karacaoğlu, S. (2016). Impacts of festivals and events on residents' well-being. *Annals of Tourism Research, 61*, 1–18.
Zheng, Y., Goh, E., & Wen, J. (2020). The effects of misleading media reports about COVID-19 on Chinese tourists' mental health: A perspective article. *Anatolia*, 1–4. https://doi.org/10.1080/13032917.2020.1747208.

Chapter 13
COVID-19 and the Stock Market: Impacts on Tourism-Related Companies

Abstract The unprecedented outbreak of the coronavirus in 2020 provided an illustration of a neglected risk that brought about an economic disaster for the world. The COVID-19 pandemic, which started as a public health emergency, rapidly transformed into an economic crisis the world had not witnessed in over a century. In particular, the COVID-19 pandemic was a source of systematic risk, which filled global stock markets with uncertainty and resulted in big moves of share prices. Using the event study methodology that significantly utilised secondary data collection and analysis, the chapter examined the impacts of COVID-19 on tourism-related stocks. Vicissitudes in stock value were used as a signature mark for the impacts. It was observed that the news of the COVID-19 outbreak and measures put in place to curb its spread dampened the stock markets and led to declines in tourism-related stock prices. The measures included travel bans, bans on mass gatherings, as well as the closure of hotels and restaurants. Tourism firms and those in their value chain became the worst performers on global stock markets with some losing up to 80% of their value within a 2-week period. This translated into billions of dollars in lost value. Governments responded to these losses by issuing stimulus and rescue packages for distressed industries. These led to notable recoveries by some firms, although most in the tourism industry remained negative. The chapter recommends an additional rescue package for the tourism industry that must be timed with visible signs that the pandemic is under control.

Keywords Tourism-related · Share price · COVID-19 · Stock markets · Volatility

13.1 Introduction

When the World Health Organization (WHO) declared COVID-19 a pandemic on 11 March 2020, it warned that the disease would potentially have devastating impacts on the various markets of the world (Ramelli and Wagner 2020). COVID-19, which emerged in Wuhan, China, as a mystery illness, quickly spiralled into a global public health and economic crisis the world had not witnessed in over a century (Hassan et al. 2020). Yan (2020) argue that when a new virus outbreak occurs, no one can fully know what its long-term impact on the economy will be, as available

G. Nhamo et al., *Counting the Cost of COVID-19 on the Global Tourism Industry*,
https://doi.org/10.1007/978-3-030-56231-1_13

models will have limitations in their predictive power. However, Ru et al. (2020) observe that the foremost factor when a pandemic occurs is not necessarily how bad the pandemic will be but how bad the market and investors perceive it to be. If investors perceive that a pandemic would disturb the value chain of certain firms, then the stock value of exposed firms would fall (Hassan et al. 2020).

Several existing works have explored the relationships between the outbreak of infectious disease and its impact on the economies of the affected countries, especially the stock markets. De Lisle (2003) quantifies the cost of the 2003 severe acute respiratory syndrome (SARS) outbreak at an estimated $3 trillion in lost gross domestic product (GDP) and $2 trillion in fallen equity in the financial markets. The research concludes that SARS had a far bigger economic shock than expected compared to its health impact. Velde (2020) highlights the negative stock market impact of the Spanish flu as being between modest to a non-event over a time span of several months. This was because information availability on the disease and its diffusion were very much constrained, and interconnections between different international economies were limited during that period. Honigsbaum (2013) observes that the first wave of the Spanish flu occurred during the First World War when news about the true extent of the outbreak was censored. Hence, it caused little panic among investors. Others such as Chev and Marinč (2018) note a significant negative reaction of stock markets to outbreak alerts and unanticipated increases in predicted infections. They also note close interrelations between stock markets in different countries, especially during times of crises. Hsieh (2013) observes that stock prices were more volatile in the trading days during crisis periods such as disease outbreaks, wars and economic crises than those during noncrisis periods.

During the COVID-19 outbreak, governments around the world responded to this biosecurity threat with a series of decisive and sweeping actions. These included travel bans, bans on mass gatherings, closure of hotels and restaurants and banning of travellers from particular countries exposed to the disease as ways to slow down the spread of the virus (Ivanov et al. 2020; Dube et al. 2020). Further, the measures imposed compulsory quarantining of people who had visited particular countries or those showing COVID-19 symptoms, partial or complete lockdowns and curfews (Hanson 2020). Ivanov et al. (2020) highlight that these government actions disturbed the normal business process and heightened the economic uncertainty caused by the pandemic. Many firms ceased operations during this period as a result of direct government orders to close, falling demand due to travel bans, voluntary precautionary measures to protect the health of employees or forced decision from disruptions in the supply chain (Pandey 2020). As a consequence, millions of employees were sent home on paid or unpaid leave or were laid off, while companies struggled to stay liquid and pay their debts (Dube et al. 2020).

Stock market returns have been noted to respond to major events such as environmental disasters, pandemics and political turbulence. Changes in share prices on stock markets allow interested and affected stakeholders to objectively decipher how different industries are impacted by pandemics such as COVID-19. Griffith et al. (2020) maintain that the changes in share prices are a reflection on market expectations about possible impacts. Such expectations and impacts include a change in demand or restrictions in the supply chain. To this end, the COVID-19 pandemic induced systematic risk, which filled global stock markets with shock and uncertainty, giving rise to big moves in stock prices (Schoenfeld 2020). This chapter examines the spillover effects of the COVID-19 pandemic on the tourism sector using changes in stock value of tourism-related firms as a signature mark for the impacts.

13.2 Literature Survey

The outbreak of the novel coronavirus will go down in history as a foremost illustration of a neglected risk. This can be noted in the World Economic Forum's (WEF) Global Risks Report 2020 released in January 2020, which ranked the theme "infectious diseases" tenth in terms of possible impact to the business, although it was considered relatively unlikely to occur (WEF 2020). The attention of global corporate decision-makers, politicians and influencers was mainly focused on traditional sources of business risk and persistent environmental concerns such as climate change (Ramelli and Wagner 2020). This suggests that global business managers systematically underestimated their exposure to pandemics (Schoenfeld 2020). The emergence and rapid spread of COVID-19 meant that the risk management models of many businesses were facing a risk that they were not prepared and had a significant blind spot for.

The tourism industry is generally very vulnerable to perceptions of risk on the part of both consumers and investors. As such, understanding how the industry responds to and rebounds from the crisis is important for its long-term management and sustainability (Morris 2020). The next parts of this section examine how previous pandemics impacted on stock markets with a special emphasis on the tourism-related stocks.

While the COVID-19 pandemic provides an extreme example of how disease outbreaks can impact the stock markets, it is not without precedent (Fernandes 2020). Much can be learnt about the behaviour and resilience of the stock markets from previous outbreaks and other catastrophic events. However, given the extraordinary nature of COVID-19, earlier experiences need to be carefully adjusted to the unique features of the current challenge because existing policy remedies and models that applied then may no longer be applicable (Barro et al. 2020).

Although there are still lessons to be learnt, Hassan et al. (2020) observe that there are more differences than similarities between the COVID-19 pandemic and the outbreak of SARS in 2003. While an industry might learn from prior experiences, ultimately, the SARS and H1N1 epidemics were much smaller in magnitude and had less severe macroeconomic consequences than the COVID-19 outbreak. Morris (2020) states that stock markets that did not have prior exposure to SARS were less volatile at the outset of the COVID-19 pandemic and were less associated with negative sentiment related to the virus compared to markets that had been exposed before. While SARS broke out in November 2002, it did not begin to affect markets until March 2003. The Hong Kong stock markets started to underperform noticeably after the WHO issued a global alert about SARS in China and other parts of Asia (Yan 2020).

The stock market reaction to the SARS outbreak was relatively moderate in Hong Kong and other Asian stock exchanges. According to Siu and Wong (2004), the Hong Kong Hang Seng Index dropped only by an average of 1.78% between 12 March and 30 April 2003. Other major market indices in the region rose over the same period, except for the Taiwan Weighted Index that dropped 7.57% and Japan's Nikkei that dropped 1.68%. Nippani and Washer (2004) examine the impacts of SARS on the stock markets of affected countries such as Canada, China, Hong Kong, Indonesia, the Philippines, Singapore, Thailand and Vietnam. They compared the mean returns for the indices during the SARS-affected period with the pre-event period. They noted that the stock markets were already underperforming before the SARS outbreak. However, they also observed that none of these indices significantly underperformed during the comparison period, implying that the SARS outbreak did not have a severe impact on any of the studied stock indices.

By the time the SARS outbreak was declared successfully contained, stock markets in Hong Kong had remained subdued and still underperformed. The most important factor, which rallied the stock market performance, was the Hong Kong government's announcement of a US$ 1.5 billion relief package to boost the domestic economy (Hassan et al. 2020). The local stock market subsequently recovered and outperformed global equities over the next several weeks (Morris 2020). At the peak of the SARS outbreak, however, share prices for the most exposed industries such as hotels and restaurants fell even deeper than the Hang Seng average (Fernandes 2020). The stock indices that performed the worst were not surprisingly those in the tourism value chain such as the Cathay Pacific Airways, which dropped by almost 15%. Globally, airline stocks also fell sharply during SARS, with the US airline stock prices dropping more than 30% (Yan 2020). However, these stocks recovered strongly once the broader market rebound began, with three out of the five worst-performing sectors moving to the top five of the best (Ru et al. 2020). These were the airlines, hotel, restaurant and leisure industries, and their recovery was fast enough not to dent the long-term profitability of the companies concerned (Morris 2020). As noted by Ramelli and Wagner (2020), during SARS, stock markets fluctuated less and even had persistent negative abnormal stock returns. The timing of the Hong Kong stimulus package to boost the markets in 2003 was also key. This came as the total number of new cases was levelling off and beginning to decline (Morris 2020).

Taiwan was also one of the territories seriously affected by SARS. This was because it was a transit country through which more than 70,000 air traffic passengers passed each day, with most of them coming from SARS-infected areas (Economist 2003). In addition to the loss of lives, the SARS outbreak had severe economic ramifications for the territory, especially the tourism industry. Thousands of business meetings and holidays were cancelled with the outbreak having almost an immediate and devastating impact on Taiwan's hotel industry (Pine and McKercher 2004). Chen et al. (2007) highlight that during the SARS outbreak, share prices of tourism-related companies dropped by as much as 28.91%. This was the heaviest drop compared to other industries such as textiles, which dropped by 14.97%, automobiles that dropped by 13.85% and construction that had a drop of 12% (Chen et al. 2007). Seven publicly traded hotel companies had significant declines in their earnings and stock prices during the SARS period. This showed that hotel stocks in Taiwan were on average exposed to above average market risk. This was consistent with the common perception that the hotel industry is most vulnerable to decreases in the number of tourists visiting an area (Vigna 2020).

Loh (2006) studied the impact of the SARS outbreak on the returns and volatility of individual airline stocks. This was after several authors observed SARS to have imposed detrimental impacts on the airline stocks listed on the stock markets of various SARS-affected nations (Harbison 2003). Loh (2006) uses data on 12 airline stocks including Air Canada (Canada), Westjet (Canada), Eastern Airlines (China), Hainan Airlines (China), Shandong Airlines (China), Shanghai Airlines (China), Cathay Pacific (Hong Kong), Singapore Airlines (Singapore), China Airlines (Taiwan), Eva Airways (Taiwan), Far Eastern Air Transport (Taiwan) and Thai Airways (Thailand). The research concludes that share prices for most airlines were lower during the SARS outbreak relative to the pre-event period. This meant that airline stocks experienced equity devaluations with the onset of SARS and investors holding these shares from the pre-event period to the SARS period were likely to realise negative returns. In terms of examining the impact of SARS on the volatility of individual airline stocks, the results indicate that the volatility of returns was significantly higher during the SARS outbreak when compared to the pre-event period for six of the 12 airline stocks that were examined. Overall, Loh (2006) concludes that negative implications of SARS on the stock markets as a whole were negligible, although its effects on airline stocks are more pronounced.

Chiang et al. (2007) argue that stock markets in a region and in the world become interlinked and interdependent, especially during crisis periods. They discovered close cross-market correlations after examining the daily stock returns for nine Asian markets for the period from 1996 to 2003. They noted a high correlation among Asian stock markets during periods of crises, with Malaysia, Vietnam and Thailand being the most financially integrated with China. This implies that a crisis in one country can easily spread to another, and events such as infectious disease outbreaks can induce negative changes in investor perceptions that can strongly affect their investment decisions and stock markets by default (Liu et al. 2020). Gates (2020) observes that in any crisis, leaders have two equally important responsibilities of solving the immediate problem and keeping it from happening again. The next section highlights the material and methods used in the collection and analysis of data. It also shows countries with the largest stock markets in the world.

13.3 Material and Methods

There are over 60 major stock exchanges around the world with different market capitalisations and monthly trading volumes. The total value of these stock markets is well in excess of $70 trillion (World Federation of Exchanges 2020). Geographically, there is a domination of the Northern Hemisphere, with North America holding 40.6% of the global stock market value (The Robust Trader 2020). These stock markets include the New York Stock Exchange, the National Association of Securities Dealers Automated Quotations (NASDAQ) and TMX Group, which operates the Toronto Stock Exchange. In Asia, which has 33.3% of the global stock value, there is the Japan Exchange Group, Shanghai Stock Exchange, Hong Kong Stock Exchange, Shenzhen Stock Exchange, Bombay Stock Exchange, National Stock Exchange of India, the Korea Exchange and the Taiwan Stock Exchange (World Federation of Exchanges 2020). Europe, which has 19.5% of global stock market value, has the Euronext, London Stock Exchange, the Deutsche Borse, SIX Swiss Exchange, the NASDAQ OMX Nordic Exchanges and the Bolsas y Mercados Espanoles (The Robust Trader 2020). Notable stock exchanges in the Southern Hemisphere include the Johannesburg Stock Exchange, Australian Securities Exchange, the Indonesia Stock Exchange and the Brazilian BM&F Bovespa. These hold a total stock value of 6.6% (World Federation of Exchanges, 2020). Figure 13.1 shows countries with the world's top 25 stock exchanges by market capitalisation.

The chapter mainly used event study methodology, dominated by secondary data generation and analysis methods. The event study methodology has its origins in finance, particularly the analysis of stocks and foreign exchange (Hayward 2018). An event study examines the impact of an event on the financial performance of a company or stock markets. In the case of this chapter, the event study was used to examine how tourism-related share prices responded to the different stages in the outbreak of COVID-19.

The research also used relevant quantitative and qualitative secondary data from authoritative sources in order to answer the objective of the chapter, which was to

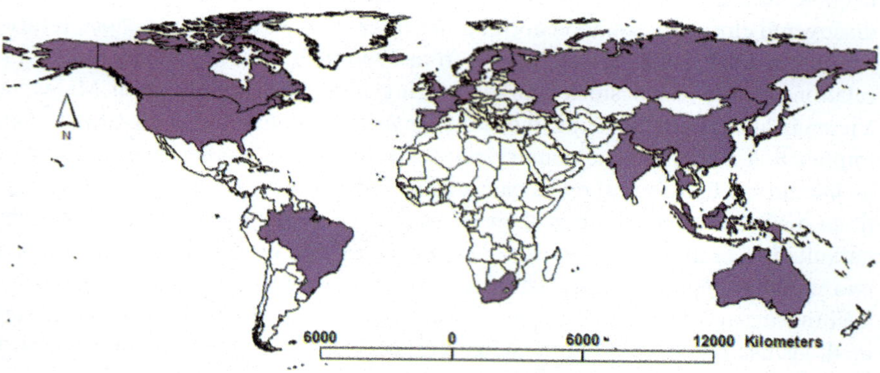

Fig. 13.1 Location of top 25 stock exchanges by market capitalisation. (Source: Authors)

document the impacts of the COVID-19 pandemic on share prices of tourism-related firms. Secondary data used in the study were derived from governmental sources, regulatory bodies, industry representative bodies, private companies and other scholarly sources addressing the research questions. Thematic content analysis was then used to thematise obtained data sets before the final analysis to derive meaning from the content was performed. Descriptive statistics, especially frequencies, were used to summarise the derived quantitative data, with the presentation being in the form of graphs. Qualitative data were analysed through thematic content analysis and presented through the use of descriptions.

13.4 Presentation of Data and Discussion of Findings

This section presents data and discusses the findings from twin perspectives: firstly, a general overview of the stock market response to COVID-19 and secondly, specific impacts of the COVID-19 on tourism-related stocks. Each of these twin subsections will now be considered as appropriate.

13.4.1 COVID-19 Impacts on the Stock Markets: An Overview

Globally, the uncertainty that characterised the period of the COVID-19 pandemic led to an unprecedented drop in stock market indices. As the pandemic rapidly spread across the world, wave after wave of bad news led to global drops in stock prices. China, the world second largest economy, for example, was hit by two waves of significant stock price declines. Firstly, stock prices dropped dramatically when the lockdown of Wuhan was announced but recovered when the virus was curbed in the country (Ozili 2020). Secondly, stock prices plunged again sharply when the global pandemic was starting to exponentiate and measures being put in place to curb the spread (Yan 2020). The timing of changes in share prices also reflected changes in market expectations. For almost every sector, the changes in share prices did not take place slowly as the coronavirus spread across the world; instead, big dips in share prices occurred from the end of February 2020, soon after Italy introduced a lockdown in Lombardy (Griffith et al. 2020). Before this event, very little change in share prices had been witnessed in global markets except in China (Ozili 2020). Yan (2020) observed that the overreaction of the stock markets to the spread of COVID-19 was mainly due to uncertainty and a lack of investor confidence given the interconnectedness of modern economies through global supply chains that stood to be disturbed during the pandemic.

Ramelli and Wagner (2020) examined how stock prices reacted to the outbreak of COVID-19. They came up with three phases, which were (1) the incubation (2 January 2020 to 17 January 2020), (2) the outbreak (20 January to 21 February) and (3) fever (24 February to at least 6 March). The events initiating the outbreak and fever periods markedly changed the attention of stock markets. This saw stock markets starting to indicate a general pessimism regarding the disruptive impact of COVID-19 on global trade (Gourinchas 2020). The emerging literature on the macro-economic impact of pandemics shows that the spread of the disease, and policy responses attempting to mitigate the spread, may result in large shocks to the value chains of different industries (Eichenbaum et al. 2020). These shocks may manifest in a variety of ways such as disruption to supply chains, shortages of labour, shutdowns of operations, the sudden drop in demand and difficulty in accessing credit lines. Those concerns, together with increased uncertainty, can lead to the decline in share stock values of the most impacted firms (Hassan et al. 2020).

The Chinese stock markets responded sensitively to the COVID-19 outbreak. Both the Shenzhen and Shanghai Composite Indexes dropped almost 3% on 23 January 2020. This represented the biggest single-day loss in almost 9 months (Ozili 2020) and was the incubation period of the pandemic. After the Wuhan lockdown was announced, investors were further unsettled by the drastic measure. Hence they sought safe havens for their investments. At the first trading day after the Chinese spring festival, market indices fell by 8% (Griffith et al. 2020). The markets then managed to slowly recover but dropped again when the WHO raised the risk level of COVID-19 to "very high" (Ozili 2020). Given that the global economy is now very interconnected and becoming more specialised with supply chains established worldwide, the COVID-19 event in China has ripple effects the world over. Once some firms along the supply chain were disrupted by COVID-19, almost all firms in the chain were affected. This was viewed by Yan (2020) as one of the factors that magnified the impact of COVID-19 on businesses and the stock markets by default. Global stock markets erased about US$6 trillion in wealth in 1 week from 24 to 28 February 2020 (Schoenfeld 2020). The S&P 500 Index also lost over $5 trillion in value in the same week in the USA. The S&P 500's largest 10 companies experienced a combined loss of over $1.4 trillion because of fear and uncertainty among investors about how COVID-19 would affect firms' profits (Ozili 2020).

There were huge losses in the stock market because of the COVID-19 panic from 2 January to 20 March 2020 (Schoenfeld 2020). The biggest losses were noted by the Russel 2000 Index, which dropped by a massive 39.2%, followed by the Dow Jones 30 Index that plunged by 32.3%. The S&P 500 stocks decreased in value by 28.6% at the onset of the pandemic (Schoenfeld 2020). This translated to an average economic loss of $18 billion per firm or close to $9 trillion in total for S&P 500 companies alone (Reinhart 2020). The companies that had the biggest declines in stock value were the Norwegian Cruises, Noble Energy, Royal Caribbean Cruises, Halliburton and Carnival. These hardest hit firms were mostly either in the tourism or energy sectors (Schoenfeld 2020). The reactions of the stock markets also suggest that pandemics were an important influence on stock markets but had not been given enough attention (Reinhart 2020).

The large drops on stock markets triggered market-wide circuit breakers, a record four times in March 2020. The circuit breakers provide safeguard pauses in trading for 15 min in the hope that the market will calm down (Fernandes 2020). The US Securities and Exchange Commission mandated the creation of these breakers to prevent a repeat of the 1987 market crash when the Dow plunged 22.6% (Funakoshi and Hartman 2020). These had only been used once in 1997. Guidelines mandate that a 15-min pause in trading on all US stock exchanges occurs if the S&P 500 Index falls more than 7% before 3:25 p.m. New York time. The same 15 min will apply for level 2, which will be a drop of 13%. However, at level 3, if a drop of more than 20% occurs, then trading will be halted for the rest of the day (Funakoshi and Hartman 2020). Figure 13.2 shows selected stock market performances by country. It shows that the most profound drop in stock markets during the COVID-19 pandemic were recorded in South America, and Brazil and Colombia, with losses of 48% and 47%, respectively. This was followed by Greece (44%) and South Africa (40%) with Japan having the lowest drop of 25%.

The implied volatility of equity markets is often used as an indicator of how risky the current environment is for investment in the stock market and how uncertain the future will be. The VIX Index, for example, is commonly referred to as the "fear index" (Cboe 2020). It is based on the traded prices of options on the S&P 500 Index and represents the market expectation of stock volatility over the next 30 days (Fernandes 2020). In Fig. 13.3, the implied volatility is given by the VIX Index, whose long-term average is approximately 20%. However, as COVID-19 spread from a regional crisis in the Chinese Hubei Province to a global pandemic, equities plummeted, and market volatility rocketed upwards around the world. It rose significantly to 85% in February 2020 because of the rise in fears over the impacts of the coronavirus on investments. The coronavirus created such uncertainty around the world that two of the largest single-day drops in the Dow Jones Industrial average were in March 2020 (Funakoshi and Hartman 2020).

Fig. 13.2 Selected stock market performances by country. (Source: Authors, based on Reuters Eikon (2020))

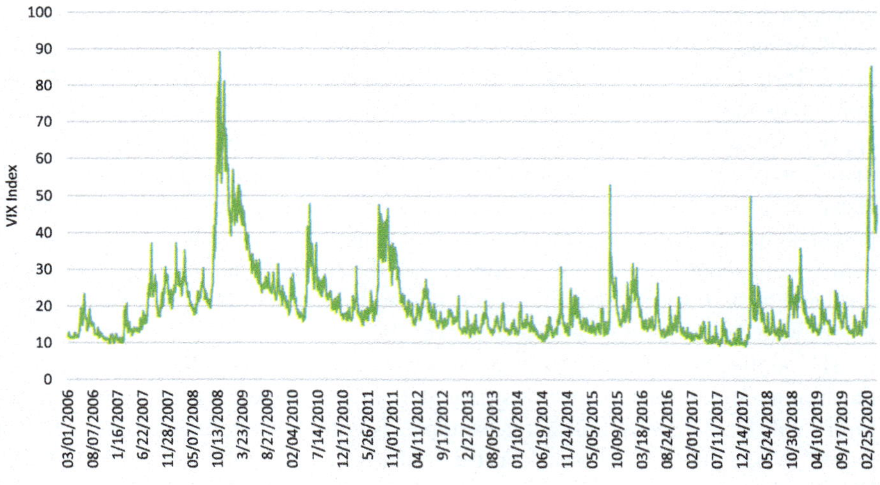

Fig. 13.3 Implied stock market volatility using the VIX Index. (Source: Authors, data from Cboe (2020))

Prior to the sudden rise, the previous rise in the volatility of such a high magnitude was in the aftermath of the 2008/9 financial crisis. This also reflected investors' uncertainty about the future (Fernandes 2020). Other smaller peaks in volatility that are shown in Fig. 13.3 depict episodes of uncertainty in the markets because of different negative news such as the American-China trade tariffs implementation, the Swine flu outbreak (H1N1) during March to May 2009, the outbreak of the Ebola virus in West Africa in October 2014 to January 2015 and many more. Of note is that when all of these past crises occurred, central banks had the firepower to prevent further damages (Fernandes 2020). None of the previous crises was in periods where the starting point of interest rates was close to zero with some in the negative. This raised concerns in the markets because there was little room for an effective policy response (Gerding et al. 2020).

After analysing daily historical jumps in the US stock markets that were greater than 2.5% up or down, Baker et al. (2020) observe 1129 stock jumps from 2 January 1900 to 24 March 2020. However, they note that before the COVID-19 pandemic, no infectious disease outbreak had ever made a sizable contribution to the USA stock market volatility. The 2003 SARS epidemic and the 2015 Ebola epidemic led to modest, short-lived spikes in volatility, and the Bird flu and Swine flu epidemics barely registered. Table 13.1 shows the historical jumps on US stock markets and the impact of the COVID-19 pandemic.

Albulescu (2020) analyses the relationships between the official COVID-19 announcements and volatility on stock markets. The analysis concluded that increases in the volatility were related positively with the increases in the infected number of people and the death ratio as of March 2020. Hence the persistence of COVID-19 had the potential to generate a new episode of international financial stress. Ru et al. (2020) argue that declines in stock market volatility would occur when the trajectory

Table 13.1 Impact of COVID-19 on US stock market jumps

Period	Number of daily US stock market moves greater than 2.5%	Number of moves attributed to pandemics	Number attributed to policy responses to pandemics
02/02/1900–21/02/2020	1116	0	0
24/02/2020–24/03/2020	18	7	8

Source: Authors, based on Baker et al. (2020)

of the pandemic becomes more certain. They also observe that stock markets reacted more quickly and strongly in countries that suffered from the SARS outbreak in 2003 compared to those that did not. This initial underreaction by countries that did not suffer from SARS lasted for several weeks until a series of severe outbreaks outside China began to occur in late February 2020, which led to the stock markets in these countries plunging. However, Gerding et al. (2020) maintain that stock price reactions were stronger in countries with a higher debt to GDP ratio.

13.4.2 COVID-19 Impacts on Tourism-Related Stocks

Tourism-related counters were observed to be some of the biggest losers during the COVID-19 outbreak. From the mobility market outlook on COVID-19 by Statista (2020), the revenue for the travel and tourism industry was set to decline globally in 2020, with Asia being the most affected region and set to decline from $225.8 billion in 2019 to roughly $165 billion in 2020. These figures refer to the cruises, hotels, package holidays and vacation rental segments of the travel and tourism industry. The tourism industry was, in particular, impacted when the travel prospects for Chinese tourists, who usually spend billions annually, were severely constrained (Yan 2020). There were increased flight cancellations, cancelled hotel bookings and cancelled local and international events worth over $200 billion (Statista 2020). Figure 13.4 shows a comparison of a global change in travel and tourism revenue because of COVID-19 by region from 2019 to 2020. This negative outlook was, therefore, also manifested in the fall of stock prices of the tourism-related companies as investors lost confidence in their ability to give a return on investment. The uncertainty in the industry was also heightened by a lack of a clear and effective roadmap to deal with the pandemic worldwide, and no specific time as to when normal operations of tourism players would be allowed to recommence.

The tourism-related stocks, also known as the BEACH stocks (BEACH – standing for booking, entertainment and live events, airlines, cruises and casinos, hotels and resorts), had over $332 billion in stock value evaporating due to COVID-19 between 19 February and 24 March 2020 (Lloyd-Jones et al. 2020). The next subsections unpack the extent of losses of by different BEACH companies during the COVID-19 pandemic.

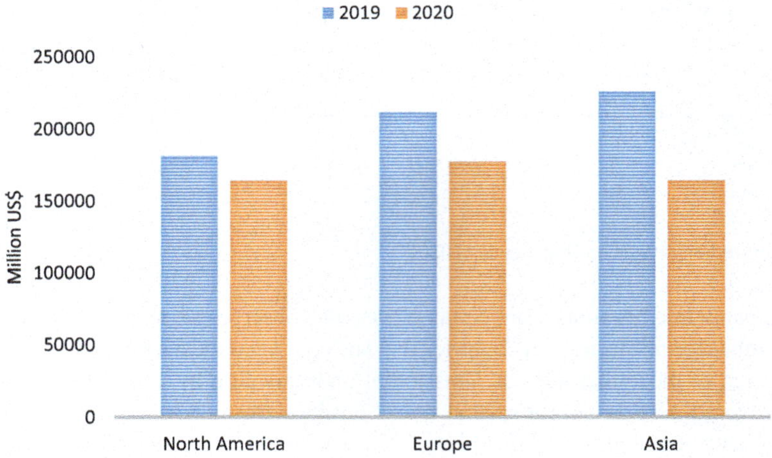

Fig. 13.4 Change in travel and tourism revenue because of COVID-19 in selected regions (2019–2020) (Source: Authors, data from Statista (2020))

13.4.2.1 Impact of Travel Counters

Due to the outbreak of the coronavirus, initially, more and more people chose to stay at home rather than travel far to avoid the disease. This was also accompanied by the general corporate travel restrictions in the face of the pandemic. In the USA, the month of March is usually full of travel as most universities are on spring breaks, and students normally go on vacation (Lloyd-Jones et al. 2020). During the COVID-19 outbreak, in fear of the disease, most people limited travel to only essential movements and cancelled their pre-arranged trips. This led to the initial decline in demand for flights. Furthermore, as the virus continued to devastate communities, travel bans were instated by different countries, including popular tourist destinations. This led to further loss of demand for airlines. Due to the unprecedented amount of cancellations and uncertain future, investors started losing confidence in travel-related stocks resulting in significant drops in value (Yan 2020). Once the airlines were grounded because of the COVID-19 induced travel bans, a negative snowball chain reaction was inevitable in the tourism value chain. Online travel agencies such as Booking were first to go out of business because they earn money from the tickets they sell. This further plunged the global travel industry into uncharted territory. Since 19 February 2020, the global airline industry alone had witnessed $157 billion wiped off stock valuations across 116 publicly traded airlines (Lloyd-Jones et al. 2020).

The United Nations World Tourism Organization (UNWTO) hints that the COVID-19 crisis is impacting on travel like no other event in history (UNWTO 2020). This resulted in almost all 96% of destinations in the world imposing travel restrictions since February 2020. The restrictions included severe measures, such as the banning all travel in some destinations. Travel restrictions were further tightened soon after 11 March 2020 when COVID-19 was declared a pandemic. Most destinations instituted complete or partial closure of borders and suspension of flights. In reaction to the announcement of the pandemic, travel tourism-related stocks further plunged on the world stock markets. As of 6 April 2020, no destination had lifted travel restrictions introduced in the context of COVID-19, further reducing the hope that the sector was likely to have a swift recovery from the drop. Figure 13.5 shows declines recorded in share values of selected travel tourism companies.

From Fig. 13.5, it emerges that the hardest hit companies in this tourism segment were cruise ship companies, with the biggest losers falling well above 80% of their pre-COVID-19 value. In less than 2 months, the biggest players in the industry including the Carnival Corporation, Royal Caribbean and Norwegian Cruise Line Holdings lost more than $42 billion in combined market capitalisation, more than half of their value (Goldman 2020). The next impact in terms of losses was witnessed by airlines. Listed airline stocks plummeted with the tightening of travel restrictions that severely impacted demand. This saw Air Canada shares dropping by 67% and the International Consolidated Airlines, commonly known as IAG stock dropping by as much as 64%. The IAG owns British Airways, Iberia of Spain and Sun Air of Scandinavia. Companies such as Delta Airlines witnessed the fall of over 50% in the value of their shares, signifying the worst performance in the history of the company. Therefore, the airline industry was likely to continue losing revenue as long as the COVID-19 pandemic persisted with worldwide airline revenue estimated to fall by as much as $113 billion in 2020 (Neufeld 2020).

This fall of the global airline industry, which employs over ten million people and supports about $2.7 trillion in global economic activity and carries an average of 12 million passengers per day (Neufeld 2020) is troublesome. There will be huge ripple effects on the tourism value chain. For example, many travel booking brands

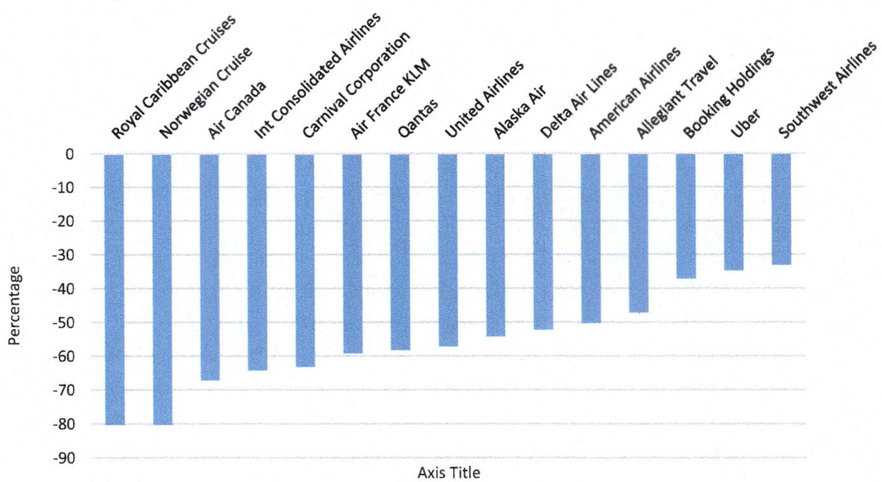

Fig. 13.5 Declines in share value of travel tourism-related firms (19 February to 24 March 2020). (Source: Authors, based on Markets Insider (2020), Statista (2020), Visual Capitalist (2020))

also plunged in value. Booking Holdings, the parent company to Booking.com, Kayak, Priceline and OpenTable, all witnessed share price drops of over 35% during this period. Companies such as the Flight Centre Travel Group suspended their earnings guidance for the 2020 fiscal year "in light of heightened coronavirus uncertainty". The travel agency had previously downgraded its full-year guidance from an underlying profit before tax of between \$310–\$350 million and \$240–\$300 million (Markets Insider 2020). This further saw its stock price drop by 60% within a month, signalling the low investors' confidence in the company. To prepare for the tough times, the company actioned cost reduction plans including shorter working weeks on proportionate pay, leave without pay, a freeze on non-essential recruitment and cuts on discretionary spending (Cullinane 2020).

13.4.2.2 Impact on Hospitality Counters

The COVID-19 pandemic and the related restrictions on travel, business activity and individual movement imposed an unprecedented impact on the hospitality industry. Hotel owners, operators, lenders and investors all faced bigger tests than ever anticipated as they had to grapple with plummeting occupancy, average rate and revenue per available room (Lloyd-Jones et al. 2020). In the end, multiple hotels globally announced temporary suspensions of normal operations in March 2020. This led to a loss of an estimated 24.3 million jobs globally, with 3.9 million losses in the USA alone (Neufeld 2020). Due to the decline in hotel occupancy during the pandemic period, the financial impact on the hotel industry was more severe than the 9/11 attack and the 2008/09 global financial recessions combined (Ozili and Arun 2020).

As this financial disaster for the hotel industry unfolded, investors were also grappling with important questions concerning risk, returns and value of hospitality shares. In reaction to the pandemic, hotel share prices realised significant declines all over the world. In the USA, an estimated \$1.4 billion in potential revenue was lost each week due to the pandemic (Vigna 2020). The Baird/STR Hotel Stock Index, which serves as a benchmark for the sector's overall health, declined by over 47% year-to-date as of March 2020 (Neufeld 2020). This shows the low confidence that investors had in this sector in terms of return on their investments during the pandemic. Figure 13.6 shows the declines in selected hospitality counter shares recorded between February and March 2020. The loss in stock value of hotels was significant, with hotels such as Park and Marriot Vacation Worldwide losing over 50% of their value, and between 38%–48% of its market value.

The damage to stock prices of the sector has ramifications for the hotel's ability to service its debts. Most of these companies had a debt that needed to be serviced. For example, Marriott had \$10.9 billion in debt, with Hilton holding approximately \$8 billion (Vigna 2020). This brought the companies to a situation where they had to choose between debt payments and operations. The major flashpoint was containing the economic damage imposed by COVID-19. Hence, Marriott cut its staff numbers, closed several hotels, slashed salaries for its executives and suspended

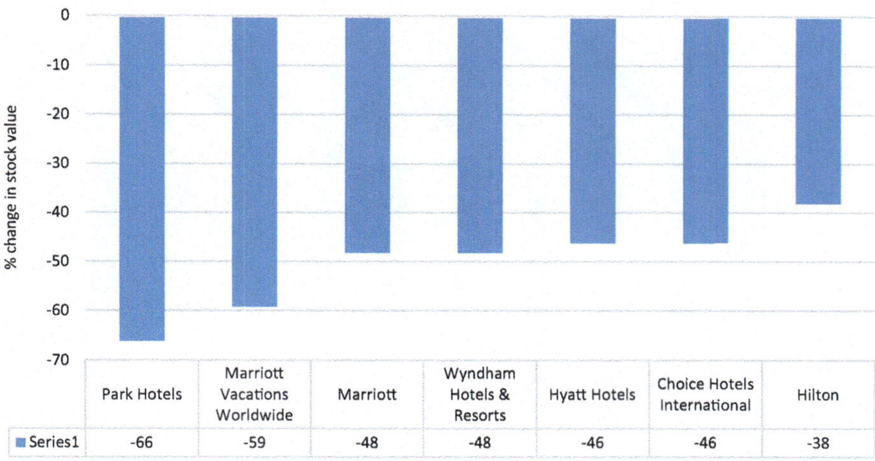

Fig. 13.6 Declines in hospitality counters. (Source: Authors, data from Statista (2020); Visual Capitalist (2020))

dividends to its shareholders. Without the crisis declared contained, it was difficult to assess with confidence how bad and permanent the damage imposed on hospitality shares and investment was and how long it would last (Vigna 2020).

13.4.2.3 Impact on Entertainment and Resort Counters

Entertainment companies and resorts also lost out during the COVID-19 pandemic. There were virtual shutdowns and almost abandonment of some of the tourist epicentres around the world. Visitors opted to stay home or were constrained in terms of travelling and social distancing measures. The chilling sights of almost deserted streets of popular resorts around the world such as the Las Vegas strip by default implied the loss in revenue and erosion of investor confidence in the sector. This led to the free fall in share prices of these companies. One of the biggest losers was the Eldorado Resorts, which reportedly lost over 76% of its value at some point. Figure 13.7 shows losses in share value of some of the world's largest resorts and entertainment companies during the pandemic.

Disney, for example, closed its theme parks in China indefinitely during the Lunar New Year. This is a known high-volume week for their theme parks. Because theme parks are places where huge volumes of people crowd, to avoid super spreading the virus, the difficult decision to close them had to be made. This had a big negative shock on Disney's revenue as its theme parks generated around 34% of its revenue. In addition, Disney planned to release the new movie "Mulan" on 27 March 2020, which was expected to be very popular in China (Owens 2020). However, as learnt from the SARS outbreak, data from the Hong Kong Box Office receipts showed that revenue dropped around 47% during the peak period and also decreased by 20% after 2 months. With

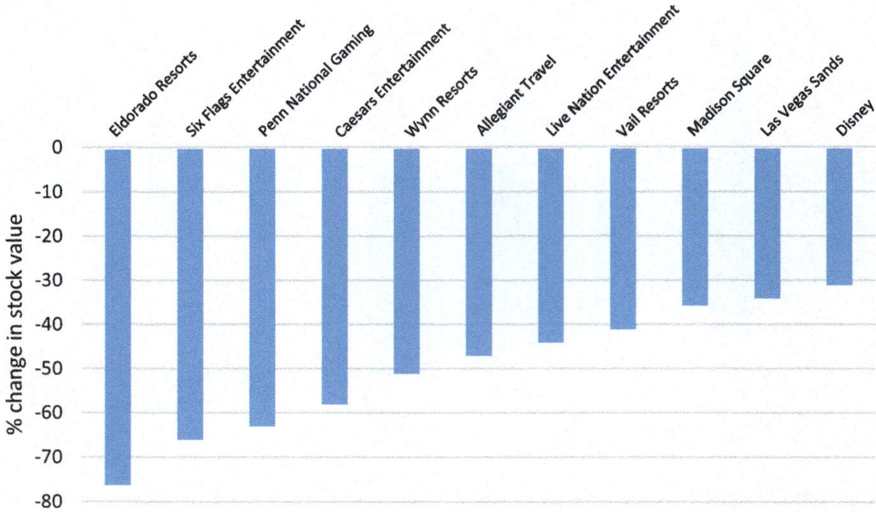

Fig. 13.7 Impact on entertainment and resort stock values. (Source: Authors, data from Visual Capitalist (2020))

this experience in mind, fewer people would go to movie theatres, and the returns from the new movie would have been suppressed (Yan 2020). This loss in potential revenue led to Disney shares at one point tumbling by as much as 31% in value reflecting investor panic. It was anticipated that the global film industry as a whole stood to lose more than $5 billion as a result from the COVID-19 pandemic (Ozili and Arun 2020).

Companies in events management also registered big losses due to the COVID-19 outbreak. Madison Square Gardens, for example, lost as much as 35% of its share value and suspended all expansion projects. Prior to 2020, the events sector shares were generally doing well. In 2018, business events hosted more than 1.5 billion participants across more than 180 countries (Oxford Economics 2018). The events industry generated more than $1.07 trillion of direct spending, representing spending to plan business events, produce business events, business events-related travel and direct spending by exhibitors. The industry also created 10.3 million direct jobs globally (Events Council 2020). During the COVID-19 pandemic, the industry was hit financially by a large number of cancellations of exhibitions, live music shows, conference, weddings, parties, corporate events, brand launches and trade shows (Ozili and Arun 2020). With no prospects of the crisis ending soon, the industry suffered negative perception from investors, hence the decline in share values.

13.4.2.4 Impact on Sports-Related Counters

The COVID-19 pandemic also hit the sports industry hard at the start of 2020. The value of the sports industry was estimated to be $471 billion in 2018, which represented a 45% increase from 2011 (Hall 2020). The spread of the coronavirus throughout the world forced the cancellations or suspensions of major global sporting events and professional leagues across the globe. Every part of the sporting value chain was being affected by the spillover effect of the cancellations (Neufeld 2020). This spelt trouble for listed sporting companies and big-name athletic companies and their stocks. Figure 13.8 shows the drop in share value of sports-related companies during the COVID-19 outbreak. Among the most affected was the Formula One (F1) season, which witnessed several of its races cancelled. The uncertainty surrounding the remaining dates in the F1 calendar led to the Formula One Group shares losing value on the stock market. The loss in value during the crisis amounted to over 44% of its value, which translated to over $5 billion (Quartz 2020).

Football-related shares were also negatively impacted by the outbreak of COVID-19. The free fall of football company shares started when they first played in empty stadiums to try and contain the virus, and it was then followed by the ultimate cancellation or postponement of matches. Among the biggest losers were Italian giants Juventus, which lost over 51% of its share value, followed by Borussia Dortmund and Manchester United, which lost 44% and 33% in value, respectively. With the football leagues postponed, the revenue streams of the clubs also declined. Hence the drop in share prices. The clubs as well as the firms in their value chain were the biggest losers. The list includes sponsors not getting value for their money, broadcasters, sportswear companies and betting companies.

Sporting is big business for gambling outfits. Betting companies such as Bet-at-home.com saw their share prices plunging by as much as 70% largely due to the cancellation of sporting events. This left their clients with nothing to bet on. This

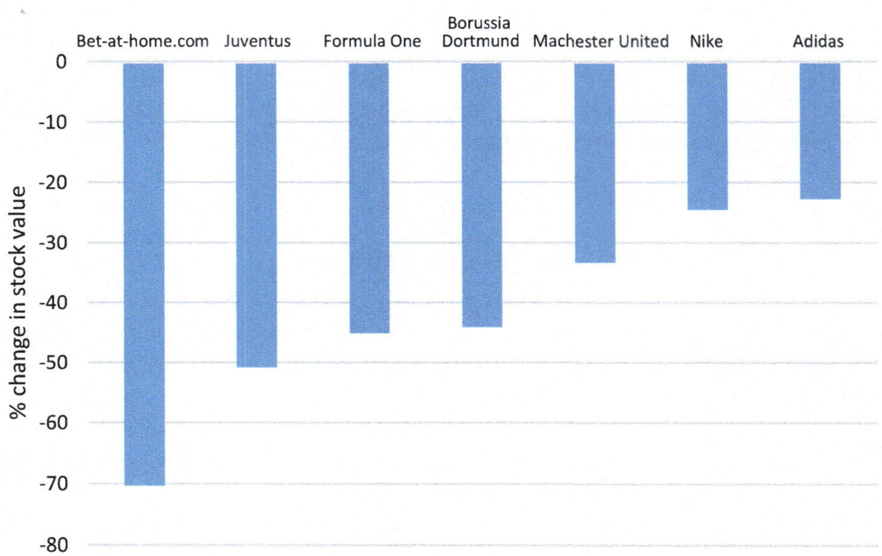

Fig. 13.8 Impact of COVID-19 on sports-related share values. (Source: Authors, Markets Insider (2020), Quartz (2020))

weighed heavily on their revenue and also desirability as investment options. Most betting companies warned their investors that the suspension of major sporting events would have significant impacts on their business. Global sportswear giants such as Nike and Adidas' share prices dropped by 12% and 14%, respectively. Most of their lucrative deals were impacted by the cancellation of global sporting events such as the Olympics and professional sporting leagues as well as the restrictions of the movement of people and closures of some stores which affected the demand for their products.

Sponsors who invest in these sporting companies were also affected by the COVID-19 pandemic. For example, Rwanda tourism invested heavily in football sponsorship to boost their tourism. They went into deals of over 40 million Euros with Arsenal and Paris Saint-Germain to have their "Visit Rwanda" slogan written on the team jerseys. Courtesy of this publicity, the country had witnessed an increase of over 25% in international arrivals (Wohlfahrt 2020). Hence, the cancellation in sporting activities and restrictions on movements meant sponsors were not getting value for their money and would in the future decide to invest their money in other areas where the return on investment was less risky.

After the meltdown in the share value of most of the sporting teams during the COVID-19 pandemic, the sustainability of some of the broadcasting and sponsorship deals was being questioned. Over the years, the collective power of professional leagues to sell media rights had been spectacular, to say the least. The USA's NBA, for example, had TV deals worth over $24 billion over 9 years, and the English Premier League had deals with broadcasters worth $12 billion over 3 years (Hall 2020). The sporting shutdown meant that the leagues were unable to meet their commitments to broadcasters, hence limiting their ability to distribute income back to the clubs, impacting on their revenue, therefore, the decline in their share values. Investors shunned their stock because no games meant no TV deals, no match-day revenue, and no income for the sporting companies. The broadcasters were also left with gaps in programming and lost attractive content for advertisers due to the cancellation or postponement of sporting activities.

13.5 Conclusion and Recommendations

The research noted that no previous infectious disease outbreak, including SARS, the Spanish flu and Ebola negatively impacted the stock markets as the COVID-19 pandemic. The COVID-19 pandemic was an unprecedented example of a neglected risk that rocked public health systems and economies around the world. In particular, stock markets reacted negatively to the emergence and tumultuous spread of the disease. Measures put in place by governments around the world to try and contain and limit the spread of the contagion included travel restrictions, imposing lockdowns and banning of public gatherings. These measures, by default, were disruptive to the global value chains of different industries that had developed over the years. The disruptions further heightened levels of uncertainty on global stock markets leading to further declines in stock values. Three major events were noted to have significantly shocked the stock markets around the world during the COVID-19 pandemic. These were the lockdown in Wuhan, China, in January 2020, the lockdown in Lombardy, Italy, in February 2020 and the declaration of COVID-19 a pandemic by the WHO on 11 March 2020. Tourism-related stocks were some of the hardest hit in terms of a decline in value. This was because most of the measures put in place to control the spread of the virus were directly disruptive to the tourism value chain, for example, travel bans were instituted in over 96% of global destinations. Hotels were forced to shut down, flights were grounded, sporting events and professional leagues were suspended or cancelled, cruise ships were forced to suspend operations and popular resorts were deserted. This resulted in an unprecedented plummeting of stock prices of counters directly linked to this industry with some dropping by as much as 80% of their value over a period of just 2 weeks. The pandemic turned a very lucrative industry in terms of stock investments into one that was bankrupt and seeking government bailout packages to survive the shock in a period of just 2 months.

Governments around the world responded to the COVID-19-induced shock on stock markets by trying to reassure the markets through policy interventions and implementing multi-trillion dollar rescue packages for distressed industries. The markets responded well to these packages with some industries making steady recoveries. However, shares in the tourism sector largely remained in the negative, and recovery was subdued. This is because a great deal of uncertainty still remained as the pandemic continued. Hence, confidence in the sector could only return once the pandemic was under control, and restrictions on travelling removed. The full depth of impacts of falling share prices on the tourism industry due to COVID-19 can only be determined once the crisis is over. However, in the short term, the drop in share value will deprive tourism firms of the much needed revenue that they could have potentially used during both the impact and recovery phases of the crisis. Unlike during the SARS outbreak when the tourism-related shares values quickly recovered after the crisis and did not dent the long-term profitability of affected firms, the COVID-19 pandemic is likely to last longer and impact the long-term profitability of tourism ventures. Recovery of share values is also likely to take longer given the gradual opening up of economies and slow return of investor confidence in the sector. In the long term, if the crisis persists or takes too long to be contained, capital-intensive firms that survived the initial shock by reducing costs may start to struggle which may put governments under further pressure for support. Some players in the tourism industry may be forced to shut down permanently, resulting in the loss of employment as well as the skills and experience of their workers. As global stocks slowly recover from the impacts of COVID-19, tourism firms must take caution on the possibilities of second and third waves of infections that may yet again plunge stock prices.

Therefore, the chapter strongly recommends another wave of stimulus packages earmarked for the tourism industry, and its value chain once signs of the pandemic declining are visible. These packages need to be synchronised major state pronouncements such as the lifting of restrictions on travelling and mass gatherings so that their effects are rapid and pronounced.

References

Albulescu, C.T. (2020). *Coronavirus and financial volatility: 40 days of fasting and fear*. Retrieved from www.ssrn.com/sol3/papers.cfm?abstract_id=3550630. Accessed 15 April 2020.

Baker, S. R., Bloom, N., Davis, S. J., Kost, K., Sammon, M., & Viratyosin, T. (2020). *The unprecedented stock market reaction to COVID-19* (Working paper). Cambridge, MA: National Bureau of Economic Research.

Barro, R. J., Urisua, J. F., & Weng, J. (2020). *The coronavirus and the great influenza pandemic: Lessons from the 'spanish flu' for the coronavirus's potential effects on mortality and economic activity* (Working Paper 26866). Cambridge, MA: National Bureau of Economic Research.

Cboe. (2020). *VIX index historical data*. Retrieved from http://www.cboe.com/micro/vix/historical.aspx. Accessed 23 April 2020.

Chen, M. H., Jang, S., & Kim, W. G. (2007). The impact of the SARS outbreak on Taiwanese hotel stock performance: An event-study approach. *Hospital Management, 26*, 200–212. https://doi.org/10.1016/j.ijhm.2005.11.004.

Chev, R., & Marinč, M. (2018). Stock prices and geographic proximity of information: Evidence from the Ebola outbreak. *International Review of Financial Analysis, 56*, 153–166.

Chiang, T. C., Jeon, B. N., & Li, H. (2007). Dynamic correlation analysis of financial contagion: Evidence from Asian markets. *Journal of International Money and Finance, 26*, 1206–1228.

Cullinane, D. (2020). Airline and travel stocks feel the impact of coronavirus. *Small Caps*. Retrieved from https://smallcaps.com.au/airline-travel-stocks-asx-feel-impact-coronavirus/. Accessed 15 April 2020.

De Lisle, J. (2003). S.A.R.S., greater China, and the pathologies of globalisation and transition. *Orbis, 47*(4), 587–604.

Dube, K., Nhamo, G., & Chikodzi, D. (2020). COVID-19 cripples global restaurant and hospitality industry. *Current Issues in Tourism*. https://doi.org/10.1080/13683500.2020.1773416.

Economist. (2003). Painful side-effects. Retrieved from http://www.economist.com/agenda/displayStory.cfm?story_id=1747241. Accessed 15 April 2020.

Eichenbaum, M. S., Rebelo, S., & Trabandt, M. (2020). *The macroeconomics of epidemics* (Working Paper 26882). Cambridge, MA: National Bureau of Economic Research.

Events Council (2020). *Coronavirus statement from CEO Amy Calvert*. https://eventscouncil.org/coronavirus. Accessed 27 April 2020.

Fernandes, N. (2020). Economic effects of coronavirus outbreak (COVID-19) on the world economy. *SSNR*. https://ssrn.com/abstract=3557504. Accessed 24 April 2020.

Funakoshi, M., & Hartman, T. (2020). Mad March: How the stock markets is being hit by COVID-19. *World Economic Forum*. www.weforum.org/agenda/2020/03/stock-market-volatility-coronavirus/. Accessed 23 April 2020.

Gates, B. (2020). Responding to Covid-19 – A once-in-a-century pandemic? *The New England Journal of Medicine*. https://doi.org/10.1056/NEJMp2003762.

Gerding, F., Martin, T., Nagler, F. (2020). *The value of fiscal capacity in the face of a rare disaster*. https://ssrn.com/abstract=3572839. Accessed 24 April 2020.

Goldman, C.M. (2020). *Air, cruise and tourism stocks whipsawed in coronavirus wake.*. Retrieved from https://www.thestreet.com/.amp/investing/coronavirus-travel-stocks-cruise-airlines. Accessed 23 April 2020.

Gourinchas, P. O. (2020). Flattening the pandemic and recession curves. *Econfip Policy Briefs, 23*. www.econfip.org/policy-brief. Accessed 23 April 2020.

Griffith, R., Levell, P., Stroud, R. (2020). The impact of COVID-19 on share prices in the UK. *Briefing note*, UKRI. Retrieved from: https://www.ifs.org.uk/publications/14773

Hall, S. (2020). This is how COVID-19 is affecting the world of sports. *World Economic Forum*. Retrieved from: www.weforum.org/agenda/2020/04/sports-covid19-coronavirus-excersise-specators-media-coverage/. Accessed 24 April 2020.

Hanson, J. (2020). *Huge hotel job losses reported due to COVID-19*. Retrieved from: https://www.law360.com/health/articles/1256131/huge-hotel-job-losses-reported-due-to-covid-19. Accessed 15 April 2020.

Harbison, P. (2003). SARS hits hard. *Flight Safety Australia, 7*, 28–29.

Hassan, T. A., Hollander, S., Van Lent, L., & Tahoun, A. (2020). *Firm-level exposure to epidemic diseases: COVID-19, SARS, AND H1N1* (NBER working paper series). Cambridge, MA: National Bureau of Economic Research.

Hayward, R. (2018). Foreign exchange speculation: An event study. *International Journal of Financial Studies, 6*, 22. https://doi.org/10.3390/ijfs6010022.

Honigsbaum, M. (2013). Regulating the 1918–19 pandemic: Flu, stoicism and the Northcliffe press. *Medical history, 57*(2), 165–185.

Hsieh, S. F. (2013). Individual and institutional herding and the impact on stock returns: Evidence from Taiwan stock market. *International Review of Financial Analysis, 29*, 175–188.

Ivanov, S., Webster, C., Stoilova, E., & Slobodskoy, D. (2020). *Biosecurity, automation technologies and economic resilience of travel, tourism and hospitality companies*. Centre for Open Science. Retrieved from www.ideas.repec.org/p/osf/socarx/2hx6f.html. Accessed 23 April 2020.

Liu, H. Y., Manzoor, A., Wang, C. Y., Zhang, L., & Manzoor, Z. (2020). The COVID-19 Outbreak and Selected Countries Stock Markets Response. *International Journal of Environmental Research and Public Health, 17*, 2800. https://doi.org/10.3390/ijerph17082800.

Lloyd-Jones, A. R., Suzanne, R., Mellen, S. R., & Pierson, T. J. (2020). *HVS report assesses COVID-19's impact on hotel values*. Retrieved from: https://www.hotelmanagement.net/own/hvs-report-assesses-covid-19-impact-hotel-values#main-content. Accessed 23 April 2020.

Loh, E. (2006). The impact of SARS on the performance and risk profile of airline stocks. *International Journal of Transport Economics, xxxiii*(3), 401–422.

Markets Insider. (2020). *Sport related stocks*. Retrieved from: www.businessinsider.com/stocks. Accessed 23 April 2020.

Morris, D. (2020). *Coronavirus: Lessons from SARS for investors*. BNP Paribas Asset Management. https://investors-corner.bnpparibas-am.com/?s=morris. Accessed 14 April 2020.

Neufeld, D. (2020). The hardest hit companies of the COVID-19 downturn: The 'BEACH' stocks. *Visual Capitalist*. Retrieved from https://www.visualcapitalist.com/covid-19-downturn-beach-stocks/. Accessed 23 April 2020.

Nippani, S., & Washer, K. M. (2004). SARS: a non-event for affected countries' stock markets? *Applied Financial Economics, 14*(15), 1105–1110. https://doi.org/10.1080/0960310042000310579.

Owens, J. C. (2020). *Disney earnings preview: 'A pivotal quarter' has Disney doing battle with coronavirus fears*. Retrieved from https://www.marketwatch.com/story/disney-earnings-preview-a-pivotal-quarter-has-disneydoing-battle-with-coronavirus-fears-2020-02-01. Accessed 24 April 2020.

Oxford Economics (2018). *Global significance of business events*. Retrieved from https://insights.eventscouncil.org/Portals/0/OEEIC%20Global%20Meetings%20Significance%20%28FINAL%29%202018-11-09-2018.pdf. Accessed 28 April 2020.

Ozili, P. (2020). Covid-19 pandemic and economic crisis: The Nigerian experience and structural causes. *SSRN*. https://ssrn.com/abstract=3567419. Accessed 23 April 2020.

Ozili, P. K., & Arun, T. G. (2020). *Spillover of COVID-19: Impact on the Global Economy*. https://ssrn.com/abstract=3562570. Accessed 24 April 2020.

Pandey, A. (2020). Coronavirus sledgehammers global job markets. *Deutsche Welle*. Retrieved from https://www.dw.com/en/coronavirus-sledgehammers-global-job-markets/a-52905245. Accessed 20 March 2020.

Pine, P., & McKercher, B. (2004). The impact of SARS on Hong Kong's tourism industry. *International Journal of Contemporary Hospitality Management, 16*(2), 139–143.

Quartz. (2020). *Impact of the coronavirus (COVID-19) pandemic on the Formula One Group's stock as of March 2020*. Retrieved from http://www.statista.com/statistics/1104366/coronavirus-f1-group-stock/. Accessed 24 April 2020.

Ramelli, S., & Wagner, A. F. (2020). *Feverish stock price reactions to COVID-19* (SSRN Working Paper). SSRN. Retrieved from https://ssrn.com/abstract=3550274. Accessed 24 April 2020.

Reinhart, C. M. (2020). *This time truly is different. Project syndicate*. Retrieved from: www.project-syndicate.org/commentary/covid19-crisis-has-no-economic-precident-by-carmen-reinhart-2020-03. Accessed 30 March 2020.

Reuters Eikon. (2020). *Global market news*. Retrieved from: www.reuters.com/finance/markets. Accessed 23 April 2020.

Ru, H., Yang, E., & Zou, K. (2020). *What do we learn from SARS-CoV-1 to SARS-CoV-2: Evidence from global stock markets*. https://ssrn.com/abstract=3569330. Accessed 24 April 2020.

Schoenfeld, J. (2020). *The invisible risk: Pandemics and the financial markets*. https://ssrn.com/abstract=3567249. Accessed 24 April 2020.

Siu, A., & Wong, Y. C. R. (2004). Economic Impact of SARS: The Case of Hong Kong. *Asian Economic Papers* 3.1. The Earth Institute at Columbia University, New York.

Statista. (2020). Forecasted change in revenue from the travel and tourism industry due to the coronavirus (COVID-19) pandemic worldwide from 2019 to 2020 (in million U.S. dollars). *Statista*. Retrieved from http://www.statista.com/statistics/1103431/covid-19-revenue-travel-tourism-industry-region-forecast/. Accessed 23 April 2020.

The Robust Trader. (2020). *Top 20 biggest stock exchanges in the world – Largest stock exchanges by market capitalisation*. Retrieved from: www.therobusttrader.com/top-20-biggest-stock-exchanges-in-the-world-largest-stock-exchanges-by-market-capitalisation. Accessed 23 April 2020.

UNWTO (United Nations World Tourism Organisation). (2020). *Supporting jobs and economies through travel & tourism: A Call for Action to Mitigate the Socio-Economic Impact of COVID-19 and Accelerate Recovery*. Geneva: UNWTO.

Velde, F. (2020). "*The Economy and Policy in the Coronavirus Crisis to Date*," presentation at a panel session for the Brookings Papers on Economics Activity, 20 March. Retrieved from: https://www.brookings.edu/blog/up-front/2020/03/20/the-economyand policy-in-the-coronavirus-crisis-to-date/. Accessed 14 April 2020.

Vigna, P. (2020). Hospitality stocks are biggest losers in week of carnage. *Wall Street Journal*. Retrieved from:https://www.wsj.com/news/collection/coronavirus0312-256f2943?mod=article_inline. Accessed 23 April 2020.

Visual Capitalist. (2020). *Market capitalisation loss of selected companies in the travel and entertainment sectors due to the coronavirus pandemic (COVID-19) worldwide in 2020*. Retrieved from http://www.statista.com/statistics/1110330/loss-beach-stock-companies-coronavirus/. Accessed 23 April 2020.

Wohlfahrt, S. (2020). Football sponsorship: Rwanda's controversial strategy to boost tourism. *France 24*. Retrieved from www.france24.com/en/africa/20200204-football-sponsorship-rwanda-s-controversial-strategy-to-boost-tourism. Accessed 24 April 2020.

World Economic Forum (WEF). (2020). *Global risk report 2020*. www.weforum.org/reports/global-risk-report-2020/. Accessed 24 April 2020.

World Federation of Exchanges. (2020). *Statistics*. www.world-exchanges.org/our-work/statistics. Accessed 28 April 2020.

Yan, C. (2020). *COVID-19 outbreak and stock prices: Evidence from China*. https://ssrn.com/abstract=3579266. Accessed 24 April 2020.

Part IV
Philanthropy and Tourism Economic Stimulus Packages

Chapter 14
Tracking of Corporate, Philanthropic and Public Donations to Dislodge COVID-19

Abstract Those with eyes have witnessed, and those with ears have heard of the coronavirus disease 2019 (COVID-19). Visible and loud are the negative impacts of the pandemic globally, threatening the attainment of both the health and other sustainable development goals (SDGs). Hence, the call of many governments to have all the stakeholders on board did not fall on deaf ears. Drawing from the events study and public announcements methodology, this chapter mainly documents the contributions from corporates, philanthropists, civil society organisations and members of the public towards COVID-19 initiatives. The findings are that the world over, many stakeholders raised their hands to help in mighty ways, particularly after the World Health Organisation's declaration of COVID-19 as a global pandemic and escalating lockdowns. This was also the case in South Africa, a country documented as a case study, where the Solidarity Fund was established with billions of rands donated. For South Africa, it was a true embracing of the *ubuntu* (I am because you are) concept post the apartheid era. It is recommended that the model followed by corporate South Africa be kept alive to enable the nation to bounce back quicker and build resilience against similar future disasters. Pandemics, epidemics and other natural disasters remain the new global and national norm. As such, working together, stakeholders can do more with less. It is also recommended that in the future, donations should be scaled up earlier.

Keywords COVID-19 · CSR · Stakeholders · Philanthropy · Coronavirus · Solidarity Fund

14.1 Introduction

On 31 December 2019, the world woke up to the announcement of the World Health Organisation (WHO) that China had notified it of an outbreak of a novel coronavirus (Amodio et al. 2020; Backer et al. 2020). The WHO later (on 11 March 2020) declared the coronavirus disease 2019 (COVID-19) a global pandemic, setting the

G. Nhamo et al., *Counting the Cost of COVID-19 on the Global Tourism Industry*, https://doi.org/10.1007/978-3-030-56231-1_14

world into a spin of fear and panic (WHO 2020a; Nishiura et al. 2020; Li et al. 2020). Some governments went into action immediately, while others adopted a wait and see attitude, anticipating that COVID-19 would go away as what happened with other recent disease outbreaks. Whatever the response mechanism, governments and their social partners had to act.

In times of pandemic, such as what the world is witnessing with COVID-19, governments, individuals, corporates, philanthropists, civil society organisations (CSOs) and non-governmental organisations (NGOs) should come to the party. For corporate entities and philanthropists, it is not time to create future business opportunities. It is a time when human life supersedes future thriving business (Dobie et al. 2018). There is also a need to build strong partnerships from the public–private and/or private–private perspectives to do the job at hand (Nguyen et al. 2017) and eradicate unnecessary barriers to success (Nguyen et al. 2018). In their study on the COVID-19 response with a focus on the Chinese experience and implications for other nations, Liu et al. (2020) provide some good insights. The authors state:

> China's bond and strict containment measures have been proved (in practice) to significantly reduce the spread of the epidemic. This was obtained through the use of emergency control measures in the epidemic areas and the integration of resources from multiple systems, including business, community, technology, education, and transportation, across the country. (Liu et al. 2020: 1)

This chapter documents corporate, philanthropic, CSO, NGO and public engagements to dislodge COVID-19 globally, with considerable space dedicated to profiling what transpired in South Africa. The processes leading to early and significant donations into the COVID-19 Solidarity Fund established by the president of South Africa on 23 March 2020 remain a good model. The authors also investigate and documents donations and initiatives outside the Solidarity Fund by corporate South Africa, philanthropists and the general public. The findings will be of value in informing global citizens in terms of the nature of financial and cooperation mechanisms that emerged to fight COVID-19 and to be better prepared in the future.

14.2 Literature Survey

Hartz (2017) highlights that disaster philanthropy is at a critical point as global disasters keep increasing due to climate change and other anthropogenic triggers. The private sector is viewed as best positioned to apply agility, expertise and resources to mitigate the effects of these disasters as it reimagines the future of disaster philanthropy and corporate social responsibility (CSR) (Owusu-Kwateng et al. 2017). India became the first country to integrate voluntary and mandatory aspects of CSR in the Companies Act 2013. Companies beyond a certain size are required to reserve and spend a certain percentage of their profits on CSR activities (Jain et al. 2020). The next sections document some of the emerging global donations and pledges towards fighting the COVID-19 pandemic outside South Africa.

14.2.1 Selected Government Initiatives

In the USA, for example, $100[1] million was made available by the Department of Labour for COVID-19 Dislocated Worker Grants (US Department of Labour 2020). Similar arrangements to cushion workers and hard-hit economic sectors such as tourism are taking place the world over. In Indonesia, an estimated $981 billion was set aside as an economic stimulus package for the tourism sector (Statista 2020). In Rwanda, on 5 April 2020, the office of the prime minister announced that all cabinet members, permanent secretaries, heads of public institutions and other senior officers would forfeit 1 month's salary towards COVID-19 social protection (Rwanda Office of the Prime Minister 2020).

The UK probably emerged as the single biggest donor towards global efforts to fight COVID-19. Calderwood (2020) states that the country reached £544 million in donations. Of this, the larger portion went to the Coalition for Epidemic Preparedness Innovations (CEPI). CEPI finances and coordinates the development of vaccines against emerging infectious diseases, of which COVID-19 became the main focus in 2020. CEPI is a partnership between public, private, philanthropic and CSOs, and was launched in Davos in 2017 (CEPI 2020). The donations that trickled into CEPI coffers to fight COVID-19, as reported by CEPI, as of 6 April 2020 are indicated in Table 14.1.

In February 2020, CEPI had placed a call for support to the tune of $2 billion. This is the basket to which the countries and philanthropists were responding. CEPI identified five funding phases as follows: Phase I ($100 million for immediate needs); Phase II ($375 million by the end of March 2020); Phase III ($400 million by end of June 2020); Phase IV ($400 million by the end of September 2020); and Phase V $500–750 million in 2021 (CEPI 2020).

The COVID-19 funding support also became a focal point post-Brexit. Even in divorce times, the European Union approved a £50 billion ($62.5) scheme for coronavirus-hit firms in the UK (News24 Team 2020). This after the UK left the European Union at the end of January 2020. In other reported news, China donated 1000 ventilators to New York State. All these initiatives show global governments in solidarity during times of need.

[1] Unless stated, all currencies in US$ will be denoted by $.

Table 14.1 Donations into CEPI coffers for COVID-19 vaccine development

Country	Pledged and/or donated amount
Belgium	Investment of €5 million (US$5.5 million) made
Canada	Pledged CA$40 million (US$28.2 million)
Germany	Announced a €140 million (US$157 million) donation
Norway	Announced a NOK 2.2 billion (US$210 million) donation Announced a NOK 36 million ($3.6 million) donation
Novavax, Inc.	CEPI announced investments worth $4.4 million in partnering agreements with Novavax Inc. and the University of Oxford to rapidly develop vaccine candidates against COVID-19
University of Hong Kong	CEPI will be investing an initial US$620,000 in a partnering agreement with the University of Hong Kong to rapidly develop a vaccine candidate against COVID-19
United Kingdom	Announced a £210 million (US$270 million) donation Announced a £50 million ($62.5) donation Announced a £20 million ($25 million) donation Announced £30 million ($37.5) donation

Source: Authors, based on CEPI (2020)

14.2.2 *WHO COVID-19 Donations Platform*

To start the ball rolling, the WHO established the COVID-19 Solidarity Response Fund with amounts of $25, $50, $100, $500, $1000 and $2500 requested on its online platforms (WHO 2020b). The WHO made it clear that COVID-19 was an unprecedented pandemic, and everybody had to come on board to assist. The WHO spelt out its mandate and the aim of the Solidarity Response Fund, which was to support nations in preventing, detecting and responding to COVID-19. All countries had to be supported, with a special focus on countries with weaker health delivery systems such as in Africa and other less developed regions of the world.

Other ways to give towards the Solidarity Response Fund, including tax deductibility and corporate and foundation options, were presented. Through a partnership with the Swiss Philanthropy Foundation, which is part of the Transnational Giving Europe network, donors in Europe and Canada were able to give via electronic transfers (WHO 2020b). Donations were made to the Facebook Fundraiser or through Google. In the USA, people donated via mobile network providers by texting WHO25 to 501501 for a $25 donation and WHO50 to 501501 for a $50 donation. In line with transparency and good governance, updates of donated amounts were posted on the WHO website (https://covid19responsefund.org/). As of 5 April 2020, a total of $129,292,984 had been raised from over 223,000 donors. As highlighted earlier, the money was to finance the WHO's 2019 Novel Coronavirus (2019-nCoV) Strategic Preparedness and Response Plan (WHO 2020c), a draft strategy, which has been in place since 3 February 2020 and focuses on three major intervention areas (Fig. 14.1).

A number of elements make up the three pillars of the Strategic Preparedness and Response Plan (WHO 2020c). For example, under international coordination and

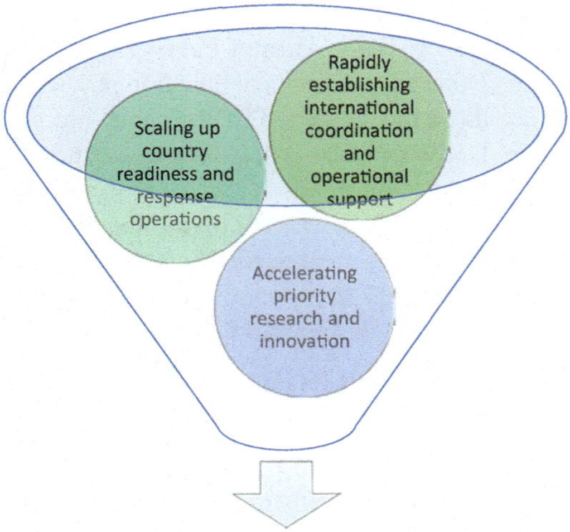

Fig. 14.1 WHO COVID-19 response strategy pillars. (Source: Authors, based on WHO (2020c))

operation support are covered matters dealing with partner coordination, epidemiological analysis and forecasting, risk communication and managing the infodemic, laboratory and diagnostics, technical expertise and guidance, pandemic supply chain coordination and travel and trade. The scaling-up country readiness and response pillar focuses on country-level coordination, risk communication and community engagement, surveillance, points of entry, rapid response teams, national laboratory system, infection prevention and control, case management and continuity of essential services, as well as logistics, procurement and supply management. The third pillar on accelerating priority research and innovation looks at enhancing global coordination of all relevant stakeholders, supporting a clear and transparent global research and innovation priority setting process and building common platforms for standardised processes, protocols and tools. Sharing of specimens, data and information also falls under this third pillar (WHO 2020c).

14.2.3 Big Corporate and Philanthropic Donations and Partnerships

On 29 March 2020, the US Department of Health and Human Services (2020) received donations of 30 million doses of hydroxychloroquine sulphate from Sandoz, the Novartis generics and biosimilar division. There were also a million doses of chloroquine phosphate from Bayer Pharmaceuticals. These were for potential use in treating COVID-19 patients or for use in clinical trials.

Nestlé (2020) announced a partnership to fight COVID-19 together with the International Federation of the Red Cross and Red Crescent Societies (IFRC) on 26 March 2020. The choice of the IFRC was based on its strength in dealing with similar disasters in the past and trust, as well as its global reach of more than a century.

Box 14.1: Four Ways of Partnership Between the IFRC and Nestlé

- Donate food, medical nutrition products and bottled water to bring relief to those most affected by the pandemic.
- Deploy available logistics capacities from the out-of-home business to support the needs of the IFRC in various countries.
- Donate CHF10 ($10.6) million for immediate deployment in countries where it is most needed.
- Match 1:1 any donations to the Red Cross and Red Crescent National Societies or Federation made by its employees.

Source: Authors, based on Nestlé (2020)

From the emergency response, the IFRC was to provide immediate help to strengthen healthcare systems, where they were needed the most. The four ways in which the partnership worked are presented in Box 14.1. For Nestlé, the time had come for the company to demonstrate its values. Given the company's presence and experience in 187 countries, the initiative to fight COVID-19 could not have come at a better time.

Biz Community (2020) reported that the Alibaba Foundation and the Jack Ma Foundation were to donate medical supplies to Africa to fight COVID-19. The donations included 20,000 testing kits, about 100,000 masks and 1000 protective suits to the 54 African countries. This donation brought the total to about 1.1 million testing kits, 6 million medical masks and 60,000 protective suits. The donated goods were to be delivered to Addis Ababa, Ethiopia. Ethiopian Prime Minister, Abiy Ahmed Ali was to take charge of the distribution logistics to nations across the continent. The foundations made additional donations that included working with African doctors online to train them on COVID-19.

On 28 February 2020, Bill Gates wrote a piece on how to respond to COVID-19 (Gates 2020). In Gates' view, crises brought two equally important responses, namely, the call to solve it immediately and the call to keep it from recurring. Such was the challenge COVID-19 brought to global leadership. Gates brought another dimension – that The Bill & Melinda Gates Foundation had invested significant resources in the past to help the world prepare for scenarios such as the COVID-19 pandemic. However, with the swift manner in which the pandemic was moving, Gates suggested that acting early to assist African and South Asian countries to be ready would save lives. This had further potential to slow the global circulation of COVID-19. To this end, a significant portion of the announced $100 million from the Bill & Melinda Gates Foundation was targeted at these countries. Bill Gates concluded his write-up by indicating that:

> Governments and industry will need to come to an agreement. During a pandemic, vaccines and antivirals won't simply be sold to the highest bidder. They'll be available and affordable for people who are at the heart of the outbreak and in greatest need. Not only is this the right thing to do, it's also the right strategy for short-circuiting transmission and preventing future pandemics. (Gates 2020)

The Bill & Melinda Gates Foundation made further pledges towards the production of the COVID-19 vaccines by pledging $1.6 billion (Alpert 2020). This was, however, amid controversy surrounding the billionaire as some sectors of the global population believed the Foundation had hidden agendas to further benefit from COVID-19 vaccines breakthroughs. These allegations were strongly refused by the Foundation (Kasprak 2020).

On 24 March 2020 the Centres for Disease Control (CDC) Foundation's Facebook Fundraiser announced a partnership where Facebook would double Facebook's donations up to $10 million (Heyns 2020). The donations would support response measures in deploying emergency staff to the USA's public health agencies and funding medical supplies. The donations would further be used to increase laboratory capacity and provide support to vulnerable communities. Another $2 million match for the CDC Foundation was announced on 19 March by Lysol (Croft 2020). Lysol donation was aimed at supporting schools, students and children.

14.2.4 Donations from the Sporting and Entertainment Fraternities

As the world continued battling COVID-19, individuals (including sporting personalities) started coming to the table. From the tennis arena, Roger Federer, Rafael Nadal and Novak Djokovic ignited the donations party. Federer donated US$1.02 million to the vulnerable in Switzerland (Razon 2020). Nadal and Djokovic joined the donation line. Nadal teamed up with Spanish basketball player Pau Gasol to launch the €11 million campaign in Spain (Tandon 2020). The fund was aimed at assisting over 1.35 million people affected by COVID-19 at the time. Spain remained the second highest affected country in Europe. Nadal also donated €1 million to Serbia to purchase hospital supplies and equipment, which included ventilators to top up Djokovic's €1 million donation. Djokovic's donation was not the first, as it followed a similar incident when Serbia was hit by floods and his foundation mobilised resources. Table 14.2 shows some of the donations from the sporting family, mainly aimed at their home countries and organisations. The list is not exhaustive, as many more were coming in at the time this chapter went into production.

Footwear News (2020) reported several personalities in the entertainment industry making COVID-19 donations. These were mixed, both in kind and cash. Through the reality star BStrong's disaster relief initiative in partnership with Global Empowerment Mission, Bethenny Frankel mobilised international businesses to deliver 500,000 medical masks and kits to frontline healthcare workers. The star further initiated a project to deliver a million full-body protective suits to the New York Health Department. Ciara and Russell Wilson donated one million meals to the Seattle Food Lifeline and urged their followers to donate to their local food banks. The next section details the methodological underpinnings of this chapter.

Table 14.2 Half a million+ mark cash donations from the sporting fraternity

Sports personality	Amount donated and possible use as applicable
Atletico Madrid Football Club	70% pay cut
Barcelona Football Club	70% pay cut to the squad donated to club employees to continue to be paid during COVID-19
British Airways	50% pay cut for 4500 pilots in April and May
Novak Djokovic	€1 million
England's contracted male cricketers	£500,000
Roger Federer	$1.02 million
Chiara Ferragni	€4 million to build a new intensive care unit at the San Raffaele Hospital in Milan, Italy
Rudy Gobert	$500,000 across three organisations
Pep Guardiola	€1 million
Zlatan Ibrahimović	Launched a GoFundMe campaign to benefit Humanitas Research Hospital in Italy with $1.07 million
Jay-Z	$2 million from the Shawn Carter Foundation, along with Rihanna's Clara Lionel Foundation ($1 million each) for marginalised communities
Robert and Anna Lewandowski	€1 million
Blake Lively	$1 million to be split between Feeding America and Food Banks Canada
Bruno Mars	$1 million to MGM employees displaced by the coronavirus
Cristiano Ronaldo and Jorge Mendes	£1 million as they finance three new intensive care units at hospitals across Portugal
Lionel Messi	€1 million
Bayern Munich, Borussia Dortmund, RB Leipzig and Bayer Leverkusen	€20 million for the neediest clubs
Rafael Nadal	€1 million
Neymar da Silva Santos Júnior	$1 million
Pink	$1 million with $500,000 to the Temple University Hospital Fund in Philadelphia and $500,000 to the City of Los Angeles Mayor's Emergency COVID-19 Crisis Fund
Rihanna's Clara Lionel Foundation	$5 million
Spanish Football Association	€500 million bailout to La Liga first and second division clubs
Cristiano Ronaldo	€1 million
Xavi	€1 million to San Raffaele Hospital in Milan, Italy

Source: Authors, based on Footwear News (2020), Razon (2020), Amitai Warehouse (2020) and Sky News (2020)

14.3 Materials and Methods

This chapter has as an objective to document corporate, philanthropic, CSO, NGO and public donations and pledges to dislodge COVID-19 globally and in South Africa specifically. To generate the relevant data and information, public and other announcements were followed – a methodology triggered by events (Bimha and Nhamo 2017). The methodology further involved document and critical discourse analysis as advised by Antwi-Agyei et al. (2018). Among the major events followed were announcements regarding big donations by foundations such as the Bill & Melinda Gates Foundation and the CDC Foundation, as well as corporate announcements like those from Facebook and Nestlé. There were also announcements made by the WHO, and the sports and entertainment industry fraternity. Further announcements came from tennis stars, including Novak Djokovic, Roger Federer and Rafael Nadal. Other announcements came from the football family, among them Pep Guardiola, Lionel Messi, Neymar da Silva Santos Júnior, Bayern Munich, Borussia Dortmund, RB Leipzig, Bayer Leverkusen and the Spanish Football Association. Announcements from the entertainment industry included those from Jay-Z, Blake Lively, Bruno Mars and Rihanna's Clara Lionel Foundation.

In South Africa, identified donors included the Motsepe Foundation, Momentum Metropolitan, Telkom, government, Rupert family and Remgro Limited, Nicky and Jonathan Oppenheimer, banks, MiWay, Liberty Mutual, OUTsurance, Naspers, Woolworths, and Mary Oppenheimer and her daughters. The lists are not in any way exhaustive. Close tracking was also made of donations coming into the South African Solidarity Relief Fund, including individual contributions. Tracking of announcements was done on a regular basis in order to keep track and not develop a backlog. The Geographical Information System was used to trace the work of the Gift of the Givers Foundation (GGF) when they opened drive-through and other test centres across the country.

Some of the websites regularly followed included those with daily global and local COVID-19 updates such as News24 (https://specialprojects.news24.com/coronavirus/). Other websites tracked were those of corporate, philanthropic and not-for-profit organisations (NPOs). This was done to minimise the risk of fake news and announcements, as there have been plenty during COVID-19. In the next section, the data are presented, and key findings are discussed, with a focus on South Africa's COVID-19 donations story.

14.4 Presentation of Data and Discussion of Findings

This section has several subsections. These include a discussion on South Africa's COVID-19 Solidarity Response Fund, donations from corporate South Africa and philanthropists and big sacrifices from the COVID-19 Patient 42, Ubuntu Beds and ventilator initiatives. Other subsections focus on interventions from mobile phone network providers, insurance and the health sector; measures from the banking sector; and contributions from the NPO space. The last subsection documents so-called smaller but significant contributions. A summary of cash donations tracked by the time of going to press is shown in Table 14.3.

From Table 14.3, there is no doubt that huge sacrifices were made. Needless to say, every donating organisation and individual made a sacrifice. What is more

Table 14.3 Selected donations towards COVID-19 relief

Date	Donor	Amount (ZAR)
18 March 2020	Momentum Metropolitan	500,000,000
22 March 2020	Telkom	1,500,000,000
23 March 2020	Government Seed Money	15,000,000,000
23 March 2020	Rupert family and Remgro Limited	100,000,000,000
23 March 2020	Nicky and Jonathan Oppenheimer	100,000,000,000
23 March 2020	Huawei	100,000,000
23 March 2020	South Africa-China Economic and Trade Association	200,000,000
24 March 2020	Shoprite	10,200,000,000
25 March 2020	Ithuba National Lottery	5,000,000,000
25 March 2020	MiWay	2,500,000,000
26 March 2020	Liberty Mutual	7,000,000,000
27 March 2020	OUTsurance	10,200,000,000
28 March 2020	Motsepe family and associated businesses	100,000,000,000
30 March 2020	Naspers	150,000,000,000
31 Mar 2020	Woolworth	50,000,000
01 April 2020	Mary Oppenheimer and daughters	100,000,000,000
09 April 2020	Vodacom	4,000,000,000
15 April 2020	The Douw Steyn Family Trust	32,000,000,000
20 April 2020	Dis-Chem	200,000,000
04 May 2020	Exxaro	2,345,000,000
08 May 2020	Economic Freedom Fighters (EFF)	600,000,000
13 May 2020	ELMA Group of Foundations	25,000,000,000
09 June 2020	The Hasso Plattner Foundation	10,000,000,000
Total		676,395,000,000

Source: Authors

important is that stakeholders came together to assist the country and its residents in ways they were able to do. This claim will become clearer in the next subsections detailing how the donations played out.

14.4.1 COVID-19 Solidarity Response Fund

The COVID-19 Solidarity Fund, interchanged with the COVID-19 Solidarity Response Fund (SRF), was established on 23 March 2020 when South African President Cyril Ramaphosa announced the escalation of measures to combat the COVID-19 pandemic (Republic of South Africa 2020a). The escalated measures entailed a country-wide lockdown that was to start on Thursday 26 March 2020. From the proclamation, it emerged that the government had provided seed funding to

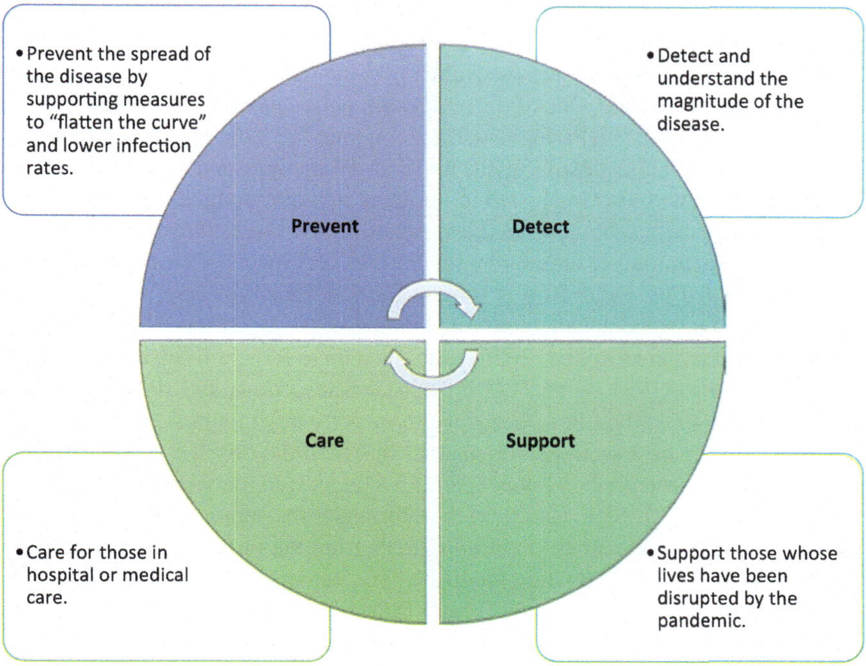

Fig. 14.2 SRF's four key aims. (Source: Authors, based on SRF (2020))

the SRF amounting to R150 million.[2] The president further announced that some pledges had been received and these ran into billions of rand. He assured the nation that the SRF would be run professionally with the openness and accountability it deserves. The SRF set a target of R4 billion and as of 8 June 2020, it had received R2.61 billion from companies, foundations, individuals and payroll giving (SRF 2020).

The SRF has four interlinked aims, as depicted in Fig. 14.2. The fund is independently administered (with an independent board) and works with other selected initiatives to achieve set objectives of fighting the spread and negative impacts of COVID-19 (SRF 2020). The SRF is supported by donations from individuals and organisations. The donations are to be secured and tax deductible. Every activity and every intervention are to be reported in a transparent manner, with the SRF administration being done through professional managers and organisations in the private sector. To this end, the president announced the appointed of Gloria Serobe as the chairperson of the fund on 18 March 2020 (Republic of South Africa 2020a). Overall, three sets of potential donors were identified (individuals, corporates and foundations), and all have dedicated email lines (SRF 2020). Donations can be made in cash and/or in kind through EFT, debit card, credit card, PayPal and SnapScan. The SRF is administered by Old Mutual. Auditing firm EY and law firm

[2] Exchange rate was \$1 = ZAR 18.20 as of 2 April 2020.

Edward Nathan Sonnenbergs (ENSafrica) developed a framework for determining how and where donations received are to be spent (Buthelezi 2020a).

In less than a fortnight, issues were already being raised regarding transparency in the operations of the SRF. One of the opposition political parties with the third highest number of members in Parliament, the Economic Freedom Fighters (EFF), asked government to disclose details of the COVID-19 pledges and donations (Mtshali 2020). The EFF was also interested in knowing how the donations would be used. The party was worried about big businesses getting into the picture as their donations could still be exploitative. Judging by the business of Business Partners Limited, an entity established by Johan Rupert (one of the R1 billion donors), the EEF's fears may have been justified. On their website landing page, the business entity highlights that Business Partners Limited (2020) had issued up to R19.5 billion in loans to business owners since 1981, with 71,000 business transactions and financing ranging from R500,000 to R50 million. Certainly, these were huge figures. However, as will be detailed in the next sections, the manner in which the donated money was to be used remained transparent and nonexploitative (at least in the view of the authors).

On 4 April 2020, SRF Chairperson Gloria Serobe announced that corporate South Africa (large and small) and individuals from the public had donated generously towards the fund. The individuals had donated amounts ranging from R1000 to R1 million, with the total fund standing at more than R2 billion (Jele 2020). Given that there were amounts from government seed money (R150 million), one corporate (R500 million) and one philanthropist (R1 billion), by implication, the individuals and small businesses had donated about R350 million. Ms. Serobe further revealed that part of the money had already been used to buy five million medical masks for frontline health workers, including doctors, nurses and community health workers. She also revealed that the future use of the SRF would focus on three main issues regarding fast-tracking and accelerating critical interventions, namely, health response, humanitarian effort and solidarity campaign (Box 14.2).

Box 14.2: SRF Proposed Future Uses

- Health response: Direct support for the healthcare system, augmenting both the private and government healthcare sectors to ensure that they have what they need, as well as enabling emergency supplies.
- Humanitarian effort: Enabling resilience augmenting the efforts of government and business to provide various forms of humanitarian aid and support to the most vulnerable households and communities to enhance their ability to cope – through sustained access to food, care and other interventions to alleviate potential loss of income.
- Solidarity campaign: Mobilising citizens in unified action to flatten the curve and manage the pandemic and its impacts on households and communities.

Source: Authors, based on Jele (2020)

There was also an announcement about the finalisation of the independent board, with Nomkhita Nqweni, former Absa CEO of Wealth Investment Management, appointed as the interim CEO of the SRF (Jele 2020). These revelations are critical in the interest of transparency and good corporate governance regarding the SRF, as demanded by stakeholders like the EFF. Accountability and transparency also remain key in attracting more donors.

14.4.2 Donations from Corporate South Africa and Philanthropists

Business Unity South Africa (BUSA) mobilised its constituency in response to COVID-19. There was a basket of interventions, some of which will not be mentioned in this paper given the limited space. What makes the initiatives great is that they came in the middle of Moody's sovereign downgrade of South Africa to junk status (Pityana 2020). Pledges and donations amounting to billions of rand were unprecedented. Several big hitters came to the fore and details are considered later in this chapter.

After a week of its existence, the SRF immediately made R100 million available for the purchase of critical medical supplies to protect health workers at the coalface of the epidemic (Buthelezi 2020a). Of the R100 million, R52 million was for 5 million surgical masks. Several big announcements were also made towards the SRF. These included R1 billion from Mary Oppenheimer and her daughters that was announced on 1 April 2020 (Omarjee 2020). In a statement following the donation, Mary Oppenheimer indicated that she and her daughters had settled on the SRF because they believed that it was the best vehicle to support the humanitarian needs of everyone in the country. The money was to be used for basic needs, food, medicine, and general care and to fight gender abuse. The same applied to a R500 million donation from Naspers (Omarjee 2020).

The donated funds fell into several categories as highlighted earlier. This section tracks some of the donations to see how the funds were allocated and used for COVID-19 interventions. The donations covered both cash and in kind (time, human resources and goods).

The donations are from those the authors tracked from the early public and other announcements. Not all the donated funds went into the SRF coffers, as other organisations announced their disbursement vehicles to use the money (Omarjee 2020). For example, Nicky and Jonathan Oppenheimer's R1 billion support to the small, medium and microenterprises (SMMEs) was administered through the South African Future Trust (SAFT). The SAFT's mandate was to extend direct financial benefits to supporting SMME employees at risk of losing their jobs or suffering loss of income due to COVID-19 (SAFT 2020). This trust would transfer money directly to deserving and participating SMMEs through interest-free loans to employees, with the companies carrying the liability (Omarjee 2020).

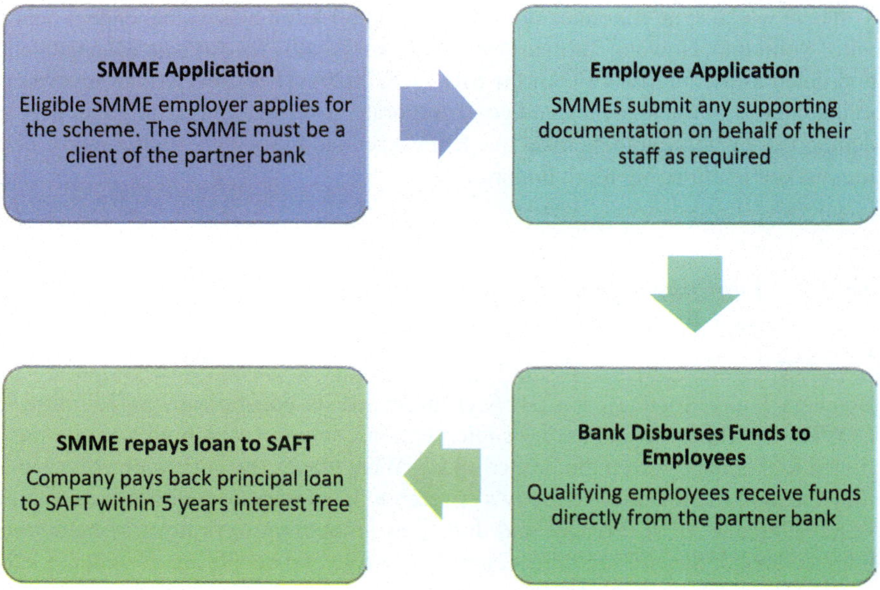

Fig. 14.3 SAFT funds disbursement mechanism. (Source: Authors, based on SAFT (2020))

The SAFT partnered with major commercial banks such as Absa, FirstRand Bank, Nedbank and Standard Bank for disbursements. However, the SMMEs had to first register with the banks starting Friday 3 April 2020, providing a list of employees "at risk" due to COVID-19 (SAFT 2020). The SAFT disbursement mechanism is highlighted in Fig. 14.3. Among the information required to access the funds are the company registration number, pay-as-you-earn (PAYE) number, valid income tax number registered with the South African Revenue Services and employment status (being confirmed permanent employment status of an employee). In addition to the business having been adversely impacted by COVID-19, additional eligibility criteria for the SMMEs include having an annual turnover of less than R25 million, trading for at least 24 months and being a sustainable business as of 29 February 2020.

What is of interest is the long-term projection of the mandate of the SAFT. The SAFT positioned itself with an ongoing role in accelerating economic growth in South Africa (SAFT 2020). To this end, any additional funds donated into the SAFT and any loans repaid by participating SMMEs would remain within the SAFT non-profit structure. The funds were to be used in ventures that prioritise initiatives focusing on employment creation. This was to be done for a period until all funds are disbursed.

The R1 billion from the Rupert family and Remgro Limited was announced on 23 March 2020 (Omarjee 2020). The money was also to assist SMMEs and to be administered by Business Partners Limited (BPL). The BPL is a leading risk finance company for SMMEs, co-owned by Johann Rupert and established in 1981. On 3 April 2020, the details on how the funds were to be disbursed emerged. Entrepreneurs – including formal sole proprietors and close corporations, companies and trusts – could get a cash payment of R25,000, and this was not to be repaid

(Business Insider SA 2020a). In addition, businesses could get the cash payment, plus a low-interest loan of up to R1 million on a 12-month repayment holiday. The newly established Sukuma Relief Fund handled other logistics, including applications. Should the SMME application have all the required documents and be approved, first payments of the relief funds were being received within 7 days, and later all the approved financial aid disbursed would be received over 3 months. Just 3 days after the announcement of logistics on applications, the Sukuma Relief Fund was oversubscribed with more than 10,000 applications (Business Tech 2020). The value of applications stood at R2.8 billion, almost three times over the available money. The application had to be stopped to allow evaluations in the next 7 days.

The R1 billion from the Motsepe family and associated businesses was announced on 28 March 2020 (Motsepe Foundation 2020a). The companies and organisations associated with the donation included the Motsepe Foundation, Sanlam, African Rainbow Capital and African Rainbow Minerals (Omarjee 2020). Part of the money was to be immediately available to purchase sanitisers, disinfectants, personal protective equipment (PPE) and other equipment. It was also to be used for any other resources needed to appropriately deal with COVID-19 (Motsepe Foundation 2020a), including the provision of water to poor rural and urban communities. This was to be facilitated by buying JoJo water tanks, drilling and equipping boreholes and building other sanitary facilities. Other projects sponsored with the medium to long-term horizons included building additional classrooms, computer centres and laboratories across the country. True to the promise, on 7 April 2020, the Motsepe Foundation donated 200,000 surgical masks to the national Department of Health (News24 Team 2020). These were to last about 8 weeks. In an earlier move, the Motsepe Foundation (2020b) donated R5 million to the University of Cape Town for its multilayered response to fight COVID-19. From the R1 billion donations, the Motsepe Foundation (2020a) lined up additional entities with which it would work (Box 14.3).

Box 14.3: Additional Partners for the R1 Billion Fund

- Traditional leaders, kings, queens and their communities that we have been working with for the past 20 years
- 34 religious and faith-based organisations that participate in the annual Motsepe Foundation National Day of Prayer
- National, provincial and local government authorities
- Trade union and other worker representative organisations
- NGOs and other local community representative organisations
- Sport organisations and entities
- Local, provincial and national business and professional organisations
- Black and white farmers, and their representative organisations
- Other organisations or structures that can assist or partner with us in dealing with the current coronavirus pandemic

Source: Authors, based on Motsepe Foundation (2020a)

During the announcement of the Motsepe family and associates' donation, the Sanlam CEO indicated that his organisation had a good history of putting people, its clients and the country first. This was reflected in the company's motto "Doing well, by doing Good" (Motsepe Foundation 2020a). Hence the donation gesture would go a long way in building sustainable societies well after COVID-19. Africa Rainbow Capital added that with the COVID-19 pandemic, South Africa was entering unchartered waters. Hence there was a need to demonstrate leadership in dealing with the disease decisively. The entire country needed to come together and fight in one accord. Only then would the pandemic be dissipated.

Naspers donated R1.5 billion towards the COVID-19 fight (Naspers 2020). However, as highlighted earlier, R500 million went straight into the SRF. The remaining R1 billion was reserve to buy PPE and other medical necessities and supplies from China. Naspers went into partnership with the Chinese government and Tencent, an entity in which it holds a 31% stake. The Group CEO highlighted that the COVID-19 pandemic demanded drastic and unconventional measures to conquer (Fin24 2020a). Naspers is 1 of the top 40 companies on the Johannesburg Stock Exchange where it has its primary listing. It also has secondary listing on the A2X Exchange in South Africa and has an ADR listing on the London Stock Exchange (Naspers 2020). The group was established in 1915 and grew into a global internet company and competes with large technology companies in the world. In South Africa, Naspers has entities including Takealot, Mr. D Food, Superbalist, OLX, Autotrader, Property24, PayU and Media24.

14.4.3 Big Sacrifices: COVID-19 Patient 42, Ubuntu Beds and Ventilators Initiatives

Although many big sacrifices were made across the corporate sector, through NPOs and by members of the public, here two initiatives are given further space. The initiatives are South Africa's COVID-19 Patient 42 and her Ubuntu Beds initiative, as well as the South Africa Emergency Ventilator Project (SAEVP).

The Ubuntu Beds (ubuntu.org) initiative was started by Ms. Kim Whitaker, who was the country's COVID-19 Patient 42 (Slabbert 2020). The idea came to her when she was in quarantine in South Africa after returning from Germany. The innovation centred around accommodating healthcare workers close to their workplaces as movements in Italy had resulted in stress and infections of health workers. Yes, hotel beds were empty in Italy as the tourism industry had come to a standstill. Patient 42 was the owner of two Whitaker's Once hotels, one in Cape Town and another in Johannesburg, with a combined total of 100 beds (50 beds each). The first 25 healthcare workers were accommodated for free in the Cape Town hotel, with the second set accommodated in the Johannesburg hotel. This move assisted in ensuring that there is a safe place for health workers, removing the COVID-19 risk from their families. By the time of completing this paper, the Ubuntu Beds initiative had 110 businesses offering assistance. In Italy, healthcare workers had self-isolated by sleeping in corners and cars. The concept of ubuntu in relation to COVID-19 and the Ubuntu Beds is presented in Box 14.4.

Box 14.4: Defining Ubuntu from the Ubuntu Beds Initiative
Globally, the COVID-19 pandemic has grown exponentially, and the risk to healthcare workers is real. In this time of need, we are committed to helping those who put their lives on the line for our community, in the spirit of ubuntu. Ubuntu is a Nguni Bantu term meaning "humanity". It is often translated as "I am because we are" or "humanity towards others", or in Xhosa "*umntu ngumntu ngabantu*" it means "the belief in a universal bond of sharing that connects all humanity". Hospitals are geared for medical emergencies and have jumped into action to provide a safe environment for their teams.

However, a great risk for healthcare workers globally, is the commute to and from the hospitals after their shifts and the fear of bringing the virus home to family and loved ones. This strategy began in China and is being adopted globally as the pandemic unfolds.

Source: Authors, based on Ubuntu Beds (2020)

From the Ubuntu Beds website, the aim of the initiative is stipulated as "to unite hospitality businesses that now stand empty with our healthcare workers who are fighting the virus on the front lines" (Ubuntu Beds 2020). Included in the list of Ubuntu Beds are hostels, hotels, bed-and-breakfasts and guesthouses hard hit by travel bans and the lockdown which can offer nurses and doctors' safe shelter and accommodation close to their workplaces. The hospitality staff are available to assist the healthcare staff with further training to handle the situation (Ubuntu Beds 2020).

The second initiative is the SAEVP. On 29 March 2020, News24's Cowan (2020a) reported that a company in the UK had refused to assist South Africa with drawings to reverse engineer a manual ventilator series, Penlon Nuffield 200. The manual ventilator looked promising to fight COVID-19 in the country and possibly in areas with limited access to electricity across the world. The company Penlon had produced the mechanical ventilator some 40 years back. A group led by businessman Justin Corbett had identified the ventilator with pans to reproduce it under the SAEVP. In response to the South African group's request for the drawings, Penlon's CEO responded that they were busy "marketing Nuffield ventilators and hence transferring all drawings etc. would be difficult" (Cowan 2020a). South Africa had only about 4000 ventilators in both the private and public health systems, and this put the country at risk.

In a follow-up article on 1 April 2020 (Cowan 2020b), it seems the momentum of the SAEVP had grown. There were offers to assist from a Canadian company, Thornhill Medical in Toronto, whose CEO is from South Africa. The company produces a US-military portable intensive care unit that includes a ventilator. It also emerged that a Dublin-based company, Medtronic, had availed its ventilator technical specifications for the PB 560 free online to companies wishing to reproduce it to supply hospitals for COVID-19 purposes. Possibly feeling the pressure from the SAEVP and its growing global partnership, Penlon made an about-turn (Cowan 2020c). On 4 April 2020, the company announced that it had revised its decision not to assist the SAEVP. Although the final outcome of the SAEVP is not known, the SAEVP deserves space in this chapter.

On 6 April 2020, the South African arms company, Denel, joined the ventilator manufacturing race. Denel announced that it was switching from making weapons to manufacturing ventilators (Business Insider SA 2020b). The report indicated that Project Sabela was being prioritised round the clock, with engineers from Denel Dynamics and Denel Aeronautics and a team of experts from Denel, Armscor, Eskom and the Council for Scientific and Industrial Research. At the time of completing this chapter, it could not be verified how far both ventilator manufacturing initiatives had progressed. South Africa was expecting to build up to 10,000 ventilators (De Wet 2020), with the capacity for a further 50,000 for export. All the ventilators were to be built with locally available material. Needless to say, both initiatives remained on everybody's radar in the country. On 13 May 2020, the South African President also announced that the USA government had donated 1000 ventilators (Republic of South Africa 2020b).

14.4.4 *Interventions from Mobile Network Providers, Insurance and Health Sector*

There were a number of interventions from the mobile phone network providers, insurance and the health sector. These initiatives are discussed briefly herein.

Telkom pledged R15 million to support healthcare workers (News24 Team 2020). Healthcare workers were identified beneficiaries given their critical role in the battle against COVID-19. Telkom's CEO made it clear that if the COVID-19 infection curve was to be "flattened", these workers had to be adequately equipped. The public was invited to contribute towards the cause from as little as R5 to R20, with three toll-free numbers provided for each of the amounts.

Insurance and the health sector remain among the worst affected. To this end, donations coming from this sector reflect huge sacrifices. As highlighted in the opening to this section, OUTsurance is one of the entities that committed R102 million to COVID-19 interventions (OUTsurance 2020). Part of this money went to the SRF. From its statement, OUTsurance indicated that it was clear that COVID-19 would have a devastating impact on the economy and all South Africans. In addition, OUTsurance presented a comprehensive campaign code-named the "Helping SA OUT" to provide relief for OUTsurance service providers who were impacted by the inability to work. Healthcare workers were to receive a share. The full list of support to beneficiaries included:

- Assistance to SMMEs claims service providers
- Assistance to clients
- Assistance to medical professionals by sourcing and distributing personal protective equipment
- Contribution to the president's Solidarity Response Fund

Further details regarding Assistance to SMMEs Claims Service Providers and clients also emerged. To qualify for the Assistance to SMMEs Claims Service Providers benefit, the service providers were to have a turnover below R50 million per year, and the rand value of work allocated by OUTsurance must have dropped by more than 50% during the period from April to June 2020 due to COVID-19. All affected service providers were to apply by sending a motivation to the OUTsurance dedicated email box covid19sphelp@out.co.za. Assistance to clients included reducing excesses for claims for the period from April to June 2020. This meant lowering amounts for business OUTsurance clients by 50% up to a maximum of R5000 and for personal lines clients by 30% up to a maximum of R5000. There would also be accelerated access to OUTbonuses vesting for 6 months from April to September 2020. This applied for missed premium payments from April to June 2020. The last intervention allowed clients to temporarily reduce premiums if they were sure their cars would not be used during the COVID-19 lockdown. This could be done through limited cover, to return to comprehensive cover afterwards.

Momentum Metropolitan (2020a) announced its first COVID-19 interventions on 6 March 2020. The company reaffirmed that its clients were to remain covered, protected and supported. This would be done in line with specific medical schemes and life insurance products they had. The medical scheme would cover COVID-19 tests and treatment. There was also provision for private hospitalisation. The group indicated that corporate employer groups under Momentum Health4Me insurance solutions would have their employees covered for treatment within the parameters of their benefits and there was 24-h free access to Hello Doctor for Momentum Health Medical Scheme members. The second set of COVID-19 interventions were announced on 18 March 2020 with Momentum Metropolitan committing R5 million to supplement resources to the public sector (Momentum Metropolitan 2020b).

Another insurance company, MiWay, announced it would give more than R25 million in premiums back to its clients (MiWay 2020). This was to be done in April 2020 in response to the COVID-19 lockdown. The company automatically applied the premium reduction to April debit orders. These were applicable to both its personal lines and business insurance clients. On average, there was a 10% reduction in car insurance premiums.

On 6 March 2020, the Discovery Health Medical Scheme extended benefits for COVID-19. The WHO Global Outbreak Benefit kicked in if a test confirmed a COVID-19 virus diagnosis and was immediately available on all Discovery Health Medical Scheme health plans (during the outbreak period) (Discovery 2020). The WHO Global Outbreak Benefit further covered out-of-hospital expenses. In partnership with mobile network provider, Vodacom and Discovery Health also provided facilities for online doctor consultations. Like all the other medical and insurance players profiled, Discovery played a role in educating its clients and those who visited the website by uploading COVID-19 educational material.

Liberty Mutual (2020) pledged $4 million (about R72 million) towards assisting communities through their philanthropy programme. Immediate grants of up to $10,000 (about R180,000) were made available to 450 community partners with which the group works. Priority was given to groups providing healthcare and/or those serving the homeless, elderly and other populations at high risk. There were further donations to 800 NPOs. Such were the initiatives coming from the sectors under consideration. The next section focuses on commitments from the banking sector.

14.4.5 Interventions from the Banking Sector

Apart from partnering with other donors to facilitate the administration of payments, banks put their own COVID-19 intervention measures in place. The early intervention from the Banking Association of South Africa was an announcement to waive Saswitch penalty fees at ATMs and the minimisation of branch closures (Van der Merwe 2020). These measures were to benefit over 11.3 million South African Social Security Agency grant beneficiaries during the lockdown. In addition, individual banks crafted their own COVID-19 intervention measures (De Villiers 2020), as shown in Table 14.4.

A follow-up was made to get further details from one of the banks, Nedbank. On 2 April 2020, Nedbank issued a document stipulating COVID-19 relief for individuals and small businesses (that would also cover the tourism sector). In terms of the bank's own arrangements, the communiqué covered the areas of payment holidays, monthly credit card repayments, Saswitch fees and penalty fees on early investment withdrawals (Nedbank 2020). It also covered arrangements from outside the bank, including the SAFT, the Debt Relief Finance Scheme, the Business Growth/Resilience Facility and Spaza Shop Grant Funding. There was also the Informal Trade Stipends, Tourism Relief Funding, Industrial Development Corporation (IDC) COVID-19 Essential Supplies Intervention, MCEP COVID-19 Programme, COVID-19 Employer/Employee Relief Scheme, COVID-19 Business Rescue Assistance and Training Layoff Scheme. Most of these schemes were from the government and would be administered through banks. Details of the total amounts available for some of the schemes are shown in Fig. 14.4.

Most of the government relief funds are hosted and administered by the National Department of Small Businesses. These include the Debt Relief Finance Scheme, the Business Growth and/or Resilience Facility and the Spaza Shop Grant Funding. The R200 million Tourism Relief Funding was hosted and administered by the National Department of Tourism, while the COVID-19 Essential Supplies Intervention and the MCEP COVID-19 Programme were both hosted and administered by the IDC, which is an agency in the National Department of

Table 14.4 Summary of initiatives from the major banks

Company/entity	Initiatives
Absa, FirstRand Bank, Nedbank and Standard Bank	Waive normal fees in managing the SAFT from the R1 billion from Nicky and Jonathan Oppenheimer, announced on 23 March 2020
Absa	Customers owing money were invited to take an up to 3-month repayment holiday, regardless of the kind of loan or how much money those customers earn Another option to reduce repayments was available Individual clients automatically qualified for the relief, but business and corporate clients were to contact the bank for a case-by-case arrangement on their debt
Standard Bank	The bank offered all its personal banking customers and business clients who earned less than R20 million a year, a 3-month debt holiday until the end of June 2020 Clients earning R7,500 or less and students automatically received the 3-month holiday, while other banking clients could apply for the relief by contacting the bank
First National Bank	The bank offered a debt repayment holiday of up to 3 months to its customers, but strict terms and conditions applied Clients had to contact the bank and have a letter from their employer to show loss of income due to the coronavirus The self-employed needed to submit financial statements
Nedbank	The bank offered clients individual solutions to cash flow challenges due to the coronavirus, including halting debt repayments (or part thereof) for a suitable period Alternatively, clients' loan periods could be extended or they could get more credit to manage short-term cash flow shortfalls Clients were encouraged to contact the bank telephonically to enquire if they qualified for the assistance
Capitec	The bank did not introduce any special initiatives to help consumers during the pandemic but said it was handling credit defaults on a case-by-case basis Clients who were facing a crisis as a result of the coronavirus were to urgently contact the bank
WesBank	The bank offered clients payment relief plans from 1 April 2020 to the end of June 2020 These included relief on financial services products such as its vehicle and asset financing, full maintenance leasing and vehicle stock funding

Source: Authors, based on De Villiers (2020) and Nedbank (2020)

Industry and Trade. The COVID-19 Employer/Employee Relief Scheme was administered by the National Disaster Benefit and UIF (Nedbank 2020). The Tourism Relief Fund announced a cap of R50,000 to deserving applicants and opened the process on 7 April 2020 until 30 May 2020 (South Africa National Department of Tourism 2020).

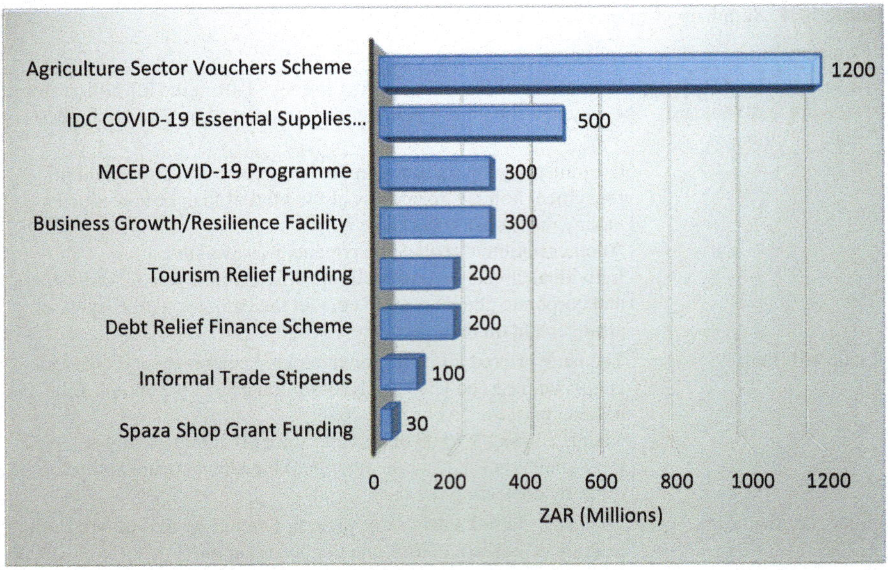

Fig. 14.4 Amounts available to selected COVID-19 relief schemes. (Source: Authors, based on Nedbank (2020) and Fin24 (2020c))

The largest relief funds were from the National Department of Agriculture, Land Reform and Rural Development. The R1.2 billion fund was distributed to financially distressed SMMEs farmers with an annual turnover of between R20,000 and R1 million (Fin24 2020c). The application processes opened on 8 April 2020. However, the scheme involved vouchers that would be exchanged for goods at designated retailers and cooperatives.

14.4.6 Contributions from the NPO Space

As is the case in most disaster situations, CSOs, NGOs and NPOs come in. A world-renowned NPO is the GGF, which has its head offices in South Africa. The organisation works in six primary focus areas: disaster response, hunger alleviation, water provision, healthcare, education and social upliftment. The GGF partnered with the government to look after more than 100 repatriated South Africans from Wuhan, China. Those repatriated were at a quarantine centre at The Ranch in Polokwane, Limpopo Province (GGF 2020). The GGF announced an Intervention Plan for COVID-19. The plan included an element of enhancing efficiency and identifying COVID-19 needs in real time. This meant a partnership had to be developed with Vula Mobile, which has 11,000 registered medical professionals on its platform nationwide. A sample of requests linked to COVID-19 across the country is shown in Box 14.5.

Box 14.5: Sample Requests for Equipment and Provisions for COVID-19

- N-95 protective masks.
- Three-tier lower cost masks for selected patients.
- Hand sanitisers and disinfectants.
- Sterile gloves and disposable sterile gowns.
- Goggles, visors, ventilators and related equipment.
- Chris Hani Baragwanath Hospital and Raheema Moosa Hospital requested the setting up of a tent, with beds, protective wear and all equipment necessary for testing and quarantine when necessary.
- The hospitals wanted the facility/tent outside existing hospital infrastructure but on the premises to reduce risk to other patients who may be immunocompromised.

Source: Authors, based on GGF (2020)

Another area that the GGF looked at closely was the provision of affordable testing. The GGF set up such a facility in Johannesburg, and this was extremely well-received. The patient paid R750, compared to R1400 in some labs (GGF 2020). The Dome in the Randburg area (Johannesburg) offered an undercover and parking facility at no cost to set up a testing site. Other testing sites eventually went up. The first delivery of 10,000 N-95 masks and 720 bottles of quality disinfectant was received by the time the first testing facility went up. Other deliveries that were in the queue included 50,000 three-tier masks. In consultation with the national Department of Health and its agencies, the GGF looked at bringing in an alternative testing kit for only R100. The kit had been approved by the US Food and Drug Administration and was being used in Europe at the time. The kit was said to be more practical and invaluable for taxi ranks. To realise much of what the GGF Plan outlined, a budget of R20 million was needed (GGF 2020).

As time progressed, the GGF established drive-through testing facilities across the country. The tests were done in partnership with Mullah Laboratories (Evans 2020). Later, Dis-Chem joined in opening drive-through testing stations. Some of the stations that were operational by the time of completing this paper are shown in Fig. 14.5.

The first set of sites that became operational on 30 March 2020 include Mullah Laboratories, Mayfair West, University of the Witwatersrand, Roshnee (all in the Johannesburg area) and the Ahmed Al Kadi Hospital in Durban (GGF 2020). More drive-through testing stations were opened at Midlands Medical Centre (Pietermaritzburg) and Momentum Metropolitan Holdings (Cape Town).

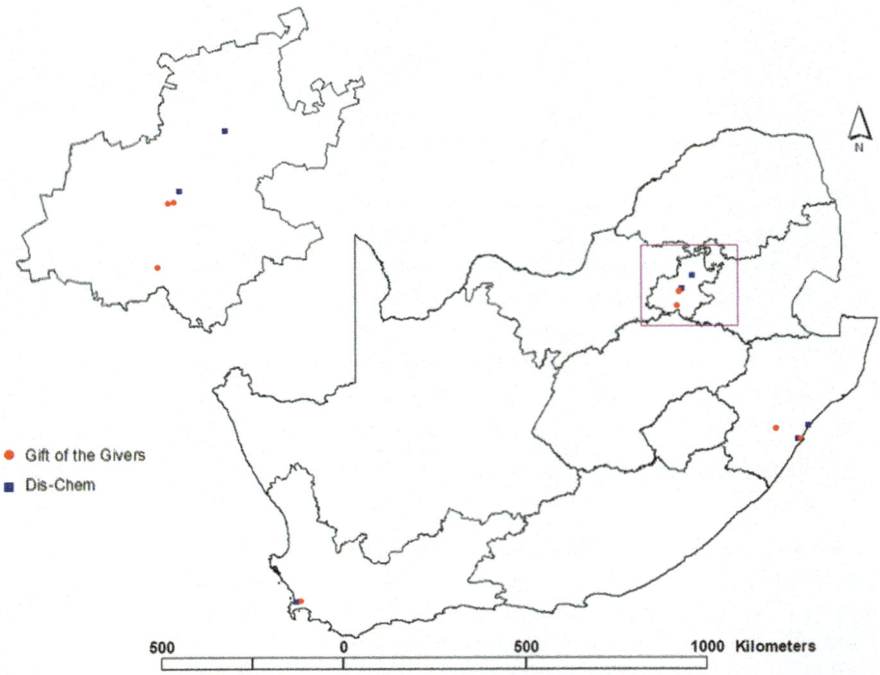

Fig. 14.5 COVID-19 drive-through testing facilities. (Source: Authors)

14.4.7 *Bringing to Light Smaller But Significant Contributions*

Often writers tend to focus on big donations, ignoring small but significant contributions that grow the overall funds. Majavu (2020) highlights Huawei's R1 million donation to the national Department of Health. Another donation to the tune of R2 million was made on the same day and function by the South Africa–China Economic and Trade Association. The funds were for testing and managing the COVID-19 outbreak. This donation was heralded by Health Minister Zweli Mkhize as an important milestone for cementing friendship and solidarity with Huawei and other Chinese businesses.

The National Film and Video Foundation – NFVF (2020) – an agency of the national Department of Arts and Culture, observed that the South African film and television industry had been severely impacted by COVID-19. The impact included halted productions and cancelled live events. As part of the state's programme that benefits young and emerging producers, the NFVF set aside R5 million for COVID-19 relief. A once-off cash injection of R500,000 was provided to 10 companies that were commissioned to assist during the epidemic.

The taxi vehicle finance provider SA Taxi announced payment assistance to taxi operators to honour their financial commitments during the lockdown (Magubane 2020). The taxi industry is huge and big business in South Africa, with millions of commuters from mainly previously disadvantaged communities using this means of transport to and from work, school, shopping venues and other places. The relief for April 2020 included monthly insurance premium payments. The two conditions stipulated to receive relief were that the account had to be in good standing and the vehicle had to be fitted with a fully functioning telematics device.

The South African Government News (SANEWS) reported on 30 March 2020 that Sasol had donated 400 litres of sanitiser solution (SANEWS 2020). The consignment was delivered to several areas surrounding Secunda (Mpumalanga province), including the Kinross Club, where a group of about 106 homeless people was accommodated. The sanitiser was also distributed to the Embalenhle Township community. The same report indicates that Spar Supermarket was going to donate foodstuffs to the homeless people for the duration of the lockdown. Other essentials such as soap, washing powder, bathroom necessities and blankets were donated to the homeless by Sabbers Supermarket and AME in Secunda.

Woolworths donated R500,000 from its loyalty programmes MySchool, MyVillage and MyPlanet (Fin24 2020b). The donation was a partnership with the GGF. The money was for the support of medical workers in terms of protective wear and equipment. Some of the equipment that the GGF was looking to buy included protective masks, hazmat suits, gowns, gloves and goggles for nurses and doctors. On 6 April 2020, it further emerged that Woolworths executives had agreed to a 30% salary cut (Buthelezi 2020b) in all 13 countries where it operates (11 in Africa as well as Australia and New Zealand).

In one of the early moves, University of Cape Town alumni donated R335,453 towards COVID-19 interventions (Motsepe Foundation 2020b). Among the 92 alumni donors was the Canadian alumnus, the Daniel Samuel Maseko Memorial Scholarship that contributed R100,000.

Responding to the announcement of the SRF, and the manner in which everyone was coming on board, BUSA President, Sipho Pitya sent praises. In his view, President Cyril Ramaphosa and key members of his Cabinet (especially Health Minister, Mkhize) had shown world-class leadership in propelling the national and government response to COVID-19 (Pityana 2020). Although limited in domestic resources, civil society had mobilised to the best of its ability through public education campaigns, volunteerism and healthcare. The general public was cited for their role in providing food and shelter and building good neighbourliness. Organised labour was keeping the workers informed of every step, particularly as COVID-19 was hitting the working class hard with retrenchments and salary cuts.

14.5 Conclusions

Documenting corporate, philanthropic and ordinary citizens' contributions to fighting COVID-19 through both financial and in-kind donations was not an easy task. Ideally, the authors needed some form of a gift and/or donations registry that could have gone to 1000 s of pages. To this end, it should be highlighted that many donations do not feature in this paper even though every donation towards the COVID-19 fight (big or small, cash or in-kind) matters. What emerged is that the WHO's declaration on 11 March 2020 elevating COVID-19 to a global pandemic ignited the scaling up of donations. Not only did the numbers of donors increase but also the magnitude of what was donated. Traditional philanthropists like Bill and Melinda Gates (through their foundation) came to the table with big donations. The sporting and entertainment industry featured greatly, with clubs and individuals donating huge sums of money and COVID-19-related equipment such as ventilators, medical masks and testing kits. Many in the sporting fraternity took huge pay cuts, up to as much as 70% sacrificed to workers and families of those running the clubs. This was touching given that the players and clubs themselves had lost billions in revenue from gate takings and advertising as games were banned during lockdowns.

From the South African perspective, all hands were on deck regarding donations. From big corporates to great philanthropists and members of the public, all sacrificed cash and in-kind giving. Within 2 weeks, the Solidarity Fund had received over R2 billion. This money was immediately put to use to source protective clothing and equipment for healthcare workers, including nurses, doctors and community healthcare workers. Other donations running over R3 billion were channelled to assist SMMEs and workers in general. The GGF worked with some donors to introduce cheaper testing facilities and drive-through testing centres across the country. Every economic sector came to the table to assist. However, two initiatives in particular are included in this paper, namely, Ubuntu Beds and the ventilator project. These remain prominent innovations for saving lives. Overall, all that donated both cash and in kind remain heroes of the time and the future.

The South African government designed a strategy to fight COVID-19 through a basket of fiscal and monetary policy instruments. Among key government interventions were tax breaks, the relaxation of certain regulations, reductions in the bank repo rate and financial contributions. These measures permitted access to both the Unemployment Insurance Fund and compensation in terms of the Compensation for Occupational Injuries and Diseases Act for employees in distress. The South African Reserve Bank (2020) lowered the repo rate by 100 basis points, putting an estimated R32 billion into the pockets of individuals to flow back into the economy.

On the downside, there seems to have been some delays in scaling up donations. Understandably so, because nobody anticipated COVID-19 spreading in the manner in which it did. By the time countries such as Italy, Spain and the USA were in great trouble, many realised COVID-19 was a disaster in its own class. The authors therefore recommend that in the future, the global leadership, corporates, philanthropists, NPOs and members of the public act quicker.

As the relief and response phases will come to an end at some point, the world will look towards whatever corner help may come from to enhance a quicker recovery and build a more pandemic-resilient world. Those with resources are called upon to extend their hearts and hands of mercy. Except for a few individuals who witnessed the 1918 influenza (Spanish flu) pandemic, the majority of the global citizenry witnessed a great pandemic for the first time.

References

Alpert, B. (2020). *Gates pledges $1.6 billion for vaccines. What that means for Covid-19 and other diseases*. Retrieved from https://www.barrons.com/articles/gates-foundation-vaccines%2D%2Dcovid-19-children-developing-countries-51591284623. Accessed 10th June 2020.

Amitai Warehouse. (2020). *REVEALED: Cristiano Ronaldo and Jorge Mendes' donation to help fight coronavirus is worth £1M as they finance THREE new intensive care units at hospitals across Portugal*. Retrieved from https://www.dailymail.co.uk/sport/sportsnews/article-8147489/Coronavirus-Cristiano-Ronaldo-Jorge-Mendes-donation-help-fight-COVID-19-worth-1million.html. Accessed 6th Apr 2020.

Amodio, E., Vitle, F., Cimino, L., Casuccio, A., & Tramuto, F. (2020). Outbreak of novel coronavirus (SARS-Cov-2): First evidences from international scientific literature and pending questions. *Health, 8*, 51. https://doi.org/10.3390/healthcare8010051.

Antwi-Agyei, P., Dougill, A. J., Agyekum, T. P., & Stringer, L. C. (2018). Alignment between nationally determined contributions and the sustainable development goals for West Africa. *Climate Policy, 18*(10), 1296–1312. https://doi.org/10.1080/14693062.2018.1431199.

Backer, J. A., Klinkenberg, D., & Wallinga, J. (2020). Incubation period of 2019 novel coronavirus (2019-nCoV) infections among travellers from Wuhan, China, 20–28 January 2020. *Euro Surveillance, 25*(5), 2000062. https://doi.org/10.2807/1560-7917.

Bimha, A., & Nhamo, G. (2017). Sustainable development, share price and carbon disclosure interactions: Evidence from South Africa's JSE 100 companies. *Sustainable Development, 25*, 400–413. https://doi.org/10.1002/sd.1670.

Biz Community. (2020). *Covid-19: Alibaba Foundation, Jack Ma Foundation to donate medical supplies to Africa*. Retrieved from https://www.bizcommunity.africa/Article/410/845/201709.html. Accessed 3rd Apr 2020.

Business Insider SA. (2020a). *Johann Rupert's R1bn: How entrepreneurs and businesses can get a slice of the money*. Retrieved from https://www.businessinsider.co.za/how-to-get-johann-rupert-money-small-business-2020-3. Accessed 5th Apr 2020.

Business Insider SA. (2020b). *Denel is making ventilators, and may turn its casspirs into ambulances*. Retrieved from https://www.businessinsider.co.za/south-african-arms-firm-denel-to-make-ventilators-in-fight-against-covid-19-2020-4. Accessed 6th Apr 2020.

Business Partners Limited. (2020). *We are business/partners*. Retrieved from https://finance.businesspartners.co.za/. Accessed 2nd Apr 2020.

Business Tech. (2020). *Johann Rupert's R1 billion coronavirus fund hits capacity in three days*. Retrieved from https://businesstech.co.za/news/business/387957/johann-ruperts-r1-billion-coronavirus-fund-hits-capacity-in-three-days/. Accessed 7th Apr 2020.

Buthelezi, L. (2020a). *SA's Solidarity Fund for coronavirus raises R500m in first week, aims to buy 5 million masks*. Retrieved from https://www.fin24.com/Economy/South-Africa/sas-solidarity-fund-for-coronavirus-raises-r500m-in-first-week-aims-to-buy-5-million-masks-20200330. Accessed 2nd Apr 2020.

Buthelezi, L. (2020b). *Woolworths announces executive pay cuts of up to 30% in bid to support staff*. Retrieved from https://www.fin24.com/Companies/Retail/woolworths-announces-executive-pay-cuts-of-up-to-30-in-bid-to-support-staff-20200406. Accessed 6th Apr 2020.

Calderwood, I. (2020). *The UK is now the single biggest contributor to the global hunt for a COVID-19 vaccine*. Retrieved from https://www.globalcitizen.org/en/content/uk-aid-biggest-contributer-covid-19-vaccine-cepi/. Accessed 6th Apr 2020.

CEPI (Coalition for Epidemic Preparedness Innovations). (2020). *Latest news*. Retrieved from https://cepi.net/news/. Accessed 6th Apr 2020.

Cowan, K. (2020a). *UK company seemingly won't help SA with ventilators marketed as simple solution for coronavirus*. Retrieved from https://www.news24.com/SouthAfrica/News/uk-company-seemingly-wont-help-sa-with-ventilators-marketed-as-simple-solution-for-coronavirus-20200329. Accessed 5th Apr 2020.

Cowan, K. (2020b). *Coronavirus crisis: Companies, engineers rally to SA's emergency ventilator cause.* Retrieved from https://www.news24.com/SouthAfrica/News/coronavirus-crisis-companies-engineers-rally-to-sas-emergency-ventilator-cause-20200401. Accessed 5th Apr 2020.

Cowan, K. (2020c). *Covid-19: UK company makes about-turn and agrees to help SA with ventilators.* Retrieved from https://www.news24.com/SouthAfrica/News/covid-19-uk-company-makes-about-turn-and-agrees-to-help-sa-with-ventilators-20200404. Accessed 5th Apr 2020.

Croft. L. (2020). *It takes all of us to combat COVID-19; thanks to Lysol for $2 million match.* Retrieved from https://www.cdcfoundation.org/blog/it-takes-all-us-combat-covid-19-thanks-lysol-2-million-match. Accessed 6th Apr 2020.

De Villiers, J. (2020). *All the help South African banks are offering consumers due to Covid-19.* Retrieved from https://www.businessinsider.co.za/all-the-help-south-african-banks-are-offering-consumers-during-covid-19-2020-4. Accessed 5th Apr 2020.

De Wet, P. (2020). *South Africa hopes to build 10 000 ventilators by end June – And up to 50 000 more if needed.* Retrieved from https://www.businessinsider.co.za/sas-national-ventilator-project-to-start-production-before-end-april-2020-4. Accessed 7th Apr 2020.

Discovery. (2020). *Discovery health medical scheme extending benefits for COVID-19.* Retrieved from https://www.discovery.co.za/corporate/covid19-extended-benefits-across-health-insurance. Accessed 6th Mar 2020.

Dobie, S., Schneider, J., Kesgin, M., & Lagiewski, R. (2018). Hotels as critical hubs for destination disaster resilience: An analysis of hotel corporations' CSR, activities supporting disaster relief and resilience. *Infrastructures, 46*(3), 1–19. https://doi.org/10.3390/infrastructures3040046.

Evans, J. (2020). *Coronavirus: Gift of the Givers' drive-thru testing sites up, but you must meet the criteria.* Retrieved at https://www.news24.com/SouthAfrica/News/coronavirus-gift-of-the-givers-drive-thru-testing-sites-up-but-you-must-meet-the-criteria-20200331. Accessed 31st Mar 2020.

Fin24. (2020a). *Naspers commits R1.5bn to fight against coronavirus.* Retrieved from https://www.fin24.com/Companies/ICT/naspers-commits-r15bn-to-fight-against-coronavirus-20200330. Accessed 2nd Apr 2020.

Fin24. (2020b). *Woolworths donates R500 000 to support medical workers.* Retrieved from https://www.fin24.com/Companies/Retail/woolworths-donates-r500-000-to-support-medical-workers-20200331. Accessed 31th Mar 2020.

Fin24. (2020c). *Government unveils R1.2bn fund to help small-scale farmers.* Retrieved from https://www.fin24.com/Companies/Agribusiness/govt-unveils-r12bn-fund-to-help-small-scale-farmers-20200406. Accessed 6th Apr 2020.

Footwear News. (2020). *Selena Gomez, Jay-Z, Taylor Swift + more celebs who are donating to the coronavirus fight.* Retrieved from https://footwearnews.com/2020/influencers/power-players/celebrity-donations-coronavirus-covid-19-live-updates-1202950323/. Accessed 5th Apr 2020.

Gates, B. (2020). *How to respond to COVID-19: And prepare for the next epidemic, too.* Retrieved from https://www.gatesnotes.com/Health/How-to-respond-to-COVID-19. Accessed 5th Apr 2020.

GGF (Gift of the Givers Foundation). (2020). *Intervention plan for Covid-19.* Retrieved from https://giftofthegivers.org/disaster-response/intervention-plan-for-covid-19/. Accessed 2nd Apr 2020.

Hartz, M. (2017). *The future of disaster philanthropy.* London/New York City: The Conference Board.

Heyns, T. (2020). *Facebook's $10 million match helps CDC Foundation fight coronavirus.* Retrieved from https://www.cdcfoundation.org/blog/facebooks-10-million-match-helps-cdc-foundation-fight-coronavirus. Retrieved on 5th Apr 2020.

Jain, A., Kansal, M., & Joshi, M. (2020). New development: Corporate philanthropy to mandatory corporate social responsibility (CSR) – A new law for India. *Public Money & Management.* https://doi.org/10.1080/09540962.2020.1714280.

Jele, N. (2020). *More than R2 billion donated to the Solidarity Fund in two weeks*. Retrieved from https://m.news24.com/SouthAfrica/News/more-than-r2-billion-donated-to-the-solidarity-fund-in-two-weeks-20200405. Accessed 5th Apr 2020.

Kasprak, A. (2020). *Are Bill Gates and the ID2020 Coalition using COVID-19 to build global surveillance state?* Retrieved from https://www.snopes.com/fact-check/bill-gates-id2020/. Accessed 10th June 2020.

Li, H., Chen, X., & Huang, H. (2020). The novel coronavirus outbreak: What can be learned from China in public reporting? *Global Health Research and Policy, 5*, 9. https://doi.org/10.1186/s41256-020-00140-9.

Liberty Mutual. (2020). *How we're responding to the coronavirus (COVID-19)*. Retrieved from https://www.libertymutualgroup.com/about-lm/corporate-information/how-were-responding-coronavirus-covid-19. Accessed 2nd Apr 2020.

Liu, W., Yue, X. G., & Tchounwou, P. B. (2020). Response to the COVID-19 epidemic: The Chinese experience and implications for other countries. *International Journal of Environmental Research and Public Health, 17*, 2304. https://doi.org/10.3390/ijerph17072304.

Magubane, K. (2020). *Coronavirus | Debt relief lifeline for taxi operators hit by lockdown*. Retrieved from https://www.fin24.com/Companies/TravelAndLeisure/coronavirus-debt-relief-lifeline-for-taxi-operators-hit-by-lockdown-20200326. Accessed 28rd Mar 2020.

Majavu, N. (2020). *Huawei hands over R1m to assist SA's Covid-19 fight*. Retrieved from https://www.sowetanlive.co.za/news/south-africa/2020-03-23-huawei-hands-over-r1m-to-assist-sas-covid-19-fight/. Accessed 23rd Mar 2020.

MiWay. (2020). *MiWay puts over R25 million back into their clients' pockets*. Retrieved from https://www.miway.co.za/blog/in-the-news/miway-puts-over-r25-million-back-into-their-clients-pockets. Accessed 25th Mar 2020.

Momentum Metropolitan. (2020a). *Coronavirus: Momentum provides support*. Retrieved from https://client.momentum.co.za/for/business/media/coronavirus-momentum-provides-support. Accessed 6th Mar 2020.

Momentum Metropolitan. (2020b). *Momentum Metropolitan Holdings partners with government to fight COVID-19*. Retrieved from https://client.momentum.co.za/wps/portal/mmiholdings-za/Home/media/2020-03-18. Accessed 30th Mar 2020.

Motsepe Foundation. (2020a). *Motsepe family and associated companies pledge R1 billion for dealing with coronavirus pandemic and related challenges*. Retrieved from http://motsepe-foundation.org/motsepe-family-and-associated-companies-pledge-r1-billion-for-dealing-with-coronavirus-pandemic-and-related-challenges/. Accessed 2nd Apr 2020.

Motsepe Foundation. (2020b). *R5 million donation for UCT'S COVID-19 offensive*. http://motsepefoundation.org/r5-million-donation-for-ucts-covid-19-offensive/. Accessed 2nd Apr 2020.

Mtshali, S. (2020). *EFF calls on government to disclose details of Covid-19 donations*. Retrieved from https://www.iol.co.za/news/politics/eff-calls-on-government-to-disclose-details-of-covid-19-donations-45891299. Accessed 2th Apr 2020.

Naspers. (2020). *Naspers commits ZAR 1,5 billion to South Africa's Covid-19 response*. Retrieved from https://www.naspers.com/home/news/naspers-commits-zar-1,5-billion-to-south-africa%e2%80%99s. Accessed 31th Mar 2020.

National Film and Video Foundation. (2020). *The NFVF to offer immediate relief to film practitioners amid COVID-19 lockdown*. Retrieved from http://www.nfvf.co.za/home/index.php?ipkMenuID=&ipkArticleID=697. Accessed 3th Apr 2020.

Nedbank. (2020). *COVID-19 relief measures: For individuals and small businesses*. Nedbank: Johannesburg.

Nestlé. (2020). *Nestlé steps up global response to COVID-19 pandemic, joins forces with the International Federation of the Red Cross and Red Crescent Societies*. Retrieved from https://www.nestle.com/media/pressreleases/allpressreleases/coronavirus-covid-19-response-partnership-ifrc. Accessed 5th Apr 2020.

News24 Team. (2020). *More than 2.6 billion worldwide told to observe lockdowns*. Retrieved from https://m.news24.com/SouthAfrica/News/coronavirus-all-the-latest-news-about-covid-19-in-south-africa-and-the-world-20200312. Accessed 6th Apr 2020.

Nguyen, D. N., Imamura, F., & Iuchi, K. (2017). Public-private collaboration for disaster risk management: A case study of hotels in Matsushima, Japan. *Tourism Management, 61*, 129–140. https://doi.org/10.1016/j.tourman.2017.02.003.

Nguyen, D. N., Imamura, F., & Iuchi, K. (2018). Barriers towards hotel disaster preparedness: Case studies of post 2011 Tsunami, Japan. *International Journal of Disaster Risk Reduction, 28*, 585–594. https://doi.org/10.1016/j.ijdrr.2018.01.008.

Nishiura, H., Jung, S., Linton, N. M., Kinoshita, R., Yang, Y., Hayashi, K., Kobayashi, T., Yuan, B., & Akhmetzhanov, A. R. (2020). The extent of transmission of novel coronavirus in Wuhan, China, 2020. *Journal of Clinical Medicine, 9*, 330. https://doi.org/10.3390/jcm9020330.

Omarjee, L. (2020). *Donations, loans & pledges: What you need to know about those billions aimed at fighting Covid-19*. Retrieved from https://m.fin24.com/Economy/South-Africa/donations-loans-pledges-what-you-need-to-know-about-those-billions-aimed-at-fighting-covid-19-20200402-2. Accessed 2nd Apr 2020.

OUTsurance. (2020). *OUTsurance commits R102 million to support COVID-19 efforts*. Retrieved from https://www.outsurance.co.za/company-news/outsurance-commits-102million-rands-to-support-covid-19-efforts/. Accessed 27th Mar 2020.

Owusu-Kwateng, K., Hamid, M. A., & Debrah, B. (2017). Disaster relief logistics operation: An insight from Ghana. *International Journal of Emergency Services, 6*(1), 4–13. https://doi.org/10.1108/IJES-10-2016-0022.

Pityana, S. (2020). *Having each other's backs is the only way through the storm*. Retrieved from https://www.fin24.com/Opinion/sipho-pityana-having-each-others-backs-is-the-only-way-through-the-storm-20200401. Accessed 30th Mar 2020.

Razon, C. (2020). *Djokovic to follow Nadal, Federer in COVID-19 donation*. Retrieved from https://www.ibtimes.com/djokovic-follow-nadal-federer-covid-19-donation-2950512. Accessed 5th Apr 2020.

Republic of South Africa. (2020a). *Statement by President Cyril Ramaphosa on escalation of measures to combat the Covid-19 epidemic, Union Buildings, Tshwane*. Retrieved from http://www.thepresidency.gov.za/speeches/statement-president-cyril-ramaphosa-escalation-measures-combat-covid-19-epidemic%2C-union. Accessed 2nd Apr 2020.

Republic of South Africa. (2020b). *Statement by President Cyril Ramaphosa on escalation of measures to combat the Covid-19 epidemic, Union Buildings, Tshwane, 13 May 2020*. Pretoria: Government Publishers.

Rwanda Office of the Prime Minister. (2020). *Communiqué of 5 April 2020*. Kigali: Rwanda.

SAFT (South Africa Future Trust). (2020). *How SAFT support works*. Retrieved from https://opp-gen.com/saft/. Accessed 2nd Apr 2020.

SANEWS (South African Government News). (2020). *Donations to fight against COVID-19 welcomed*. Retrieved from https://www.sanews.gov.za/south-africa/donations-fight-against-covid-19-welcomed. Accessed 30th Mar 2020.

Sky News. (2020). *Coronavirus: British Airways pilots to take 50% pay cut in April and May*. Retrieved from https://news.sky.com/story/coronavirus-british-airways-pilots-to-take-pay-cut-over-pandemic-turmoil-11960718. Accessed 6th Apr 2020.

Slabbert, A. (2020). *Covid-19 patient offers hotel for healthcare workers*. Retrieved from https://city-press.news24.com/News/covid-19-patient-offers-hotel-for-healthcare-workers-20200404. Accessed 5th Apr 2020.

South Africa National Department of Tourism. (2020). *Tourism relief fund for SMMEs funding criteria and the application process*. Cape Town: National Department of Tourism.

South Africa Reserve Bank. (2020). *Statement of The Monetary Policy Committee: 19 March 2020*. https://www.resbank.co.za/Lists/News%20and%20Publications/Attachments/9790/March%202020%20MPC%20statement.pdf. Accessed 23rd Mar 2020.

SRF (Solidarity Response Fund). (2020). *Solidarity fund.* Retrieved from https://www.solidarityfund.co.za/. Accessed 10 June 2020.

Statista. (2020). *Breakdown of the economic stimulus package for the tourism sector to combat the impact of SARS-CoV-2 in Indonesia 2020.* Retrieved from https://www.statista.com/statistics/1103513/indonesia-coronavirus-economic-incentives-for-tourism-sector/. Accessed 6 Apr 2020.

Tandon, K. (2020). *United in this fight: Nadal, Djokovic extend generous COVID-19 pledges.* Retrieved from https://www.tennis.com/pro-game/2020/03/rafael-nadal-novak-djokovic-join-fight-against-covid-19-pau-gasol-federer-muguruza/88156/. Accessed 5th Apr 2020.

U.S. Department of Health and Human Services. (2020). *HHS accepts donations of medicine to Strategic National Stockpile as possible treatments for COVID-19 patients.* Retrieved from https://www.hhs.gov/about/news/2020/03/29/hhs-accepts-donations-of-medicine-to-strategic-national-stockpile-as-possible-treatments-for-covid-19-patients.html. Accessed 2nd Apr 2020.

U.S. Department of Labour. (2020). *$100M in dislocated worker grants available.* New York: U.S. Department of Labour.

Ubuntu Beds. (2020). *Ubuntu Bes.* Retrieved from https://www.ubuntubeds.org/. Accessed 5th Apr 2020.

Van der Merwe, M. (2020). *Banks to waive ATM penalty fees, offer customers a helping hand.* Retrieved from https://www.fin24.com/Companies/Financial-Services/lockdown-banks-to-waive-atm-penalty-fees-offer-customers-a-helping-hand-20200326. Accessed 26 Mar 2020.

WHO (World Health Organisation). (2020a). *The first few X cases and contacts (FFX) investigation protocol for coronavirus disease 2019 (COVID-19).* Geneva: WHO.

WHO (World Health Organisation). (2020b). *COVID-19 solidarity response fund for WHO.* Retrieved from https://covid19responsefund.org/. Accessed 5th Apr 2020.

WHO (World Health Organisation). (2020c). *2019 novel coronavirus (2019-nCoV): Strategic preparedness and response plan.* Geneva: WHO.

Chapter 15
Tourism Economic Stimulus Packages as a Response to COVID-19

Abstract Economic stimulus packages in times of global crises and pandemics are nothing new. Following the 2008 global financial meltdown, many countries announced fiscal interventions to stimulate their economies. Before the COVID-19 pandemic, global tourism has never been in need of a bailout. However, due to the pandemic, hotel occupation and restaurant visits plummeted, and some airlines and related businesses were liquidated. By means of secondary data and public pronouncements, as well as document and critical discourse analysis, the chapter profiles stimulus packages to save the tourism industry. From small island developing states (SIDS) in the Pacific and the Caribbean to major global economies and in countless African countries, COVID-19 devastated the tourism sector. The findings are that stimulus packages included measures such as the postponement and restructuring of tourism business loan payments; the postponement of tourism surcharges, levies and taxes; delayed tax declarations and deductions for the sector; the introduction of special tourism sector loans and credit lines; exemption from utility payments and delayed tax collection; cuts and subsidies in interest rates and guarantees on loans for small, medium and microenterprises; the marketing of domestic tourism and discounts on air tickets and jet fuel. The most popular stimulus packages were income tax and corporate tax deferrals. The chapter recommends that wealthier countries, development banks and donors urgently consider topping up fiscal packages for SIDS and least developed countries because the demands occasioned by COVID-19 are bigger than domestic resources (if there are any). It also argues for the escalation of national stimulus packages already in place and quick peer learning and assistance across the world. Since some major companies are able to remain afloat, corporations who are to be bailed out with taxpayers' money have to be selected carefully.

Keywords COVID-19 · Coronavirus · Tourism · Stimulus packages · Subsidies · Loans

G. Nhamo et al., *Counting the Cost of COVID-19 on the Global Tourism Industry*, https://doi.org/10.1007/978-3-030-56231-1_15

15.1 Introduction

Tourism remains one of the leading economic sectors across the world. The industry accounts for about 5% of the global gross domestic product (GDP) and an estimated 8% of world employment (Abuhjeeleh 2019). With the number of arrivals steadily growing, the industry became more visible in many global economies' growth and development radars. However, epidemics and pandemics such as COVID-19 result in curtailed travel and even discrimination against people from countries where many cases have been reported (Jamal and Budke 2020; Rodríguez-Morales et al. 2020). The Australian Tourism Industry Council (ATIC) estimates that 302,000 tourism businesses were affected by COVID-19 (ATIC 2020) and more than 90% of these were small, medium and microenterprises (SMMEs) that employed one million Australians. Such global downturns force countries to put in place fiscal stimulus packages to prevent their economies from free falling into a recession or depression (Ramey 2019).

Deloitte (2020) notes that several economic sectors that recovered remarkably well from the 2008 financial meltdown such as food and drink, and tourism, borne the brunt of COVID-19. Travel bans, lockdowns and the closure of many international and domestic ports brought the tourism industry to a standstill, with the sea, air, rail and road transport carrying only essential commodities. The Scottish tourism sector was at its peak at the time of the COVID-19 outbreak. Up to 13% of the Scottish economy and around 19% of employment come from distribution, hotels and restaurants. Tourism remains the main source of employment in rural Scotland (Ibid.).

In response to COVID-19, the United Nations (2020) came up with a report in March 2020 forecasting the effect of the pandemic on the 2030 Agenda for Sustainable Development and the 17 Sustainable Development Goals (SDGs). The report presented measures to reduce COVID-19 transmissions, explained the social, economic and multidimensional impacts and proposed sustainable solutions. It ended by calling for stronger partnerships to accelerate global response and singling out two regions where the tourism sector was highly vulnerable, namely, Asia and Central America, including the Caribbean.

Several stimulus issues were considered. Weak local currencies constrain the governments' ability to stabilise their economies (United Nations 2020). At a global level, the United Nations advocated for a coordinated stimulus package in the region of two-digit percentage points of the global GDP. Such a stimulus package had to be extended to the business sector, including lead companies, workers and SMMEs. One of the key stakeholders it identified was the tourism sector. For the stimulus package to be effective, directed and targeted programmes to transfer resources to the most vulnerable sectors and households were essential. There had to be "social protection, tax abatement, low-interest rates, access to credit, insurance and wage support schemes" (United Nations 2020: 14). Furthermore, specific interventions were suggested to boost the economies of developing countries, including small island developing states (SIDS), in response to the 2030 Agenda for Sustainable Development's call to leave no one behind. This would result in national solidarity.

There are hardly better examples of global economic stimulus packages than those occasioned by the 2008 global financial meltdown and the severe acute respiratory syndrome (SARS) of 2003. Several stimulus packages were introduced to resuscitate economies (Li et al. 2020). One sector that benefitted from them was tourism. Saha and Von Weizsäcker (2009) affirm that a significant part of the European Union's 2008 economic stimulus of about €200 billion went to the tourism sector. During the SARS outbreak, Hai et al. (2004) observe that tourism was the hardest hit sector in China. By the end of 2003, its total loss was estimated at $16.8 billion. If the multiplier effect was applied to the Chinese economy, the loss was as high as $25.3 billion. Given its enormous losses, the tourism sector had to receive bailouts. A further literature review on tourism economic stimulus packages will be conducted in the section dedicated to it.

This chapter draws the reader's attention to economic stimulus packages aimed at saving the tourism industry following the COVID-19 pandemic. While acknowledging other generic stimulus packages for the tourism sector, such as those aimed at alleviating unemployment and reducing repo rates, this chapter restricts itself to stimulus packages for the tourism sector. In many instances, they included schemes for SMMEs. In the SIDS of the Pacific and the Caribbean, the least developed countries in Africa and in the larger economies of China, the USA, the UK and Germany, COVID-19 had a devastating effect on the tourism sector. The chapter will illustrate how countries responded to save the tourism sector and how, in the future, negative impacts could be diminished through similar economic stimulus packages. The chapter also calls for bailout selection criteria since taxpayers' money cannot be channelled to big corporates with considerable assets that are able to remain in operation.

15.2 Literature Survey

Gámez et al. (2014) focus on tourism and the 2008 financial crisis, which led to a 4.8% slump in global growth in 2009. The authors found that the tourism sector rebounded remarkably in 2010. International arrivals increased by 6.6% to hit 940 million, resulting in $919 billion in receipts. International tourism grew by about 5% during the first half of 2011. Only the Middle East and North Africa could not recover as quickly because of the ongoing conflict. Compagnoli (2020) notes that the US tourism sector took longer, between 3 and 6 years, to recover from the September 11 and 2008 crises.

Following the global financial crisis in 2008, the International Labour Organization (ILO) reviewed sector-specific stimulus packages and policy responses. Some of the measures undertaken by certain countries for the hotel, catering and tourism sector are shown in Table 15.1. In Germany, huge credit line

Table 15.1 2008 Economic stimulus packages for the hotel, catering and tourism sector

Country	Hotel, catering and tourism sector stimulus
China	CNY1.8 billion to refund the quality guarantee deposit for travel agencies
	Travel agencies could claim a temporary refund of 70% of the quality guarantee deposit they paid to CNTA when they set up a business. This policy applied for a period of 2 years starting in January 2009
	CNY1 billion investments to improve infrastructure
Indonesia	The policy of zero unemployment after 1 year of graduation from tourism schools supervised by the Ministry of Culture and Tourism
	Capacity building and technical assistance for tourism industries in 15 priority destinations
Philippines	Reduction of airfare and hotel rates to attract tourists
	Tourism Department planned to create new jobs in support of President Arroyo's programme to create one million jobs in the first 6 months of 2009
Singapore	Government deferred for 1 year an increase in assessment rates for hotel rooms, which were to grow to 25% in January 2009
Thailand	Tax reductions and exceptions for the private sector
	A microcredit scheme was initiated to provide special loans for SMMEs
	Lending period for microcredit scheme was set at 2 years with a −3% of the median low rate of interest, −1% from cooperated banks and −2% from the budget allocated by government
Vietnam	Implementation of preferential programmes to reduce VAT and to extend the tax payment duration
Argentina	Soft credit line of 300 million pesos to encourage refurbishment and extension works in the hotel sector through its national bank
Uruguay	Tourists could buy national products tax-free
Barbados	Government provided $15 million to assist failing enterprises and to maintain employment in the tourism sector
	Tourism Industry Relief Product together with $20 million was made available to the Barbados Tourism Authority
Jamaica	Tax cuts and low-cost loans for business, manufacturing and tourism activities
Belgium	VAT reduction from 21% to 6% on restaurant services
Czech Republic	VAT reduction from 19% to 9% on restaurant services
Cyprus	€51 million for tourism, which included tax and airport landing fee reductions
	Introduction of a scheme to encourage domestic holidays
France	Government stimulated job creation through an incensement of 40,000 places
	Reduction in VAT to 5.5% for the catering sector
Germany	The German government extended a €500 million credit line for tourism, a €50 million credit line for restaurants and a €400 million credit line for SMMEs
Netherlands	Revoked airport departure tax of €11.25 for short-haul flights and €45 for long-haul flights that the Dutch government had introduced in July 2008
Slovenia	The government subsidised 50% of the labour costs of each worker. Wages had to be at least equal to the minimum guaranteed wage of €589, and support lasted for a maximum of 6 months
Egypt	New formula for reduced takeoff and landing fees for new operators was put in place
Jordan	Reduced the Royal Jordanian ticket prices and hotel rate sales tax from 14% to 8%

Source: Authors, based on ILO (2009: 7–38)

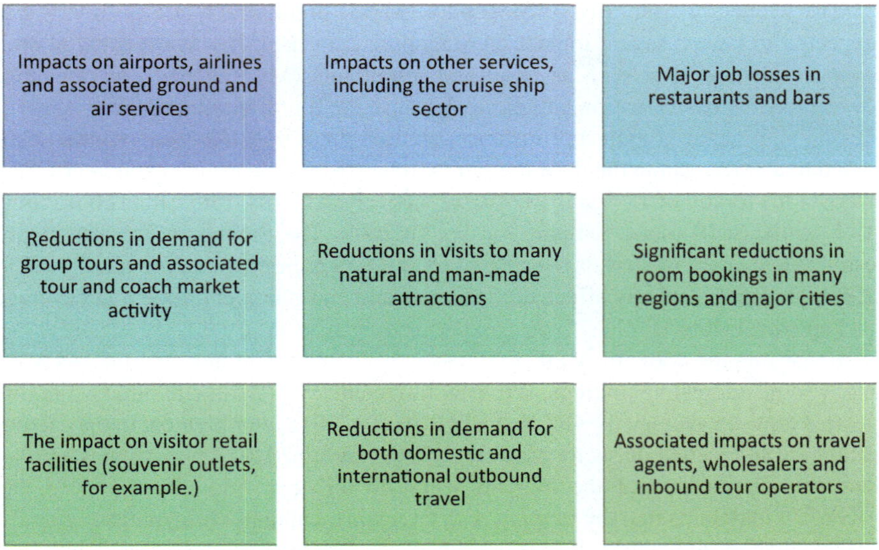

Fig. 15.1 Tourism segments impacted by COVID-19. (Source: Authors, based on TTFA (2020: 8))

facilities of up to €950 million were created for the hotel, catering and tourism sector. *The Economist* (2009) reports that, to stimulate the tourism sector in, among others, the premier exporting province of Guangdong, China, set up a fund of CNY20 million to encourage senior citizens to join tour groups. In Zhenjiang and Nanjing in the Jiangsu province and in Ningbo in the Zhejiang province, tourism vouchers were issued for local scenic spots (Ibid.).

During the SARS outbreak in 2003, the Asian Development Bank (ADB) reported that several Asian economies that had close ties with China experienced a significant reduction in tourist arrivals. Among them were Indonesia, Thailand and Korea. In China, the United Nations World Tourism Organisation (UNWTO) indicated that arrivals dipped −7.7% in 2003 (ADB 2020). In 1994 a plague outbreak in Surat (India) led to an export loss of $420 million (at 1994 prices), and tourist bookings declined by 2.2 million (World Bank 2020). The total economic loss stood at US$2 billion. Italy was severely affected by COVID-19. The tourism sector came to a standstill. Badiani (2020) notes that the tourism sector contributed up to 13.2% of the country's GDP in 2018 and employed 3.5 million workers. Key segments of the tourism sector that were impacted by COVID-19 and required stimulus packages are shown in Fig. 15.1.

The Monetary Authority of Singapore (MAS) highlights that COVID-19 was expected to have a severe impact on Singapore's tourism. It was expected to spill over to other sectors, such as retail and restaurants (MAS 2020). The downturn owing to COVID-19 was an anticlimax because the fourth quarter of 2019 had witnessed significant growth. For instance, arrivals grew by 6.9% year on year compared to a 3.5% gain in the third quarter. Arrivals from Indonesia and Greater China were in the region of 11% and 10.6%, respectively. Arrivals from China contributed 19% to the 2019 totals compared to 9% in 2003. The Singapore Tourism Board estimates the dent in arrivals because of COVID-19 at between −25% and −30% in 2020. This drop severely affected hotels, arts and entertainment, food and beverage and air travel (Ibid.).

Regarding the 2008 financial crisis's effect on the tourism industry, the UNWTO made some interesting findings. While employment levels across all other economic sectors grew on average by 11% between 2010 and 2018, employment in the accommodation and restaurant sector grew by 35% (UNWTO 2020). These figures emphasise the significance of the tourism sector in all economies. Furthermore, the UNWTO maintains that the tourism sector remains a major employment engine for disadvantaged and vulnerable groups in society such as women, the youth and unskilled foreigners. In 2019 the tourism sector took up 30% of the global export services totalling $1.5 trillion and up to 45% of total export services in developing countries. This explains why governments trembled at the outbreak of the COVID-19 pandemic. In France, for example, up to 160,000 catering businesses, 140,000 non-food trade and 100,000 tourism businesses closed their doors (WTTC 2020).

Experts believe that the airline industry should not be bailed out with taxpayers' money. de Rugy and Leff (2020) are of the opinion that airlines' requests for a bail-out were inappropriate. Instead of grants, loans needed to be extended. Airlines were advised to simply apply for bankruptcy before they get any help. In the past, US airlines such as American, Delta and United survived the bankruptcy.

The next section presents the materials and methods applied to generate and analyse the data.

15.3 Materials and Methods

Every country on earth benefits from tourism. Therefore, tourism requires substantial stimulus packages in times of a pandemic or economic meltdown. Figure 15.2 shows 80 countries with significant annual tourism earnings, including economic powerhouses such as Brazil, South Africa, China, India, Australia, Spain, Portugal, Italy, England, Germany, Russia, the USA, Canada, France and Japan. Smaller countries with significant annual tourism earnings are Bahrain, Barbados, Bhutan, Costa Rica, Fiji, Hong Kong, Taiwan, Singapore, Macau, the Maldives, Mauritius, Panama, Rwanda and Zimbabwe.

The research method was a critical document review and analysis (Linton et al. 2019; Bowen 2009). This method was appropriate, given the considerable amount

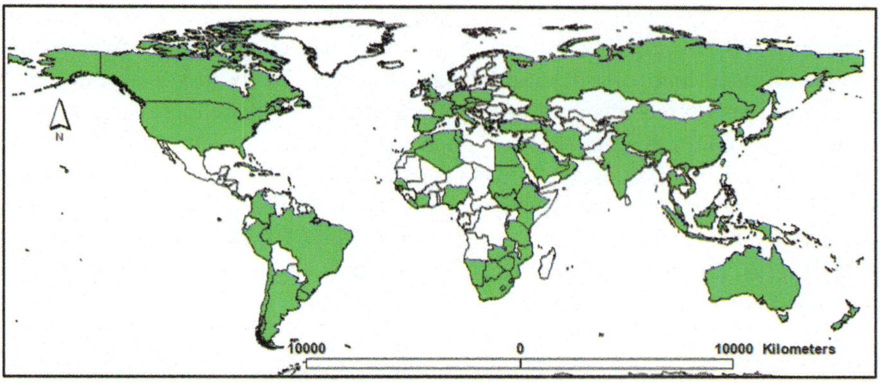

Fig. 15.2 Countries with significant annual tourism earnings. (Source: Authors)

of data about COVID-19 in general and the impacts of COVID-19 on the tourism and hospitality sector in particular. Documents were analysed, sources were verified, and triangulation was executed as advised by Wood et al. (2020). This included case studies. Among the principal sources of data were the UNWTO (2020), the World Travel and Tourism Council (2020) and the International Monetary Fund (2020). Announcements on COVID-19 stimulus packages across the world were tracked, and, after a careful selection, general stimulus packages and specific allocations for the tourism sector were analysed. The analysis was done to, first, get an overview of tourism stimulus packages by using selected examples. The second part focused on selected country case studies and deeper presentations. The data and findings are presented in the next section.

15.4 Presentation of Data and Discussion of Findings

The data and key findings are presented in two parts. The first focuses on an overview of the COVID-19 impact on the tourism sector and examples of economic stimulus packages and the second on selected case studies.

15.4.1 COVID-19 Impacts and Examples of Tourism Economic Stimulus Packages

The Asian Development Bank (ADB) and the United Nations Economic and Social Commission for Asia and the Pacific (ESCAP) identified lower tourism and business travel as manifestations of the economic disruption Asia experienced owing to COVID-19 (ADB 2020; ESCAP 2020). International tourism accounted for up to

40% of GDP in the economies of Palau, and the Maldives and local tourism contributed up to 10% of GDP in more than 50% of Asian countries. The World Bank (2020) affirmed that some Pacific Island States were likely to see poverty levels double as tourism receipts go down. For example, Fiji (where tourism contributed up to 38% of GDP), Kiribati, Palau, Samoa and Vanuatu remain the most vulnerable to a tourism decline. China, the initial epicentre of COVID-19, accounted for 9% of global tourism. In light of these figures, the UNWTO called for urgent interventions to mitigate the socio-economic impacts of COVID-19 and accelerate recovery (UNWTO 2020). It also noted the tourism industry's ability to bounce back quicker and inject the necessary regrowth stimuli in economies.

The United Nations Development Programme (UNDP) predicted that China's domestic tourism would drop by 56% in the first quarter of 2020 and overall by 15.5% due to COVID-19 (UNDP 2020). The drop constituted about 932 million tourists year on year, and domestic tourism revenue would go down by 20.6% ($167 billion). The impact of COVID-19 on the tourism and related sectors was expertly captured by the World Bank (Box 15.1).

Box 15.1: Unprecedented Impacts of COVID-19 on Tourism, Hospitality and Recreation Sectors

- Since 20 April the accommodation and lodging sector's quarterly revenues were down 75%
- Travel agents witnessed a slowdown in bookings of 50% in March of 2020
- Airlines worldwide were expected to lose $113 billion in revenues for 2020
- At the peak of the outbreak, 70% of scheduled flights in China were cancelled
- As of mid-March 2020, international travel grounded to a halt, and the World Travel and Tourism Council anticipated global travel to decline by at least 25% in 2020

Source: Authors, based on World Bank (2020: 49)

The fact that Chinese tourists increased eightfold from less than 11 million in 2003 to about 87 million in 2018 posed an additional challenge to the Asian region (ADB 2020). In Myanmar, Thailand, Mongolia, the Republic of Korea, Vietnam, Cambodia, Palau and Hong Kong, Chinese tourists accounted for more than 25% of arrivals in 2018. Hence, when China banned outward-bound travel on 24 January 2020, the Asian tourism sector was completely taken aback. Many airlines in the region suspended travel to China (Congressional Research Services 2020). The World Bank (2020) indicated that an estimated 30% of Cambodia and Thailand's arrivals were from China.

The ADB estimates that, in a moderate-case scenario, the impact of COVID-19 on the tourism sector could cost the world $156 billion (ADB 2020). Other scenarios put the figures between $77 billion and $347 billion (0.1–0.4% of global GDP). Anecdotal evidence suggested that tourism arrivals would drop in developing Asian economies by between 50% and 90% in February 2020 compared to the same time in the previous year. With this in mind, Hong Kong, India, China, Japan and Singapore had already rolled out stimulus packages to subsidise tourism and other affected sectors (ESCAP 2020). Arezki and Nguyen (2020) focus on the COVID-19 impact on arrivals in the Middle East and North Africa. Because of China's travel ban and Saudi Arabia's suspension of pilgrimages to Mecca, arrivals were set to drop drastically. Examples of tourism stimulus measures announced at the time of writing the chapter are shown in Table 15.2.

The UNWTO proposed nine stimulus packages for the tourism sector (Box 15.2). It further made some recommendations for building future resilience. The recommendations included the diversification of markets, products and services; investment in marketing intelligence systems and digital transformation; the reinforcement of tourism governance at all levels; improved preparation for a crisis and the assurance that the tourism sector forms part of national emergency mechanisms; investment in human capital and talent development; assurance that sustainable tourism is on the national agenda and, lastly, a just transition to the circular economy that embraces the Sustainable Development Goals or SDGs (UNWTO 2020). The SDGs are highlighted because tourism is embedded in the targets of several SDGs which will remain on the global radar until 2030.

The Organisation for Economic Cooperation and Development (OECD) conducted a survey to determine how the SMMEs were impacted by COVID-19. Of 6 000 responding Italian firms, 72% indicated that they were directly affected due to a drop in demand or problems in the supply chain and logistics (OECD 2020). The worst-hit SMMEs were in transport (98.9%), tourism (89.9%), fashion (79.9%) and agrofood (77.7%). Details about selected COVID-19 stimulus packages across the world that were audited by the OECD, including the tourism sector, are presented in Fig. 15.3. The most popular of the eight stimulus packages are income tax and corporate tax deferrals. A total of 54 out of 54 countries have implemented it. This was followed by direct lending (39 out of 54 countries have put it into effect). The two least popular interventions were reduced VAT (13 out of 54 countries instituted it under deferrals) and social security (16 out of 54 countries resorted to it) (OECD 2020). Although no explanation for the popularity of the top two stimuli was given, it could be because they are easy and quicker to administer from a government perspective. Furthermore, a 500 million euro fund to mitigate damage in the aviation industry was set up (WTTC 2020).

What emerges from the OECD study is that all SMMEs, including transport, fashion and agrofood SMME, have either direct or indirect links with tourism SMMEs. This illustrates the significant role the tourism sector plays in the economy in general by driving jobs and creating wealth in the tourism global value chain. Every economic stimulus package has to take the tourism sector into consideration in one way or another. The next section is dedicated to case study countries.

Table 15.2 Examples of stimulus packages for the tourism industry

Country	Economic stimulus measures for the tourism industry as of 17 April 2020
Bahrain	Exemption of tourist facilities from tourism fees
Belize	Reducing risk weights for banks on loans in the tourism sector from 100% to 50%
Brunei Darussalam	Businesses in the tourism, hospitality/event management, restaurant/cafe and air transport sectors ("Affected Sectors") were to be given a 6-month deferment of their principal repayments of financing/loans
Cabo Verde	Measures were taken to support the private sector, such as loan guarantees and tax obligation facilities of up to 80% for companies in the tourism and transport sectors (CVE1 billion)
Colombia	Authorities announced faster direct contracting for services associated with the emergency response, a new credit line providing liquidity support to all tourism-related companies and delayed tax collection for the tourism and air transportation sectors
Cyprus	Support for affected businesses to maintain jobs, support for the tourism sector and deferral of VAT payments due in 2 months
Egypt	$3.2 billion allocated to the tourism sector
El Salvador	A 3-month extension for income tax payments for individuals and firms operating in the tourism sector with a taxable income lower than US$25 000
	A 3-month exemption from the special tourism tax for companies operating in the tourism industry
Georgia	GEL1 billion (2% of GDP) support package on 13 March
	The package includes suspension of property and income taxes for the tourism sector until November 2020, provision of interest subsidy to small- and medium-sized hotels, an increase in credit guarantee scheme, an acceleration of VAT refunds and higher capital spending
Hungary	When the 2020 budget was revised, the tourism development contribution was temporarily cancelled
Iceland	ISK4.6 billion injection into the travel and tourism sector
	Residents over 18 years were to collectively receive ISK1.5 billion worth of travel vouchers from government for domestic travel
	Special ISK15 billion investment for acceleration initiatives, including infrastructure development in national parks, airports extensions, harbour and roads improvement as well as tourism-related technology improvements
	Key measures to restart the economy (1.1 percent of GDP) include public investment, tax incentives for real estate improvement, temporary tax relief for the tourism sector and marketing efforts to encourage domestic tourism
Indonesia	US$725 million stimulus package to support the tourism, airline and property industries (about 0.2% of GDP)
	This first package comprises support to the tourism sector (tax cuts and discounts on airplane tickets and jet fuel) and low-income households (social assistance and subsidies for home buyers)
Lao P.D.R.	Tax relief and extensions of interest payments are being discussed for tourism and agriculture companies
Latvia	50% cut in interest rates on loans for SMMEs in the tourism sector and a 15% cut for large enterprises

(continued)

Table 15.2 (continued)

Country	Economic stimulus measures for the tourism industry as of 17 April 2020
Myanmar	COVID-19 fund worth MMK100 billion (S$70 million, 0.1% of GDP) has been established at the Myanmar Economic Bank to provide soft loans to affected business (particularly the garment and tourism sectors and SMMEs) at a 1% per annum interest rate for a 1-year period, and terms are to be reassessed as needed
Palau	The parliament authorised additional funding (up to $6 million or 2.1% of GDP) to maintain government services in the face of declining tourism revenue
Philippines	The aid of $271 million (PHP14 billion) from the Tourism Infrastructure and Enterprise Zone Authority (TIEZA) has been earmarked for various programmes and projects of the Department of Tourism
Portugal	€9.2 billion package to support workers and provide liquidity for companies affected by the coronavirus outbreak
	Just over half of the 3 billion euros in credit lines is aimed at companies working in tourism, hotels and restaurants
Qatar	QAR75 billion ($20.6 billion or about 13% of GDP) package to reduce the effects of COVID-19 was announced on 16 March 2020
	The programme aims to shore up small businesses and hard-hit sectors (hospitality, tourism, retail, commercial complexes and logistics) through 6-month exemptions on utilities payments (water, electricity)
Russian Federation	Tax deadlines extended for the tourism and aviation industries as well as for SMMEs and other affected industries
	Tourism companies not to pay contributions to tourist sector reserve funds
Rwanda	IMF approved SDR80.1 million (about US$109.4 million) under the Rapid Credit Facility
	This money will alleviate COVID-19 pressures on tourism, trade and foreign exchange reserves and provide resources for health expenditure and households and firms affected by the crisis
	The notice period for park permits for both tourists and tour operators was removed
	Tourists who need to postpone park visits due to cancelled flights or travel bans or who are infected with COVID-19 were given up to 2 years to postpone bookings at no extra cost. Tourists with low season-discounted permits for mountain gorilla trekking can postpone and use them within 2 years
Spain	Introduction of a special credit line for the tourism sector through the ICO (€400 million) with a limit of €500,000
	The qualifying self-employed and tourism companies included transport operators, taxis, hotels, restaurants, car rentals, travel agencies and museums
	Deferred repayment of loans granted to businesses by the Ministry of Industry, Trade and Tourism
Sri Lanka	The President announced a wide-ranging debt repayment moratorium which included a 6-month moratorium on bank loans for the tourism and SMMEs and reduced rate working capital loans
Sweden	SEK1 billion for culture and sport
	Airlines received credit guarantees of a maximum of SEK5 billion

(continued)

Table 15.2 (continued)

Country	Economic stimulus measures for the tourism industry as of 17 April 2020
Tonga	Restructuring loans to businesses that have reduced business hours in affected sectors such as tourism and related industries such as transportation and to individuals who have been laid off
Turkey	Reduced and postponed taxes for affected industries, particularly tourism
United Arab Emirates	Dubai announced a US$400 million economic stimulus for the tourism and hospitality sector
	Tourism and hospitality tax cut from 7.5% to 3.5%
	The new initiatives provided for water and electricity subsidies as well as credit guarantees and liquidity support to small and medium enterprises. In addition, the government of Abu Dhabi has announced a reduction or suspension of various government fees and penalties as well as a rebate on commercial lease payments in the tourism and hospitality sectors
UK	12-month business rates holiday for all retail, hospitality and leisure businesses
	Grant funding of £25,000 for retail, hospitality and leisure businesses with property valued at between £15,000 and £51,000
USA	In terms of the CARES Act, $2.2 trillion was set aside as relief for individuals, businesses, state and local governments as well as the healthcare system
	$10 billion in airport and tourism grants for maintaining airport operations
	$5 billion in community development block grants to state and local governments to mitigate disruptions, including grants to tourism businesses
	$46 billion in direct loans to airlines, cargo airlines and businesses critical to national security such as Boeing
	$32 billion in grants to passenger airlines, cargo airlines and air service contractors (including airline catering) for payroll costs but on strict conditions
Uzbekistan	The authorities announced a temporary reduction of social contributions for individual entrepreneurs, postponing surcharges on tourism, extending the moratorium on tax audits and delaying tax declarations for 2019 income taxes (until August 2020)

Source: Authors, data from IMF (2020); Congressional Research Services (2020: 25–45); World Bank (2020: 32–35); IATA (2020: 1–11); WTTC (2020: 2–29)

15.4.2 *Tourism and Economic Stimulus Packages of Selected Countries*

Given the central role China plays in the regional and global tourism industry, it is inevitable that this country's stimulus package for the tourism industry would come under scrutiny. The Chinese government temporarily returned part of the tourism service quality deposit to travel agencies. It also extended enterprises' carry-forward period for losses from 5 to 8 years. Each travel agent was due to receive about $28,000, bringing the total package for the sector to $1.1 billion (UNDP 2020). A total of 39,000 registered travel agents operated in China prior to COVID-19. At the time of finalising this chapter, 35,200 of them had already submitted applications for a temporary refund of guarantees – in other words, 90% of all registered travel agents. The Tourism and Transport Forum Australia (TTFA) acknowledged that

Box 15.2: Providing Stimulus and Accelerating Recovery in the Tourism Sector

(i) Provide a financial stimulus for tourism investment and operations
(ii) Review taxes, charges and regulations impacting travel and tourism
(iii) Advance travel facilitation
(iv) Promote new jobs and skills development in particularly the digital world
(v) Mainstream environmental sustainability in stimulus and recovery packages
(vi) Understand the market and act quickly to restore confidence and stimulate demand
(vii) Boost marketing, events and meetings
(viii) Invest in partnerships
(ix) Mainstream tourism in national, regional and international recovery programmes and in development assistance

Source: Authors, based on UNWTO (2020: 12)

China remained the largest inbound market, accounting for about 1.5 million arrivals in 2019 for Australia (TTFA 2020). This represented about AU$12.3 billion in annual tourism receipts, nearly 33% of all tourism receipts since May 2019. Each month an estimated 120,000 Chinese visit the country. The Chinese travel ban of January 2020, therefore, had severe impacts on the Australian tourism industry.

When the Australian government announced its initial economic stimulus package of AU$17.6 billion, it emphasised that COVID-19, on top of the bushfires in the summer of 2019, worsened the countries' economic difficulties (Australian Government 2020). Because of the bushfires, many travellers to the country cancelled or delayed their trips. The Australian government set aside AU$1 billion to support regions and communities affected by the economic impacts of the pandemic. Three industries were to benefit, namely, tourism, agriculture and education. A waiver of the fees payable by tourism businesses in the Great Barrier Reef Marine Park region and another of entry fees for Commonwealth National Parks were proclaimed. A further AU$66 billion was to be spent until September 2020 (WTTC 2020).

The TTFA estimated a 40% decrease in international arrivals during the first half of 2020 compared to the same period in 2019 (TTFA 2020). This translated in a loss of 1.8 million visitors. In addition, about AU$2 billion post-March 2020 monthly averages were expected in the sector. Other statistics of interest involved staff reductions of between 15% and 20% for 2020. These figures translated to a loss of between 99,000 and 133,200 mostly part-time jobs. In the broader economy, lost salaries and wages would range between AU$3.9 billion and AU$5.3 billion. The drop in income from inbound tourism from China, other Asian countries and other markets by June 2020 was estimated at AU$4.01 billion (Ibid.).

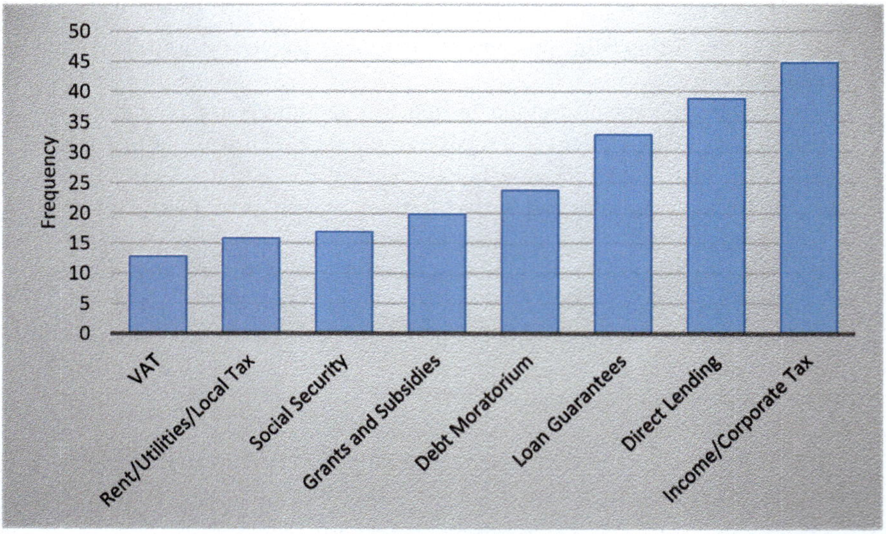

Fig. 15.3 SMMEs COVID-19 economic stimulus packages (n = 54). (Source: Authors, data from OECD (2020: 10–11))

In imitation of the Australian government, state governments also announced stimulus packages. Queensland, for example, allocated AU$27.25 million of which AU$7 million was to be spent on international tourism promotion (KPMG 2020). Furthermore, Queensland deferred tourism lease payments. In Tasmania, AU$100,000 was voted for the Tourism Industry Council of Tasmania and the Tasmanian Hospitality Association to train and support workers in the tourism and hospitality sectors over the winter months. The state government of Tasmania also waived tourism operator lease, license and entry fees from April to June 2020. This benefitted close to 300 tourism businesses. The Tasmanian government allocated a further AU$20 million towards free business loans for SMMEs with an annual turnover of less than AU$5 million (Tasmanian Government 2020). For 4 months, a payroll tax waiver was put in place as well as small grants of AU$5000 for SMMEs to hire an apprentice/trainee in the tourism sector. In Victoria, AU$500 million was set aside to establish a Business Support Fund for hospitality, tourism, accommodation, arts and entertainment and retail. Lastly, some AU$12.85 million tourism boost was voted for domestic tourism in Western Australia to subsidise, among other things, regional airfares (Ibid.).

Malaysia was among the first countries to proclaim economic and tourism stimulus packages because of the central role tourism plays in the national economy. The Malaysian government announced on 27 February 2020 that the tourism, airline and travel and tourism-dependent retailers were among the most affected (Crowe 2020). Four packages were announced (Fig. 15.4). Businesses such as travel agencies, hotels and airlines were regarded as part of the tourism industry. A 15% discount on electricity bills applied to hotel operators, travel agencies, local airline offices,

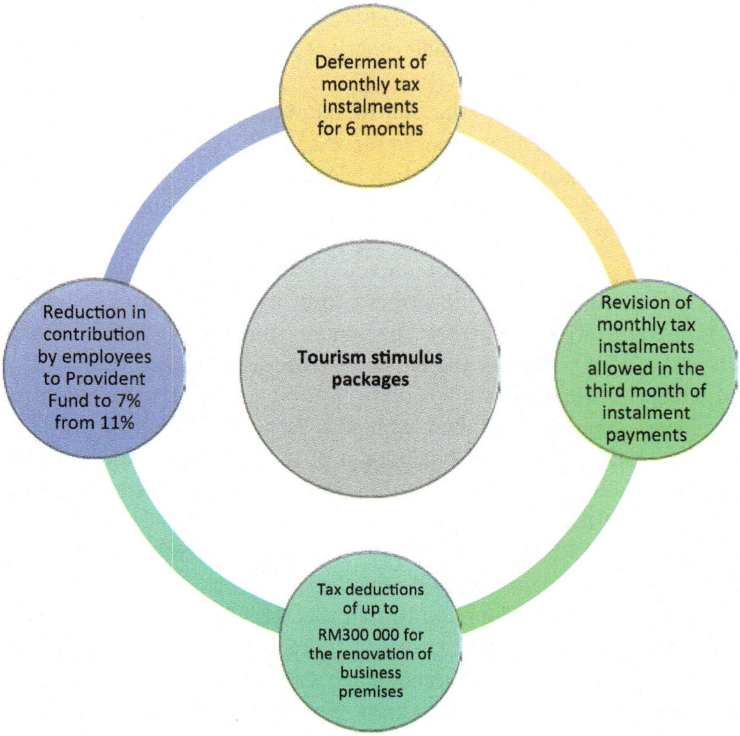

Fig. 15.4 Four key interventions in the tourism and hospitality-related sector. (Source: Authors, based on Crowe (2020: 3))

shopping malls, convention centres and theme parks from 1 April 2020 to 30 September 2020. Digital discount travel vouchers of up to RM100 per person were issued and could be used for domestic flights, railway travel and hotel accommodation. The digital vouchers and matching grant amounted to RM500 million. In April 2020 a once-off cash incentive of RM600 was paid to taxi drivers, tour bus drivers, tour guides and trishaw drivers who were registered since 31 December 2019 (Bakertilly 2020) to supplement their diminished revenues because of slow business during the lockdown.

The Luther Services (2020) announced that the Malaysian government-owned Bank Simpanan Nasional was to pay MYR200 million into a microcredit facility for SMMEs in affected sectors and in particular the tourism industry. The loan would be at an interest rate of 4%, and repayment was to start after 6 months of the loan disbursement. Furthermore, individual income tax relief of up to MYR1000 for expenditure in certain segments of domestic tourism was implemented. A double deduction would be made for expenses incurred for approved tourism-related training. The Malaysia Airport Holdings Berhad facilitated rebates on the rental of airport properties and landing and parking charges (Bakertilly 2020). All of these packages were welcomed. Lastly, More Advent (2020) pointed out the Human

Resource Development Fund's (HRDF) matching grant of up to RM100 million to 40,000 employees in the tourism and other sectors to increase productivity during the economic slowdown.

Like the economy of Malaysia, that of Indonesia depends heavily on tourism. The Indonesian government also put in place an early COVID-19 tourism stimulus package. After manufacturing, trade, hotels and restaurants contribute second most to their GDP. In 2008 its share stood at 17.72%, whereas manufacturing contributed 26.9% (Titiheruw et al. 2009). These percentages remained unchanged in 2020. Medina (2020) reports that Indonesia voted a hefty $725 million stimulus package to support tourism, airlines and the property industry. Other sources put the stimulus package figure at $750 (IATA 2020). The announcement came on 25 February 2020, well before the WHO declared the pandemic (Medina 2020). Additional measures included waivers on taxes for hotels and restaurants in selected local destinations, including Denpasar, Batam, Bintan, Manado, Yogyakarta, Labuan Bajo, Belitung, Lombok and Malang. The total breakdown of the package is shown in Fig. 15.5. Mufti (2020) cited the Chairman of the Indonesia Hotel and Restaurant Association, who indicated that the tourism sector had suffered losses of at least $1.5 billion between January and 12 March 2020. This was due to cancellations associated with COVID-19. Hotel occupancy rates dropped to as low as 20%. The capital Bali as well as Jakarta, Manado, Batam and Bintan was the worst affected.

South Africa also put in place several tourism stimulus packages. A tourism relief fund of R200 million was announced, and applications could be submitted between 7 April 2020 and 30 May 2020 (South African Department of Tourism 2020). It provided for a once-off capped grant of R50 000 to SMMEs in the sector

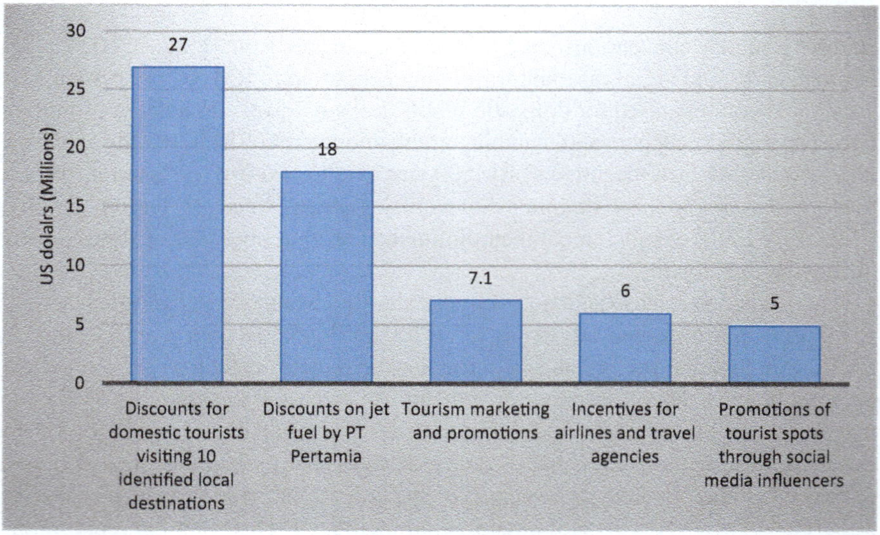

Fig. 15.5 Indonesian COVID-19 tourism and economic stimulus packages. (Source: Source: Authors, data from Medina (2020))

to subsidise fixed, operational, supplies and other pressure cost items. Businesses in three tourism categories, including accommodation establishments, hospitality and related services and travel and related services, received a grant. Under accommodation establishments came hotels, lodges, B&Bs, guest houses and backpacker hostels. Hospitality and related services included restaurants and conference venues that are not part of hotels, professional caterers and attractions. Travel and related services comprised tour operators, travel agents, tourist guides, car rental companies and coach operators. The qualification scoring with 11 indicators is shown in Box 15.3.

Box 15.3: South Africa Tourism Relief Fund Qualification Matrix

(i) Proof of registration with the Companies and Intellectual Property Commission
(ii) An applicant must be an Exempted Micro Enterprise (EME) defined in terms of the Amended Tourism Broad Based Black Economic Empowerment Sector Code, 2015
(iii) A valid tax clearance certificate or PIN
(iv) Proof of compliance with the minimum wage requirements
(v) Proof of UIF registration of employees
(vi) An applicant must be an existing tourism-specific establishment as outlined in the scope of application (suppliers and intermediaries are not eligible)
(vii) An applicant must be in operation for at least 1 financial year
(viii) Proof that the relief is required as a result of the impact of COVID-19
(ix) An applicant must submit his latest statements of financial position, financial performance and cash flows
(x) An applicant must submit 6 months' bank statements
(xi) A grading certificate or proof of application to be graded as an accommodation establishment

Source: Authors, based on South Africa National Department of Tourism (2016: 1)

The stipulations excluded many enterprises such as fast-food and take-away restaurants, nightclubs, bars and gaming and gambling venues (South Africa National Department of Tourism 2016). Franchise restaurants, as well as restaurants attached to tourism facilities or fully or partly owned by the government, also did not qualify. SMMEs had to apply online, and details were provided accordingly. Up to 12 documents had to accompany the application, such as a completed application form, a company registration certificate, certified ID copies of the director(s), 6 months' bank statements and the latest financial statements.

In Hong Kong, HK$700 million was allocated to the Hong Kong Tourism Board and HK150 million to the Anti-epidemic Fund for travel agent relief (WTTC 2020). At the time of finalising this chapter, some 1350 or 98% of the licenced travel agents had already received payments. A once-off subsidy of HK80,000 was paid to eligible applicants. In addition, the Hong Kong Airport Authority provided a relief package of HK$1.6 billion ($206 million) for airport communities. This was for waivers on arrivals and air navigation fees and charges. Part of the money was used to subsidise certain licensing fees, reduced rents and other services. Furthermore, a budget of HK$400 million was approved by the Hong Kong Tourism Board to launch a recovery plan for local and international travel agencies, hotels, airlines and attractions. It would have included retail, dining and meetings.

To build confidence in the country's hotels, the Singapore government subsidised their professional cleaning and disinfection costs. The subsidy included up to S$20,000 for a hotel with confirmed COVID-19 cases and up to S$10,000 for hotels with suspected cases (WTTC 2020). In addition, Singapore established a clean certification programme to protect the health of locals and visitors to tourism venues. Other measures included, among other things, training subsidies and the waiving of tourist guides' fees for multilanguage proficiency tests. A rebate on aircraft landing and parking fees as well as airport shops and cargo rental fees was introduced at Changi Airport. The rebate was valued at S$112 million (US$82 million) (Ibid.).

The stimulus packages that were instituted globally and in case study countries are in line with packages identified by the OECD (2020) and the UNWTO (2020). These packages included grants and subsidies; vouchers and domestic tourism promotions; drastic repo rate reductions; direct lending; tax waivers and reductions; social security payments; rent and utility payments; loan guarantees; debt moratoriums and other debt repayment arrangements. However, as highlighted in the literature section, not all sectors of the tourism value chain qualified for a bailout. The aviation sector, for instance, had the financial muscle to weather the COVID-19 crisis without assistance (de Rugy and Leff 2020).

Breuss et al. (2009) affirm that Australia set aside 4.2% of its 2008 GDP for an economic stimulus package aimed at, among other things, the tourism industry. Simulation models indicated that in 2010 Australia's economic stimulus package slowed down the decrease in its GDP by 2.1 percentage points. An estimated 41 500 jobs were preserved, and the stimulus package reduced the rate of unemployment by 0.7 percentage points (Ibid.).

Although general tax reductions often form part of economic stimulus packages, they are said to be expensive and to result in losses (Belke 2009). Besides, it is difficult to scale back after a financial crisis or natural disaster, and households and companies are accused of hoarding the additional money instead of using it for the intended purpose. Economists are sceptical of temporary VAT cuts such as those the UK government introduced after the 2008 financial meltdown. Bailed-out enterprises could channel the money to where they deem it is needed most. For example, if the family of a microenterprise owner goes hungry, it is likely that the owner would, instead of stimulating his business, divert the funds to put food on the table. For this reason, complementary economic stimulus packages are needed as a social safety net for individuals and households because many workers in the tourism industry lose their jobs or have to accept a salary reduction. Furthermore, minor VAT cuts of 2% or so do not stimulate the economy and, for the purposes of this chapter, the tourism sector. Economists have reservations about the transmission mechanism of reduced VAT from the government and the ways in which retailers pass it on to the consumer. Immediately after a crisis, governments who fell back on reduced taxation have to lessen their countries' deficit and debt ratios (Köhler-Töglhofer and Reiss 2009).

If firms are heavy borrowers (Belke 2009) and many firms were in countries such as Spain, France and Italy during the 2008 financial crisis, economic stimulus packages should be aimed at improving the liquidity of companies. Postponing the payment of corporate taxes for a year or 2 is a suitable way to accomplish it. This was evident during the COVID-19 pandemic. However, this type of intervention may benefit only firms that make a profit. Walsh (2010: 1) is of the view that economic stimulus "measures were successful when they were sufficiently large, properly focused, and implemented promptly". Because Thailand, one of the tourism-dependent economies, did not meet these conditions, the results were negligible. When GDPs and tax revenues dwindled globally subsequent to the 2008 financial crisis, several countries abandoned fiscal stimulus packages in favour of fiscal consolidation measures (Ramey 2019). What comes out of this discussion is that, because there isn't a one-size-fits-all economic stimulus package for governments and the tourism industry, in particular, fit-for-purpose and closely monitored economic stimulus packages remain on the table for the near future.

15.5 Conclusion

Tourism establishments could not anticipate the COVID-19 meltdown. The spate of cancellations took them by surprise after the WHO had declared COVID-19 a pandemic on 11 March 2020. The declaration leads to a global economic slowdown because of lockdowns and port closures. Economic stimulus packages were one of the best ways to bring immediate relief. It will take the global economy a long time and a great deal of effort to recover. Revenue in many subsectors of the tourism industry dwindled to zero during the lockdown. Many airlines have grounded their planes. Big businesses ran huge overheads, and small businesses experienced cash flow problems. Hence, all stakeholders, including governments, the tourism industry and organised labour, had to come to the table.

A number of economic stimulus packages for the tourism sector were introduced as immediate and intermediate interventions. They include grants and subsidies; voucher systems and domestic tourism promotions; drastic repo rate reductions; direct lending; tax waivers and reductions; social security payments; delayed rent and utility payments; loan guarantees; debt moratoriums and other debt repayment arrangements. Eight popular types of economic stimulus packages implemented by 54 countries worldwide were examined. They were (from the least to the most preferred): VAT; rent, utilities and local tax; social security, grants and subsidies; debt moratoriums; loan guarantees; direct lending and reduced income and corporate tax. Monetary policies that resulted in a drastic repo rate reduction by central banks were not included in the tracking of the 54 countries. Although effective in developing countries where repo rates remained high, it had little impact in developed economies where repo rates were, in many instances, close to zero.

The chapter recommends that the tourism industry rethinks its current business model and puts the right fundamentals in place. Obstacles in visa regimes have to be eliminated to make it easy to travel. Local recovery should come first, as this brings quick and big wins. Chinese arrivals are another challenge. Since some Asian economies depend on tourists from China, current models should be revised. The revamp of the tourism industry will necessitate fresh minds to lead innovation. Digital marketing has to be enhanced. Investment in hygiene in hotels, restaurants and related establishments will regain the trust of tourists. In addition, product pricing will be critical. Needless to say, blanket economic stimulus packages are unfeasible as some subsectors such as airlines may dispose of assets to cover their COVID-19 losses.

The promotion of regional and continental markets will bring in quick and big earnings for global tourism. In Africa, a Buy Africa campaign could be the way to go. However, Africa's tourism industry needs to keep an eye on global travel sentiment. To market travel, countries that came off well from COVID-19 have to be identified. The entire tourism industry supply chain must be reviewed. It is like a puzzle: one missing piece and the set is incomplete. Lastly, exchange rates that tanked against the dollar offered opportunities to persuade tourists to buy holidays for 2021 at lower rates. The South African rand, for example, lost close to 40% of its value and hit $1 to 20 ZAR in April 2020.

References

Abuhjeeleh, M. (2019). Rethinking tourism in Saudi Arabia: Royal vision 2030 perspective. *African Journal of Hospitality, Tourism and Leisure, 8*(5), 1–16.

ADB (Asian Development Bank). (2020). *The economic impact of the COVID-19: Outbreak on developing Asia*. Kuala Lumpur: ADB.

Arezki, R., & Nguyen, H. (2020). Novel coronavirus hurts the Middle East and North Africa through many channels. In R. Baldwin & B. W. di Mauro (Eds.), *Economics in the time of COVID-19* (pp. 53–58). London: Centre for Economic Policy Research Press.

ATIC (Australian Tourism Industry Council). (2020). *Media statement: Tourism industry body thumbs up to Federal stimulus combatting COVID-19*. Sydney: ATIC.

Australian Government. (2020). *Economic response to the Coronavirus*. Sydney: Australia Government.

Badiani, R. (2020). *Italy: Complete shutdown to contain Covid-19 set to deepen Italian GDP losses in 2020*. London: HIS Markit.

Bakertilly. (2020). *Malaysia economic stimulus package 2020*. Kuala Lumpur: Bakertilly.

Belke, A. (2009). Fiscal stimulus packages and uncertainty in times of crisis: Economic policy for open economies. *Economic Analysis & Policy, 39*(1), 25–46.

Bowen, G. A. (2009). Document analysis as a qualitative research method. *Qualitative Research Journal, 9*(2), 27–40.

Breuss, F., Kaniovski, S., & Schratzenstaller, M. (2009). Macro-economic effects of the Fiscal stimulus measures in Austria. *Austrian Economic Quarterly, 4*, 205–2016.

Compagnoli, R. (2020). *Competitive Agility: How the hospitality industry can win Amidst new challenges*. Retrieved from https://www.entrepreneur.com/article/349354. Accessed 17 Apr 2020.

Congressional Research Services. (2020). *Global economic effects of COVID-19: Updated March 26, 2020*. New York: Congressional Research Services.

Crowe. (2020). *Malaysia Economic Stimulus Package 2020: Announced on 27 February 2020*. Kuala Lumpur: Crowe.

de Rugy, V., & Leff, G. (2020, 25 March). *The case against Bailing out the Airline Industry*. SSRN Special Edition Policy Brief, https://doi.org/10.2139/ssrn.3571441.

Deloitte. (2020). *Economic Commentary: Vol 44 No 1 – Coronavirus* (Special edn). London: Deloitte.

ESCAP (Economic and Social Commission for Asia and the Pacific). (2020). *The impact and policy responses for COVID-19 in Asia and the Pacific*. Kuala Lumpur: ESCAP.

Gámez, A., Ivanova, A., & Campiranon, K. (2014). *Tourism and economic crisis management within APEC*. Phuket/Los Cabos: Cases.

Hai, W., Zhao, Z., Wang, J., & Hou, Z. G. (2004). *The short-term impact of SARS on the Chinese Economy* (Asian Economic Papers 3:1). New York: The Earth Institute at Columbia University and the Massachusetts Institute of Technology.

IATA (International Air Transport Association). (2020). *Overview on COVID-19 Announced relief measures*. Geneva: IATA.

ILO (International Labour Organisation). (2009). *Review of sector-specific stimulus packages and policy responses to the global economic crisis*. Geneva: ILO.

IMF (International Monetary Fund). (2020). *Policy responses to COVID-19: Policy tracker*. Retrieved from https://www.imf.org/en/Topics/imf-and-covid19/Policy-Responses-to-COVID-19. Accessed 31 Mar 2019.

Jamal, T., & Budke, C. (2020). Tourism in a world with pandemics: Local-global responsibility and action. *Journal of Tourism Futures*. https://doi.org/10.1108/JTF-02-2020-0014.

Köhler-Töglhofer, W., & Reiss, L. (2009). The effectiveness of fiscal stimulus packages in times of crisis. *Economic Analysis & Policy, 39*(1), 78–99.

KPMG. (2020). *Federal Government Second Economic Response Package: A review of measures announced as of 22 March 2020*. Sydney: KPMG.

Li, H., Chen, X., & Huang, H. (2020). The novel coronavirus outbreak: What can be learned from China in public reporting? *Global Health Research and Policy, 5*(9). https://doi.org/10.1186/s41256-020-00140-9.

Linton, M. J., Coast, J., Williams, I., Copping, J., & Owen-Smith, A. (2019). Developing a framework of quality indicators for healthcare business cases: A qualitative document analysis consolidating insight from expert guidance and current practice. *BMC Health Services Research, 19*, 433. https://doi.org/10.1186/s12913-019-4269-9.

MAS (Monetary Authority of Singapore). (2020). *Recent economic developments in Singapore*. Pulau Ujong: MAS.

Medina, A. F. (2020). *Indonesia Unveils Stimulus Package to combat Coronavirus impact*. Retrieved from https://www.aseanbriefing.com/news/indonesia-unveils-stimulus-package-to-combat-coronavirus-impact/. Accessed 7 Apr 2020.

More Advent. (2020). *Malaysia 2020 Economic Stimulus Package*. Kuala Lumpur: Luther Services.

Mufti, R. R. (2020). *COVID-19: Government calls for limits to all tourist activities*. Retrieved from https://www.thejakartapost.com/news/2020/03/17/covid-19-government-calls-for-limits-to-all-tourist-activities.html. Accessed 7 Apr 2020.

OECD (Organisation for Economic Cooperation and Development). (2020). *SME Policy Responses to COVID-19*. London: OECD.

Ramey, V. A. (2019). Ten years after the financial crisis: What have we learned from the renaissance in fiscal research? *Journal of Economic Perspectives, 33*(2), 89–114.

Rodríguez-Morales, A. J., MacGregor, K., Kanagarajah, S., Patel, D., & Schlagenhauf, P. (2020). Going global: Travel and the 2019 novel coronavirus. *Travel Medicine and Infectious Disease*. https://doi.org/10.1016/j.tmaid.2020.101578.

Saha, D., & Von Weizsäcker, J. (2009). *EU Stimulus Packages: Estimating the size of the European stimulus packages for 2009*. Brussels: Breuegel.

Services, L. (2020). *The Malaysian Economic Stimulus Package 2020 – Support for the Malaysian Economy in the Times of COVID-19*. Kuala Lumpur: Luther Services.

South Africa National Department of Tourism. (2016). *National Tourism Sector Strategy*. Cape Town: National Department of Tourism.

Tasmanian Government. (2020). *Tasmanian support & Stimulus Package*. Retrieved from www.health.tas.gov.au/coronavirus. Accessed 16 Apr 2020.

The Economist. (2009). *China's stimulus package: A six-month report card*. London: The Economist.

Titiheruw, I. S., Soesastro, H., & Atje, R. (2009). *Global Financial Crisis Discussion Series Paper 6: Indonesia*. Paris: Overseas Development Institute.

TTFA (Tourism and Transport Forum Australia). (2020). *Estimated impact of coronavirus on the Australian visitor economy*. Sydney: TTFO.

UNDP (United Nations Development Programme). (2020). *Assessment Report on Impact of COVID-19 Pandemic on Chinese Enterprises*. Beijing: UNDP.

United Nations. (2020). *Shared responsibility, global solidarity: Responding to the socio-economic impacts of COVID-19*. New York: United Nations Secretariat.

UNWTO (United Nations World Tourism Organisation). (2020). *Supporting jobs and economies through travel & tourism: A call for action to mitigate the socio-economic impact of COVID-19 and accelerate recovery*. Geneva: UNWTO.

Walsh, J. (2010). Thai Khem Kaeng: An inadequate response by the Thai State to the impact of the 2008 economic crisis. *Journal of Economics and Behavioural Studies, 1*(1), 1–8.

Wood, L. M., Sebar, B., & Vecchio, N. (2020). Application of rigour and credibility in qualitative document analysis: Lessons learnt from a case study. *The Quality Report, 25*(2), 456–470. https://nsuworks.nova.edu/tqr/vol25/iss2/11.

World Bank. (2020). *East Asia and Pacific in the Time of COVID-19: Update April 2020*. Washington, DC: World Bank.

WTTC (World Travel & Tourism Council). (2020). *COVID-19 related policy shifts supportive of travel and tourism sector*. Geneva: WTTC.

Part V
Conclusion and Policy Recommendations

Chapter 16
Conclusions and Policy Recommendations: Building Back Better Global Tourism Systems Post-COVID-19

Abstract The tourism industry remains one of the largest economic sectors in the world in terms of both employment creation and contribution to the global economy. However, the industry is sensitive to external shocks such as natural and/or human-induced disasters. This book documented the impact of the COVID-19 pandemic on the global tourism industry. This chapter (also serving as a policy brief) presents emerging key findings and conclusions, as well as policy recommendations that are embedded within the Sendai Framework of Disaster Risk Reduction (DRR) and management. Drawing from past disasters, the emphasis is placed on building back better (BBB) global tourism systems post-COVID-19. The decision to pivot the chapter on the DRR and management framework that embraces the BBB concept is informed by the inevitable new global normal. The new global normal is characterised by increasing and more intense disasters across all spatial levels, from a local to a global scale, which normally results in severe negative impacts on all economic sectors as experienced during the Indian Ocean earthquake and tsunami in 2004 and the COVID-19 pandemic in 2019/2020. This orientation is further informed by the futuristic thinking that incorporates matters of building resilience and adaptation to disaster risks within the tourism industry. Therefore, the chapter provides the tourism industry and its key stakeholders with both immediate ideas for quick and big wins post-COVID-19 recovery pathways and future ideas to build resilience that permits adaptation of the tourism industry to future disasters. The development of globally accepted protocols enhancing health and safety, especially for the tourist and those working in the tourism and related industries, is recommended. Should these measures be fully upheld, then international travel, which is critical in the tourism sector, may resume sooner rather than later, with spill over effects on exciting domestic tourism too.

Keywords COVID-19 · Tourism industry · "New normal" · DRR · Build back better · Stakeholders

G. Nhamo et al., *Counting the Cost of COVID-19 on the Global Tourism Industry*,
https://doi.org/10.1007/978-3-030-56231-1_16

16.1 Introduction

The impacts of previous epidemics, pandemics and/or other disasters on the tourism industry have been observed by Page et al. (2012) to differ according to the extent to which a country depends on the sector and its contribution to the gross domestic product (GDP). The authors also note that pandemics often lead to the shifting of demand from locations deemed to be unsafe to those perceived to be safer. However, unlike previous pandemics and disasters, the impact of COVID-19 has been global, with almost 100% of world destinations affected (UNWTO 2020a). This left very few (if any) alternative destinations deemed to be safe. In terms of recovery of the tourism industry from COVID-19, some like Peters et al. (2020) see the impossibility of the industry returning to the pre-COVID-19 normal. This is mainly attributed to the widespread negative impacts of the pandemic concerning issues such as job insecurity; higher unemployment rates in both tourism and other sectors across the economy; reduced savings, which will lead people to prioritise necessities; and tourism suffering as a luxury spend.

The book started by providing an overview of tourism, COVID-19 and global development agendas, including the politics of COVID-19 and the overarching methodological framework. Detailed chapters were developed to highlight the plight and responses of the global tourism system, giving attention to the following industries: aviation, hoteliers, restaurants, cruise ships, airports, entertainment and car rentals. Space was also allocated for chapters dealing with sports and religious tourism, preparedness and response to COVID-19, responses from stock markets and tourism-related shares, philanthropic activities and the emerging economic stimulus packages. Some of the emerging key findings and conclusions from the book are presented in the next section.

16.2 Emerging Key Findings and Conclusions

Overall, the key findings show that the COVID-19 pandemic slowed down and in some cases reversed the gains already made in the implementation of global development agendas, particularly the 2030 Agenda for Sustainable Development (AfSD) and the 17 Sustainable Development Goals (SDGs). The book also reveals that the unthinkable happened in the tourism industry, a near total shutdown of all international passenger travel at some point due to the pandemic leading to some tourism operations completely closing down in many places. Almost universally, there were national regulations put in place to subdue the rapid spread of COVID-19. The measures include social distancing, isolations, quarantines, lockdowns and curfews, among others. All these accelerated the rate at which the tourism and related industries were collapsing. Broadly, the following key impacts were witnessed: (1) death, illness and COVID-19-induced fear to potential tourists; (2) loss of revenue and income; (3) loss of employment and reduced work hours; (4) postponed sporting and religious tourism events; (5) elements of unpreparedness by stakeholders across the tourism value chain; (6) a historical global financial and other resources mobilisation movement; (7) blame and COVID-19 witch-hunting; (8) institution of economic stimulus packages, mostly at a national level; and (9) concerted efforts by key tourism stakeholders to accelerate the relaxation of lockdowns through lobbying and the development of appropriate voluntary global start-up protocols.

With initial responses to the COVID-19 pandemic, the findings in this book reveal that generally, Asian countries were prepared to react quicker compared to countries in other regions. This was mainly due to a regular epidemic and pandemic outbreaks in the region. After several months of lockdowns and the implementation of social distancing measures, world leaders (under pressure from citizens and residents) started implementing step-wise relaxations of these restrictions. The main aim of relaxing the restrictions was to "flatten" the plummeting economic curve. This was inevitable since by the end of March, and during April 2020, the main noise in global economic systems was to focus on "flattening" both the COVID-19 and economic curves. Both these curves were of interest to the tourism industry. In the next paragraphs, additional details on the overall impact of the COVID-19 pandemic on the global economy and selected tourism sectors are provided.

With the COVID-19 pandemic, it emerged that global international tourist arrivals were likely to drop by between 20% and 30% compared to 2019 (Gössling et al. 2020). In South Africa, which had one of the toughest lockdowns in the world, the CEO of the Tourism Business Council highlighted that there was likely to be nothing left if the industry was left to reopen only in December 2020 (Smith 2020). From a survey on the impact of COVID-19 by the Cape Town Tourism Department, it emerged that 56% of companies had no recovery plans in place, 36% partially paid staff, while 31% put staff on unpaid leave and 18% retrenched employees. About 83% of businesses sampled revealed that they were not likely to survive beyond 6 months under prevailing lockdowns conditions (Ibid.).

The first line of defence by governments to the coronavirus was to prepare for and prevent the spread of COVID-19 after the World Health Organisation (WHO) declared a global pandemic on 11 March 2020. The precautionary measures included the closure of ports (land, sea and air) and screening and testing for infected persons. Precautionary measures also included the promotion of some measures learnt from the 1918 influenza pandemic, as improved upon over time through experiences of other outbreaks such as SARS, Ebola and MERS. The measures included handwashing, good sneezing etiquette and social and physical distancing. These measures led to negative chain impacts as many tourism establishments and businesses closed, among such were restaurants, hotels, B&Bs, cinemas, live shows and amusement and national parks. Port closures also resulted in the total grounding of airlines and an almost full closure of airports as these were now only servicing repatriations and the movement of essential goods and services like medicines, as well as medical and security personnel.

The port closures and other COVID-19 preventive measures highlighted herein, however, proved insufficient as the pandemic continued to spread rapidly across the world. Nearly every country on earth recorded COVID-19 infections. From China as the first epicentre, COVID-19 ravaged Europe, with Italy as its new global and European epicentre of infections. However, other European countries were badly hit by the pandemic, including Spain, the UK, France, Russia and Germany. In no time the pandemic spread to North America, with the USA becoming the new global epicentre. As this book was being finalised in June 2020, the USA was still the global epicentre both in terms of COVID-19 infections and deaths, although there were serious worries regarding Brazil, Russia and South Africa.

As the COVID-19 pandemic spiralled globally, many countries moved into response mode. This phase involved the continuation of awareness-raising, although the main activities were testing, contact tracing, self-isolation, quarantining of infected or suspected individuals and treatment. This meant huge investments in procuring COVID-19 specific quarantine centres, hospital beds, ventilators, testing kits, gloves, face masks and other personal protective equipment (PPE). Since many governments had not prepared for this huge expenditure, there was a need to set up national platforms to mobilise financial, technical and other resources. In countries such as South Africa, a COVID-19 Solidarity Fund was established. Earlier, the WHO had also established its own Solidarity Response Fund to assist in building COVID-19 response capacity in developing countries. Many donations came in from individuals, companies, nations and philanthropists. Governments and central banks came in with a variety of economic stimulus packages and repo rate cuts. From the preliminary findings, economic stimulus packages were generally much higher than the 2008/2009 global financial crisis economic stimulus packages. In certain instances, some central banks cut repo rates three times in less than 2 months as the gravity of the pandemic kept growing bigger and clearer. These countries include Canada, Pakistan and the Philippines. The tourism industry was among the priority sectors to receive the economic stimulus packages. However, the cruise ship industry was having challenges to access these packages as many companies were registered outside their countries of origin to avoid strict labour laws and taxes such as those from the USA. Debates were also ongoing as to whether airlines should benefit from economic stimulus packages as these should have been prepared financially for such pandemics (Rugy and Leff 2020).

Although such economic stimulus packages are a good gesture, governments the world over were left in a dilemma regarding which sectors to prioritise and which planed programmes to put on a standstill or cancel altogether. As budgets were being shifted and repo rates cut, economies were also shrinking, and some countries projected huge shortfalls in tax revenues. Some statistics regarding the first quarter of 2020 GDP in selected countries are shown in Fig. 16.1. The worst affected country was China, which experienced an almost 10% GDP decline. This decline was likely to have ripple effects on other countries that have close trade links with China. In particular, the tourism industry was likely to suffer the most given that outbound Chinese visitors have the highest presence in destinations worldwide.

Debates were also raging regarding COVID-19 status in Africa as some still believed the peak was yet to be reached given the limited testing and case isolation capacity. A diplomatic war also erupted between the USA and Chinese administration as the USA continued building an unproven case that China had released the

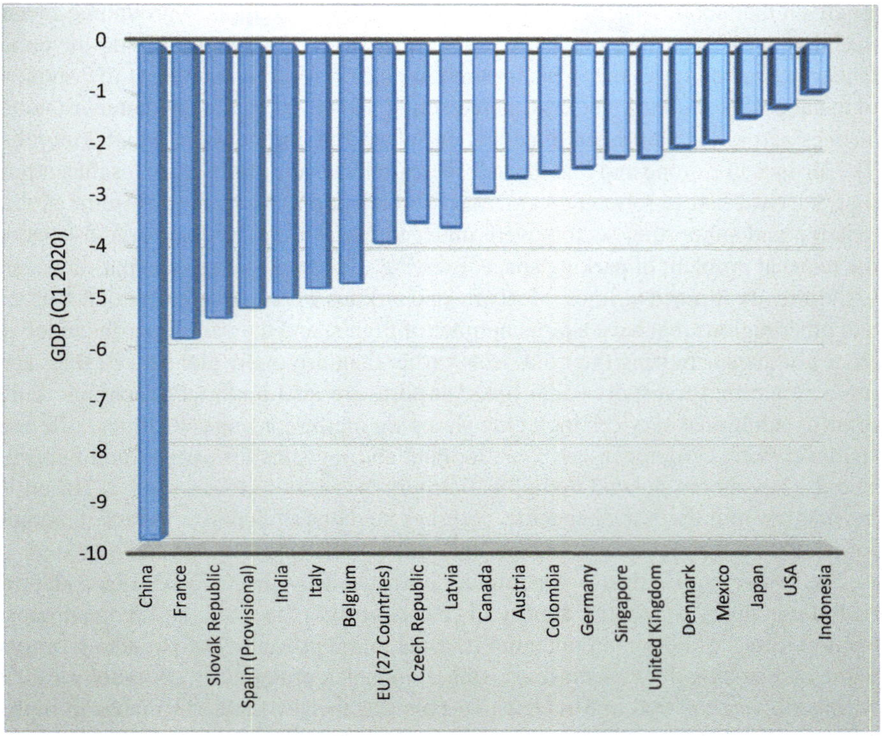

Fig. 16.1 Reduction in GDP growth due to COVID-19 (other sources put the figure at 6.8% [Financial Times, 2020. China's GDP fall is a chance to ditch its national growth targets, Retrieved from https://www.ft.com/content/983d5074-805d-11ea-8fdb-7ec06edeef84 (accessed 11 May 2020)]). (Source: Authors, data from OECD (2020), MTI (2020) and The Guardian (2020))

coronavirus causing COVID-19 from its laboratory in Wuhan (deliberately or otherwise), rather than it being a natural zoonotic disease. Debates continued concerning the competence of the WHO Chief, with the USA also accusing the WHO head of conniving with China to hide the COVID-19 outbreak leading to a slow global response and preparedness. The manner in which China handled the notification of the coronavirus outbreak to the WHO remains worrying. This is so against the background that China was not a first-time offender. The country had been in a similar situation as it took months to notify the WHO of the SARS outbreak in 2002/2003. Although China accepted responsibility and managed to rebuild trust with the global community, the extent of the COVID-19 outbreak is beyond human imagination, and the backlash was growing. To this end, Japan joined the USA in the backlash by announcing huge incentives for its companies to relocate from China back to Japan or any other country of choice. However, it is important that the said tension was possibly fuelled by the raging demand for global economic and political dominance, particularly between China and the USA that was raging on at the time of this pandemic.

Given that about 49% of the airlines' expenses are fixed, their grounding meant that they were incurring costs even when they were not generating any revenue. Under normal circumstances, airlines require a cash reserve equivalent to 3 months of usual costs for them to be considered sound. However, as the pandemic hit, most airlines across the world had a median cash flow ranging between 1 and 2 months. The airlines in Europe and Pacific Asia were under severe strain, and this meant that their survival had to rely on an external cash injection. As such, jobs in the global aviation and supporting sectors were under threat. Airlines and airports also faced the unusual problem of parking space, exerting additional storage and parking costs to the already strained airlines. Most affected airlines were in North America, Europe and other regions that have a large number of flights, and the parking problem led to some companies retiring their old fleets earlier than originally planned. Besides acting as important transport nodal links, airports are also important economic hubs offering additional services including shopping malls, car rental facilities, ride and share services, conference facilities, lodging and logistics for exports and imports. Initial assessments showed that global airports were set to lose over $76 billion in the first few months of the pandemic outbreak, and this spilled over to loss of income for related air navigation companies and others in the airport value chain.

The cruise ship industry was almost devastated by the COVID-19 pandemic, with the industry cancelling almost all cruises around the world and in the process losing billions of dollars in potential revenue in the process. The pandemic further witnessed stock values of most cruising companies plunging by as much as 80% within a 2-week period in March 2020. This was their worst performance in living memory. In addition, cruise passengers were traumatised as they could not disembark in many jurisdictions, with some getting infected and also dying on board as documented in this book. The Diamond Princess is the commonly cited case in this space. However, the industry was poised to make a recovery post the COVID-19 crises, although it would not be a quick rebound. Rays of hope for the revival emerged with some regular passengers expressing a willingness to return to the cruise ships once the virus was under control and with some even making future bookings during this period. Investors were also showing faith in the industry by injecting funds into it during the COVID-19 lockdown. With the "no sail" order imposed by the Centre for Disease Control and Prevention (CDC) likely to be lifted after 24 July 2020 (Froneman 2020), Carnival Cruise opened bookings for August 2020. The bookings surged 600% within 3 days, 200% compared to the same period in 2019. This act shows a ray of hope for some recovery in the industry. The only challenge was that the bookings were conditional to sailing orders being lifted.

It emerged that there were unprecedented cancellations in accommodation bookings for hotels, B&Bs and other establishments due to travel restrictions, lockdowns and direct orders from governments for the sector to close. Faced with imminent collapse, hotels resorted to massive and systematic job layoffs and furloughs with most hotels temporarily shutting down. There was also a noted realist possibility of most hotels defaulting on their financial obligations such as loan repayments and servicing of rates. Most global hotel groups deferred or cancelled paying dividends to shareholders and suspended capital expenditure as highlighted in the book. Initial assessment showed that for the hotel industry alone, billions of dollars in potential revenue was lost. The figure could translate into trillions of dollars if one considers the knock-on effect of hotel shutdowns on the entire tourism value chain.

COVID-19 also adversely affected the sports tourism industry. Most sporting events around the world were either cancelled and/or postponed. The events included professional football matches in major European, African and South American leagues, as well as sports mega events lined up in the first half of 2020. The highest profile sporting event that fell victim to COVID-19 was the 2020 Tokyo Olympics. The cancellations of sporting activities led to losses for the individual teams, broadcasters and sponsors, leaving most of them in financial distress. From an athlete's perspective, the COVID-19 pandemic resulted in the loss of income as most sporting clubs significantly cut salaries with some reducing this by as much as 70%. The athletes also faced a major battle to remain fit during the period of the pandemic. This was a difficult task given the fact that they had no access to training facilities. In the early days of the COVID-19 pandemic, there were visible commercial interests as employers seemed to have buried their heads in the sand, continuing with scheduled events even when it was no longer safe to do so. This facilitated the spread of the disease, risking the health and wellbeing of sports personalities and spectators. It took the contraction of the virus by several athletes for action to be taken. Like all the other sectors, the impacts of this pandemic on sports tourism was not only direct. Consequently, also affected were stakeholders in the supply chain, including batting syndicates and catering firms.

Religious activities, and by default religious tourism, were also negatively affected by the COVID-19 pandemic, resulting in them being either being scaled down, cancelled or postponed. Such major events included Saudi Arabia's Hajj and Umrah for Moslems as well as the Easter festivities for the Christian churches. There were loses made by stakeholders from both the hosts and receiving places. With some economies such as Saudi Arabia cashing in on the Hajj and Umrah, government revenue was reduced. Other industries in the value chain such as travel agencies, hotels, transport and restaurants also took a knock. Unfortunately, pilgrimages such as the Hajj and Easter are calendar-specific and cannot be rescheduled, and thus there were no possibilities to reschedule these in 2020. Consequently, there were instances of confrontations between governments and religious leaders when the latter ignored the lockdowns and social distancing measures. In some cases, ignoring government regulations resulted in many infections and in other instances deaths. Such instances were reported in Iran, South Africa, South Korea and India.

The book also investigated the impact of COVID-19 on the stock market, especially tourism-linked counters. It emerged that no previous infectious disease outbreak, including SARS, the Spanish Flu and Ebola, negatively impacted the stock markets the way COVID-19 did. Generally, stock markets reacted negatively to the emergence and tumultuous spread of the disease as measures were put in place by governments around the world to try and contain and limit the spread of the contagion, and by default, disrupted many (if not all) industries. Three major events were noted to have significantly shocked the stock markets, namely, (1) the lockdown in Wuhan, China, in January 2020; (2) the lockdown in Lombardy, Italy, in February 2020; and (3) the declaration of COVID-19 as a pandemic by the WHO on 11 March 2020. Tourism-related stocks were hardest hit as travel bans and lockdowns were

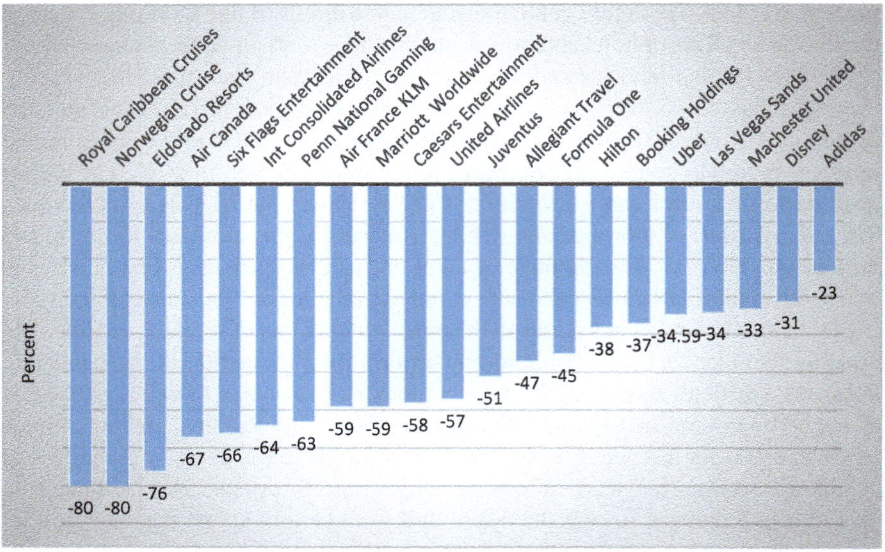

Fig. 16.2 Drop in share value of tourism-related firms. (Source: Authors, data from Statista (2020))

instituted in all global destinations. Figure 16.2 summarises the drop in share value (as a percent) of selected tourism-related firms during the COVID-19 pandemic.

Governments around the world responded to the COVID-19-induced shock on the stock markets by trying to reassure the markets through policy interventions and implementing multi-trillion-dollar rescue packages for distressed industries. The markets responded well to these packages with some industries making steady recoveries. However, shares in the tourism sector largely remained in the negative, and recovery was subdued. This is because a deep uncertainty remained as the pandemic continued to spread and intensify. Hence, confidence in the sector could only return once the pandemic was under control and restrictions on travelling removed.

The impact of the COVID-19 on the entertainment industry was another aspect investigated, with revelations that the pandemic devastated the industry to levels also never witnessed before. There was a loss of production time as well as cancellation of productions and film and music festivals. This led to the untold suffering of artists and losses of revenue by entertainment companies. In some instances, livelihoods changed overnight! Global giants in entertainment such as Box Office lost billions of dollars in potential revenue due to the postponement of major film launches and the shutting down of studios. These losses were likely to have long-term impacts on revenue. The television and entertainment sector found itself compromised. This was in terms of providing new content, an element that came at a time when there was large demand for entertainment during the lockdown periods where the audience was mostly available, as many people were not going to work or school. As a consequence, there was a rare growth in share price for the video on demand and streaming companies like Netflix and Showmax. As was witnessed from other line industries in the tourism sector, firms in entertainment had to contain costs through rescheduling capital investments, laying off and furloughing employees. Unfortunately, the entertainment industry was likely to remain subdued as many lockdown relaxation levels were putting the industry in the last segments due to social distancing requirements that the industry cannot guarantee.

Given the magnitude of COVID-19 generally, and its foregoing impacts on the tourism sector, specifically, governments had to intervene through economic stimulus packages, including monetary policy measures. The economic stimulus packages remained one of the major immediate relief and response measures. However, it emerged that the road to recovery was never going to be quick or painless. No tourism business, small or big, was immune to the crisis. Big businesses run huge overheads, while small businesses usually have cash flow challenges. As such, solutions post, COVID-19 had to bring all stakeholders on board including governments, tourism industry, philanthropists, individuals and organised labour. Several typologies of economic stimulus packages emerged for the tourism sector, and these included grants and subsidies, use of voucher systems and domestic tourism promotions, drastic repo rate reductions, direct lending, tax waivers and reductions, social security payments, rent and utility payments, loan guarantees and debt moratorium and other debt repayment arrangements. Among eight economic stimulus packages tracked across 54 countries worldwide, the most popular included VAT, rent/utilities/local tax, social security, grants and subsidies, debt moratorium, loan guarantees, direct lending and income and corporate tax breaks in varied forms and degrees.

In the absence of a COVID-19 cure and/or vaccination, tight controls on human mobility are likely to remain in place the world over, ushering in the so-called new normal (WTTC 2020a). Johnson and Boone (2020) observe that as the world exited lockdowns, it was entering another challenging period of lock-ins. These lock-ins were likely to severely constrain international travel, a key pillar in the tourism value chain and recovery. Players in the tourism industry, therefore, needed to abandon their old ways of doing things and reinvent new strategies of operation that are in harmony with the new reality. The new normal in international travel although difficult to fully envision, was likely to witness every port of entry ramping up COVID-19 testing, quarantines or traveller, entry/departure declines and returns (whether arriving by air, sea or road). However, given the incubation period of COVID19, such measures were likely to deter travellers, forcing them to stay home for a while. Besides the health aspect, the global economy was also likely to feel the effects of the pandemic for years to come and was likely to trigger a worldwide economic depression, which would further hurt the prospects of a quick recovery in the tourism industry. Figure 16.3 shows the projected revenue loss by the tourism industry in the year 2020.

The next and last section in this chapter focuses on policy recommendations. The authors hope to present a concise tabulation of action points and protocols that the tourism industry can utilise to realise quick and big wins in post-COVID-19 recovery. Although the main focus is on recovery, space is also reserved to reflect on preparedness as well as relief and response as countries across the world are at different stages in experiencing and addressing the pandemic.

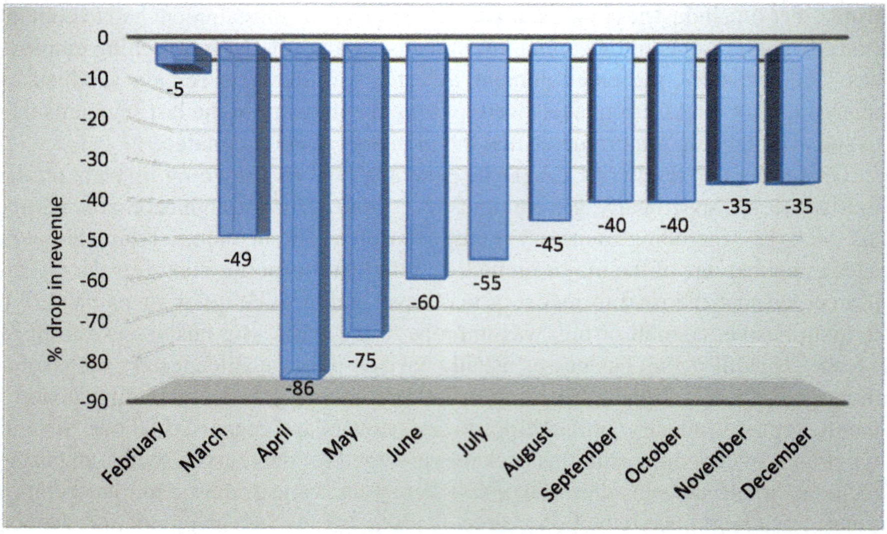

Fig. 16.3 Projected revenue loss by the tourism industry in 2020. (Source: Authors, data from Statista (2020))

16.3 Policy Recommendations: BBB Global Tourism Post-COVID-19

This policy recommendations section is structured around the DRR and management framework and cycle focusing on (1) preparedness, (2) relief and response as well as (3) recovery and reformation (embedding the BBB concept in rehabilitation and reconstruction). The BBB concept implies becoming more resilient, bouncing back quicker, and more efficiently as well as leaving no one behind after disasters (World Bank 2018). The perspectives on building resilience (including adaptive capacity) across economic sectors are well-articulated in current major global development agendas, especially the Sendai Framework and the Paris Agreement (Fig. 16.4). In the Sendai Framework, the scope of DRR was broadened to focus on both natural and human-made hazards, as well as related environmental, technological and biological hazards and risks, with health resilience strongly promoted throughout (UNDRR 2015). To this end, investing in DRR for resilience is one of the four main priorities for action. Resilience (adopted in the Sendai Framework) is then defined as:

> The ability of a system, community or society exposed to hazards to resist, absorb, accommodate and recover from the effects of a hazard in a timely and efficient manner, including through the preservation and restoration of its essential basic structures and functions. (UNDRR 2009: 24)

Resilience is also supposed to be promoted at workplaces that include tourism industries through both structural and non-structural measures (UNDRR 2015). It was due to be enhanced in national health systems, through the integration of disaster risk management, into all forms of the healthcare system at a local level.

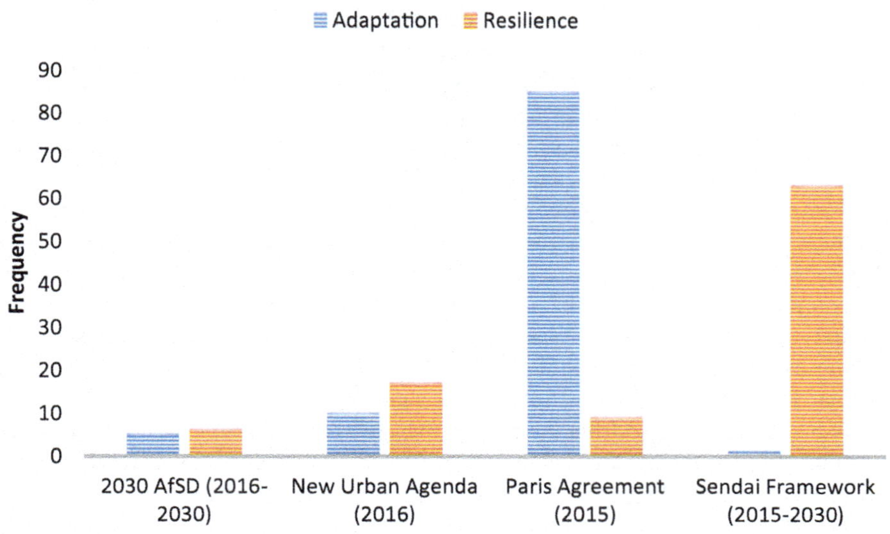

Fig. 16.4 Resilience and adaptation mentions across current global development agendas. (Source: Authors, based on UNODRR (2015), United Nations (2015), UN Habitat (2016) and UNFCCC (2015))

Concerning tourism, the Sendai Framework makes direct reference to such under Priority 2 that focuses on strengthening disaster risk governance to mitigate the impact of the disaster. As such, the industry is supposed "to promote and integrate disaster risk management approaches throughout the tourism industry, given the often-heavy reliance on tourism as a key economic driver" (UNDRR 2015: 20). From the 2030 AfSD and the New Urban Agenda, the world committed to promoting sustainable tourism (UN Habitat 2016) and embedded this into selected targets, namely, 8.9, 12.b and 14.a, as presented in this book (United Nations 2015).

16.3.1 Preparedness

To holistically grasp the perspectives on disaster preparedness, the Sendai Framework for Disaster Risk Reduction (2015–2030) should be recognised and elevated to be the overarching DRR framework explicitly covering the tourism sector as well. Although the WHO has a mandate over global health matters, pandemics such as COVID-19 remain one of the disasters listed under the Sendai Framework. For some reason, it seems the world has prioritised climate change-related disasters, many of which are not pandemics, with huge annual gatherings taking place to review progress. To this end, key business and development players, including those in the tourism sector, bought into this narrative. The climate change narrative also found its way into many other key global development agendas including the 2030 AfSD where SDG 13 (Climate Action) is somehow elevated among other SDGs with the United Nations Framework Convention given the mandate to strictly oversee it. Climate change also featured strongly in the United Nations Habitat III's New Urban Agenda. Hence, the COVID-19 pandemic somehow reminded the world and the tourism industry to refocus and place all potential global disasters on an equal footing regarding preparedness. Like climate change, the COVID-19 pandemic became global and notably acute beyond anybody's imagination.

There seems to be epidemic, pandemic and/or disaster fatigue settling in across countries. The same could be true for economic sectors such as tourism and related firms. However, the best way to be ready for any such eventualities is for countries to continuously learn from past disasters and refine their national, regional and global response protocols. Disasters come like a thief in the night! To this end, countries and regions that regularly experience epidemics and pandemics such as China, Hong Kong, Taiwan, South Korea and Asia, in general, were ready for COVID-19. This was so from the perspectives of existing comprehensive measures including having testing kits and protocols in place, contact tracing, general hygiene in terms of handwashing and other practices, ventilators, masks and mask-wearing, other PPE, self-isolation and quarantine. Hence the pandemic was quickly contained. The level and speed of response also seemed to be linked to the manner in which past epidemics and pandemics impacted economies, with Hong Kong, South Korea and Taiwan being good examples.

Time has also come to rethink WHO disease notification protocols and impact determinations. Drawing from improvements and emerging efficiencies in impact-based weather forecasting approaches, which has moved away from merely alerting stakeholders of certain measured and quantified millimetres of rainfall, amount of snow or wind speeds, etc., the sector now informs users of the nature of impacts from such. The users are more interested in the potential impact the measured millimetres of rainfall, amount of snow or wind speed will have on certain economic sectors. Hence it was not good enough for the WHO to indicate the various stages in the spread of COVID19, going from being a global health emergency to being declared a pandemic. The global citizens needed to understand the likely health, social, economic and other impacts across sectors, including tourism. Such information becomes more relevant and useful for preparedness and future planning. With the assistance of other United Nations agencies, professional bodies, academia and businesses, the WHO can start to draw up matrices indicating three to five levels of possible impacts of epidemics and pandemics across economic sectors. Known and past epidemics and pandemics can be used as baselines, as well as the COVID-19 experience.

The global tourism industry has weathered many storms in the recent past. These include the 2004 Indian Ocean earthquake and tsunami that hit the Asian region; terrorist attacks such as the 9/11 event in the USA; the London, Bali and Paris bombings; and previous disease outbreaks such as SARS in 2003, swine flu in 2009 and the West African Ebola in 2014 (Higgins-Desbiolles 2020). Lessons from these crises led the tourism industry to advance toolkits of risk management, risk mitigation, crisis recovery and resilience to future challenges within its value chains (Hall et al. 2018). From a disaster risk reduction (DRR) perspective, the industry tried to devise strategies for the prevention and preparation of incidences that it had faced in the past in accordance with the magnitude of the disasters. However, and despite these previous experiences, nothing could prepare the industry for the unprecedented impact of the COVID-19 pandemic. The pandemic turned out to be a perfect storm as it was of much higher intensity, magnitude, temporal and spatial resolution compared to past events. As a result, the tourism industry found itself in unchartered waters.

Although global preparedness for emerging infections with pandemic potential has improved by building and learning from past experiences, the focus has mainly been on the action taken when the infection had emerged, and not on prevention. It is therefore recommended that more emphasis needed to be put on prevention, which is less costly than taking action when there is already an outbreak, which is more expensive to control in the long term. Adopting the prevention health approach could substantially improve the situation, manage the related costs and reduce the threat of future incidents.

Generally, the tourism sector, like any other industry, seemed to have been ill-prepared for the COVID-19 pandemic. The pandemic came across like a level five global hurricane. It came across like a level five global cyclone. It was like a global earthquake of the highest magnitude. It was a global tsunami, indeed! Hence the tourism sector lacked the desired understanding in terms of the magnitude of the COVID-19 pandemic. In addition, the capacity to deal with disaster risk governance remained weak. Furthermore, there was no convincing evidence from the tourism sector regarding investments in DRR for resilience building and adaptation. Yes, as highlighted earlier, global tourism systems had encountered isolated major disasters such as the 9/11 terrorist attacks, London bombings and the 2004 Indian Ocean earthquake and tsunami and had worked overtime to come up with disaster plans aligned to such. However, COVID-19 took the industry by surprise, such as no other modern pandemic had been witnessed, apart from the 1918 Spanish flu pandemic. Although similar, advances in travel, especially the airlines and cruise ships, meant the COVID-19 pandemic spread across the world literally overnight.

Given that pandemics such as COVID-19 can easily spread in confined and crowded spaces, the focus on improving preparedness in airlines, cruise ships and all ports of entry must be the priority focus for the future. Ports of entry failed to detect the coronavirus. As for the cruise industry, there is a need to improve its state of medical preparedness on board so that it relies less on ground-based facilities in the case of emergencies and future outbreaks. There is also the need to improve sanitation procedures on cruise ships, with guests being educated on the importance of good hygiene. Additional work on better evacuation and quarantine procedures from the cruise industry in the event of future outbreaks needs to be done too. Generally, the cruise industry did not respond quickly to the COVID-19 outbreak as voyages could have been suspended sooner. Modern forms of disease screening prior to boarding remain a possible solution to be enforced in the industry.

Since major gatherings, especially sporting events and pilgrimage, are known disease and pandemic super-spreaders, ongoing awareness-raising and education by the relevant authorities and religious leaders need ratcheting. Equally important is to engage religious leaders in terms of roles they could play to build community resilience to epidemics and pandemics. This could be done without unilaterally impeding their right to worship. The use of social influencers, religious and community leaders have been shown to bear good results in the fight against COVID-19 in many places. Hence scaling it up in other places impacted strongly recommended.

Many travellers would have been saving for a long time to embark on a holiday. This made it difficult to follow travel advice and go ahead with cancellations. With this in mind, the tourism sector should work with the insurance industry on a long-term plan to scale up travel insurance in preparedness for likely repeats of disasters of the COVID-19 magnitude or worse. This is likely to be an uphill task given that the insurance industry was already revising clauses in order to limit liabilities. Additional measures for individuals may include governments revamping unemployment and general health insurance.

Lastly, the tourism industry needs to come up with reformed DRR protocols, particularly revamping preparedness. The Sendai Framework for DRR and management remains the main guideline document for such. Both small and big businesses should develop scenarios on potential impacts and map out how particular industries in the tourism value chain must ready themselves.

16.3.2 Relief and Response

With virtually all global destinations still under some form of COVID-19 travel restrictions as per the United Nations World Tourism Organisation (UNWTO) as of the 8 May 2020 report, relief and response measures remained a reality (UNWTO 2020a). The UNWTO further details that 25% of all destinations worldwide had undergone at least 3 months of travel restrictions, with 40% having such for at least 2 months. About 156 destinations had completely closed their borders for international travel (57% in Africa, 80% in the Americas, 70% in Asia and the Pacific, 83% in Europe and 62% in the Middle East). This resulted in a huge business transaction burden in the tourism sector.

While it remained a huge challenge for many countries to have prepared for COVID-19, relief and response measures depended heavily on timely warnings from the WHO. Unfortunately, the WHO, in turn, relies on timely member states notifications, which China allegedly might not have done. Overall, governments and private entities should continue with economic stimulus packages and favourable monetary policies, especially those aimed at bailing out SMMEs. Although some jurisdictions such as Canada, Pakistan and the Philippines cut their repo rates three times, and many others did so twice, there remains room for more cuts where these are still above zero basis points. Where there is still a leeway to institute additional rounds of economic stimulus packages, urgency in this regard is paramount. Social safety nets, including unemployment benefits and insurance, were also triggered. However, if room to manoeuvre still existed, more should have been or should still be done. All commonly applicable COVID-19 economic stimulus packages – including relief on income generation constraints and relief from corporate taxes obligations, direct lending, loan guarantees, debt moratorium, grants and subsidies, social security as well as rent/utility measures – must be brought fully on board. These may also be extended in the recovery phase.

One of the tourism sectors that was difficult to deal with in terms of relief was arts and sports. There were vulnerable members from these industries that governments did not place on the radar early enough. This was made more complex by certain huge clubs in other countries that were filing for relief, only to be pressurised to withdraw the claims. While it may be rather late to consider the affected members for relief, measures to assist the sector to recover quicker were more appropriate, including bailouts for certain organisations and individuals.

16.3.3 Recovery and Reformation

To assist member states with gradual recovery from COVID-19, the UNWTO designed the "COVID-19 Tourism Recovery Technical Assistance Package" (UNWTO 2020b). From the package, the UNWTO emphasises that all stakeholders in the tourism sector must come to terms with the new reality in the hospitality, travel and tourism sector. The package is structured around three pillars, namely, economic recovery, marketing and promotion as well as institutional strengthening and building resilience. The concept of building resilience draws the reader back to the Sendai Framework's DRR and management framework and the BBB concept. From the economic recovery front, both quantitative and qualitative impacts of COVID-19, particularly on SMMEs, should be assessed, with appropriate economic measures implemented. As for marketing, appropriate promotion strategies covering the identification and targeting of markets that can aid quicker recovery, product diversification and coming up with attractive pricing and packaging guidelines were necessary. Lastly, institutional strengthening and resilience building would focus on enhancing working conditions in terms of health, safety and restoring consumer confidence. It would also focus on skills enhancement and the development of other relevant protocols. Some of these matters were picked up by the World Travel and Tourism Council (WTTC). The UNWTO further mapped the SDGs that the COVID-19 Tourism Recovery Technical Assistance Package would impact on (Table 16.1).

The WTTC also came up with a supporting initiative to the UNWTO assistance package. The WTTC's "Safe Travels Global Protocols" initiative was designed for the new normal in the tourism sector (WTTC 2020a, b, c). The Safe Travel Global Protocols was gaining momentum and acceptance from key stakeholders post-COVID-19 recovery. The global protocols (fully endorsed by authors of this work)

Table 16.1 UNWTO COVID-19 recovery package and impact on SDGs

COVID-19 recovery pillar	Sustainable Development Goal (SDG) impact
Economic recovery	SDG 1: No poverty SDG 8: Decent work and economic growth SDG 10: Reduced inequalities SDG 12: Responsible consumption and production SDG 17: Partnership for the Goals
Marketing and promotion	SDG 9: Industry, innovation and infrastructure SDG 11: Sustainable cities and communities SDG 10: Reduced inequalities SDG 17: Partnership for the Goals
Institutional strengthening and building resilience	SDG 1: No poverty SDG 4: Quality education SDG 5: Gender equality SDG 8: Decent work and economic growth SDG 9: Industry innovation and infrastructure SDG 10: Reduced inequalities SDG 11: Sustainable cities and communities SDG 12: Responsible consumption and production SDG 17: Partnership for the Goals

Source: Authors, based on UNWTO (2020b: 6–26)

act as common protocols and procedures for the tourism sector and were being created for the following: hospitality, outdoor retail, aviation, airports, cruise, tour operators, convention centres and MICE, car rental and insurance industries. The protocols are segmented into four pillars highlighted in Fig. 16.5. The first two global protocols for the hospitality and outdoor retail industries were released on 12 May 2020. The reader is directed to the WTTC website for future Safe Travel Global Protocols that were still to be developed.

Linked to the WTTC global protocols tourism de-risking model, South African Tourism (2020) developed a three-phase recovery model focusing on the "local first" (domestic), regional and international tourism. The three phases each had twin sub-phases (I and II) aided by transition points. The transition point within domestic tourism includes the implementation of health protocols by the tourism sector. To move from domestic into regional recovery and spaces, the transition point involves the lifting of regional country border closures and the institution of globally accepted health protocols such as those proposed by the WTTC. The transition point from regional into international tourism covers elements such as the lifting of long-haul country air travel bans, the institution of globally accepted health protocols and practices.

Permitting day trips, overnight and local air travel are envisaged as quick and big wins to recover domestic tourism. Self-driving and regional/continental air travel will facilitate recovery of regional tourism, while the resumption of international carriers including flights and cruises will signal progress towards full recovery. However, given that COVID-19 lockdown relaxations are going to be gradual and varied, rather than a once-off worldwide announcement, global tourism recovery

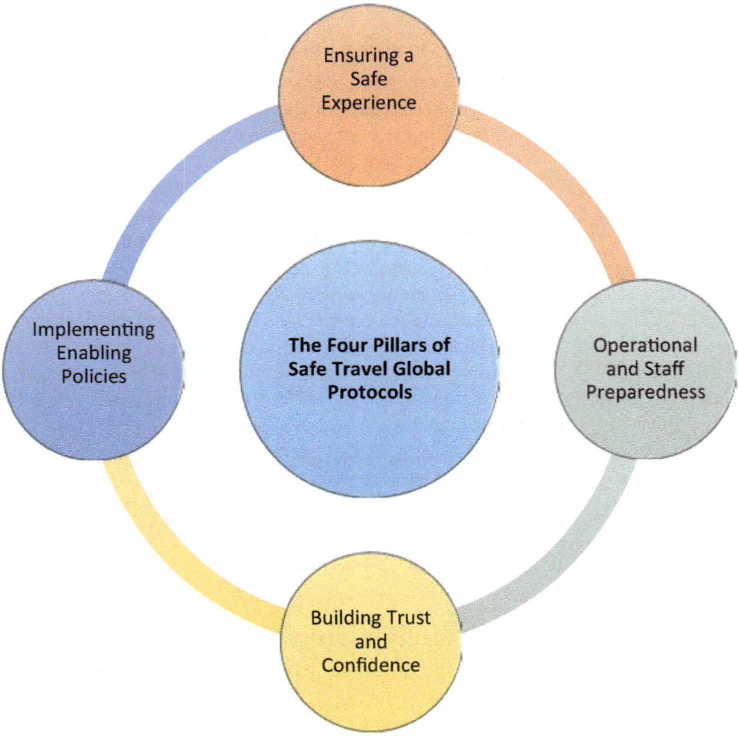

Fig. 16.5 Safe Travel Global Protocols pillars. (Source: Authors, based on WTTC (2020b: 2))

will take longer. Apart from promoting domestic and regional tourism, the industry needs to be on the lookout as to when positive global travel sentiments will start emerging. There will also be a need to cleverly identify countries that were not badly hit by COVID-19 in order to market these as early start-up locations and/or sources of travellers.

Although the entire tourism industry was battered by COVID-19, some other players and sectors will not manage to recover without large cash injections and other forms of support. The SMMEs, in particular, remain at the coalface of struggling operations with many facing permanent closures, hostile takeovers, forced mergers and/or significant scaling down. Given the critical role of SMMEs in employment creation, there is a need for governments to continue systematically bailing this sector out through incentives that include tax deferments, loan guarantees and other such appropriate low risk measures. This should be done quickly as the sector was already in severe stress. Many government economic stimulus packages had been rolled out and were being utilised by SMMEs, but these were proving to be insufficient, particularly in developing economies.

The hotel and restaurant industries remain causalities of the COVID-19-induced fear (real or perceived). To this end, measures focused on bringing back the trust from patrons should be designed, including drawing back local patrons. Rebuilding

Table 16.2 Ensuring safe experience in hospitality and outdoor shopping industries

Hospitality industry	Outdoor shopping destinations/retail establishments
Implement processes focused on enhanced sanitation, disinfection and deep cleaning practices as well as increase cleaning/disinfection frequency Implement customer processes, including guest information and social distancing Enhance food safety and hygiene at restaurants Implement processes focused on enhanced cleaning, disinfection and social distancing for meetings and events	Enhance sanitation, deep cleaning practices and reinforce mitigation measures, including revisiting guidance for cleaning staff relating to washrooms and baby-changing facilities and private client spaces. Within the scope of sanitation and hygiene, provide hand sanitiser dispensers with 70% alcohol content in washrooms, near entrances and exits and at regular intervals throughout Introduce guest wellbeing procedures, such as thermal scanning, social distancing as well as queue management Minimise touch contact by digitising guest services such as digital maps, digital queue management, e-menus, virtual personal shopping and roving concierges and promote contactless payment wherever possible. Provide complimentary WIFI to encourage the use of digital services Suspend services that cannot support social distancing and health and safety requirements, such as playgrounds, drinking fountains and ATMs, as well as items that may have been previously lent to guests, which may include umbrellas and pushchairs

Source: Authors, based on WTTC (2020b: 3, c: 4)

trust and confidence in the hospitality industry could involve measures such as enhanced communication (with clear and consistent messaging) with customers regarding new health and hygiene protocols (WTTC 2020b). The front offices should play a central role in responding to guest queries on COVID-19 and other possible likely disasters. There will also be a need for hotels, B&Bs and other accommodation establishments to erect clear signage informing guest on revamped cleaning protocols and social distancing measures. Additional guidelines may include the need to wear face masks, as well as regular handwashing and maintaining sneezing and coughing etiquette, which is also applicable to outdoor shopping and other industries in the sector (WTTC 2020c).

Some customers are likely to feel uneasy and fearful touching menu boards and making physical contact with automated payment machines. The alternatives could include using electronic display screens and QR codes to limit physical contact with the material. Perceptions of the possibility of getting infected in a restaurant or hotel were likely to persist. Other innovations could include riding on and scaling up the already existing use of drones and robots in the restaurant space. Drones could be used for home deliveries, while sit-in customers could be enticed with reduced costs and other promotional materials and deals. However, there remains a catch-22 situation. This massive technology roll-out may result in technology-induced joblessness. Nonetheless and drawing from the WTTC Safe Travel Global Protocols on hospitality and outdoor retail industries, a sample on how to ensure a safe experience for staff and guests is summarised in Table 16.2.

One of the focus areas of recovery should be destination attractions such as national parks, waterfalls, mountains and rivers. While there was evidence of these attractions becoming less polluted and the wild animals emerging from their usual hideouts in conservation areas (which is a plus), the state of these natural resorts could have been degraded due to limited maintenance during the lockdown periods. This is so given that many are under the care of governments through parastatals. There was also a real chance that many developing countries would redirect development and maintenance budgets for these resorts to the health sector and other COVID-related pressing needs. The recommendation, therefore, is that governments should still prioritise tourism as one of the quick and big wins in terms of recovery, creating jobs and incomes to communities as well as revenue for the state.

For future travel, the airline industry has been placed in the spotlight in terms of the models of aircraft needed and immediate future seating arrangements. While the 70% capacity ratio is being used as an indicative figure, more innovation will be needed in this regard. For travellers (business or leisure), it is going to take courage for one to be back in a plane with all the headaches of social distancing, food handling and other stringent COVID-19 screening measures at both ports of departure and entry. Anticipated long delays at airports will add to the equation. Furthermore, the unintended consequence of possible reductions in air travel for business is due to the newly discovered working-from-home scenario that may disrupt the business travel sector. In South Africa, for example, government officials, parliamentarians and ministers regularly shuttle between Pretoria and Cape Town, as well as other main cities. This practice stopped completely during lockdowns and with budgetary constraints, so there are likely to review that may negatively impact business travel. Working from home (the new normal) is also widespread across sectors globally, and similar negative impacts on air travel demand are likely to be witnessed in short to medium term. One must not rule out the possibility of this new normal becoming permanent in certain organisations.

As air travel resumes, governments may also wish to discount jet fuel (although this should be done with further assessments on oil price). At the time of completing the book, crude oil prices had at some point in April 2020 plummeted to below $17 per barrel due to COVID-19 and ongoing oil price war between Saudi Arabia and Russia. As of 12 May 2020, crude oil was trading at exactly $30 per barrel (21:54 CAT). The airlines will need to offer incentives to excite travel again. This had already started with Qatar Airways offering 100,000 free tickets to medical staff as a measure to say "thank you" for their work on fighting the COVID-19 pandemic (Bitmez and Kosak 2020).

On 13 May 2020, the International Air Transport Association (IATA) and the Airports Council International Europe (ACI Europe) released the Safely Restarting Aviation guide. The safe restart of the industry meant ensuring that aviation "is not a meaningful vector for the spread of COVID-19" (IATA and ACI Europe 2020: 2). Restarting the industry also meant that there had to be significant changes throughout the end-to-end passenger journey map. Governments were encouraged to disrupt current port practices of managing the risk of communicable diseases by investing in advanced technologies. Passengers were also encouraged to take control of their travel safety considerations, including assessing personal levels of health risks prior to travel. The details concerning some of the proposed new normal measures for aviation are presented in Table 16.3.

Table 16.3 Proposed new normal measures on end-to-end passenger journeys

Stage in the travel journey	Proposed new normal measures
Preflight	Passenger contact tracing (COVID-19 status information in electronic format and provided in advance prior to airport arrival, including through e-visa and electronic travel authorisation)
Airport terminal	Continue with revamped entry/exit temperature screening measures Social/physical distancing Use of PPE as guided by local authorities Cleaning and disinfection Application of reliable and fast COVID-19 testing COVID-19 immunity passports for recovered patients Self-check-ins prior to airport arrival, although this requires governments to institute new regulations Airlines to promote self-bag drops where such facilities exist and look for ways to enhance them across airports Boarding reforms will be necessary, including redesigns of boarding areas by the relevant authorities as well as use of automation (self-scanning and biometrics)
Inflight	Although COVID-19 transmission from passenger to passenger has been assessed to be low, physical distancing will be necessary, and passengers can use face coverings Physical distancing from blocked seats may also be necessary, depending on aircraft Cabin crew is encouraged to use comprehensive guidelines already in place that include the management of suspected COVID-19 cases and simplified service and pre-packaged catering Sanitisation wipes could also be provided to clean spaces around them Limiting movement inflight is also advised Revised aircraft cleaning guidelines have been issued
Arrival airport	Revised border and customs control is necessary, including considerations for electronic options (mobile and QR codes), use of green/red lanes for self-declarations, contactless passport reading and redesigning of immigration halls Efficiencies in baggage collection will be needed to avoid passengers unnecessarily waiting for long. Social/physical distancing should also be encouraged and other regulatory measures in the jurisdiction like face mask-wearing need to be observed if in place Security and health screening for transferring passengers should take maximum advantage of "one-stop security arrangements" – Relying on mutual recognition of screening measures at the originating airport and eliminates re-screening in the transfer process In the event that health and safety transfer screening is required, this should follow appropriate sanitary requirements outlined earlier

Source: Authors, based on IATA and ACI Europe (2020: 3–6)

The car rental as well as the ride and share service companies need to ensure the comfort and safety of customers. Ensuring the cleanliness and sanitisation of vehicles will be the focus of many tourists. Education and training of drivers is crucial, and the development of safety protocols will be needed to protect the travellers and the drivers alike. Training the drivers on how to clean vehicles religiously and use of certified chemicals in fumigating and cleaning the vehicles should be obligatory to ensure passenger safety. Some companies had already instituted measures to instil a sense of customer safety. For example, from its 2020 first-quarter report, Hertz indicated that it had started implementing stringent measures in line with the US CDC guidelines to safeguard personnel and customers. In addition to following social distancing best practices at its locations, every vehicle was sealed and certified "Hertz Gold Standard Clean" after undergoing a 15-point cleaning and sanitisation process. This process follows US CDC guidelines and uses Environmental Protection Agency-approved products.

Regarding the sports sector, recovery measures started popping up in May 2020, with some major football leagues, rugby and cricket tests scheduled to start between the end of May and July 2020. The New Zealand rugby, German Bundesliga and English Premier League were among the early pacesetters to getting back to the fields, mainly to play in empty stadiums. The dilemma from the leagues was to find an agreed method to conclude the leagues' and award titles to deserving teams. The French Football League 1 points-per-game (PPG) system was applied to conclude the league (Howard 2020). However, there was still no consensus across the football leagues on the best way to conclude the leagues, although this work supports the PPG to be used to conclude the leagues without endangering the players from COVID-19 infections. Given the scenario, vulnerable sports personalities and administrators should be covered by off-field disaster insurance, and if this is not available currently, this should be planned for the future. Match and/or sports insurance is common in cricket as some matches are insured against bad weather. A disaster relief fund could also be established, ready for disbursement using applicable evaluation criteria when disasters such as COVID-19 strike. The fate of sports personalities and administrators was similar to those in the arts sector.

Targeted marketing to open up new age cohort-related tourism should be perused. Evidence revealed that the older generation dominated cruises, yet they were also among the highly impacted in terms of COVID-19 deaths both on and off-board the ships. Gear and McGorian (2020) present two scenarios from the cruise ships, The Diamond Princess and The Naval Craft. The Diamond Princess on average had passengers aged 68 years compared to presumably young and healthy passenger demographics on The Naval Craft. The rate of infections between the two cruise ships was 18.9% and 15.1%, respectively, with the population on board also similar at 3711 and 4500 people, respectively. What was intriguing was the mortality rates of 0.32% (12 deaths) for Diamond Princess compared to 0.02% (a single death) on The Naval Craft. Hospitalisation rates were at 20% and 1.03% for the two ships, respectively. Moving to countries, as of 4 May 2020, there were 25,635 deaths in Italy with 95.3% of these being 60 years or older. In the USA, with 34,343 deaths as of 1 of May 2020, 92.05% were 55 years or older (Ibid.). These emerging figures are the reason why the tourism industry should rethink demographics in its marketing strategies as it recovers from the COVID-19 pandemic.

Given that the findings on the impact of COVID-19 on falling share prices on tourism-related firms remain provisional, there is a need to undertake additional work to determine the full impact once the crisis is over. The same applies to many other aspects raised across the tourism industries in this book. Unlike during the SARS outbreak in 2002/2003, the values of the tourism-related shares quickly

Table 16.4 Implementation of enabling policies in hospitality and outdoor shopping industries

Hospitality industry	Outdoor shopping destinations/retail establishments
Work collaboratively with industry and other governments as new rules for cross-border travel are developed Provide financial relief to the sector through the reduction of taxes, fees and charges to stimulate demand Create incentives and provide direct support to boost stays in hotels as well as offering tax incentives Enhance destination promotion to boost demand both domestically and internationally	Enhance destination promotion to boost demand both domestically and internationally Provide financial relief to the sector through the reduction of taxes and charges to stimulate demand Create incentives to promote tourism as well as tax incentives Implement visa facilitation measures, making visas more affordable and simpler to obtain Implement public transport measures to and from the destination, which observe social distancing

Source: Authors, based on WTTC (2020b: 5, c: 6)

recovered after the crisis and did not dent the long-term profitability of affected firms the COVID-19 pandemic is likely to last longer. Therefore, governments need to consider another wave of stimulus packages earmarked for the tourism industry and synchronised with the lifting of restrictions on travelling and mass gatherings. In addition, future share prices for these firms should factor such risks for greater reflection of the true value of the related stocks. This needs new thinking in the stock market share pricing approaches.

Overall, drawing from the BBB concept, recovery in the tourism industry cannot be a one-size-fits-all scenario. Furthermore, all the key stakeholders in the entire global tourism value chain must come on board. For example, an individual dealing with souvenirs at the corner of a street needs tourist to be able to sleep somewhere, be transported, eat and travel within the locality to increase the visibility of his/her products. This is the value chain in which many actors are at play. The Chinese outbound tourism market cannot be ignored. Whatever resentments are emerging regarding the origins and the handling of COVID-19, the world needs China in as much as China needs the world. Hence, China remains a key player in tourism recovery. BBB in the tourism sector cannot be dissociated from drawing up enabling policies. The WTTC (2020b, c) provides samples of how these enabling policies could look like in the hospitality and outdoor shopping industries. Details are presented in Table 16.4.

The new reality (normal) requires that the tourism industry should first re-establish trust with the public after the initial closures meant to control the rapid spread of the disease. Operators needed to show their customers that their facilities were safe to use. Tourism firms need to reimagine their businesses as to safety companies first and hospitality providers second (Klasko et al. 2020). As highlighted in the book, some sectors like restaurants remain closed or are operating at reduced capacity due to restrictions on their operations. In addition, the new normal meant shifting from sit-ins to takeaways and having new strict hygienic standards. In

Lebanon, for example, the previous model of operation was hinged on having more numbers to make profits now they find themselves in a situation where they were initially allowed to work at 30% capacity before gradually scaling up to 50% and a maximum of 70% (Azhari 2020). This phased opening was designed to minimise the risk of a second wave of COVID-19 infections.

On 13 May 2020, the European Union became the first trading block to provide comprehensive guidelines and recommendations for the safe resumption of travel and consequently the rebooting of its tourism sector. The guidelines and recommendations were to assist member states to gradually lift restrictions and permit tourism business to reopen (European Union 2020). This package was to enable citizens and residents to get some rest, refresh, relax and get fresh air post-COVID-19. The opening coincides with the summer season – a typical travel and holiday time in Europe. As highlighted earlier in this chapter and elsewhere in the book, the implementation of the necessary health and safety measures remained a top priority to resume business. The package that addresses tourists and travellers, as well as business stakeholders, include the following that the reader is free to follow-up for more details:

- An overall strategy towards recovery in 2020 and beyond
- A common approach to restoring free movement and lifting restrictions at EU internal borders in a gradual and coordinated way
- A framework to support the gradual re-establishment of intra-EU transport arrangements while ensuring the safety of passengers and personnel
- A recommendation that aims to make travel vouchers an attractive alternative to a cash reimbursement for consumers
- Criteria for restoring tourism activities safely and gradually and for developing health protocols for hospitality establishments such as hotels (European Union 2020)

The tourism industry must also join hands with governments and other organisations on COVID-19 psychosocial support activities. Mental health must be prioritised at firm level, especially, as there will be individuals transporting passengers in various modes of transport including aeroplanes, cruise ships, motorcycles, trains, cars, buses and others. Mental health matters will also be of concern to the security clusters that have most of their personnel carrying firearms to protect tourists and other global citizens. To this end, mental health services were identified as an essential part of all government responses to the COVID-19 by the United Nations Secretary General (Guterres 2020), as mental health remains central to humanity and economic advancements.

Lastly, should the tourism industry take up the emerging policy recommendations and build back better, then there will still be sustainable tourism in the future. The following take-home messages, therefore, stand out from this chapter for both the tourism industry, governments and all other stakeholders for as long as the COVID-19 treatment and vaccines have not been found:

1. Gradual and science-based lockdown relaxations that are fully negotiated and agreed between governments and their citizens, residents and key stakeholders in the business.
2. Enhancing general health and safety of the tourist, workers in the tourism industry and the general public, particularly through maintaining social/physical distancing.
3. Continued and future handwashing, mask-wearing (surgical, homemade, scarf or bandana) as well as practising and adapting to cough and sneezing etiquette.
4. Continued swift and widespread COVID-19 testing, contact tracing, isolation, quarantine and treatment.
5. Appropriately packaged government messaging and COVID-19 awareness-raising.
6. Continued and future enhancement of epidemic and pandemic preparedness through investments in healthcare systems, including mental health, health insurance and instituting disaster-related frontline staff monetary incentives.
7. Prioritisation of the tourism and hospitality industry in economic and other stimulus packages. This includes the allocation of maintenance budgets to existing natural attractions such as national parks, ocean, pollution reduction (including climate change mitigation), waterfalls, mountains and rivers.
8. The restaurant industry requires some special dispensation as there are many SMMEs that are either on a franchise or standalone operations. Those on franchise business continued feeling the heat as they face the triple challenge of paying franchise fees, rentals and loans from financing institutions.
9. The development, negotiation and application of globally accepted (tourism) recovery protocols cross the tourism industry subsectors.
10. Engage in urgent WHO and its epidemic/pandemic notification protocols reforms and not proxy trade wars that emerged between the USA and China as well as the USA withholding its $400 million contributions to the World Health Organization.
11. Document as many good practice case studies in tourism epidemic, pandemic and/or disaster preparedness, response and recovery.

References

Azhari, T. (2020). *Lebanon's restaurants 'sinking' after coronavirus lockdown.* Retrieved from https://www.aljazeera.com/ajimpact/lebanons-restaurants-sinking-coronavirus-lockdown-200506095822226.html. Accessed 13th May 2020.

Bitmez, S., & Kosak, C. (2020). *Qatar Airways offers 100,000 medical staff free tickets.* Retrieved from https://www.aa.com.tr/en/middle-east/qatar-airways-offers-100-000-medical-staff-free-tickets/1837412. Accessed 13th May 2020.

de Rugy, V., & Leff, G. (2020, March 25). The case against bailing out the airline industry. *SSRN Special Edition Policy Brief.* https://doi.org/10.2139/ssrn.3571441.

European Union. (2020). *Tourism and transport: Commission's guidance on how to safely resume travel and reboot Europe's tourism in 2020 and beyond.* Retrieved from https://europeansting.com/2020/05/13/tourism-and-transport-commissions-guidance-on-how-to-safely-resume-travel-and-reboot-europes-tourism-in-2020-and-beyond/. Accessed 14th May 2020.

Froneman, A. (2020). *Carnival Cruises see 600% increase in bookings*. Retrieved from https://www.getaway.co.za/travel-news/carnival-cruises-see-600-increase-in-bookings/. Accessed 13th May 2020.

Gear, J., & McGorian, I. (2020). *Lockdown in South Africa can no longer be justified*. Retrieved from https://www.dailymaverick.co.za/article/2020-05-13-lockdown-in-south-africa-can-no-longer-be-justified/. Accessed 13th May 2020.

Gössling, S., Scott, D. and Hall, C. (2020). Pandemics, tourism and global change: A rapid assessment of COVID-19. Journal of Sustainable Tourism, https://doi.org/10.1080/09669582.2020.1758708.

Guterres, A. (2020). *Mental health services are an essential part of all government responses to COVID-19*. Retrieved from https://www.un.org/en/coronavirus/mental-health-services-are-essential-part-all-government-responses-covid-19. Accessed 15th May 2020.

Hall, C. M., Prayag, G., & Amore, A. (2018). *Tourism and resilience: Individual, organisational and destination perspectives*. Bristol: Channelview.

Higgins-Desbiolles, F. (2020). Socialising tourism for social and ecological justice after COVID-19. *Tourism Geographies*. https://doi.org/10.1080/14616688.2020.1757748.

Howard, D. (2020). *The final Premier League table: If they follow French Ligue 1 points-per-game system*. Retrieved from https://www.brightonandhoveindependent.co.uk/sport/football/brighton-and-hove-albion/final-premier-league-table-if-they-follow-french-ligue-1-points-game-system-2661508. Accessed 12th May 2020.

IATA (International Air Transport Association) and ACI Europe (Airports Council International Europe). (2020). *Safely restarting aviation: ACI and IATA joint approach*. Geneva/Brussels: IATA/ACI Europe.

Johnson, S., & Boone, P. (2020). *From lockdown to lock in, here's what post-pandemic travel could look like*. World Economic Forum. www.weforum.org

Klasko, S., Gleanson, J. L., & Hoad, M. (2020). *What hotels must learn from hospitals for the new reality of tourism*. World Economic Forum. www.weforum.org

Ministry of Trade and Industry Singapore. (2020). *Singapore's GDP contracted by 2.2 per cent in the first quarter of 2020*. Retrieved from https://www.mti.gov.sg/Newsroom/Press-Releases/2020/03/Singapore-GDP-Contracted-by-2_2-Per-Cent-in-the-First-Quarter-of-2020. Accessed 13th May 2020.

OECD (Organisation for Economic Cooperation and Development). (2020). *Quarterly national accounts*. Retrieved from https://stats.oecd.org/index.aspx?queryid=350. Accessed 13th May 2020.

Page, S., Song, H., & Wu, D. C. (2012). Assessing the impacts of the global economic crisis and swine flu on inbound tourism demand in the United Kingdom. *Journal of Travel Research, 51*(2), 142–153.

Peters, K., Peters, J., & Peters, N. (2020). *Tourism recovery after disaster*. Visit people. KPPM strategy.

Smith, C. (2020). *There will be 'nothing left' of tourism industry if it can only reopen in December, warns CEO*. Retrieved from https://www.fin24.com/Companies/TravelAndLeisure/there-will-be-nothing-left-of-tourism-industry-if-it-can-only-reopen-in-december-warns-ceo-20200512. Accessed 12th May 2020.

South African Tourism. (2020). *Tourism risk rating – A PowerPoint presentation*. Pretoria: South African Tourism.

Statista. (2020). *Forecasted change in revenue from the travel and tourism industry due to the coronavirus (COVID-19) pandemic worldwide from 2019 to 2020 (in million U.S. dollars)*. Retrieved from https://www.statista.com/forecasts/1103426/covid-19-revenue-travel-tourism-industry-forecast. Accessed 23rd Apr 2020.

The Guardian. (2020). *UK economy shrinks by 2% as Covid-19 slump begins*. Retrieved from https://www.theguardian.com/business/live/2020/may/13/uk-economy-gdp-q1-2020-march-covid-19-recession-business-live. Accessed 13th May 2020.

UNDRR (United Nations Office for Disaster Risk Reduction). (2009). *2009 UNISDR terminology on disaster risk reduction*. Geneva: UNDRR.

UNDRR (United Nations Office for Disaster Risk Reduction). (2015). *Sendai framework on disaster risk reduction (2015–2030)*. New York: UNDRR Secretariat.

UNFCCC (United Nations Framework Convention on Climate Change). (2015). *Paris agreement*. Bonn: UNFCCC Secretariat.

United Nations. (2015). *Transforming our world: The 2030 agenda for sustainable development*. New York: United Nations Secretariat.

United Nations Habitat. (2016). *Habitat III: New urban agenda*. New York: United Nations Habitat Secretariat.

UNWTO (United Nations World Tourism Organisation). (2020a). *COVID-19 related travel restrictions: A global review for tourism – Third report as of 8 May 2020*. Geneva: UNWTO.

UNWTO (United Nations World Tourism Organisation). (2020b). *COVID-19 tourism recovery technical assistance package*. Geneva: UNWTO.

World Bank. (2018). *Building back better: Achieving resilience through stronger, faster and more inclusive post-disaster reconstruction*. Washington, DC: World Bank.

WTTC (World Travel & Tourism Council). (2020a). *Leading global protocols for the new normal*. London: WTTC.

WTTC (World Travel & Tourism Council). (2020b). *Leading global protocols for the new normal: Hospitality*. London: WTTC.

WTTC (World Travel & Tourism Council). (2020c). *Leading global protocols for the new normal: Outdoor shopping destinations/retail establishments*. London: WTTC.

Index

G. Nhamo et al., *Counting the Cost of COVID-19 on the Global Tourism Industry*, https://doi.org/10.1007/978-3-030-56231-1